Quantization of Gauge Systems

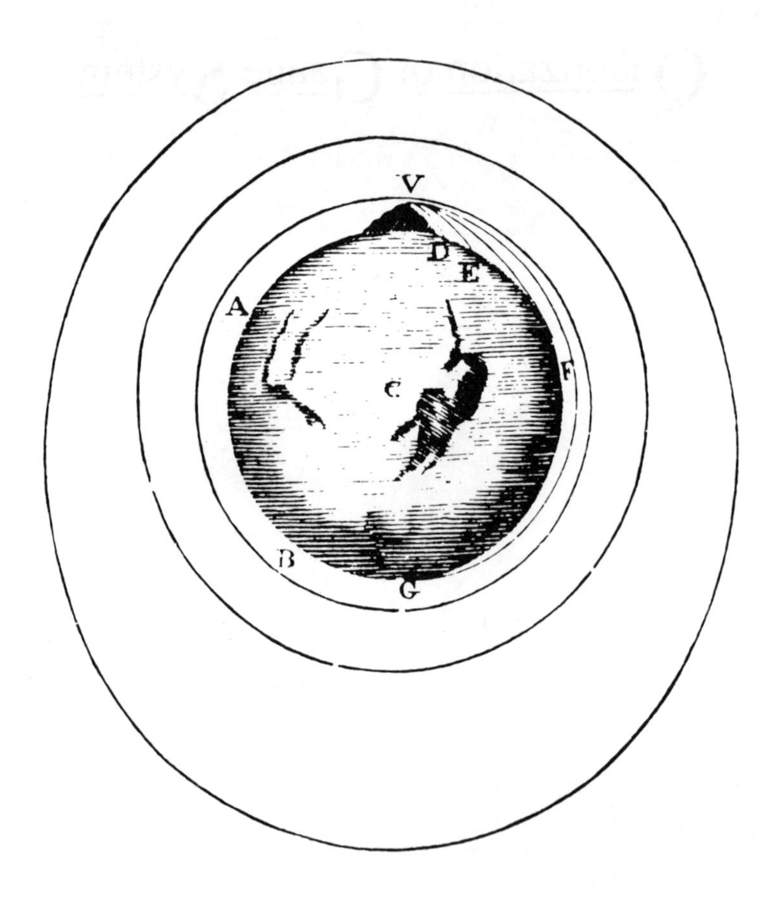

Published by Princeton University Press in association
with the Centro de Estudios Científicos de Santiago

Quantization of Gauge Systems

Marc Henneaux and Claudio Teitelboim

PRINCETON UNIVERSITY PRESS
PRINCETON, NEW JERSEY

Published by Princeton University Press, 41 William Street,
Princeton, New Jersey 08540

Library of Congress Cataloging-in-Publication Data

Henneaux, Marc.
Quantization of gauge Systems / Marc Henneaux and Claudio Teitelboim.
p. cm.
Includes bibliographical references and index.
ISBN 0-0691-08775-X
1. Gauge fields (Physics) 2. Quantum theory. I. Teitelboim, Claudio. II. Title.
QC793.3.F5H46 1992
530.1'435—dc20 92-11585
CIP

Princeton University Press books are printed on acid-free paper, and meet the guidelines for permanence and durability of the Committee on Production Guidelines for Book Longevity of the Council on Library Resources

Printed in the United States of America

10 9 8 7 6 5 4 3 2 1

The amount of theoretical work one has to cover before being able to solve problems of real practical value is rather large, but this circumstance is an inevitable consequence of the fundamental part played by transformation theory and is likely to become more pronounced in the theoretical physics of the future.

— P.A.M. Dirac

(from the preface to the first edition of
The Principles of Quantum Mechanics,
Oxford, 1930)

CONTENTS

PREFACE

Physical theories of fundamental significance tend to be gauge theories. These are theories in which the physical system being dealt with is described by more variables than there are physically independent degrees of freedom. The physically meaningful degrees of freedom then reemerge as being those invariant under a transformation connecting the variables (gauge transformation). Thus, one introduces extra variables to make the description more transparent and brings in at the same time a gauge symmetry to extract the physically relevant content.

It is a remarkable occurrence that the road to progress has invariably been toward enlarging the number of variables and introducing a more powerful symmetry rather than conversely aiming at reducing the number of variables and eliminating the symmetry.

This book is devoted to the general theory of gauge systems both classical and quantum. It starts from the classical analysis of Dirac, showing that gauge theories are constrained Hamiltonian systems, and works its way up to ghosts and the Becchi–Rouet–Stora–Tyutin symmetry and its cohomology, including the formulation in terms of antifields. The quantum mechanical analysis deals with both the operator and path integral methods.

We have attempted to give a fully general and unified treatment of the subject in a form that may survive future developments. To our knowledge, such a treatment was not previously available.

Applications are not included except for a chapter on the Maxwell field and on two-form gauge fields, which are used as an example of how to apply many parts of the general formalism to a specific system. Any attempt to cover a reasonably complete list of applications would have ended up inevitably in a treatise on theoretical physics at large. Exercises are, however, provided with each chapter.

Marc Henneaux
Claudio Teitelboim
Santiago de Chile, April 1991

ACKNOWLEDGMENTS

We are grateful to many colleagues for helpful discussions. Among them, Laurent Baulieu, Jean Fisch, Tullio Regge, Christiane Schomblond, James Stasheff, Claude Viallet, John Wheeler, Edward Witten, and Jorge Zanelli deserve special mention.

Warm thanks are due to the Centro de Estudios Científicos de Santiago, the Institute for Advanced Study at Princeton, the Istituto di Fisica Teorica dell'Università di Torino, Princeton University, the Université Libre de Bruxelles, and the University of Texas at Austin for hospitality during the writing of this book.

For assistance in the research that went into this volume, we are especially grateful to the European Community, the Fonds National de la Recherche Scientifique (Belgium), the Fondo Nacional de Ciencia y Tecnología (Chile), the International Centre for Theoretical Physics, the John D. and Katherine T. MacArthur Foundation, the National Science Foundation (USA), and the Tinker Foundation.

Finally, we extend our warm thanks to Mrs. Elizabeth Baker for her wonderful and patient work in producing the book in TeX.

NOTATIONS

First-class constraints	$\gamma_a \approx 0$ or $G_a \approx 0$
Multipliers for first-class constraints	u^a or λ^a
Second-class constraints	$\chi_\alpha \approx 0$
Momentum conjugate to λ^a	b_a
Grassmann parity of A	$\varepsilon_A = 0, 1 \pmod 2$
Ghost conjugate pairs	$(\eta^a, \mathcal{P}_a)$
Antighost conjugate pairs	$(\bar{C}_a, \rho^a)$
BRST generator	Ω
BRST symmetry	s
Poisson bracket of phase space functions A, B	$[A, B]$
Dirac bracket of phase space functions A, B	$[A, B]^*$
Poisson bracket of phase space coordinates z^A	$[z^A, z^B] = \sigma^{AB}(z)$
Symplectic 2-form in coordinates z^A	$\sigma_{AB}(z), \sigma^{AB}\sigma_{BC} = \delta^A{}_C$
(Graded) commutator $AB - (-)^{\varepsilon_B \varepsilon_A} BA$ of operators A, B	$[A, B]$

Remark. The summation convention over repeated indices is used throughout, except when the index is solely repeated in a sign factor. For instance, there is a summation over a in $\lambda^a \mu_a (-)^{\varepsilon_a}$ but none in $\lambda^a (-)^{\varepsilon_a}$.

Quantization of Gauge Systems

CHAPTER ONE

CONSTRAINED HAMILTONIAN SYSTEMS

1.1. GAUGE INVARIANCE—CONSTRAINTS

A gauge theory may be thought of as one in which the dynamical variables are specified with respect to a "reference frame" whose choice is arbitrary at every instant of time. The physically important variables are those that are independent of the choice of the local reference frame. A transformation of the variables induced by a change in the arbitrary reference frame is called a gauge transformation. Physical variables ("observables") are then said to be gauge invariant.

In a gauge theory, one cannot expect that the equations of motion will determine all the dynamical variables for all times if the initial conditions are given because one can always change the reference frame in the future, say, while keeping the initial conditions fixed. A different time evolution will then stem from the same initial conditions. Thus, it is a key property of a gauge theory that *the general solution of the equations of motion contains arbitrary functions of time.*

The most thorough and foolproof treatment of gauge systems is that which proceeds through the Hamiltonian formulation. Once that formulation is understood, one can go back to the Lagrangian. One can even often shortcut the Hamiltonian—at least to a great extent, but to do so correctly, it is of great help to have a solid understanding of the Hamiltonian.

Therefore, we will start the analysis of gauge systems by studying their Hamiltonian formulation. Even though one may rightly regard the Hamiltonian formulation as the more fundamental one, we will begin the discussion by assuming that the action principle is given in Lagrangian form, and we will proceed to pass to the Hamiltonian. We do this only because it is the situation most often found in practice.

It will emerge from the discussion given below that the presence of arbitrary functions of time in the general solution of the equations of motion implies that the canonical variables are not all independent. Rather, there are relations among them called constraints. *Thus, a gauge system is always a constrained Hamiltonian system.* The converse, however, is not true. Not all conceivable constraints of a Hamiltonian system arise from a gauge invariance. The analysis developed below covers, nevertheless, all types of constraints.

1.1.1. The Lagrangian as a Starting Point: Primary Constraints

The starting point for discussing the dynamics of gauge systems will be the action principle in Lagrangian form.

The classical motions of the system are those that make the action

$$S_L = \int_{t_1}^{t_2} L(q, \dot{q})\, dt \tag{1.1}$$

stationary under variations $\delta q^n(t)$ of the Lagrangian variables $q^n (n = 1, \ldots, N)$, which vanish at the endpoints t_1, t_2.

The conditions for the action to be stationary are the Euler–Lagrange equations

$$\frac{d}{dt}\left(\frac{\partial L}{\partial \dot{q}^n}\right) - \frac{\partial L}{\partial q^n} = 0, \qquad n = 1, \ldots, N. \tag{1.2}$$

Equations (1.2) can be written in more detail as

$$\ddot{q}^{n'} \frac{\partial^2 L}{\partial \dot{q}^{n'} \partial \dot{q}^n} = \frac{\partial L}{\partial q^n} - \dot{q}^{n'} \frac{\partial^2 L}{\partial q^{n'} \partial \dot{q}^n}. \tag{1.3}$$

We immediately see from (1.3) that the accelerations $\ddot{q}^n$ at a given time are uniquely determined by the positions and the velocities at that time if and only if the matrix $\partial^2 L/\partial\dot{q}^{n'}\partial\dot{q}^n$ can be inverted; that is, if the determinant

$$\det\left(\frac{\partial^2 L}{\partial\dot{q}^n\partial\dot{q}^{n'}}\right) \tag{1.4}$$

does not vanish.

If, on the other hand, the determinant (1.4) is zero, the accelerations will not be uniquely determined by the positions and velocities and the solution of the equations of motion could then contain arbitrary functions of time. So, the case of interest for systems having gauge degrees of freedom is the one where $\partial^2 L/\partial\dot{q}^{n'}\partial\dot{q}^n$ cannot be inverted. We must, therefore, allow for that possibility.

The departing point for the Hamiltonian formalism is to define the canonical momenta by

$$p_n = \frac{\partial L}{\partial\dot{q}^n}, \tag{1.5}$$

and we see that the vanishing of the determinant (1.4) is just the condition for the noninvertibility of the velocities as functions of the coordinates and momenta. In other words, the momenta (1.5) are not all independent in this case, but there are, rather, some relations

$$\phi_m(q,p) = 0, \qquad m = 1,\ldots,M, \tag{1.6}$$

that follow from the definition (1.5) of the momenta. Thus, when the p's in (1.6) are replaced by their definition (1.5) in terms of the q's and $\dot{q}$'s, Eq. (1.6) reduces to an identity. The conditions (1.6) are called *primary constraints* to emphasize that the equations of motion are not used to obtain these relations and that they imply no restriction on the coordinates q^n and their velocities $\dot{q}^n$.

We assume for simplicity that the rank of the matrix $\partial^2 L/\partial\dot{q}^n\,\partial\dot{q}^{n'}$ is constant throughout $(q,\dot{q})$-space and that Eqs. (1.6) define a submanifold smoothly embedded in phase space. This submanifold is known as the *primary constraint surface.* If the rank of $\partial^2 L/\partial\dot{q}^n\,\partial\dot{q}^{n'}$ is equal to $N - M'$, there are M' independent equations among (1.6), and the primary constraint surface is a phase space submanifold of dimension $2N - M'$. We do not assume that the constraints (1.6) are independent so that M may be strictly greater than M'. However, we shall impose on (1.6) regularity conditions to be detailed in the next subsection.

It follows from (1.6) that the inverse transformation from the p's to the $\dot{q}$'s is multivalued. Given a point (q^n, p_n) that fulfills the constraints (1.6), the "inverse image" $(q^n, \dot{q}^n)$ that solves (1.5) is not unique, since (1.5) defines a mapping from the $2N$-dimensional manifold of the q's and the $\dot{q}$'s to the smaller manifold (1.6) of dimension $2N - M'$. Therefore, the inverse images of a given point of (1.6) form a manifold of dimension M' (see Fig. 1). In order to render the transformation single-valued, one needs to introduce extra parameters, at least M' in number, that indicate the location of $\dot{q}$ on the inverse manifold. These parameters will appear as Lagrange multipliers when we define the Hamiltonian and study its properties.

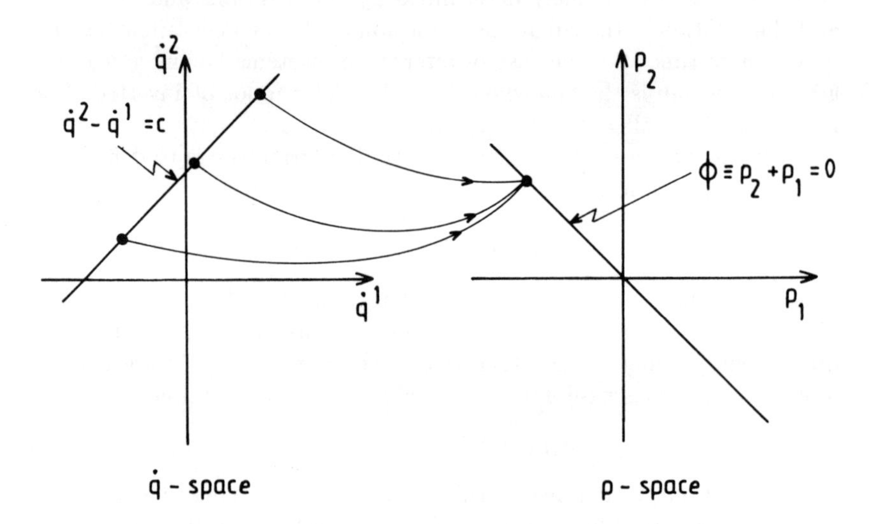

Figure 1: The figure shows the example of a system with two q's and Lagrangian $\frac{1}{2}(\dot{q}^1 - \dot{q}^2)^2$. The momenta are $p_1 = \dot{q}^1 - \dot{q}^2$ and $p_2 = \dot{q}^2 - \dot{q}^1$. There is one primary constraint $\phi = p_1 + p_2 = 0$. All of $\dot{q}$-space is mapped on the straight line $p_1 + p_2 = 0$ of p-space. Moreover, all the $\dot{q}$'s on the straight line $\dot{q}^2 - \dot{q}^1 = c$ are mapped on the same point $p_1 = -c = -p_2$ belonging to the constraint surface $\phi = 0$. The transformation $\dot{q} \rightarrow p$ is thus neither one-to-one nor onto. To render the transformation invertible, one needs to adjoin extra parameters to the p's (see below).

1.1.2. Conditions on the Constraint Functions

There exist many equivalent ways to represent a given surface by means of equations of the form (1.6). For instance, the surface

$$p_1 = 0 \tag{1.7a}$$

can equivalently be written as

$$p_1{}^2 = 0 \tag{1.7b}$$

or as

$$\sqrt{|p_1|} = 0 \tag{1.7c}$$

or, redundantly, as

$$p_1 = 0, \qquad {p_1}^2 = 0. \tag{1.7d}$$

To pass to the Hamiltonian formalism, it turns out to be necessary to impose some restrictions on the choice of the functions ϕ_m, which represent the primary constraint surface. These conditions play an important role in the theory and are referred to in the sequel as the *regularity conditions.*

They can be stated as follows. The $(2N - M')$–dimensional constraint surface $\phi_m = 0$ should be coverable by open regions, on each of which ("locally") the constraint functions ϕ_m can be split into "independent" constraints $\phi_{m'} = 0$ $(m' = 1, \ldots, M')$, which are such that the Jacobian matrix $\partial(\phi_{m'})/\partial(q^n, p_n)$ is of rank M' on the constraint surface, and "dependent" constraints $\phi_{\bar{m}'} = 0$ $(\bar{m}' = M' + 1, \ldots, M)$, which hold as consequences of the others, $(\phi_{m'} = 0 \Rightarrow \phi_{\bar{m}'} = 0)$.

The condition on the Jacobian matrix $\partial(\phi_{m'})/\partial(q^n, p_n)$ can be alternatively reformulated as:

(*i*) the functions $\phi_{m'}$ can be locally taken as the first M' coordinates of a new, regular, coordinate system in the vicinity of the constraint surface; or

(*ii*) the gradients $d\phi_1, \ldots, d\phi_{M'}$ are locally linearly independent on the constraint surface; *i.e.*, $d\phi_1 \wedge \ldots \wedge d\phi_{M'} \neq 0$ ("zero is a regular value of the mapping defined by $\phi_1, \ldots, \phi_{M'}$"); or

(*iii*) the variations $\delta\phi_{m'}$ are of order ε for arbitrary variations δq^i and δp_i of order ε (Dirac's terminology).

Returning to the example $p_1 = 0$, we see that the descriptions of the constraint surface by means of (1.7a) and (1.7d) are both admissible. Indeed, $\partial(p_1)/\partial(q^n, p_n)$ is of rank one, while ${p_1}^2 = 0$ is a clear consequence of $p_1 = 0$. However, neither (1.7b) nor (1.7c) is admissible because $\partial({p_1}^2)/\partial(q^n, p_n)$ vanishes when ${p_1}^2 = 0$, whereas $\partial(\sqrt{|p_1|})/\partial(q^n, p_n)$ is singular there. Another example that is excluded by the regularity conditions is ${p_1}^2 + {p_2}^2 = 0$. In that case, an admissible description of the constraint surface is, for instance, $p_1 = 0$, $p_2 = 0$.

It should be emphasized that although we assume that the above split of the contraint functions can locally be performed, it is by no means necessary to explicitly perform this separation in order to develop the theory. The subsequent formulas will not be based on any such split. All that is required is to choose the functions ϕ_m in such a way that the split can in principle be achieved.

When the constraint functions ϕ_m fulfill the required regularity conditions, the following useful properties, which will be repeatedly used in the sequel, are easily seen to hold.

Theorem 1.1. *If a (smooth) phase space function G vanishes on the surface $\phi_m = 0$, then $G = g^m \phi_m$ for some functions g^m.*

Theorem 1.2. *If $\lambda_n \delta q^n + \mu^n \delta p_n = 0$ for arbitrary variations $\delta q^n, \delta p_n$ tangent to the constraint surface, then*

$$\lambda_n = u^m \frac{\partial \phi_m}{\partial q^n},$$

$$\mu^n = u^m \frac{\partial \phi_m}{\partial p_n}$$

for some u^m. The equalities here are equalities on the surface (1.6).

The proof of the first theorem is based on the fact that one can locally choose the independent constraint functions $\phi_{m'}$ as first coordinates of a regular coordinate system $(y_{m'}, x_\alpha)$, with $y_{m'} \equiv \phi_{m'}$. In these coordinates one has, since $G(0, x) = 0$,

$$G(y, x) = \int_0^1 \frac{d}{dt} G(ty, x)\, dt$$

$$= y_{m'} \int_0^1 G_{,m'}(ty, x)\, dt,$$

and thus

$$G = g^m \phi_m$$

with $g^{m'} = \int_0^1 G_{,m'}(ty, x)\, dt$ and $g^{\bar{m}'} = 0$. This yields a local proof of Theorem 1.1. It is straightforward to extend the proof to the whole of phase space. In order not to obscure the discussion by technical considerations, the global argument is given in Appendix 1.A.

The proof of the second theorem is based on the observation that the constraint surface is of dimension $2N - M'$, and therefore the tangent variations $\delta q^n, \delta p_n$ at a point form a $(2N - M')$-dimensional vector space. Hence, there exist exactly M' independent solutions of $\lambda_n \delta q^n + \mu^n \delta p_n = 0$. By the regularity assumptions, the M' gradients $(\partial \phi_{m'}/\partial q^n, \partial \phi_{m'}/\partial p_n)$ of the independent constraints are linearly independent. Since these gradients clearly solve $\lambda_n \delta q^n + \mu^n \delta p_n = 0$ for tangent variations, they yield a basis of solutions and Theorem 1.2 holds. Note that in the presence of redundant constraints, the functions u^m exist but are not unique.

1.1.3. The Canonical Hamiltonian

The next step in the Hamiltonian analysis is to introduce the canonical Hamiltonian H by

$$H = \dot{q}^n p_n - L. \tag{1.8}$$

As defined by (1.8), H is a function of the positions and the velocities. However, the remarkable fact is that the $\dot{q}$'s enter H only through the combination $p(q, \dot{q})$ defined by (1.5). This general property of the Legendre transformation is what makes H interesting. It is verified by evaluating the change δH induced by arbitrary independent variations of the positions and velocities:

$$\begin{aligned} \delta H &= \dot{q}^n \delta p_n + \delta \dot{q}^n p_n - \delta \dot{q}^n \frac{\partial L}{\partial \dot{q}^n} - \delta q^n \frac{\partial L}{\partial q^n} \\ &= \dot{q}^n \delta p_n - \delta q^n \frac{\partial L}{\partial q^n}. \end{aligned} \tag{1.9}$$

Here, δp_n is not an independent variation but is regarded as a linear combination of δq's and $\delta \dot{q}$'s. We see, thus, that the $\delta \dot{q}$'s appear in (1.9) only through that precise linear combination and not in any other way. This means that H is a function of the p's and the q's.

The Hamiltonian defined by (1.8) is not, however, uniquely determined as a function of the p's and the q's. This may be understood by noticing that the δp_n in (1.9) are not all independent but are restricted to preserve the primary constraints $\phi_m \approx 0$, which are identities when the p's are expressed as functions of the q's and $\dot{q}$'s via (1.5).

We arrive then at the conclusion that the canonical Hamiltonian is well defined only on the submanifold defined by the primary constraints and can be extended arbitrarily off that manifold. It follows that the formalism should remain unchanged by the replacement

$$H \to H + c^m(q, p)\phi_m,$$

and we will see below that this is indeed the case.

Equation (1.9) can be rewritten as

$$\left(\frac{\partial H}{\partial q^n} + \frac{\partial L}{\partial q^n}\right)\delta q^n + \left(\frac{\partial H}{\partial p_n} - \dot{q}^n\right)\delta p_n = 0,$$

from which one infers, using Theorem 1.2, that

$$\dot{q}^n = \frac{\partial H}{\partial p_n} + u^m \frac{\partial \phi_m}{\partial p_n}, \tag{1.10a}$$

$$-\left.\frac{\partial L}{\partial q^n}\right|_{\dot{q}} = \left.\frac{\partial H}{\partial q^n}\right|_{p} + u^m \frac{\partial \phi_m}{\partial q^n}. \tag{1.10b}$$

The first of these relations is particularly important because it enables us to recover the velocities $\dot{q}^n$ from the knowledge of the momenta p_n (obeying $\phi_m = 0$) and of extra parameters u^m. These extra parameters can be thought of as coordinates on the surface of the inverse images of a given p_n.

If the constraints are independent, the vectors $\partial\phi_m/\partial p_n$ are also independent on $\phi_m = 0$ because of the regularity condition [Exercise 1.1(a)]. Hence, no two different sets of u's can yield the same velocities in (1.10a). This means that the u's can be expressed, in principle, as functions of the coordinates and the velocities by solving the equations

$$\dot{q}^n = \frac{\partial H}{\partial p_n}(q, p(q, \dot{q})) + u^m(q, \dot{q}) \frac{\partial \phi_m}{\partial p_n}(q, p(q, \dot{q})).$$

If we define the Legendre transformation from $(q, \dot{q})$-space to the surface $\phi_m(q, p) = 0$ of (q, p, u)-space by means of

$$\begin{cases} q^n = q^n, \\ p_n = \dfrac{\partial L}{\partial \dot{q}^n}(q, \dot{q}), \\ u^m = u^m(q, \dot{q}), \end{cases} \tag{1.11a}$$

we see that this transformation between spaces of the same dimensionality $2N$ is invertible, since one has

$$\begin{cases} q^n = q^n, \\ \dot{q}^n = \dfrac{\partial H}{\partial p_n} + u^m \dfrac{\partial \phi_m}{\partial p_n}, \\ \phi_m(q, p) = 0. \end{cases} \tag{1.11b}$$

Hence, Eqs. (1.11b) imply Eqs. (1.11a), and vice versa. Invertibility of the Legendre transformation when $\det(\partial^2 L/\partial \dot{q}^n \partial \dot{q}^{n'}) = 0$ can thus be regained at the price of adding extra variables.

It should be mentioned that the preceding discussion is only of local validity. We will assume from now on that (1.11) is also globally correct. This implies, in particular, that a Hamiltonian H can be globally defined as a function of q, p by means of (1.8) and is not, say, multivalued.

The only modification that arises in the analysis when some constraints depend on others is that the variables u^m are no longer determined by q and $\dot{q}$. Rather, one should view them as functions of $q, \dot{q}$ and of extra parameters u^α $(\alpha = 1, \ldots, M' - M)$ in number equal to the degree $M' - M$ of redundancy. The formulas (1.11a)–(1.11b) are otherwise unchanged.

1.1.4. Action Principle in Hamiltonian Form

The relations (1.10) enable one to rewrite the original Lagrangian Eqs. (1.2) in the equivalent Hamiltonian form

$$\dot{q}^n = \frac{\partial H}{\partial p_n} + u^m \frac{\partial \phi_m}{\partial p_n}, \tag{1.12a}$$

$$\dot{p}_n = -\frac{\partial H}{\partial q^n} - u^m \frac{\partial \phi_m}{\partial q^n}, \tag{1.12b}$$

$$\phi_m(q, p) = 0. \tag{1.12c}$$

That Eqs. (1.12) follow from (1.2) is a direct consequence of (1.10) and of the definition of the momenta in terms of the velocities. That, conversely, Eqs. (1.12) imply (1.2) results from the fact that (1.12a) and (1.12c) lead, as we have just shown, to $p_n = \partial L / \partial \dot{q}^n$. When this relation is inserted in (1.12b) and (1.10b) is taken into account, one gets the original Lagrangian equations of motion.

The Hamiltonian equations (1.12) can be derived from the variational principle

$$\delta \int_{t_1}^{t_2} (\dot{q}^n p_n - H - u^m \phi_m) = 0 \tag{1.13}$$

for arbitrary variations $\delta q^n, \delta p_n, \delta u_m$ subject only to the restriction $\delta q^n(t_1) = \delta q^n(t_2) = 0$. The new variables u^m, which were introduced to make the Legendre transformation invertible, appear now as Lagrange multipliers enforcing the primary constraints (1.12c). One can alternatively fix the p's, rather than the q's, at the endpoints. In that case, the $p\dot{q}$ term in (1.13) should be replaced by $-q\dot{p}$. Yet another variational principle, in which the p's and the q's are treated symmetrically, is analyzed in §7.1.3 below.

It is clear from the form of the action principle that the theory is invariant under $H \to H + c^m \phi_m$, since this change merely results in a renaming $u^m \to u^m + c^m$ of the Lagrange multipliers. The variational principle (1.13) is also equivalent to the alternative variational principle with fewer variables in which the constraints are solved, namely,

$$\delta \int_{t_1}^{t_2} (\dot{q}^n p_n - H)\, dt = 0 \tag{1.14a}$$

for independent variations of the coordinates and the momenta subject to the conditions

$$\phi_m = 0, \qquad \delta \phi_m = 0. \tag{1.14b}$$

This follows from the standard Lagrange multiplier method. The regularity condition on the constraints plays again a key role here, since otherwise (1.14) would, in general, not be equivalent to (1.13). (See Exercise 1.3 in this context.)

The equations of motion derived from (1.13) can be written as

$$\dot{F} = [F, H] + u^m [F, \phi_m]. \tag{1.15}$$

Here, $F(q,p)$ is an arbitrary function of the canonical variables, and the Poisson bracket (P.B.) is defined as usual by

$$[F, G] = \frac{\partial F}{\partial q^i} \frac{\partial G}{\partial p_i} - \frac{\partial F}{\partial p_i} \frac{\partial G}{\partial q^i}. \tag{1.16}$$

1.1.5. Secondary Constraints

Let us now examine some of the consequences of the equations of motion (1.15). A basic consistency requirement is that the primary constraints be preserved in time. Thus, if we take F in (1.15) to be one of the ϕ_m, we should have $\dot{\phi}_m = 0$. This gives rise to the consistency conditions,

$$[\phi_m, H] + u^{m'} [\phi_m, \phi_{m'}] = 0. \tag{1.17}$$

Equation (1.17) can either reduce to a relation independent of the u's (thus involving only the q's and the p's) or it may impose a restriction on the u's. In the former case, if the relation between the p's and the q's is independent of the primary constraints, it is called a *secondary constraint.* Secondary constraints differ from the primary ones in that the primary constraints are merely consequences of Eq. (1.5) that defines the momentum variables, while for the secondary constraints one has to make use of the equations of motion as well. If there is a secondary constraint—$X(q,p) = 0$, say—coming in, we must impose a new consistency condition,

$$[X, H] + u^m [X, \phi_m] = 0. \tag{1.18}$$

Next, we must again check whether (1.18) implies new secondary constraints or whether it only restricts the u's, and so on. After the process is finished, we are left with a number of secondary constraints, which will be denoted by

$$\phi_k = 0, \qquad k = M+1, \ldots, M+K, \tag{1.19}$$

where K is the total number of secondary constraints. The reason for the notation (1.19) is that the distinction between primary and secondary constraints will be of little importance in the final form of the theory, and it is thus useful to be able to denote all constraints (primary and secondary) in a uniform way as

$$\phi_j = 0, \qquad j = 1, \ldots, M + K = J. \tag{1.20}$$

We make the same regularity assumptions on the full set of constraints ϕ_j as on the primary constraints. Namely, we assume not only that (1.20) defines a smooth submanifold but we also take the constraint functions ϕ_j to obey the regularity conditions described in §1.1.2. It will be further assumed below that the rank of the matrix of the brackets $[\phi_j, \phi_{j'}]$ is constant throughout the surface (1.20) where the constraints hold.

1.1.6. Weak and Strong Equations

It is useful at this stage to introduce the *weak equality symbol* "$\approx$" for the constraint equations. Thus, (1.20) is written as

$$\phi_j \approx 0$$

to emphasize that the quantity ϕ_j is numerically restricted to be zero but does not identically vanish throughout phase space. This means, in particular, that it has nonzero Poisson brackets with the canonical variables.

More generally, two functions F, G that coincide on the submanifold defined by the constraints $\phi_j \approx 0$ are said to be *weakly equal,* and one writes $F \approx G$. On the other hand, an equation that holds throughout phase space and not just on the submanifold $\phi_j \approx 0$ is called *strong,* and the usual equality symbol is used in that case. Thus (by Theorem 1.1 with ϕ_m replaced by ϕ_j),

$$F \approx G \quad \Leftrightarrow \quad F - G = c^j(q,p)\phi_j. \tag{1.21}$$

1.1.7. Restrictions on the Lagrange Multipliers

Assuming now that we have found a complete set (1.20) of constraints, we can go over to study the restrictions on the Lagrange multipliers u^m. These restrictions are

$$[\phi_j, H] + u^m[\phi_j, \phi_m] \approx 0, \tag{1.22}$$

where m is summed from 1 to M and j takes on any of the values from 1 to J. We can consider (1.22) as a set of J nonhomogeneous linear equations in the $M \leq J$ unknowns u^m, with coefficients that are functions of the q's and the p's. These equations should possess solutions, for otherwise the system described by the Lagrangian (1.1) would be inconsistent.

The general solution of (1.22) is of the form

$$u^m = U^m + V^m, \tag{1.23}$$

where U^m is a particular solution of the inhomogeneous equation (1.22) and V^m is the most general solution of the associated homogeneous system

$$V^m[\phi_j, \phi_m] \approx 0. \tag{1.24}$$

Now, the most general V^m is a linear combination of linearly independent solutions $V_a{}^m$, $a = 1, \ldots, A$, of the system (1.24). The number A of independent solutions $V_a{}^m$ is the same for all q, p on the constraint surface because we assume the matrix $[\phi_j, \phi_m]$ to be of constant rank there. We thus find that the general solution of (1.22) is

$$u^m \approx U^m + v^a V_a{}^m \tag{1.25}$$

in terms of coefficients v^a, which are *totally arbitrary.* We have thus explicitly separated that part of u^m that remains arbitrary from the one that is fixed by the consistency conditions derived from the requirement that the constraints be preserved in time.

A more detailed analysis of these consistency conditions and of how (1.19) and (1.25) explicitly arise is given in §1.6.3 and §3.3.2.

1.1.8. Irreducible and Reducible Cases

If the equations $\phi_j = 0$ are not independent, one says that the constraints are "reducible" (or "redundant") and that one is in the "reducible case." One is in the irreducible case when all the constraints are independent.

By dropping the dependent constraints, one does not lose any information. In that sense, one can always assume that one is (locally) in the irreducible case. However, the separation of the constraints into "dependent" and "independent" ones might be awkward to perform, might spoil manifest invariance under some important symmetry, or might even be globally impossible because of topological obstructions. For that reason, it is preferable to construct the general formalism in both the irreducible and reducible contexts. The reducible case arises, for example, when the dynamical coordinates include p-form gauge fields (see Sec. 19.2).

It should be added that, conversely, any irreducible set of constraints can always be replaced by a reducible one by introducing constraints that are consequences of the ones already at hand. The formalism should (and will) be invariant under such replacements.

1.1.9. Total Hamiltonian

We now return to the equations of motion (1.15) and use expression (1.25) for u^m to rewrite those equations in the equivalent form,

$$\dot{F} \approx [F, H' + v^a \phi_a], \tag{1.26}$$

where we have defined

$$H' = H + U^m \phi_m, \tag{1.27}$$

$$\phi_a = V_a{}^m \phi_m. \tag{1.28}$$

In arriving at (1.26) we have used

$$[F, U^m \phi_m] = U^m [F, \phi_m] + [F, U^m]\phi_m \approx U^m [F, \phi_m] \tag{1.29}$$

and similar expressions for $[F, V_a{}^m \phi_m]$.

The function

$$H_T = H' + v^a \phi_a, \tag{1.30}$$

which appears in (1.26), is called the *total Hamiltonian.* So in terms of the total Hamiltonian, the equations of motion read simply

$$\dot{F} \approx [F, H_T]. \tag{1.31}$$

These equations contain A arbitrary functions v^a and are equivalent, by construction, to the original Lagrangian equations of motion (1.2).

1.1.10. First-Class and Second-Class Functions

We have mentioned before that the distinction between primary and secondary constraints is of little importance in the final form of the Hamiltonian scheme. A different classification of constraints—and, more generally, of functions defined on phase space—plays, however, a central role. This is the concept of *first-class* and *second-class* functions.

A function $F(q, p)$ is said to be first class if its Poisson bracket with every constraint vanishes weakly,

$$[F, \phi_j] \approx 0, \qquad j = 1, \ldots, J. \tag{1.32}$$

A function of the canonical variables that is not first class is called second class. Thus, F is second class if there is at least one constraint such that its Poisson bracket with F does not vanish weakly.

An important feature of the first-class property is that it is preserved under the Poisson bracket operation. In other words the Poisson bracket of two first-class functions is first class. This is proved as follows: if F and G are first class, then

$$[F, \phi_j] = f_j{}^{j'} \phi_{j'}; \qquad [G, \phi_j] = g_j{}^{j'} \phi_{j'}. \tag{1.33}$$

Now by the Jacobi identity we have

$$\begin{aligned}
[[F, G], \phi_j] =& [F, [G, \phi_j]] - [G, [F, \phi_j]] \\
=& [F, g_j{}^{j'} \phi_{j'}] - [G, f_j{}^{j'} \phi_{j'}] \\
=& [F, g_j{}^{j'}]\phi_{j'} + g_j{}^{j'} f_{j'}{}^{j''} \phi_{j''} \\
& - [G, f_j{}^{j'}]\phi_{j'} - f_j{}^{j'} g_{j'}{}^{j''} \phi_{j''} \approx 0.
\end{aligned} \tag{1.34}$$

As a first application of the first-class concept we note that H' and ϕ_a, respectively defined by (1.27) and (1.28), are first class. This follows from (1.22) and (1.24). Moreover, the ϕ_a are a complete set of first-class primary constraints, *i.e.*, any first-class primary constraint is a linear combination of the ϕ_a (with coefficients that are functions of the q's and the p's and modulo squares of second-class constraints). This is so because $v^a V_a{}^m$ is the most general solution of (1.24) on the surface $\phi_j = 0$.

Thus, we learn that the total Hamiltonian (1.30) is the sum of the first-class Hamiltonian H' and the first-class primary constraints multiplied by arbitrary coefficients. It should be pointed out here that the splitting of H_T into H' and $v^a \phi_a$ is not unique because U^m appearing in (1.27) can be any solution of the inhomogeneous equation (1.22). This means that by merely renaming the arbitrary functions v^a, we can admit into H' in (1.30) any linear combination of the ϕ_a without changing the total Hamiltonian.

1.2. FIRST-CLASS CONSTRAINTS AS GENERATORS OF GAUGE TRANSFORMATIONS

1.2.1. Transformations That Do Not Change the Physical State. Gauge Transformations

The presence of arbitrary functions v^a in the total Hamiltonian tells us that not all the q's and p's are observable. In other words, although the physical state is uniquely defined once a set of q's and p's is given, the converse is not true—*i.e.*, there is more than one set of values of the canonical variables representing a given physical state. To see how this conclusion comes about, we notice that if we give an initial set of canonical variables at the time t_1 and thereby completely define the physical state at that time, we expect the equations of motion *to fully determine the physical state at other times.* Thus, by definition, any ambiguity in the value of the canonical variables at $t_2 \neq t_1$ should be a physically irrelevant ambiguity.

Now, the coefficients v^a are arbitrary functions of time, which means that the value of the canonical variables at t_2 will depend on the choice of the v^a in the interval $t_1 \leq t \leq t_2$. Consider, in particular, $t_2 = t_1 + \delta t$. The difference between the values of a dynamical variable F at time t_2, corresponding to two different choices $v^a, \tilde{v}^a$ of the arbitrary functions at time t_1, takes the form

$$\delta F = \delta v^a [F, \phi_a] \tag{1.35}$$

with $\delta v^a = (v^a - \tilde{v}^a)\delta t$. Therefore, the transformation (1.35) does not alter the physical state at time t_2. We then say, extending a terminology used in the theory of gauge fields, that *the first-class primary constraints generate gauge transformations.* The gauge transformations (1.35) are independent if and only if the constraints $\phi_a = 0$ are irreducible. When these constraints are reducible, some of the gauge transformations (1.35) lead to $\delta F \approx 0$.

In general, the transformations (1.35) are not the only ones that do not change the physical state. In fact, the following two results hold:

1. The Poisson bracket $[\phi_a, \phi_{a'}]$ of any two first-class primary constraints generates a gauge transformation.

Proof. Applying to a generic dynamical variable F four successive transformations of the form (1.35) with parameters δv^a given by $(\varepsilon^a, \eta^a, -\varepsilon^a, -\eta^a)$ we obtain by virtue of the Jacobi identity

$$\delta F = \varepsilon^a \eta^{a'} \left[F, [\phi_a, \phi_{a'}]\right] + 0(\varepsilon^2) + 0(\eta^2). \tag{1.36}$$

Since ε^a and η^a are arbitrary, $\varepsilon^a \eta^{a'}$ is also arbitrary and the result follows.

2. The Poisson bracket $[\phi_a, H']$ of any first-class primary constraint ϕ_a with the first-class Hamiltonian H' generates a gauge transformation.

Proof. We compare the values of the dynamical variable F at time $t + \varepsilon$ obtained by (i) first making a gauge transformation (1.35) of parameter $\delta v^a = \eta^a$ and then evolving the system with H'; and (ii) doing the same operations in reverse order. The net difference must be a gauge transformation. Repeated application of (1.31) and (1.35) yields for the change in F (we keep only terms up to $\varepsilon\eta^a$ and neglect $(\eta^a)^2$ and ε^2. This suffices for the argument):

$$\begin{aligned}\delta F &= +\big([[F,\phi_a],H'] - [[F,H'],\phi_a]\big)\,\varepsilon\eta^a \\ &= +[F,[\phi_a,H']]\varepsilon\eta^a. \end{aligned} \tag{1.37}$$

This shows that $[\phi_a, H']$ generates gauge transformations.

The two results obtained above indicate that in general we may expect at least some secondary first-class constraints to act also as gauge generators. In fact, we know that since ϕ_a and H' are first class, the brackets $[\phi_a, \phi_{a'}]$ and $[\phi_{a'}, H']$ will also have that property, which means that they will be linear combinations of the first-class constraints. There is, however, no reason to expect this linear combination to contain only primary constraints, and in practice a good many secondary first-class constraints do show up in this way.

It is not possible to infer from these considerations that every first-class secondary constraint is a gauge generator ("Dirac conjecture"). One can actually construct counterexamples (see the next subsection and subsection 1.6.3). Nevertheless, *one postulates, in general, that all first-class constraints generate gauge transformations.* This is the point of view adopted throughout this book. There are a number of good reasons to do this. First, the distinction between primary and secondary constraints, being based on the Lagrangian, is not a natural one from the Hamiltonian point of view. On the contrary, the division of the constraints into first class and second class relies only on the fundamental structure of the Hamiltonian theory, the Poisson bracket. Second, the scheme is consistent in that: (*i*) the transformation generated by a first-class constraint preserves all the constraints (first class and second class) and thus maps an allowed state onto an allowed state, and (*ii*) the Poisson bracket of two gauge generators remains a gauge generator (the Poisson bracket of two first-class constraints is again a first-class constraint). Third, as we shall see later, the known quantization methods for constrained systems put all first-class constraints on the same footing, *i.e.,* treat all of them as gauge generators. It is actually not clear if one can at all quantize otherwise. Anyway, since the conjecture holds in all physical applications known so far, the issue is somewhat academic. (A proof of the Dirac conjecture under simplifying regularity conditions that are generically fulfilled is given in subsection 3.3.2.)

Finally, a word of caution. The arguments leading to the identification of ϕ_a and $[\phi_a, H']$ as generators of transformations that do not change the physical state at a given time implicitly assume that the time t (the integration variable in the action) is observable. That is information brought in from the outside. One may also take the point of view that some of the gauge arbitrariness indicates that the time itself is not observable. This is done in the so-called generally covariant theories (Chapter 4). One of the arbitrary functions is then associated with reparametrizations $t \to f(t)$ of the time variable. Which function is chosen is also based on additional information. One may ask and answer the same questions within both interpretations of the formalism (see Chapter 4 and §16.2.3).

1.2.2. A Counterexample to the Dirac Conjecture

To illustrate the above considerations, it is of interest to analyze a system that violates the conjecture. This system is described by the Lagrangian

$$L = \tfrac{1}{2}\, e^{y}\, \dot{x}^2. \tag{1.38}$$

The equations of motion leave y arbitrary but restrict x to being constant in time, $x = x_0$. The variable y is, therefore, pure gauge. A "physical state" of the system is completely specified by a single constant x_0, the initial value of x.

The passage to the Hamiltonian is straightforward. One finds

$$\phi \equiv p_y \approx 0 \tag{1.39a}$$

as a primary constraint. The Hamiltonian reads

$$H = \tfrac{1}{2}\, e^{-y}\, p_x^2. \tag{1.39b}$$

There is one secondary constraint, namely,

$$\dot{p}_y \approx 0 \Rightarrow p_x^2 \approx 0 \Rightarrow p_x \approx 0. \tag{1.39c}$$

The constraints are both first class. However, only the first one generates a gauge transformation. The second one generates shifts in x, but these shifts do not correspond to any arbitrariness in the general solution of the equations of motion following from (1.38). Therefore, the property conjectured by Dirac does not hold for the model (1.38).

However, it appears necessary to adopt p_x as a gauge generator. Otherwise, one runs into difficulties. Indeed, the space of physically distinct initial data for (1.38) is then one-dimensional. That space has no bracket structure, and it is not clear how to pass to quantum mechanics. The way out is to postulate that the secondary first-class constraint $p_x = 0$ generates gauge transformations, even though this is not exhibited explicitly by the original Lagrangian. If x is postulated to be a pure gauge variable, the physical phase space of (1.38) is zero-dimensional and the system has no physical degree of freedom. The quantization is then straightforward: the physical Hilbert space contains a single state.

Once this point of view is adopted, *as it will be throughout this book,* the proof of the "Dirac conjecture" is somewhat of marginal interest. Its sole purpose is to determine whether the time evolution derived from the original Lagrangian exhibits explicitly all the transformations that do not change the physical state of the system at a given time.

1.2.3. The Extended Hamiltonian

We argued above that the really important classification of constraints from the Hamiltonian point of view is the one that distinguishes between first- and second-class constraints. It is therefore useful to introduce a new notation to distinguish these two kinds of constraints. We denote the first-class constraints by the letter γ—and, subsequently, by G—(for "generator" or "gauge") and the second-class ones by χ. The set of all constraints (first and second class) will be denoted by $\{\phi_j\}$ as before.

Now, the most general physically permissible motion should allow for an arbitrary gauge transformation to be performed while the system is dynamically evolving in time. The motion generated by the total Hamiltonian H_T contains only as many arbitrary gauge functions as there are first-class primary constraints. We thus have to add to H_T the first-class secondary constraints multiplied by additional arbitrary functions. The first-class function obtained in this way has the form

$$H_E = H' + u^a \gamma_a \tag{1.40}$$

and is called the *extended Hamiltonian.* (Here the index a runs over a complete set of first-class constraints.)

For gauge-invariant dynamical variables (variables such that their Poisson brackets with the gauge generators γ_a vanish weakly), the evolution predicted by H', H_T, and H_E is of course the same. For any other kind of variable we must use H_E to account for all the gauge freedom.

It should be emphasized here that strictly speaking, the need for the extended Hamiltonian does not follow from the Lagrangian theory. It is rather the total Hamiltonian H_T that generates the original Lagrangian equations of motion, since H_E contains more arbitrary functions of time than does H_T. The introduction of H_E is a new feature of the Hamiltonian scheme, which truly extends the Lagrangian formalism by making manifest all the gauge freedom. A precise comparison between the Hamiltonian equations generated by H_T and H_E will be given in Chapter 3 below.

1.2.4. Extended Action Principle

It has been shown in §1.1.4 that the equations of motion derived from the original action (1.1) are equivalent to the Hamiltonian equations of motion derived from the action (1.13),

$$S_T = \int (p_n \dot{q}^n - H' - u^m \phi_m)\, dt, \tag{1.41}$$

in which the sum $u^m \phi_m$ runs over the primary constraints only. The Hamiltonian equations of motion that follow from (1.41) are those of the nonextended formalism.

On the other hand, the equations of motion for the extended formalism can be derived from the "extended action principle,"

$$S_E = \int (p_n \dot{q}^n - H' - u^j \phi_j)\, dt, \tag{1.42a}$$

where the sum contains *all* the constraints and not just the primary ones. Indeed, the equations of motion that follow from (1.42a) imply that $u^j = u^a A_a{}^j$, where $A_a{}^j$ is such that the first-class constraints are $\gamma_a = A_a{}^j \phi_j$ and where the u^a's are arbitrary. They then reduce to

$$\dot{F} \approx [F, H_E], \tag{1.42b}$$

$$\phi_j \approx 0, \tag{1.42c}$$

with H_E given by (1.40).

1.3. SECOND-CLASS CONSTRAINTS: THE DIRAC BRACKET

1.3.1. Separation of First-Class and Second-Class Constraints

Let us now turn to second-class constraints, which are present whenever the matrix $C_{jj'} = [\phi_j, \phi_{j'}]$ does not vanish on the constraint surface. To keep the discussion simple, let us assume that the constraints are irreducible. Remarks concerning the reducible case will be gathered in §1.3.4. We also assume that the rank of the matrix $C_{jj'}$ of the brackets of *all* the constraints is constant on the constraint surface.

Theorem 1.3. *If* $\det C_{jj'} \approx 0$, *there exists (at least) one first-class constraint among the* ϕ_j*'s.*

Proof. If $\det C_{jj'} \approx 0$, one can find a nonzero solution λ^j of $\lambda^j C_{jj'} \approx 0$. The constraint $\lambda^j \phi_j$ is then easily seen to be first class, which proves the theorem.

By redefining the constraints as $\phi_j \to a_j{}^{j'} \phi_{j'}$, with an appropriate invertible matrix $a_j{}^{j'}$, one can use the constraint $\lambda^j \phi_j$ as the first constraint of an equivalent representation of the constraint surface. In that representation $C_{1j} = -C_{j1} \approx 0$.

Upon repeated use of Theorem 1.3, one finally arrives at an equivalent description of the constraint surface in terms of constraints $\gamma_a \approx 0$, $\chi_\alpha \approx 0$, whose Poisson bracket matrix reads *weakly*

$$\begin{array}{cc} & \begin{array}{cc} \gamma_a & \chi_\alpha \end{array} \\ \begin{array}{c} \gamma_b \\ \chi_\beta \end{array} & \begin{pmatrix} 0 & 0 \\ 0 & C_{\beta\alpha} \end{pmatrix}, \end{array} \tag{1.43}$$

where $C_{\beta\alpha}$ is an antisymmetric matrix that is everywhere invertible on the constraint surface.

In this representation, the constraints are completely split into first and second classes. No combination of the χ_α is first class and the γ_a's exhaust all first-class constraints, while any second-class constraint must have a component along χ_α. Note that the number of second-class constraints must be even, since otherwise the antisymmetric matrix $C_{\beta\alpha}$ would possess zero determinant. This feature will not be maintained, however, in the presence of fermionic degrees of freedom.

The separation (1.43) is not unique. It is preserved by the redefinitions

$$\gamma_a \to a_a{}^b \gamma_b, \quad \chi_\alpha \to a_\alpha{}^\beta \chi_\beta + a_\alpha{}^a \gamma_a \tag{1.44}$$

with $\det a_a{}^b \neq 0$, $\det a_\alpha{}^\beta \neq 0$. Also, one can add squares of second-class constraints to γ_a without changing the first-class property, $\gamma_a \to \gamma_a + t_a^{\alpha\beta} \chi_\alpha \chi_\beta$.

We will assume that the second-class functions χ_α are such that $\det C_{\alpha\beta} \neq 0$ everywhere on the surface $\chi_\alpha = 0$ and not just on $\chi_\alpha = 0$, $\gamma_a = 0$. This is necessary to properly handle second-class constraints.

1.3.2. Treatment of Second-Class Constraints: An Example

Second-class constraints cannot be interpreted as gauge generators, or, more generally, as generators of any transformation of physical significance. The reason is that by definition, the contact transformation generated by a second-class constraint χ does not preserve all the constraints $\phi_j \approx 0$ and thus maps an allowed state onto a nonallowed state.

How, then, should second-class constraints be treated? Considerable insight into this question is obtained by examining the simplest example of a theory with second-class constraints: one with N pairs of canonical coordinates where the first pair (q^1, p_1) is constrained to be zero. The constraints are then

$$\chi_1 = q^1 \approx 0, \tag{1.45a}$$

$$\chi_2 = p_1 \approx 0. \tag{1.45b}$$

These constraints are second class because

$$[\chi_1, \chi_2] = 1 \not\approx 0. \tag{1.45c}$$

It is rather obvious what we have to do in this case: Equations (1.45a)–(1.45b) tell us that the first degree of freedom is not important, and consequently we just discard q^1 and p_1 and work with a modified Poisson bracket:

$$[F, G]^* = \sum_{n=2}^{N} \left(\frac{\partial F}{\partial q^n} \frac{\partial G}{\partial p_n} - \frac{\partial G}{\partial q^n} \frac{\partial F}{\partial p_n} \right). \tag{1.46}$$

The modified bracket (1.46) of each of the two constraints (1.45) with an arbitrary dynamical variable is identically zero, which means that when working with $[\ ,\ ]^*$ we can set the χ_α equal to zero before evaluating the bracket. Thus, if in this example we use the star bracket instead of the Poisson bracket, we can set the second-class constraints strongly equal to zero. It is also clear that the equations of motion for the other ($n \geq 2$) degrees of freedom remain unchanged if we replace the original Poisson bracket by the modified bracket. Moreover, the bracket (1.46) clearly satisfies all the good properties of a Poisson bracket (antisymmetry, derivation property $[F, GR]^* = [F, G]^* R + G[F, R]^*$, and the Jacobi identity).

1.3.3. Dirac Bracket

The generalization of (1.46) for an arbitrary set of second-class constraints was invented by Dirac.

Since the matrix $C_{\alpha\beta}$ is invertible, it possesses an inverse $C^{\alpha\beta}$,

$$C^{\alpha\beta} C_{\beta\gamma} = \delta^\alpha{}_\gamma. \tag{1.47}$$

The Dirac bracket is now defined as

$$[F, G]^* = [F, G] - [F, \chi_\alpha] \, C^{\alpha\beta} \, [\chi_\beta, G]. \tag{1.48}$$

A constructive way to arrive at (1.48) is discussed in Exercise 1.12. Here, we shall simply point out that (1.48) has all the good properties it should have, namely,

$$[F, G]^* = -[G, F]^* \tag{1.49a}$$

$$[F, GR]^* = [F, G]^* R + G[F, R]^*, \tag{1.49b}$$

$$\left[[F, G]^*, R\right]^* + \left[[R, F]^*, G\right]^* + \left[[G, R]^*, F\right]^* = 0, \tag{1.49c}$$

$$[\chi_\alpha, F]^* = 0 \qquad \text{for any } F, \tag{1.50}$$

$$[F, G]^* \approx [F, G] \qquad \text{for } G \text{ first class and } F \text{ arbitrary,} \tag{1.51a}$$

$$\left[R, [F, G]^*\right]^* \approx \left[R, [F, G]\right]$$

$$\text{for } F \text{ and } G \text{ first class and } R \text{ arbitrary.} \tag{1.51b}$$

The proof of all the above equations except the Jacobi identity (1.49c) is quite simple and straightforward. One merely uses the definition (1.48) and the fact that a quadratic combination of constraints is always first class, even if the original constraints were second class. The proof of (1.49c) is more elaborate and is discussed in the exercises.

It follows from (1.50) that the second-class constraints can be set equal to zero either before or after evaluating a Dirac bracket. Furthermore, since the extended Hamiltonian (1.40) is first class, we see from (1.51a) that the H_E still generates the correct equations of motion in terms of the Dirac bracket, *i.e.*,

$$\dot{F} \approx [F, H_E] \approx [F, H_E]^*, \qquad \text{for any } F. \tag{1.52}$$

In particular, the effect of a gauge transformation can also be evaluated by means of the Dirac bracket:

$$[F, \gamma_a] \approx [F, \gamma_a]^*, \qquad \text{for any } F. \tag{1.53}$$

The general situation at this stage is then the following. The original Poisson bracket is discarded after having served its purpose of distinguishing between first-class and second-class constraints. All the equations of the theory are formulated in terms of the Dirac bracket, and the second-class constraints merely become identities expressing some canonical variables in terms of others (strong equations). In simple cases [such as (1.45)], the second-class constraints can actually be used to eliminate entirely some canonical variables from the formalism. However, in more complicated situations, the elimination of some degrees of freedom in favor of others may be very difficult, even though it can always be achieved in principle.

As a final point, we note that the formalism remains unchanged under the replacement (1.44) of the second-class constraints χ_α by $\bar{\chi}_\alpha = a_\alpha{}^\beta \chi_\beta + a_\alpha{}^a \gamma_a$ in the sense that the Dirac brackets of the gauge-invariant functions among themselves are not modified on the surface $\gamma_a = 0$.

1.3.4. Reducible First-Class and Second-Class Constraints

The previous considerations can be extended to cover the reducible case.

We will say that the reducible constraints $\phi_j = (\gamma_a, \chi_\alpha)$ are separated into first-class constraints (γ_a) and second-class constraints (χ_α) when they obey the following conditions:

(*i*) The reducibility conditions are split into pure first-class and pure second-class sets as

$$Z_{\bar{a}}{}^a \gamma_a = 0 \qquad (a = 1, \ldots, A;\ \bar{a} = 1, \ldots, \bar{A}); \tag{1.54a}$$

$$Z_{\bar{\alpha}}{}^\alpha \chi_\alpha = 0 \qquad (\alpha = 1, \ldots, B;\ \bar{\alpha} = 1, \ldots, \bar{B}); \tag{1.54b}$$

where the reducibility functions $Z_{\bar{a}}{}^a$ and $Z_{\bar{\alpha}}{}^\alpha$ may depend on the q's and the p's;

(*ii*) The brackets $[\gamma_a, \gamma_b]$ and $[\gamma_a, \chi_\alpha]$ weakly vanish,

$$[\gamma_a, \gamma_b] \approx 0, \qquad [\gamma_a, \chi_\alpha] \approx 0; \tag{1.54c}$$

(*iii*) The matrix $[\chi_\alpha, \chi_\beta]$ is of maximal rank $B - \bar{B}$ on the constraint surface

$$\text{rank}\left([\chi_\alpha, \chi_\beta]\right) = B - \bar{B}. \tag{1.54d}$$

(We assume all the conditions (1.54b) to be independent, so that there are exactly $B - \bar{B}$ independent second-class constraints.) It is easy to see that one can always reach locally the separation (1.54) by appropriate redefinitions of the constraints. This can be done, for example, by first choosing an independent subset of constraints $\phi_u = 0$ to which one applies the results of the previous sections. One then redefines the dependent constraint functions ϕ_v so as to fulfill (1.54) (take, *e.g.*, $\phi_v \equiv 0$).

Because of (1.54), the constraints $\gamma_a = 0$ are all first class, and furthermore there is no combination of the constraints $\chi_\alpha = 0$ that yields a nontrivial first-class constraint.

Once the separation (1.54) has been achieved, one can consistently set equal to zero all the second-class constraints, as in the irreducible case. This can be seen by again choosing a maximum subset of $B - \bar{B}$ independent second-class constraints, say, χ_Λ $(\Lambda = 1, \ldots, B - \bar{B})$, in terms of which all the χ_α are expressible, *i.e.*, $\chi_\alpha = m_\alpha{}^\Lambda \chi_\Lambda$ for appropriate $m_\alpha{}^\Lambda$. The matrix $C_{\Lambda\Gamma}$ of the brackets of this subset is invertible by assumption; otherwise, (1.54d) would not be of rank $B - \bar{B}$. One can thus use the Dirac bracket (1.48) associated with χ_Λ. Since $\chi_\Lambda = 0$ implies $\chi_\alpha = 0$, this procedure consistently enforces all the second-class constraints. (By "consistently," it is meant that $[A, F]^*$ vanishes as a consequence of $\chi_\alpha = 0$ for all functions F that are zero on the surface $\chi_\alpha = 0$.)

One can directly write down the appropriate Dirac brackets without having to explicitly display a complete, independent subset of second-class constraints. Indeed, it follows from (1.48) and our above discussion that $[A, B]^*$ takes the form

$$[A, B]^* = [A, B] - [A, \chi_\alpha]\, D^{\alpha\beta}\, [\chi_\beta, B], \tag{1.55a}$$

where the matrix $D^{\alpha\beta} = -D^{\beta\alpha}$ obeys on $\chi_\alpha = 0$

$$D^{\alpha\beta}\, [\chi_\beta, \chi_\rho] = \delta^\alpha{}_\rho + Z_{\bar{\alpha}}{}^\alpha\, \lambda^{\bar{\alpha}}{}_\rho \tag{1.55b}$$

for some $\lambda^{\bar{\alpha}}{}_\rho$.

Even though Eq. (1.55b) leaves an ambiguity in $D^{\alpha\beta}$, given by

$$D^{\alpha\beta} \to D^{\alpha\beta} + Z_{\bar{\alpha}}{}^{[\alpha}\, n^{\beta]\bar{\alpha}} + d^{\alpha\beta\gamma}\chi_\gamma, \tag{1.55c}$$

the expression (1.55a) is well defined on the surface $\chi_\alpha = 0$. This is because $Z_{\bar{\alpha}}{}^\alpha \chi_\alpha = 0$, so that the ambiguous terms in (1.55c) do not contribute to (1.55a) on $\chi_\alpha = 0$. Hence, Eqs. (1.55a) and (1.55b) completely characterize the Dirac bracket.

Finally, we mention that it is essential here that the reducibility conditions (1.54b) on the second-class constraints do not involve the first-class ones. If $Z_{\bar{\alpha}}{}^\alpha \chi_\alpha = 0$ were to be replaced by $Z_{\bar{\alpha}}{}^\alpha \chi_\alpha + d^a{}_{\bar{\alpha}}\, \gamma_a = 0$, then setting $\chi_\alpha = 0$ would also amount to setting some first-class constraints equal to zero. This would lead to inconsistencies.

As an example, consider the system of constraints

$$\chi_1 = q^1, \quad \chi_2 = p_1, \quad \chi_3 = p_1 + p_2 + q_1, \quad \gamma = p_2.$$

The constraint γ is first class. The constraint functions χ_1, χ_2, and χ_3 are all second class, since $[\chi_1, \chi_2] = 1$, $[\chi_1, \chi_3] = 1$, and $[\chi_2, \chi_3] = -1$. One may thus superficially think that it is possible to consistently enforce $\chi_1 = \chi_2 = \chi_3 = 0$ by defining an appropriate bracket. However, it is easy to see that p_2 vanishes on $\chi_1 = \chi_2 = \chi_3 = 0$, and yet there is no way to choose $D^{\alpha\beta}$ in the Dirac bracket (1.55a) so that $[q^2, p_2]^* = [q^2, p_2] - [q^2, \chi_\alpha]\, D^{\alpha\beta}\, [\chi_\beta, p_2] = 1$ vanishes. The problem arises because the constraints have been incompletely separated: the reducibility condition on the second-class constraints χ_1, χ_2, and χ_3—namely, $\chi_1 + \chi_2 - \chi_3 = -\gamma$—involves also the first-class constraint γ.

1.4. GAUGE FIXATION—INDEPENDENT DEGREES OF FREEDOM

1.4.1. Canonical Gauges

As we have seen, the presence of first-class constraints and the associated gauge freedom indicates that there is more than one set of canonical variables that corresponds to a given physical state. Sometimes it is desirable in practice to eliminate this ambiguity by imposing further restrictions on the canonical variables (canonical gauge conditions) so that there is a one-to-one correspondence between physical states and values of those canonical variables that are left independent after the additional conditions are imposed. The gauge conditions are therefore not a consequence of the theory, but rather they are ad-hoc equations brought in from the outside to avoid "multiple counting of states." It is permissible to bring in these additional conditions because they merely remove the unobservable arbitrary elements of the theory and do not affect the observable (gauge-invariant) properties.

To simplify the discussion, we assume in this section that the constraints are all independent (irreducible case). We leave it as an exercise to generalize the criteria for a good set of gauge-fixing conditions to the reducible case.

There are two properties that a satisfactory set

$$C_b(q,p) \approx 0 \tag{1.56}$$

of gauge conditions must satisfy:

(a) The chosen gauge must be accessible. That is, given any set of canonical variables there must exist a gauge transformation that maps the given set of q's and p's onto a set that satisfies (1.56). This transformation must be obtained by iteration of infinitesimal transformations of the form $\delta u^a[F, \gamma_a]$.

This requirement guarantees that (1.56) does not affect the physically relevant (gauge-invariant) properties of the system but merely restricts the gauge freedom. Since the number of independent parameters δu^a is equal to the number of independent first-class constraints γ_a, we conclude that the number of independent gauge conditions (1.56) cannot be greater than the number of independent γ_a's.

(b) The conditions (1.56) must fix the gauge completely. This means that there must exist no gauge transformation other than the identity that preserves (1.56). In other words the equations

$$\delta u^a[C_b, \gamma_a] \approx 0. \tag{1.57}$$

must imply

$$\delta u^a = 0. \tag{1.58}$$

Equations (1.57) can imply (1.58) only if the number of independent equations is equal to or greater than the number of unknowns δu^a. Thus (a) and (b) together imply that in order to fully fix the gauge, *the number of independent gauge conditions must be equal to the number of independent first-class constraints.* The Poisson brackets $[C_b, \gamma_a]$ form then a square matrix. In order for (1.57) to imply (1.58) this matrix should be invertible. Thus, we have the condition

$$\det [C_b, \gamma_a] \neq 0. \tag{1.59}$$

But this condition precisely expresses that the constraints C_b, γ_a form together a second-class set. Thus, we see that after complete gauge-fixing, no first-class constraint is left.

It is quite reasonable that after bringing in a complete set of gauge-fixation conditions, we are left with second-class constraints only. In fact, if any first-class constraint were left, we would still have some gauge freedom remaining corresponding to the transformations generated by that constraint.

After fixing the gauge we can pass to the Dirac bracket. We then have a theory that is effectively free of constraints in the sense that all constraints may be regarded at that point as identities expressing some dynamical variables in terms of others ("operator equations" in the sense of quantum mechanics).

One can give a geometric description of the process of gauge-fixing. The surface $C_a = 0$ of the gauge conditions should intersect the gauge orbits, which lie on the constraint surface, once and only once. The condition (1.59) guarantees this property locally (see Fig. 2).

Even though (1.59) guarantees that the gauge is locally fixed—*i.e.*, that both conditions (a) and (b) above are locally fulfilled—they do not imply that the conditions (1.56) still meet this requirement globally. The geometry of the constraint surface and of the gauge orbits may actually be such that it forbids the existence of global gauge conditions. In that case, any choice of C_b obeying the local criterium (1.59) would either intersect some orbits at least twice or would not intersect at all some others. One usually refers to this problem as the "Gribov obstruction."

Because of the Gribov obstruction, gauge conditions of the form (1.56) may be of limited validity. This is one of the reasons for developing a theory of first-class constraints without having to fix the gauge in the sense of (a) and (b).

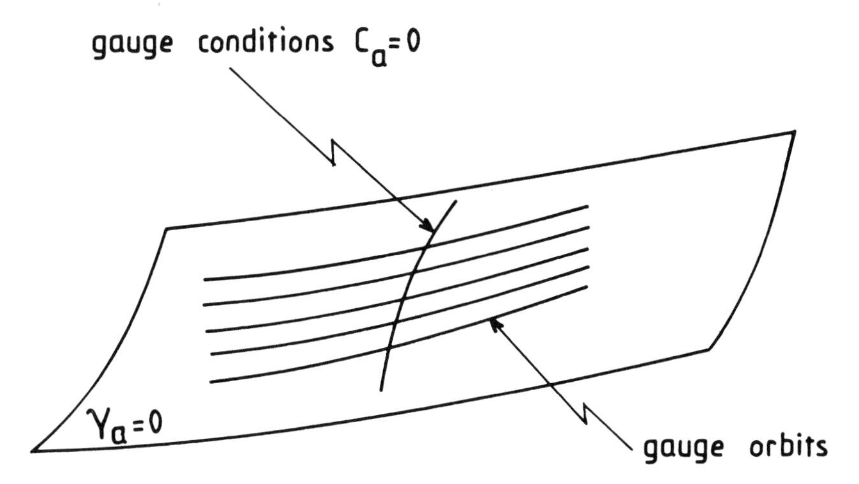

Figure 2: A good set of gauge conditions should determine a surface $C_a = 0$ that intersects the gauge orbits once and only once.

1.4.2. Counting of Degrees of Freedom

In a theory possessing only second-class constraints no arbitrary functions appear in the Hamiltonian. A set of canonical variables that satisfies the constraint equations determines then one and only one physical state. Since after fixing the gauge there are only second-class constraints left, we arrive at the following counting of physical degrees of freedom:

$$
\begin{aligned}
2\times &\begin{pmatrix}\text{Number of physical}\\ \text{degrees of freedom}\end{pmatrix} = \begin{pmatrix}\text{Number of independent}\\ \text{canonical variables}\end{pmatrix}\\
&= \begin{pmatrix}\text{Total number of}\\ \text{canonical variables}\end{pmatrix} - \begin{pmatrix}\text{Number of original}\\ \text{second-class constraints}\end{pmatrix}\\
&\quad - \begin{pmatrix}\text{Number of}\\ \text{first-class constraints}\end{pmatrix} - \begin{pmatrix}\text{Number of}\\ \text{gauge conditions}\end{pmatrix} \qquad (1.60)\\
&= \begin{pmatrix}\text{Total number of}\\ \text{canonical variables}\end{pmatrix} - \begin{pmatrix}\text{Number of original}\\ \text{second-class constraints}\end{pmatrix}\\
&\quad - 2\times\begin{pmatrix}\text{Number of}\\ \text{first-class constraints}\end{pmatrix}.
\end{aligned}
$$

In the reducible case, the counting proceeds in the same way, provided one only counts the number of independent constraints.

As the number of second-class constraints is always even, we see from (1.60) that the number of independent canonical variables is also even corresponding to an integer number of physical degrees of freedom.

The above counting of degrees of freedom is well defined and unambiguous for a finite (or perhaps countable) number of degrees of freedom. When one turns over to a continuous number of degrees of freedom, subtleties arise. These subleties are not only of a mathematical nature, they also are of physical importance. They bear on the problem of distinguishing gauge transformations from transformations that change the physical state.

In a typical continuous case the "gauge part" $u^a \gamma_a$ of the extended Hamiltonian becomes an integral of the form

$$\int dx\, u^a(x)\, \gamma_a(x), \tag{1.61}$$

and the question that arises is: How arbitrary can the functions $u^a(x)$ be in order that

$$\delta F = \int dx\, u^a(x) \big[F, \gamma_a(x)\big] \tag{1.62}$$

be a gauge transformation? There is no general answer to this question, which requires a case-by-case analysis. One important element of the analysis is that $u^a(x)$ can be properly thought of as Lagrange multipliers for the constraints $\gamma_a(x) \approx 0$. Thus, the functions u^a must be just general enough so that the vanishing of (1.61) for all u^a implies $\gamma_a(x) \approx 0$. In many cases, this amounts to u^a and γ_a being in the same function space.* For example, if γ_a is periodic, then also u^a must be periodic.

For functions u^a that do not belong to the function space determined by the above argument, the transformation (1.62) still maps one allowed state onto another whenever it is well defined. However, it does not follow from the general analysis that the transformation does not change the physical state. That demand would amount to an additional assumption. For example, for gauge fields that vanish at infinity, the use of a $u(x)$ that tends to a nonvanishing constant at infinity gives rise to an overall "charge rotation." Invariance under such a transformation picks up states of zero total charge. The total charge becomes then a new constraint, which generates the additional gauge invariance.

An even more extreme case may arise. There may exist finite transformations of the same functional form as the iterated transformations of the form (1.62) but which are not continuously connected with the identity

---

* The function space to which $\gamma_a(x)$ belongs is determined by the requirement that phase space should contain all physically interesting solutions $q(x,t)$, $p(x,t)$ of the equations of motion.

("large-gauge transformations"). They again map allowed states on allowed states, but regarding them as proper gauge transformations again amounts to an additional assumption.

1.4.3. Do All Second-Class Constraints Arise from Gauge Fixation?

Upon gauge-fixation, the first-class constraints become, together with the gauge conditions, second class. This suggests the possibility of viewing any set of second-class constraints as resulting from gauge-fixation of an equivalent system with a gauge invariance and no second-class constraints.

This can, indeed, be done, although the removal of the second-class constraints is not unique. Rather than trying to develop here the general procedure, we will merely illustrate the method on two simple examples. These examples show explicitly that the equivalent gauge system may either contain, in the simplest cases, the same number of variables or need extra (gauge) degrees of freedom.

Example 1: Constraints $q_1 = 0$, $p_1 = 0$. One can view $q_1 = 0$ as a gauge-fixing condition for the invariance generated by $p_1 = 0$, $q_1 \rightarrow q_1 + u$. The two second-class constraints $q_1 = 0$, $p_1 = 0$ are thus equivalent to the single first-class constraint $p_1 = 0$.

In that simple example, the second-class constraints split into two equal groups, with the constraints of one group being all first class. This separation cannot always be achieved in practice, and one may then need extra variables in order to remove the second-class constraints. How this is done can be illustrated on the same model.

Example 2: Same set of constraints. One extends the system by introducing another pair of canonically conjugate variables (q^2, p_2), and one replaces the constraints $q_1 = 0$, $p_1 = 0$ by

$$\phi_1 = q_1 + q_2 = 0,$$
$$\phi_2 = p_1 - p_2 = 0.$$

These constraints are clearly first class and generate the gauge invariance

$$q_1 \rightarrow q_1 + \varepsilon_2, \qquad p_1 \rightarrow p_1 - \varepsilon_1,$$
$$q_2 \rightarrow q_2 - \varepsilon_2, \qquad p_2 \rightarrow p_2 - \varepsilon_1.$$

If one imposes the gauge conditions $q_2 = 0$, $p_2 = 0$, which are permissible, q^2 and p_2 drop out and one is left with the original system. Hence, the enlargement of the system by the addition of extra variables does not change its dynamical content.

By enlarging the original system, we have traded a pair of second-class constraints for a pair of first-class ones. The price paid is the introduction of new degrees of freedom. There is an advantage, however, in having only first-class constraints in that it is no longer necessary to introduce Dirac brackets, since one can keep working with Poisson brackets as long as one does not fix the gauge by means of canonical gauge conditions.

Bypassing the Dirac brackets may be a substantial advantage. Indeed, the quantum realization of Dirac brackets that depend on the canonical variables may be highly nontrivial and is by no means guaranteed.

The examples studied here are quite simple. However, they illustrate well the general idea of viewing second-class constraints as resulting from gauge-fixing of a larger system with a gauge invariance. A more complete discussion is given in Exercise 1.22.

It results from our analysis of gauge-fixing that one can say that second-class constraints cover the general situation, since one can always fix the gauge. But one can equally say that first-class constraints also cover the general case, since one can always "undo" the gauge fixation if second-class constraints are present.

1.5. GAUGE-INVARIANT FUNCTIONS

1.5.1. Functions on the Constraint Surface

It is useful at this stage to introduce a series of concepts that turn out to play an important role in the analysis of constrained systems. We denote phase space by P and, as is conventional, the space of smooth phase space functions by $C^\infty(P)$.

The vector space $C^\infty(P)$ is endowed with two algebraic structures: one is the ordinary pointwise multiplication, for which $C^\infty(P)$ is an associative algebra; the other is the Dirac bracket operation, for which $C^\infty(P)$ is a Lie algebra. These two operations are related by

$$[F_1 F_2, F_3]^* = [F_1, F_3]^* F_2 + F_1 [F_2, F_3]^*.$$

Because the dynamical system is constrained to be on the constraint surface, which we denote by Σ, two phase space functions that coincide on Σ cannot be distinguished. In other words, the relevant functions are not all the smooth phase space functions but, rather, the smooth functions on Σ.

The space $C^\infty(\Sigma)$ of smooth functions on Σ can be algebraically characterized as follows. The functions that vanish on Σ form an ideal in $C^\infty(P)$ for the ordinary multiplication. That is, the product of an arbitrary phase space function with a function that vanishes on Σ also vanishes on Σ.* We denote this ideal by $\mathcal{N}$. By virtue of the Theorem 1.1, the elements in $\mathcal{N}$ are of the form $\lambda^a\gamma_a + \lambda^\alpha\chi_\alpha$, *i.e.*, $\mathcal{N}$ is generated by γ_a and χ_α.

Now, given $\mathcal{N}$, one can consider the quotient algebra $C^\infty(P)/\mathcal{N}$ containing the equivalence classes of phase space functions whose difference is an element of $\mathcal{N}$. This quotient algebra is just $C^\infty(\Sigma)$—and the induced product is just the ordinary multiplication in $C^\infty(\Sigma)$—because any equivalence class of $C^\infty(P)/\mathcal{N}$ defines a function on Σ. Conversely, a function on Σ can be extended off Σ to a smooth phase space function. Two extensions differ by an element of $\mathcal{N}$, and so a function in $C^\infty(\Sigma)$, indeed, defines an equivalence class of $C^\infty(P)/\mathcal{N}$.

1.5.2. Classical Observables

A (classical) "observable" is, by definition, a function on the constraint surface that is gauge invariant. Alternatively, an observable can be described as a phase space function that has a weakly vanishing Dirac bracket with the first-class constraints,

$$[F, \gamma_a]^* \approx 0, \tag{1.63}$$

and one should identify two such functions that coincide on the constraint surface. The concept of observables involves, thus, two steps: (*i*) first, the restriction to the constraint surface; (*ii*) second, the gauge-invariance condition (1.63). This second condition is the first-class condition with respect to the Dirac bracket.

Although the terminology "observable" is used, it will not be attempted in this book to give a direct experimental significance to that concept by means of appropriate apparatuses, as this is not necessary for the development of the theory. Indeed, no information other than the action principle was required in the determination and classification of the constraints. Thus, the action itself enables one to decide what are the observables. It should also be mentioned that some classically observable functions may no longer be quantum-mechanically observable. This would occur, for instance, in the presence of superselection rules.

---

* The definitions of algebra, subalgebra, ideal, ..., are collected in Chapter 8.

1.5.3. Algebraic Characterization of the Observables

The observables can be algebraically characterized in a different manner that suggests itself once one analyzes the following question: the quotient algebra $C^\infty(\Sigma) \equiv C^\infty(P)/\mathcal{N}$ inherits a well-defined multiplication from $C^\infty(\Sigma)$. What about the bracket structure?

It is clear that there is no well-defined bracket in $C^\infty(\Sigma)$, except in the case where the constraints are all second class. This is because $\mathcal{N}$, even though an ideal for the ordinary multiplication, is not an ideal for the Dirac bracket: if $F \in C^\infty(P)$ and $G \in \mathcal{N}$ ($\Leftrightarrow G = \lambda^a \gamma_a + \mu^\alpha \chi_\alpha$), then $[F, G]^* \approx [F, \gamma_a]^* \lambda^a \neq 0$, even on the constraint surface, unless F happens to be gauge invariant, $[F, \gamma_a]^* \approx 0$. The gauge-invariant functions define, therefore, the maximum subalgebra $\mathcal{O}$ containing $\mathcal{N}$ for which $\mathcal{N}$ is an ideal, $[\mathcal{O}, \mathcal{N}]^* \subset \mathcal{N}$. For these functions—and only for these—there is a well-defined bracket because of the equality

$$[F + \lambda^a \gamma_a + \mu^\alpha \chi_\alpha, F' + \lambda'^a \gamma_a + \mu'^\alpha \chi_\alpha]^* \approx [F, F']^*.$$

Therefore, the necessity to get a well-defined bracket—a mandatory step towards quantization—forces one to consider functions that are invariant under the transformations generated by *all* (primary and secondary) first-class constraints.

1.5.4. Gauge-Invariant Extensions

One way to construct observables is based on the fact that these are constant under gauge transformations. Hence, if one knows a function f in a specific canonical gauge, one can extend it off that gauge in a gauge-invariant way. This yields a function $F(q, p)$, which is gauge invariant and called the gauge-invariant extension of f.

Furthermore, if G is an arbitrary phase space function, one can consider its restriction g to a given canonical gauge. One can then extend g in a gauge-invariant way and get a gauge-invariant function G^*, which coincides with G in the chosen gauge. In that sense, one can say that the functions one manipulates when the gauge is fixed by canonical methods are all gauge invariant. It should be kept in mind, however, that the relation between the function G and the gauge-invariant function G^* depends on the gauge-fixing conditions unless G happens to be gauge invariant, in which case $G \approx G^*$. Also worth mentioning is the fact that $[G^*, F^*] \approx [G, F]^*$ for any two phase space functions G, F, where $[\ ,\ ]^*$ is the Dirac bracket in the chosen gauge.

1.6. EXAMPLES

1.6.1. System with n Generations of Constraints

In order to show that generations of constraints of arbitrarily high order can occur, we take as our first example a model with n coordinates q_i $(i = 1, \ldots, n)$ described by the Lagrangian

$$L = \sum_{i=1}^{n-1} \frac{1}{2}\left(q_i - \frac{dq_{i+1}}{dt}\right)^2. \tag{1.64a}$$

For $n = 2$, the Lagrangian reduces to

$$L = \frac{1}{2}\left(q_1 - \frac{dq_2}{dt}\right)^2 \tag{1.64b}$$

and can be thought of as a truncation of electromagnetism at a fixed wave vector $\vec{k}$, in which the transverse components are set equal to zero $(q_1 \sim A_0/|\vec{k}|, q_2 \sim A^{\text{longitudinal}})$. The gauge transformations read

$$\delta q_i = \frac{d^{n-i}\varepsilon}{dt^{n-i}} \qquad \left(\text{with } \frac{d^0\varepsilon}{dt} \equiv \varepsilon\right) \tag{1.65a}$$

and leave the Lagrangian invariant, since

$$\delta\left(q_i - \frac{dq_{i+1}}{dt}\right) = \frac{d^{n-i}\varepsilon}{dt^{n-i}} - \frac{d}{dt}\left(\frac{d^{n-i-1}\varepsilon}{dt^{n-i-1}}\right) = 0. \tag{1.65b}$$

The Hamiltonian analysis is straightforward. The momenta are given by

$$\pi_i = \frac{\partial L}{\partial \dot{q}^i} = \dot{q}^i - q^{i-1} \qquad (i \geq 2), \tag{1.66a}$$

and

$$\pi_1 = 0. \tag{1.66b}$$

The canonical Hamiltonian is

$$H = \pi_i \, \dot{q}^i - L \tag{1.67a}$$

$$= \tfrac{1}{2}\sum_{i\geq 2} \pi_i^2 + \sum_{i\geq 2} \pi_i \, q_{i-1}, \tag{1.67b}$$

and the time evolution is generated by

$$H_T = H + u\,\pi_1,$$

since (1.66b) is the only primary constraint.

By requiring $\dot{\pi}_1 = 0$, one finds as successive secondary constraints

$$\pi_2 = 0, \pi_3 = 0, \ldots, \pi_n = 0. \tag{1.68}$$

The constraints are all first class. The extended Hamiltonian is

$$H_E = H + u^i \pi_i \approx 0 \tag{1.69}$$

and is a combination of the constraints. The theory possesses no physical degree of freedom.

The gauge transformations (1.65a) of the Lagrangian (1.64a) are canonical transformations whose generator G is a combination of the constraints (1.66b) and (1.68),

$$G = \sum_i \frac{d^{n-i}\varepsilon}{dt^{n-i}}\, \pi_i. \tag{1.70}$$

The general rule for constructing the gauge transformations of the original Lagrangian out of the first-class constraints will be given in Chapter 3.

1.6.2. $L = 0$ and $L = -\frac{1}{2}\sum_i (q^i)^2$

The simplest Lagrangian that one can devise is $L = 0$ ("topological classical mechanics"). It can be thought of as being a function of n coordinates q^i, and it is invariant under arbitrary shifts of the q's. In that case, the constraints are

$$\pi_i = 0, \tag{1.71a}$$

and the Hamiltonian vanishes. The extended Hamiltonian is a combination of the constraints,

$$H_E = u^i \pi_i. \tag{1.71b}$$

The gauge transformations are generated by

$$G = \sum_i \varepsilon^i\, \pi_i, \tag{1.71c}$$

where the ε^i's are arbitrary functions of time.

If one compares this example with the previous one, one finds that both systems are described by exactly the same set of constraints and the same extended Hamiltonian. Hence, these examples actually describe the same system. One can view the first example as a partly gauge-fixed version of the second one, in which the conditions $\varepsilon^i = d\varepsilon^{i+1}/dt$ are imposed on the functions ε^i occurring in (1.71c).

It should also be observed that in these two different descriptions of the same theory, the constraints do not arise in the same order. While they are all primary for $L = 0$, they are successively primary, secondary, ..., up to n-ary for the Lagrangian (1.64a). This indicates, as stressed previously, that the distinction between primary, secondary, etc., constraints does not possess a fundamental significance.

Yet another description of the same theory is obtained by starting from

$$L = -\sum_i \tfrac{1}{2}\,(q^i)^2. \tag{1.72a}$$

The primary constraints are in that case

$$\pi_i = 0, \tag{1.72b}$$

and the Hamiltonian reads

$$H = \tfrac{1}{2}\sum_i (q^i)^2. \tag{1.72c}$$

The secondary constraints are

$$q^i = 0, \tag{1.72d}$$

and the set (1.72b)–(1.72d) is second class. The Hamiltonian weakly vanishes. The conditions (1.72d) can be viewed as gauge-fixing conditions for the transformations generated by (1.72b). If one unfreezes the gauge, which is completely fixed by (1.72d), one gets the same constrained Hamiltonian system as in the previous two cases.

1.6.3. More on the Consistency Algorithm

The consistency algorithm that determines the secondary constraints and the functional dependence of the Lagrange multipliers u^m is notoriously subtle. This is because all conceivable possibilities can arise in principle (first-class primary constraints generating second-class secondary constraints or vice versa ...).

We end this section by giving a series of examples that constitute a nonexhaustive sample of the possibilities.

Example 1:

$$H = e^{\lambda} q^2 + e^{-\mu} p^2 + {\pi_\nu}^2 + q\pi_\nu, \tag{1.73a}$$

$$\phi_1 \equiv \pi_\lambda = 0, \qquad \phi_2 \equiv \pi_\mu = 0. \tag{1.73b}$$

The secondary constraints are

$$\dot{\pi}_\lambda = 0 \Rightarrow q^2 = 0 \Rightarrow q = 0, \tag{1.74a}$$

$$\dot{\pi}_\mu = 0 \Rightarrow p^2 = 0 \Rightarrow p = 0. \tag{1.74b}$$

One then finds

$$\dot{q} = [q, H + u^m \phi_m] \approx 0, \tag{1.75a}$$

but

$$\dot{p} = [p, H + u^m \phi_m] \approx -\pi_\nu, \tag{1.75b}$$

and thus

$$\pi_\nu = 0 \tag{1.75c}$$

arises as a tertiary constraint. The final step is

$$\dot{\pi}_\nu = [\pi_\nu, H + u^m \phi_m] \approx 0. \tag{1.76}$$

There is no further constraint.

The primary and tertiary constraints are first class. The secondary constraints are second class. Hence, first-class constraints can generate second-class ones [(1.73b) ⇒ (1.74)] and vice versa [(1.74b) ⇒ (1.75c)]. The first-class constraint $\pi_\nu = 0$ does not generate any gauge invariance (ν is not arbitrary), and the system is a counterexample to Dirac conjecture.

Example 2:

$$H = (q_2)^2 e^{q_1}, \tag{1.77}$$

$$\phi_1 \equiv p_1 = 0, \qquad \phi_2 \equiv p_2 = 0. \tag{1.78}$$

The secondary constraints are

$$\dot{p}_1 = 0 \Rightarrow q_2 = 0, \tag{1.79a}$$

$$\dot{p}_2 = 0 \Rightarrow q_2 = 0. \tag{1.79b}$$

There is no further constraint. The primary constraint $p_1 = 0$ is first class. The other constraints are second class, and the Lagrange multiplier u^2 is determined by the condition $\dot{q}_2 = 0$. This example shows that the same second-class constraint ($q_2 = 0$) is generated simultaneously by a first-class and a second-class constraint.

Example 3:

$$H = \tfrac{1}{2}(p_3)^2 + \tfrac{1}{2}(p_5)^2 + q_4 p_5 + q_5 p_3, \tag{1.80}$$

$$\phi_1 \equiv p_1 = 0, \qquad \phi_2 \equiv p_2 + p_3 e^{q_1} = 0, \qquad \phi_3 \equiv p_4 = 0. \tag{1.81}$$

The requirement $\dot{\phi}_m = 0$ implies

$$\dot{\phi}_1 = 0 \Rightarrow u_2 e^{q_1} p_3 = 0, \tag{1.82a}$$

$$\dot{\phi}_2 = 0 \Rightarrow u_1 e^{q_1} p_3 = 0, \tag{1.82b}$$

$$\dot{\phi}_3 = 0 \Rightarrow p_5 = 0. \tag{1.82c}$$

From (1.82a)–(1.82b), two possibilities should be considered: (i) $p_3 \neq 0$ and $u_1 = 0 = u_2$; or (ii) $p_3 = 0$, no condition on u_1 or u_2. The constraint algorithm "bifurcates."

However, the first possibility is incompatible with the preservation in time of the secondary constraint,

$$\dot{p}_5 = 0 \Rightarrow p_3 = 0. \tag{1.83}$$

Hence, $p_3 = 0$ arises as a tertiary constraint, and only the second possibility (ii) is consistent.

All the constraints are first class. What should be remarked is that the tertiary constraint $p_3 = 0$ appears not only in the bracket of H with the secondary constraint but also in the algebra of the primary constraints,

$$[\phi_1, \phi_2] = -e^{q_1} p_3, \tag{1.84a}$$

$$[\phi_2, \phi_3] = 0, \qquad [\phi_3, \phi_1] = 0. \tag{1.84b}$$

For this reason, it is not clear that $p_3 = 0$ should be called a tertiary constraint, since it already appears in the first consistency equation $\dot{\phi}_1 = 0$. This is actually why we could not exclude the possibility (i) (ϕ_1 and ϕ_2 second class with u^1 and u^2 equal to zero) *a priori.*

Example 4:

$$H = 0, \tag{1.85}$$

$$\phi_1 \equiv p_1 = 0, \qquad \phi_2 \equiv p_3 + q^1 q^2 = 0. \tag{1.86}$$

The preservation in time of the constraints leads to

$$\dot{\phi}_1 = 0 \Rightarrow u^2 q^2 = 0, \tag{1.87a}$$

$$\dot{\phi}_2 = 0 \Rightarrow u^1 q^2 = 0. \tag{1.87b}$$

Hence, two possibilites should be considered:

(i) $q^2 \neq 0$. In that case, u^1 and u^2 are required to vanish by (1.87) and the constraints are second class.

(ii) $q^2 = 0$. The condition $\dot{q}^2 = 0$ is then automatically fulfilled. The constraints are first class; there is no condition on either u^1 or u^2.

In this example, there is no global separation of the constraints into first or second class because the rank of the matrix $[\phi_i, \phi_j]$ is not constant on the constraint surface. This system is somewhat pathological, since one cannot give to it a uniform treatment. There is a bifurcation in the analysis of the consistency conditions leading to physically distinct consistent possibilities. It is precisely to eliminate such examples that the assumption of constant rank for the matrix $[\phi_j, \phi_{j'}]$ was made in the text (§1.3.1.)

Because of all the various possibilities that exist, a general, systematic study of the consistency algorithm appears to be rather involved. It is, therefore, fortunate that the distinction between primary and secondary constraints turns out to be of no physical importance in the final form of the theory and that, consequently, the detailed form of the consistency algorithm is not needed in its development.

APPENDIX 1.A: GLOBAL PROOF OF $G \approx 0 \Rightarrow G = g^j \phi_j$

We have shown in the text that any function vanishing on the constraint surface $\phi_j = 0$ can be written as $G = g^j \phi_j$ with regular coefficients g^j, at least in an open neighborhood of any point of the constraint surface.

We extend here the argument to provide a global proof of this property. We adopt a method that will be very useful later in the study of the BRST symmetry, since it will enable us to prove along the same steps acyclicity of appropriate differentials occurring in the formalism.

Consider a covering of a neighborhood of the constraint surface by open sets $O_{(i)}$, on each of which the property $G \approx 0 \Rightarrow G = g^j{}_{(i)}\phi_j$ holds, with functions $g^j{}_{(i)}$ that are defined on $O_{(i)}$. Complete the covering of phase space by adding open sets V_k that do not intersect the constraint surface. Choose the V_k so that on each of them, at least one constraint, denoted by $\phi_{j(k)}$, does not vanish ($\phi_{j(k)} = 0$ at no point of V_k). [Example: Take V_j to be the open set containing all points where $\phi_j \neq 0$.]

Now, on each V_k the property $G \approx 0$ implies $G = g^j{}_{(k)}\phi_j$. In fact, any function on V_k can be written as $g^j{}_{(k)}\phi_j$; take, *e.g.*, $G = (G/\phi_{j(k)})\phi_{j(k)}$ ($\phi_{j(k)} \neq 0$ on V_k).

Therefore, we have a covering of phase space by open sets $O_\alpha = (O_{(i)}, V_k)$ such that on each O_α the weakly vanishing function G is given by $G = g^j_\alpha \phi_j$. The coefficients g^j_α depend on the open set O_α, and there is no guarantee at this point that $g^j_\alpha = g^j_\beta$ on $O_\alpha \cap O_\beta$. Thus, it is not clear at this stage that G can be written as $G = g^j \phi_j$ with coefficients g^j that are globally defined throughout phase space.

The functions g^j_α are, however, not unique, and one can redefine them by equivalent functions g^j in such a way that the expansion $G = g^j \phi_j$ has global validity. This is done with the help of a set of nonnegative smooth functions f_α with the following properties:

(i) supp $f_\alpha \subset O_\alpha$ (supp f_α is the support of f_α, *i.e.*, the closure of the set on which $f_\alpha \neq 0$);
(ii) $\sum_\alpha f_\alpha = 1$ everywhere; and

(*iii*) each point possesses a neighborhood in which the series $\sum_\alpha f_\alpha$ reduces to a finite sum (and thus only a finite number of $f_\alpha \neq 0$ at each point).

The functions f_α define a "partition of unity."

Now, consider the function $g^j = \sum_\alpha f_\alpha g^j_\alpha$. This function is well defined, since (*i*) the sum $\sum_\alpha f_\alpha g^j_\alpha$ is actually finite (so no convergence problem arises); and (*ii*) the function $f_\alpha g^j_\alpha$ goes smoothly to zero as one leaves the open set O_α because supp $f_\alpha \subset O_\alpha$, so that one can consistently interpret $f_\alpha g^j_\alpha$ as zero out of O_α, where g^j_α may not be defined.

By definition, one finds $g^j \phi_j = \sum_\alpha f_\alpha g^j_\alpha \phi_j = \sum_\alpha f_\alpha G = G$, since $\sum_\alpha f_\alpha = 1$. This proves the theorem because the coefficients g^j are well defined and regular everywhere.

EXERCISES. CHAPTER ONE

1.1. Consider a Lagrangian $L(q^n, \dot{q}^n)$ with primary constraints $\phi_m(q,p) = 0$, $n, n' = 1, \ldots, N$, $m = 1, \ldots, M$.

(a) Show that the M vectors $\partial\phi_m/\partial p_n(q, p(q,\dot{q}))$ provide a complete set of null vectors of the Hessian matrix $\partial^2 L/\partial\dot{q}^n \partial\dot{q}^{n'}$. [Hint: Differentiate the identity $\phi_m(q, p(q,\dot{q})) = 0$ with respect to $\dot{q}^n$ to verify that these vectors are annihilated by $\partial^2 L/\partial\dot{q}^n\, \partial\dot{q}^{n'}$; differentiate with respect to q^n and use the regularity conditions to show that they are complete (*i.e.*, contain M' independent combinations).]

(b) Infer from (a) that two inverse images $\dot{q}^n_1$ and $\dot{q}^n_2$ of a given p_n must be related as $\dot{q}^n_2 - \dot{q}^n_1 = u^m \partial\phi_m/\partial p_i$ for some u^m. (Hint: Use the fact that $\partial\phi_m/\partial p_n$ is constant along the null surfaces of $\partial^2 L/\partial\dot{q}^n \partial\dot{q}^{n'}$, since it only depends on $\dot{q}^n$ through p_n.)

1.2. Same notations as for Exercise 1.1. This exercise provides an alternative derivation of the equivalence between the Lagrangian and Hamiltonian equations of motion.

(a) Let $\partial^2 L/\partial\dot{q}^{n'} \partial\dot{q}^n$ be of rank $N - M'$. Order the variables in such a way that the $(N - M') \times (N - M')$ submatrix $\partial^2 L/\partial\dot{q}^\alpha\, \partial\dot{q}^\beta$ $(\alpha, \beta = M' + 1, \ldots, N)$ is invertible. (This can be done because $\partial^2 L/\partial\dot{q}^n\, \partial\dot{q}^{n'}$ is symmetric and of rank $N - M'$.) The velocities $\dot{q}^\alpha$ can be expressed in terms of p_α and $\dot{q}^{m'}$ $(m' = 1, \ldots, M')$, *i.e.*, the transformation $\dot{q}^n \to \dot{q}^{m'}, p_\alpha$ is invertible. Show that the primary constraints can be taken to be $p_{m'} - P_{m'}(q^n, p_\alpha) = 0$, $(m' = 1, \ldots, M')$, where $P_{m'}$ does not depend on $\dot{q}^{m'}$.

(b) Check explicitly that

$$H = (p_\alpha \dot{q}^\alpha + P_{m'}\dot{q}^{m'} - L)\big|_{\dot{q}^\alpha = \dot{q}^\alpha(q^n,\, \dot{q}^{m'},\, p_\alpha)}$$

depends only on q^n and p_α and not on $\dot{q}^{m'}$.

(c) Show that the transformation inverse to $p_\alpha = \partial L/\partial\dot{q}^\alpha$, $\dot{q}^{m'} = \dot{q}^{m'}$ is

$$\dot{q}^\alpha = \partial H/\partial p_\alpha - \dot{q}^{m'} \partial P_{m'}/\partial p_\alpha\,, \quad \dot{q}^{m'} = \dot{q}^{m'}.$$

(d) Prove that the Lagrangian equations of motion are equivalent to

$$\delta \int (p_\alpha \dot{q}^\alpha + P_{m'}(q^n, p_\alpha)\dot{q}^{m'} - H) = 0$$

under arbitrary variations of p_α, q^n.

(e) Show, finally, that this last variational principle is equivalent to

$$\delta \int (p_i \dot{q}^i - H - u^m \phi_m)$$

under arbitrary variations of q^i, p_i and u^m. Relate the velocities $\dot{q}^{m'}$ to the Lagrange multipliers.

1.3. Consider a system with a single coordinate q and Lagrangian $L \equiv 0$.

(a) Compute the constraint ϕ and the Hamiltonian. Check explicitly that the equations of motion in Lagrangian and Hamiltonian forms are equivalent. What happens if one replaces the constraint function ϕ by ϕ^2 in the equations of motion? (Answer: The equations are no longer correct.)

(b) Repeat the exercise with $L = 1/2(\dot{q}^1 - \dot{q}^2)^2$.

1.4. Consider an action $S[y^i, z^A]$ that depends on some coordinates y^i, as well as "auxiliary fields" z^A. That is, assume that the variational equations of motion

$$\frac{\delta S}{\delta z^A} = 0$$

can be solved to yield the z^A as functions of y^i and their time derivatives,

$$\frac{\delta S}{\delta z^A} = 0 \Leftrightarrow z^A = z^A(y^i, \dot{y}^i).$$

(a) Show that the variational principle based on $S[y^i, z^A]$ is equivalent to the variational principle $S'[y^i]$ obtained by eliminating the auxiliary fields z^A from S, using their own equations of motion.

(b) Use this property to show that the action principle (1.13) is equivalent to the original action principle. (Hint: Take p_i and u^m for the auxiliary z^A.)

1.5. The issue discussed in this exercise arises in general relativity when one treats the cosmological constant as a dynamical variable.

Consider the action

$$S[q, A] = \int dt \frac{1}{2}\left(\dot{q}^2 + \frac{\dot{A}^2}{q^2}\right).$$

(a) Show that the equations of motion imply that

$$\dot{A} = cq^2$$

and that q obeys the harmonic oscillator equation

$$\ddot{q} + c^2 q = 0,$$

where c is a constant of integration.

(b) Show that it is not legitimate to replace $\dot{A}$ by cq^2 in the variational principle if one treats c in the new action as a given constant.

(c) Is there a contradiction with the previous exercise?

(d) This exercise shows the care that must be involved in order to take proper account of the boundary conditions. The apparent contradiction above results from the fact that the constant of integration c cannot be treated as independent from q. Rather, it is a functional of q. Indeed, in the given variational principle the proper boundary conditions are obtained by specifying q and A at the endpoints t_1, t_2 (since one must use $\delta q = \delta A = 0$ at the endpoints in order to derive the equations). This means that c is not a given constant but must be adjusted so as to fulfill $A(t_2) = A_2$, $A(t_1) = A_1$. In doing so, one finds that c is a functional of $q(t)$,

$$c = \frac{A_2 - A_1}{\int_{t_1}^{t_2} q^2(u)\, du}.$$

Only if one takes that functional dependence into account does one recover the correct equations of motion for q from the new action principle.

In order to be able to treat c as a constant, one would need a variational principle in which $\dot{A}$ is given at the endpoints (and then one would have $c = \dot{A}_1/{q^2}_1$). Such a variational principle exists and simply differs from the original one by the boundary term $-[(\dot{A}A)/q^2]_{t_1}^{t_2}$,

$$S' = \int dt \left[\frac{1}{2}\left(\dot{q}^2 + \frac{\dot{A}^2}{q^2}\right) - \frac{d}{dt}\left(\frac{\dot{A}A}{q^2}\right)\right].$$

It is easy to see that S' is an extremum along the classical trajectory for histories $q(t)$, $A(t)$, which obey $q(t_1) = q_1$, $q(t_2) = q_2$, $\dot{A}(t_1) = \dot{A}_1$, $\dot{A}(t_2) = \dot{A}_2$ at the endpoints. Consequently, it is legitimate to treat c as a constant in the action principle that one obtains upon elimination of A from S' using its own equation of motion. As a final point, we note that the terminology "auxiliary field," which we introduced in Exercise 1.4 above, usually covers only the case where the fields that can be eliminated obey nondifferential equations of motion, for which there is then no boundary condition at t_1 or t_2 to be given. In that sense, the field A is not an auxiliary field, but the p_n's and the u^m's of the Hamiltonian formalism are. The concept of auxiliary field is extremely useful, and examples are given in the exercises to Chapters 3 and 18.

1.6. Consider a Lagrangian $L(q, \dot{q})$ invariant under the following gauge transformations

$$\delta L = 0 \qquad \text{for} \qquad \delta q^i = a^i{}_A(q)\varepsilon^A,$$

where ε^A are arbitrary functions of time. Assume that these transformations "close," *i.e.*,

$$[a_A, a_B]^i \equiv a^j{}_A\, \partial a^i{}_B/\partial q^j - a^j{}_B\, \partial a^i{}_B/\partial q^j = C^C{}_{AB}\, a^i{}_C.$$

(a) From $\delta L = 0$ and the independence of ε^A and $\dot{\varepsilon}^A$ at any given time, infer the following identities

$$\frac{\partial L}{\partial q^i} a^i{}_A + \frac{\partial L}{\partial \dot{q}^i} \frac{\partial a^i{}_A}{\partial q^j} \dot{q}^j = 0 \tag{i}$$

$$\frac{\partial L}{\partial \dot{q}^i} a^i{}_A = 0. \tag{ii}$$

(b) Assume that (ii) are the only primary constraints,

$$\phi_A \equiv p_i \, a^i_A = 0,$$

i.e., that the degeneracy of the Lagrangian is only due to the above gauge invariance. Show from (i) that the primary constraints are preserved in time by the equations of motion, so there are no secondary constraints.

(c) Compute $[q^i, u^A \phi_A]$ and compare the result with the gauge transformations.

1.7. Consider the Lagrangian

$$L = (\dot{q} - x)(\dot{y} - q)$$

invariant under

$$\delta x = \ddot{\varepsilon}, \qquad \delta q = \dot{\varepsilon}, \qquad \delta y = \varepsilon.$$

(a) Compute the primary constraints.
(b) Compute the canonical Hamiltonian.
(c) Show that one gets both secondary and "tertiary" constraints.

1.8. Let ϕ_j be reducible constraints obeying the following reducibility equations,

$$Z^j{}_v \, \phi_j = 0.$$

Show that any function λ^j obeying

$$\lambda^j \, \phi_j = 0$$

can be written as

$$\lambda^j = \lambda^{jj'} \, \phi_{j'} + k^v \, Z^j{}_v$$

with $\lambda^{jj'} = -\lambda^{j'j}$. Accordingly, the equation $\lambda^j \phi_j = 0$ does not imply $\lambda^j = 0$, even in the irreducible case. (More on the the equations $\lambda^j \phi_j = 0$ in Chapters 9 and 10.)

1.9. Show that if the constraints

$$p_a + K_a(q^i, p_j, q^a) \approx 0$$

$(a = 1, \ldots, A; \; i, j = A+1, \ldots, N)$ are first class, and if there is no other constraint in the theory, then the Poisson bracket $[p_a + K_a, \, p_b + K_b]$ vanishes *strongly*.

1.10. Consider a system with constraints

$$\pi_k = 0, \qquad \phi_\alpha(q^i, p_i, \lambda_k, \pi_k) = 0,$$

where (q^i, p_i) and (λ_k, π_k) are canonically conjugate pairs. Assume that the constraints are all first class.

(a) Show that one can redefine the constraints $\pi_k = 0$, $\phi_\alpha = 0 \to \pi_k = 0$, $\psi_\alpha = 0$ in such a way that the new constraint functions ψ_α do not depend on either π_k or λ_k, $\psi_\alpha = \psi_\alpha(q^i, p_i)$. [Hints: (*i*) prove first that one can eliminate the π_k-dependence; (*ii*) show next, using the first-class property, that if $\phi_\alpha(q^i, p_i, \lambda^k_0) = 0$ for q^i, p_i, λ^k_0, then $\phi_\alpha(q^i, p_i, \lambda^k)$ vanishes for any λ^k and the same q^i, p_i; (*iii*) use finally this information to show that the equations $\phi_\alpha(q^i, p_i, \lambda_k) = 0$ and $\psi_\alpha(q^i, p_i) = 0$ are equivalent. Here, $\psi_\alpha(q^i, p_i) = \phi_\alpha(q^i, p_i, \lambda_k = C_k)$, where C_k are arbitrary fixed values of λ_k, *e.g.*, $C_k = 0$.]

(b) Prove that any first-class function is weakly equal to a function that depends on q^i and p_i only.

1.11. Check the Jacobi identity for the Dirac bracket by direct computation.

1.12. Consider a system with second-class constraints $\chi_\alpha \approx 0$. Let F be an arbitrary phase space function.

(a) Show that one can define another function F^*, equal to F on the surface of the second-class constraints $\chi_\alpha \approx 0$,

$$F^* = F + \nu^\alpha \chi_\alpha$$

such that

$$[F^*, \chi_\alpha] \approx 0.$$

[Throughout this exercise, "weak equality" means "equality modulo the second-class constraints only."] In particular, one finds that the function H^* is just the first-class Hamiltonian.

(b) Show that

$$[F^*, G] \approx [F^*, G^*].$$

(c) Verify that the Poisson bracket $[F^*, G^*]$ of F^* and G^* is weakly equal to the Dirac bracket $[F, G]^*$ of F and G.

(d) Show that

$$[[F, G]^*, H]^* \approx [[F^*, G^*], H^*].$$

(e) Infer from (c) and (d) that the Jacobi identity for the Dirac bracket holds weakly.

1.13. Same notations as for Exercise 1.12, with χ_α split as $\chi_\alpha = (\chi_{\bar{\alpha}}, \chi_a)$. Check that if the subset of constraints $\chi_{\bar{\alpha}} \approx 0$ is also second class by itself, *i.e.*, $\det [\chi_{\bar{\alpha}}, \chi_{\bar{\beta}}] \neq 0$, then the Dirac bracket associated with the full set of second-class constraints $\chi_\alpha = 0$ can be computed recursively by first eliminating the constraints $\chi_{\bar{\alpha}} = 0$ and then by eliminating the remaining second-class constraints $\chi_a = 0$ (*i.e.*, $[\ ,\ ]^*\big|_{\chi_\alpha} = [\ ,\ ]^*\big|^*_{\chi_{\bar{\alpha}}}\big|_{\chi_a}$).

1.14. Find the relation between the Dirac brackets corresponding to two different choices χ_α and $\bar{\chi}_\alpha = m_{\alpha\beta}\, \chi_\beta$ of second-class constraints where $m_{\alpha\beta}$ is an invertible matrix. Discuss.

1.15. Show that the brackets $[\gamma_a, \gamma_b]$ of two first-class constraints must be at least quadratic in the second-class constraints,

$$[\gamma_a, \gamma_b] = C_{ab}{}^c(q, p)\gamma_c + T_{ab}^{\alpha\beta}(q, p)\, \chi_\alpha \chi_\beta .$$

1.16. Consider the first-order action $S[q^i(t), p_i(t)] = \int (p_i \dot{q}^i - H)\, dt$, which is already in canonical form.

(a) Define momenta for both p_i and q^i, $\Pi^i = \partial L / \partial \dot{p}_i$, $P_i = \partial L / \partial \dot{q}^i$, and compute the constraints and Hamiltonian.

(b) Show that the Dirac brackets read $[q^i, q^j]^* = 0$, $[p_i, p_j]^* = 0$, and $[q^i, p_j]^* = \delta^i{}_j$, as expected.

1.17. In this exercise and the next two, $\gamma_a = 0$ are first-class constraints, and $C_a = 0$ are canonical gauge conditions. Assume that $\gamma_a = p_a + k_a(z^\Delta, q^a)$, where

(q^a, p_a) are canonically conjugate pairs possessing vanishing brackets with z^Δ (which stand for the other canonical variables).

(a) Show that $C^a \equiv q^a = 0$ are good gauge conditions.

(b) Compute the Dirac brackets in the gauge $q^a = 0$.

1.18. Let $[F, G]$ be two gauge-invariant functions, $[F, \gamma_a] \approx 0$, $[G, \gamma_a] \approx 0$. Prove that $[F, G]^* \approx [F, G]$, no matter which (good) gauge conditions are adopted.

1.19. Let F be the (gauge-invariant) generator of a global symmetry of the theory.

(a) Observe that if $[C_a, F] \neq 0$, the transformation generated by F does not preserve the gauge conditions. One must thus "improve" the symmetry transformation generated by F by adding to $\delta z^A = [z^A, F]$ a gauge transformation that brings one back to the gauge $C_a = 0$.

(b) Show that this improved symmetry is just generated by F acting through the *Dirac* bracket, $\delta^{(\text{improved})} z^A = [z^A, F]^*$.

1.20. Let $L = \frac{1}{2} a_{ij}(q) \dot{q}^i \dot{q}^j + b_i(q) \dot{q}^i + C(q)$. Show that the primary constraints $\phi_m = 0$ (if any) are linear in the momenta. Prove that if $[\phi_m, \phi_{m'}] = C_{mm'}{}^{m''} \phi_{m''}$, then one can make a change of coordinates $q^i \to Q^i(q)$ and add a total derivative to L, $L \to L' = L + (dV/dt)$, so that the primary constraints become $P_m \equiv \partial L' / \partial \dot{Q}_m = 0$.

1.21. Assume that the second-class constraints $\chi_\alpha = 0$ split as $\chi_\alpha \equiv (\gamma_a, C_a)$, where the subset $\gamma_a = 0$ is first class by itself, $[\gamma_a, \gamma_b] = C_{ab}{}^c \gamma_c$. Let F be an arbitrary function. Prove the existence of an equivalent function $\bar{F} = F + \lambda^\alpha \chi_\alpha$, which is first class with respect to the γ_a's, $[\bar{F}, \gamma_a] = f^b{}_a \gamma_b$. Show that the first-class system with constraints $\gamma_a = 0$ and Hamiltonian $\bar{H}$ is equivalent to the original second-class system.

1.22. (a) Show that any set of mixed second- and first-class constraints can locally be replaced by equivalent constraints that have brackets equal to 0 or ± 1. [Hint: The proof closely parallels the proof of the Darboux theorem. For more information, see Exercise 2.10.]

(b) Infer from (a) that one can always replace locally any set of constraints by an equivalent pure first-class one containing the same number of constraints and involving as many new conjugate pairs as half the number of second-class constraints. Furthermore, the equivalent first-class constraints and Hamiltonian differ from the original ones by terms that can be assumed to be linear in the new variables. [Hint: Apply to the constraints of (a) the method of §1.4.3 and rewrite the result in terms of the original constraints.]

(c) Show that if the original constraints are all second class, the construction of the equivalent first-class constraints $\gamma_\alpha = \chi_\alpha^{\text{original}}$ + "terms linear in the new variables" cannot bypass step (a). That is, one can find the constraints of (a) having brackets equal to 0 or ± 1 from the first-class constraints γ_α by mere algebraic manipulations..

1.23. Consider the following system of second-class constraints:

$$\chi_1 = 1 + q(e^{2qP-Q} + p + P^2),$$
$$\chi_2 = p + P^2,$$

where (q, p) and (Q, P) are conjugate pairs. Construct a first-class system equivalent to it.

1.24. (a) Let γ_a be constraints forming an "open algebra," $[\gamma_a, \gamma_b] = C_{ab}{}^c(q,p)\gamma_c$, with $C_{ab}{}^c$ depending on the canonical variables (§3.2.5). Show that, in general, the (strong) equations $[F, \gamma_a] = 0$ imply that $F = \text{const}$.

(b) Assume now that the γ_a's form a Lie algebra, $[\gamma_a, \gamma_b] = C_{ab}{}^c\gamma_c$ with $C_{ab}{}^c$ constant. Show that within each class $F \to F' = F + k^a\gamma_a$ of gauge-invariant functions F ($[F, \gamma_a] \approx 0$), one can find a function that is strongly gauge invariant.

(c) Let $F_a = 0$ be abelian constraints ($[F_a, F_b] = 0$) and let Q_α be gauge-invariant functions obeying

$$[Q_\alpha, F_a] \approx 0,$$
$$[Q_\alpha, Q_\beta] \approx D_{\alpha\beta}{}^\gamma Q_\alpha$$

with $D_{\alpha\beta}{}^\gamma$ constant. Show that one can locally redefine $Q_\alpha \to Q'_\alpha = Q_\alpha + k^a_\alpha F_a$ so as to have $[Q'_\alpha, F_a] = 0, [Q'_\alpha, Q'_\beta] = D_{\alpha\beta}{}^\gamma Q'_\gamma$ (strongly).

1.25. Consider a system with two canonical pairs (q^1, p_1) and (q^2, p_2) and one constraint $p_1 = 0$. Construct the gauge-invariant extensions of q^1 in the gauges $q^1 = 0$ and $q^1 = q^2$.

1.26. Extend the analysis of Dirac to higher-order Lagrangians depending on the accelerations, $\ddot{q}, \dddot{q}, \ldots$ up to order k. [Hint: Treat the velocities and the accelerations as independent variables by means of Lagrange multipliers in order to get a Lagrangian that contains only first-order time derivatives. Note: The extra added variables can be eliminated by means of their own equations and are thus auxiliary fields (Exercise 1.4).]

1.27. Consider a system with a single primary constraint $\phi_1 \approx 0$. Let $\phi_2 = [\phi_1, H]$, $\ldots, \phi_k = [\phi_{k-1}, H] \ldots.$ Assume that $[\phi_k, \phi_1]$ vanishes on $\phi_S = 0$ $(S \leq k)$ for $k \leq L$ and that $[\phi_L, \phi_1] \neq 0$ (even weakly).

(a) Show that the consistency conditions successively lead to the constraints $\phi_2 \approx 0, \phi_3 \approx 0, \ldots, \phi_L \approx 0$.

(b) Prove that $[\phi_k, \phi_s] \approx 0$ for $k + s \leq L$ and that $[\phi_L, \phi_1] \approx -[\phi_{L-1}, \phi_2] \approx [\phi_{L-2}, \phi_3] \approx - \cdots$. (Hint: Show that $[\phi_i, \phi_j] = [[\phi_{i-1}, H], \phi_j] \approx -[\phi_{i-1}, \phi_{j+1}]$ by using the Jacobi identity and $[H, \phi_k] \approx 0$ for $k < L$. Here, $i, j < L$.) Conclude that L is even and that the constraints are all second class.

CHAPTER TWO

GEOMETRY OF THE CONSTRAINT SURFACE

The distinction between first-class and second-class constraints was formulated in the previous chapter in terms of the matrix of the Poisson brackets of the contraints.

To apply the criterion for deciding whether the constraint surface is first class or second class, it is necessary that the constraints obey the regularity conditions that were formulated in §1.1.2. This appears strikingly in the second-class case, where $\det[\chi_\alpha, \chi_\beta] \neq 0$. If one replaces $\chi_\alpha \approx 0$ by $\bar{\chi}_\alpha = \chi_\alpha^2 \approx 0$, say, one gets $[\bar{\chi}_\alpha, \bar{\chi}_\beta] \approx 0$. It could thus seem that the same constraint surface ($\chi_\alpha = 0 \Leftrightarrow \chi_\alpha^2 = 0$) can be viewed as either second class or first class. By imposing the regularity condition on the constraints, the paradox is avoided (Exercise 2.1). What happens is that the functions χ_α^2, which are indeed first class, do not obey the regularity condition.

The purpose of this chapter is to provide an intrinsic characterization of the constraint surface that does not rely on the regularity condition and that is, on this ground, more satisfactory. This alternative reformulation is based on the geometrical features of the constraint surface. It sheds considerable light on the analysis of the previous chapter.

2.1. INDUCED TWO-FORM ON THE CONSTRAINT SURFACE

2.1.1. An Analogy: Surfaces in Minkowski Space

The geometrical description of the constraint surface is based on the structure that this surface inherits from its embedding in phase space. A good analogy is provided by the study of surfaces embedded in a pseudo-riemannian space. Since this topic is probably more familiar, we first briefly recall what can occur to a surface that is embedded, say, in Minkowski space.

From its embedding, a surface in Minkowski space inherits a metric. If $\eta_{\lambda\mu}$ is the Minkowski metric and if $x^\lambda = x^\lambda(y^i)$ are the parametric equations of the surface (with $x^\lambda_{,i} \neq 0$ in order for y^i to be a good coordinate system on the surface), then the induced metric is given by

$$g_{ij} = \eta_{\lambda\mu}\, x^\lambda_{,i}\, x^\mu_{,j} \qquad (,i \equiv \partial/\partial y^i). \tag{2.1}$$

The signature of the induced metric g_{ij} characterizes the surface as spacelike when the induced metric is of euclidean signature, null when it possesses tangent vectors orthogonal to all tangent directions, and timelike (mixed signature) otherwise. These features are intrinsic to the given embedded surface and do not depend on how it is represented.

It is worth noticing that the induced metric of an embedded surface cannot contain an arbitrary number of timelike or null directions, since this may conflict with the signature of Minkowski space. For instance, a hypersurface cannot contain two orthogonal timelike directions, since one cannot find two orthogonal timelike vectors in Minkowski space.

One can develop similar considerations for surfaces embedded in phase space and characterize them by means of the rank of an appropriate induced covariant tensor. The only difference with surfaces in Minkowski space is that the induced tensor is now antisymmetric. (There is in general no natural metric in phase space, but there is a natural nondegenerate antisymmetric tensor $\sigma_{\lambda\mu}$, known as the "symplectic two-form.")

2.1.2. Geometry of Phase Space (Symplectic Geometry)

The study of the geometrical properties of phase space is known as "symplectic geometry." We only give here an elementary presentation of those of its basic ideas that are relevant to our purposes. These basic ideas will be developed for convenience in local coordinates, but all the concepts introduced are intrinsic.

The characteristic feature of phase space is the existence of a Poisson bracket structure. If we denote the phase space coordinates as x^λ, this structure is completely captured by a rank-two, contravariant, antisymmetric tensor, whose components in the coordinate system x^λ are

$$\sigma^{\lambda\mu} = [x^\lambda, x^\mu]. \tag{2.2}$$

One easily verifies that $\sigma^{\lambda\mu}$ is indeed a tensor, *i.e.*,

$$\sigma'^{\lambda\mu} = (\partial x'^\lambda / \partial x^\alpha)(\partial x'^\mu / \partial x^\beta)\sigma^{\alpha\beta}$$

under the change of coordinates $x'^\lambda = x'^\lambda(x^\alpha)$. The tensor $\sigma^{\lambda\mu}$ obeys

$$\det \sigma^{\lambda\mu} \neq 0, \tag{2.3a}$$

$$\sigma^{[\lambda\mu}{}_{,\rho}\, \sigma^{\nu]\rho} = 0 \qquad \text{(Jacobi identity)}. \tag{2.3b}$$

Furthermore, the bracket of two functions F, G is given by

$$[F, G] = F_{,\lambda}\, \sigma^{\lambda\mu}\, G_{,\mu}. \tag{2.3c}$$

The proof of (2.3) is immediate in canonical coordinates $(x^\lambda) = (q^i, p_i)$, since then

$$\sigma^{\lambda\mu} = \begin{pmatrix} 0 & I \\ -I & 0 \end{pmatrix}. \tag{2.4}$$

Because $\det \sigma^{\lambda\mu} \neq 0$, one can invert $\sigma^{\lambda\mu}$. The inverse two-form is denoted by $\sigma_{\lambda\mu}$,

$$\sigma_{\lambda\mu}\, \sigma^{\mu\rho} = \delta^\rho_\lambda, \tag{2.5}$$

and the basic properties (2.3a)–(2.3b) of $\sigma^{\lambda\mu}$ are equivalently expressed in terms of $\sigma_{\lambda\mu}$ as

$$\det \sigma_{\lambda\mu} \neq 0, \tag{2.6a}$$

$$\sigma_{[\lambda\mu,\nu]} = 0 \qquad (\sigma_{\lambda\mu} \text{ is closed}). \tag{2.6b}$$

A two-form with the properties (2.6) defines a "symplectic structure." We thus see that phase space is equipped with a symplectic structure.

2.1.3. Induced Two-Form

It turns out to be more convenient to work with the covariant tensor $\sigma_{\lambda\mu}$ than with its contravariant inverse $\sigma^{\lambda\mu}$. This is because covariant tensor fields always induce a (covariant) tensor field on a surface embedded in a manifold, while contravariant tensors do not without additional structure.

Let $x^\lambda = x^\lambda(y^i)$ be the parametric equations of a surface Σ embedded in phase space. Then, the functions σ_{ij} given by

$$\sigma_{ij} = \sigma_{\lambda\mu} \frac{\partial x^\lambda}{\partial y^i} \frac{\partial x^\mu}{\partial y^j} \tag{2.7}$$

are the components of an antisymmetric tensor field on the embedded surface. We will refer to this tensor field as the "induced two-form."

Now, σ_{ij} inherits from $\sigma_{\lambda\mu}$ the property (2.6b),

$$\sigma_{[ij,k]} = 0 \qquad (\sigma_{ij} \text{ is closed}), \tag{2.8}$$

but it will fail, in general, to be invertible. This is just as in Minkowski space, where the metric induced on a hyperplane may possess null eigenvectors. The rank of σ_{ij}—which we assume constant over the surface—provides an intrinsic characterization of the embedded surface.

Let N and $N - M$ be the dimensions of phase space and of the surface Σ, respectively. (Thus, $N_{\text{here}} = 2N_{\text{Chapter 1}}$.) Then, σ_{ij} is an antisymmetric $(N-M)\times(N-M)$ square matrix. Its maximum possible rank is $N - M$,

$$\operatorname{rank} \sigma_{ij} \leq N - M. \tag{2.9}$$

What is its minimum possible rank? To answer this question, let us take a basis of vectors at each point of the surface as follows: take the first vectors X_a^λ $(a = 1, \ldots, A)$ to be a basis of the space of the null eigenvectors of σ_{ij}; they are tangent to Σ; complete X_a^λ to a basis of the tangent space to Σ by adding appropriate vectors $Y_{\bar{a}}^\lambda$ $(\bar{a} = A+1, \ldots, N-M)$. Finally, add M vectors Z_α^λ $(\alpha = N - M + 1, \ldots, N)$ to form a basis of the full vector space. By construction, one has

$$\sigma_{\lambda\mu} X_a^\lambda X_b^\mu = \sigma_{\lambda\mu} X_a^\lambda Y_{\bar{a}}^\mu = 0, \tag{2.10}$$

from which it results that $\sigma_{\lambda\mu} X^\lambda Y^\mu = 0$ whenever X^λ is a linear combination of the null vectors X_a^λ and Y^μ is a vector tangent to Σ.

Consider now the linear, homogeneous equations

$$A^a X_a^\lambda \sigma_{\lambda\mu} Z_\alpha^\mu = 0. \tag{2.11}$$

These are M homogeneous equations for the A unknown coefficient A^a. If $A > M$, these equations would possess nontrivial solutions no matter what $\sigma_{\lambda\mu} X_a^\lambda Z_\alpha^\mu$ is (more unknowns than equations). But in that case, one would find a nonzero vector $X^\lambda = A^a X_a^\lambda$ that has a vanishing skew-product $\sigma_{\lambda\mu} X^\lambda Z^\mu$ with *any* vector Z^μ, contradicting the fact that $\det \sigma_{\lambda\mu} \neq 0$. Hence, $A \leq M$ and the lower bound on rank σ_{ij} is given by $N - 2M$. Combining this result with (2.9), we have thus proved

$$N - 2M \leq \operatorname{rank} \sigma_{ij} \leq N - M. \tag{2.12}$$

We will now relate the rank of σ_{ij} to the first-class versus second-class properties of the constraint surface.

2.2. FIRST-CLASS CONSTRAINT SURFACE

2.2.1. Rank of Induced Two-Form

We first consider the case where all the constraints are first class (no second-class constraint). For simplicity, we also assume them to be independent, since including the reducible case would only unnecessarily cloud the issue without changing the main points of the discussion.

The constraint functions γ_a $(a = 1, \ldots, M)$ define vectors X_a^λ through

$$X_a^\lambda = \sigma^{\lambda\mu}\, \partial_\mu\, \gamma_a. \tag{2.13a}$$

By construction, one has

$$X_a^\lambda\, \partial_\lambda\, F \equiv \partial_a\, F = [F, \gamma_a], \tag{2.13b}$$

and, furthermore, the M vectors X_a^λ are linearly independent at each point of the constraint surface because of the irreducibility and regularity conditions. The vectors X_a^λ are sometimes called "Hamiltonian vector fields" associated with the functions γ_a. They generate the infinitesimal gauge transformations.

Theorem 2.1. *The induced two-form σ_{ij} is maximally degenerate,*

$$\text{rank}\ \sigma_{ij} = N - 2M, \tag{2.14}$$

and the null directions are spanned by the vectors X_a^λ.

Proof. Since we know that the $(N - M) \times (N - M)$ matrix σ_{ij} is of rank $\geq N - 2M$, all we have to do is show that it possesses M linearly independent null vectors.

Now, a vector Y^λ is tangent to $\gamma_a = 0$ if and only if $Y^\lambda\, \partial_\lambda\, \gamma_a \approx 0$. This implies:

(*i*) that the vectors X_a^λ associated with the (first-class) constraints γ_a are tangent to $\gamma_a = 0$, since $X_b^\lambda\, \partial_\lambda\, \gamma_a = [\gamma_a, \gamma_b] \approx 0$; and

(*ii*) that $\sigma_{\lambda\mu}\, X_a^\lambda\, Y^\mu \approx 0$ whenever Y^μ is tangent to the constraint surface, since $\sigma_{\lambda\mu}\, X_a^\lambda\, Y^\mu = -Y^\mu\, \partial_\mu\, \gamma_a$. Therefore, the M linearly independent vectors X_a^λ are null eigenvectors of the induced two-form, which proves the theorem.

The demonstration of the converse property, namely, that a surface with rank $\sigma_{ij} = N - 2M$ is described by constraints $\gamma_a = 0$, which are first class, $[\gamma_a, \gamma_b] \approx 0$, is sketched in Exercise 2.2.

A final comment concerning the terminology: First-class constraint surfaces—which contain, as we have just seen, all directions orthogonal to them in the skew-product determined by $\sigma_{\lambda\mu}$—are called by mathematicians "co-isotropic surfaces."

2.2.2. Null Surfaces and Gauge Orbits

Theorem 2.2. *On the constraint surface, the vectors X_a^λ are "surface-forming," i.e., generate M-dimensional manifolds.*

Proof. The proof amounts to checking Frobenius' integrability condition, namely,

$$[X_a, X_b]^\lambda = \text{combination of the } X_c^\lambda \text{ on } \Sigma \tag{2.15}$$

with

$$[X_a, X_b]^\lambda = X_b^\mu \, X_{a,\mu}^\lambda - X_a^\mu \, X_{b,\mu}^\lambda \qquad \text{(notice sign).} \tag{2.16}$$

Expression (2.16) is known as the "Lie bracket" of the vector fields X_a and X_b. We have included a minus sign in (2.16), as compared to the standard definition of the Lie bracket, so that we can use the following convenient property: If X_a^λ correspond to γ_a and X_b^λ to γ_b, then $[X_a, X_b]^\lambda$ corresponds to $[\gamma_a, \gamma_b]$, with γ_a and γ_b in the same order.

The verification of (2.15) is straightforward. One has

$$\begin{aligned}[X_a, X_b]^\lambda &= \sigma^{\lambda\mu}(C_{ab}{}^c \gamma_c)_{,\mu} \\ &= C_{ab}{}^c X_c^\lambda + \gamma_c \, \sigma^{\lambda\mu} \, \partial_\mu \, C_{ab}{}^c && \text{(2.17a)} \\ &\approx C_{ab}{}^c X_c^\lambda, && \text{(2.17b)}\end{aligned}$$

since $[\gamma_a, \gamma_b] = C_{ab}{}^c \gamma_c$.

It results from (2.17b) that the M vector fields X_a^λ generate M-dimensional surfaces on the constraint surface, as stated by the theorem.

Off the constraint surface, however, the second term in the right-hand side of (2.17a) does not vanish unless $\partial_\mu C_{ab}{}^c = 0$. It may not even be a combination of the vectors X_c^λ, in which case the transformations generated by the X_a^λ's fail to be integrable off $\gamma_a = 0$. In a terminology often used, "for an algebra that only closes on shell, gauge orbits only exist on shell."

By construction, any vector tangent to the M-dimensional surfaces spanned by the vector fields X_a^λ on the constraint surface is annihilated by the induced two-form; for that reason, these surfaces will be called the "null surfaces" from now on. Because the vector fields X_a^λ also generate the gauge transformations, we see that the *null surfaces coincide with the gauge orbits.* We have thus obtained an intrinsic characterization of the gauge orbits in terms of the induced geometrical structure.

The identification of the gauge orbits with the null surfaces of the induced two-form relies strongly on the postulate made throughout the book that all first-class constraints generate gauge transformations. If this were not the case, the gauge orbits would be strictly smaller than the null surfaces, and there would be null directions not associated with any gauge transformation.

2.2.3. Reduced Phase Space

Because the induced two-form σ_{ij} cannot be inverted, there is no induced Poisson bracket on the constraint surface. However, if one factors out the null surfaces, i.e, if one takes the quotient of the constraint surface by the gauge orbits, one gets a smaller space with an invertible two-form and a well-defined Poisson bracket (see Appendix 2.A for the details). This smaller space, obtained by identifying all points on the same orbit, is known as the *reduced phase space.*

Functions defined over the reduced phase space are constant along the gauge orbits, *i.e.,* are the "observables" defined in §1.5.2. Not surprisingly, one can easily show (see Appendix 2.A) that the induced bracket in the reduced phase space is simply the original Poisson bracket evaluated for gauge-invariant functions. For that reason, and because the Hamiltonian is gauge invariant, the projected motion in the reduced phase space is generated by standard, unconstrained Hamiltonian equations with a Poisson bracket structure.

It might appear from our discussion that one should reformulate the theory in the reduced phase space and drop all the other degrees of freedom, which either vanish on the constraint surface or fail to commute with the constraints.

However, when one takes the quotient of the constraint surface by the gauge orbits, one loses, in practice, desirable features such as manifest Lorentz invariance or, in the case of field theory, polynomiality in the fields and locality in space. Furthermore, it is usually technically impossible to reformulate the theory in terms of gauge-invariant quantities only, since this requires a complete set of solutions to the differential equations $[A, \gamma_a] \approx 0$.

For these reasons, we will not develop the theory of first-class constraint systems in terms of the reduced phase space. Rather, we will carry along all the dynamical variables. Actually, as we will see, a much greater simplicity is achieved by extending even more the phase space instead of reducing it. This will be the object of BRST theory.

2.3. SECOND-CLASS CONSTRAINTS

2.3.1. Rank of Induced Two-Form

Let us now turn to the case of pure second-class constraints $\chi_\alpha = 0$. We again define the Hamiltonian vector fields X^λ_α associated with the constraints as

$$X^\lambda_\alpha = \sigma^{\lambda\mu}\, \partial_\mu\, \chi_\alpha, \qquad \alpha = 1, \dots, M. \tag{2.18}$$

These vectors are linearly independent by the regularity and irreducibility assumptions.

In the first-class case, we found that the vectors (2.18) were all tangent to the constraint surface. Here, we have the opposite situation.

Theorem 2.3. *No linear combination of the vectors X^λ_α is tangent to the constraint surface.*

Proof. A vector Y^λ is tangent to $\chi_\alpha = 0$ if and only if $Y^\lambda\, \partial_\lambda\, \chi_\alpha \approx 0$. If $Y^\lambda \approx a^\alpha X^\lambda_\alpha$, one finds that $Y^\lambda\, \partial_\lambda\, \chi_\alpha$ is weakly equal to $a^\beta C_{\alpha\beta}$. Since $C_{\alpha\beta}$ is invertible, $a^\beta C_{\alpha\beta}$ weakly vanishes if and only if the coefficients a^α are all weakly zero, *i.e.*, $Y^\lambda \approx 0$. This establishes the theorem.

An immediate consequence of Theorem 2.3 is that the two-form σ_{ij} induced on $\chi_\alpha = 0$ is nondegenerate.

Theorem 2.4. *A surface $\chi_\alpha = 0$ is second class if and only if σ_{ij} is of maximum rank $N - M$,*

$$\det \sigma_{ij} \neq 0. \tag{2.19}$$

Proof. The vectors X^λ_α and their linear combinations are the only ones for which the skew-product $\sigma_{\lambda\mu}\, Y^\lambda X^\mu_\alpha$ is equal to zero for arbitrary tangent vectors Y^λ. This is because a generic vector tangent to $\chi_\alpha = 0$ is subject only to the conditions $Y^\lambda\, \partial_\lambda\, \chi_\alpha \equiv \sigma_{\lambda\mu}\, Y^\lambda X^\mu_\alpha = 0$ and to no others. On these grounds, the induced two-form σ_{ij} is degenerate if and only if there exists at least one nonzero linear combination $a^\alpha X^\mu_\alpha$ of the vectors X^μ_α that is tangent to $\chi_\alpha = 0$; *i.e.*, if and only if there exists at least one nonzero solution of $a^\alpha X^\mu_\alpha\, \partial_\mu\, \chi_\beta \equiv a^\alpha C_{\beta\alpha} = 0$. Theorem 2.3 states that there is no such solution when the constraints are second class. Conversely, if $a^\alpha C_{\beta\alpha} = 0$ implies $a^\alpha = 0$, then the matrix $C_{\beta\alpha}$ is invertible and the constraints are second class.

Theorem 2.4 provides an intrinsic characterization of a second-class constrained surface. It may be used to define a second-class surface without reference to its explicit representation by constraint equations $\chi_\alpha = 0$.

2.3.2. Dirac Bracket Revisited

Because $\det \sigma_{ij} \neq 0$, there is no null submanifold on the constraint surface, and one does not need to perform a further reduction to get an invertible induced two-form. It is because this second step is not required that second-class constraints only kill half as many degrees of freedom as do first-class ones.

Since σ_{ij} is invertible and closed, there is a well-defined induced Poisson bracket structure on the constraint surface: given two functions $f(y^i)$ and $g(y^i)$ on $\chi_\alpha = 0$, one can define

$$[f, g]^* = \frac{\partial f}{\partial y^i} \sigma^{ij} \frac{\partial g}{\partial y^j} \tag{2.20a}$$

with

$$\sigma^{ij} \sigma_{jk} = \delta^i{}_k. \tag{2.20b}$$

The situation is then the following: if F and G are phase space functions, one can consider their restrictions f, g on $\chi_\alpha = 0$ and compute their bracket $[f, g]^*$ according to (2.20). Or one can first compute the Dirac bracket $[F, G]^*$ by means of formula (1.48) and then consider the restriction of $[F, G]^*$ on $\chi_\alpha = 0$. Are the answers so obtained the same?

The answers indeed coincide, as was anticipated by our notations. This is the content of the next theorem.

Theorem 2.5. *The Dirac bracket associated with the second-class constraints $\chi_\alpha = 0$ is simply the bracket induced on the surface $\chi_\alpha = 0$. More precisely,*

$$[F, G]^* \big|_{\chi_\alpha = 0} = [f, g]^* \tag{2.21a}$$

where

$$F \big|_{\chi_\alpha = 0} = f, \qquad G \big|_{\chi_\alpha = 0} = g. \tag{2.21b}$$

The statement expressed by Eqs. (2.21a)–(2.21b) is meaningful because $[F, G]^*$ and $[F', G]^*$ coincide on the constraint surface whenever $F - F' \approx 0 \big([\chi_\alpha, G]^* = 0\big)$. Also, the equality (2.21a) provides another proof of the Jacobi identity for the Dirac bracket (1.48). Indeed, $[f, g]^*$ obeys (2.3b), since σ_{ij} is a closed two-form. (Only the Jacobi identity on $\chi_\alpha = 0$ is verified in this manner, but this is all that matters when one sets the constraints $\chi_\alpha = 0$ strongly equal to zero.)

Proof. Both the left- and the right-hand sides of (2.21a) are coordinate-independent. Hence, it is sufficient to check (2.21a) in a convenient system of local coordinates. To that end, we extend off $\chi_\alpha = 0$, the coordinates y^i defined on $\chi_\alpha = 0$, by requiring $[y^i, \chi_\alpha] \approx 0$. This is possible because if $\bar{y}^i$ is an arbitrary extension of y^i, then $\bar{y}^i + \lambda^{i\alpha}\chi_\alpha$ with $\lambda^{i\alpha} C_{\alpha\beta} = -[\bar{y}^i, \chi_\beta]$ fulfills the necessary requirement. Next, we take as phase space coordinates y^i and χ_α. The matrix of the Poisson brackets of these coordinates is given, on $\chi_\alpha = 0$, by

$$\sigma^{\lambda\mu} = \begin{pmatrix} [y^i, y^j] & 0 \\ 0 & C_{\alpha\beta} \end{pmatrix}, \tag{2.22a}$$

so that $\det [y^i, y^j] \neq 0$. The symplectic two-form is obtained by inversion,

$$\sigma_{\lambda\mu} = \begin{pmatrix} \sigma_{ij} & 0 \\ 0 & C^{\alpha\beta} \end{pmatrix} \tag{2.22b}$$

with

$$\sigma_{ij}[y^j, y^k] = \delta_i^k, \tag{2.22c}$$

$$C^{\alpha\beta} C_{\beta\rho} = \delta_\rho^\alpha. \tag{2.22d}$$

Since σ_{ij} are the components of the induced two-form, one infers from (2.20b) and (2.22c) that

$$[y^j, y^k]\big|_{\chi_\alpha=0} = [y^i, y^k]^*, \tag{2.23a}$$

where $[y^i, y^k]^* = \sigma^{ik}$ is the induced bracket. This proves the Theorem 2.5 for the functions y^i because $[y^i, \chi_\alpha] = 0$.

Now, consider an arbitrary pair of functions F, G. One has

$$\begin{aligned} [F, \chi_\alpha]\big|_{\chi_\beta=0} &= \left(\sigma^{\lambda\mu}\, \partial_\lambda\, F\, \partial_\mu\, \chi_\alpha\right)\big|_{\chi_\beta=0} \\ &= \left(\partial_\beta\, F\, C_{\beta\alpha}\right)\big|_{\chi_\beta=0} \end{aligned}$$

with $\partial_\beta F = \partial F / \partial \chi_\beta$. Therefore,

$$\begin{aligned} [F, G]\big|_{\chi_\beta=0} &= \left(\sigma^{\lambda\mu}\, \partial_\lambda\, F\, \partial_\mu G\right)\big|_{\chi_\beta=0} \\ &= \left(\sigma^{ij}\, \partial_i\, F\, \partial_j\, G\right)\big|_{\chi_\beta=0} + \left(\partial_\alpha\, F\, C_{\alpha\beta}\, \partial_\beta\, G\right)\big|_{\chi_\beta=0} \\ &= \sigma^{ij}\, \partial_i\, f\, \partial_j\, g + [F, \chi_\alpha]\, C^{\alpha\beta}\, [\chi_\beta, G]\big|_{\chi_\beta=0}. \end{aligned} \tag{2.23b}$$

By passing the second term in the right-hand side of this equality to the left-hand side, one gets the required Eq. (2.21a).

Since σ_{ij} is an intrinsic object, (2.21) provides an intrinsic definition of the Dirac bracket. From this point of view, a violation of the regularity conditions for second-class constraints, such as the one arising when one replaces χ by χ^2, simply renders ill-defined the particular representation (1.48) of the Dirac bracket. ($C^{\alpha\beta}$ becomes infinite while $[F, \chi_\alpha]$ and $[\chi_\beta, G]$ vanish.)

2.3.3. Solving the Constraints inside the Action

Once the second-class constraints have been eliminated, the equations of motion on the constraint surface, expressed in terms of the coordinates y^i, take the form of ordinary, unconstrained Hamiltonian equations,

$$\dot{y}^i = \left[y^i, h(y)\right]^* \tag{2.24a}$$

with

$$h(y) = H\big|_{\chi^\alpha=0}. \tag{2.24b}$$

If $a_i(y)$ is a one-form potential for the closed two-form σ_{ij},

$$\sigma_{ij} = \frac{\partial a_j}{\partial y^i} - \frac{\partial a_i}{\partial y^j}, \tag{2.25}$$

which locally exists by Poincaré lemma, Eqs. (2.24) derive from the variational principle $\delta S = 0$, with

$$S\left[y^i(t)\right] = \int (a_i\, \dot{y}^i - h)\, dt. \tag{2.26}$$

This can be checked by direct calculation.

What is the relationship between this variational principle and the original first-order variational principle

$$\delta S_E = 0, \qquad S_E\left[q^n(t), p_n(t), u^\alpha(t)\right] = \int (p_n \dot{q}^n - H - u^\alpha \chi_\alpha)\, dt, \tag{2.27}$$

in which the second-class constraints $\chi_\alpha = 0$ are enforced by means of the Lagrange multipliers u^α?

The answer is that (2.26) is simply obtained by solving all the (second-class) constraints inside the action (2.27). This is because

(*i*) Upon substituting $x^\mu = x^\mu(y)$ [with $x^\mu \equiv (q^n, p_n)$] and discarding a total time derivative, the kinetic term in (2.27) becomes,

$$p_n \dot{q}^n \equiv a_\mu(x)\dot{x}^\mu = a_i(y)\dot{y}^i \tag{2.28a}$$

with

$$a_\mu = (p_n, 0), \tag{2.28b}$$

since the exterior derivative of the induced one-form

$$a_i(y) = a_\mu\big(x(y)\big)\frac{\partial x^\mu}{\partial y^i} \tag{2.28c}$$

is equal to the induced two-form σ_{ij}. ("The exterior derivative operator d commutes with the pull-back of forms"); and

(*ii*) the last terms $H + u^\alpha \chi_\alpha$ in (2.27) reduce to $h(y)$ on the constraint surface.

The observation that (2.27) leads to (2.26) provides a new method for computing the Dirac brackets: one solves the constraints inside the action and reads the Dirac brackets from the kinetic term of the unconstrained action, using formulas (2.25) and (2.20). In many applications, this method turns out to be the most convenient.

Finally, a word concerning first-class constraints. *Provided the constraints* $\gamma_a = 0$ *obey the regularity condition,* it is certainly legitimate to solve them inside the action

$$S_E[q, p, u^a] = \int (p\dot{q} - H - u^a \gamma_a)\, dt. \tag{2.29}$$

This is just the content of the Lagrange multipler method, which guarantees that the implementations of the constraints $\gamma_a = 0$, either directly or by means of Lagrange multipliers, are equivalent. However, although technically correct, this procedure is not as useful as for second-class constraints because the equations of motion on the constraint surface do not possess thc canonical form: there is no induced bracket on the constraint surface. To get Hamiltonian equations, one needs to go to the reduced phase space. This can be done by adding canonical gauge conditions $C_a = 0$.

One may implement the gauge condition by adding C_a with a new Lagrange multiplier μ^a to (2.29), thus achieving the form (2.27),

$$S[q, p, u^a] \to S[q, p, u^a, \mu^a] = \int (p\dot{q} - H - u^a \gamma_a - \mu^a C_a)\, dt. \tag{2.30}$$

The action (2.30) yields the correct equations of motion because $\dot{\gamma}_a = 0$ implies $\mu^a = 0$. The equations for q, p are therefore unchanged and include the particular u^a fixed by $\dot{C}_a = 0$. The kinetic term of the reduced action yields the Dirac brackets in the gauge $C_a = 0$.

Canonical gauge conditions are quite peculiar in this regard. The noncanonical gauge conditions studied in Sec. 3.4 are such that one cannot insert them inside the action without losing some equations of motion.

2.4. MIXED CASE

The geometrical structure of the constraint surface in the mixed case with both first-class and second-class constraints is worked out in a straightforward manner.

One finds that the rank of σ_{ij} is equal to $N - M - k$

$$\text{rank } \sigma_{ij} = N - M - k, \tag{2.31a}$$

$$0 < k < M, \tag{2.31b}$$

where k is the number of independent first-class constraints. Moreover, the orbits generated by the gauge transformations are again the null surfaces of the induced two-form σ_{ij}. The reduced phase space is obtained by quotienting out the constraint surface by the orbits generated by the first-class constraints. It is clearly invariant under the redefinitions (1.44) of the constraints, since these redefinitions modify neither the constraint surface nor the gauge orbits. Finally, the dynamical equations on the surface $\chi_\alpha = 0$ of the second-class constraints, written in terms of the Dirac bracket, are generated by the extended action in which the second-class constraints have been solved for.

APPENDIX 2.A. MORE ON THE STRUCTURE OF THE REDUCED PHASE SPACE

In this appendix, we analyze in more detail the structure of the reduced phase space that arises when one factors out the gauge orbits generated by the first-class constraints $\gamma_a = 0$.

Through each point y^i on the constraint surface, there is one and only one gauge orbit. Therefore, the quotient space of the constraint surface by the orbits is well defined. We assume that this quotient space is a smooth differentiable manifold. To study its structure in the vicinity of an arbitrary point, we locally introduce $N-2M$ independent functions $z^\alpha(y^i)$ $(\alpha = 1, \ldots, N-2M)$ such that the equations of the gauge orbits are

$$z^\alpha(y^i) = \text{const.}$$

The functions z^α can be thought of as local coordinates on the reduced phase space and obey

$$X_a^i \frac{\partial z^\alpha}{\partial y^i} = 0 \qquad (a = 1, \ldots, M), \tag{2.A.1}$$

where X_a^i are the components in the y^i-coordinate system of the vectors tangent to the orbits.

We want to show explicitly that the reduced phase space inherits a nondegenerate two-form field from the degenerate two-form field $\sigma_{ij}(y)$. To that end, we must prove that $\sigma_{ij}(y)$ can be written as

$$\sigma_{ij}(y) = \sigma_{\alpha\beta}\big(z(y)\big)\,\frac{\partial z^\alpha}{\partial y^i}\,\frac{\partial z^\beta}{\partial y^j}, \tag{2.A.2}$$

where $\sigma_{\alpha\beta}$ is such that (i) it only depends on z; (ii) it is closed, $\sigma_{[\alpha\beta,\gamma]} = 0$; and (iii) it is invertible, $\det \sigma_{\alpha\beta} \neq 0$.

The functions $z^\alpha(y^i)$ define a map from the constraint surface to the reduced phase space. Equation (2.A.2) expresses that σ_{ij} (which is known) can be viewed as the pull-back under this map of a well-defined, invertible two-form $\sigma_{\alpha\beta}$ on the reduced phase space (which is to be determined).

Since (2.A.2) is covariant, we may analyze it by choosing the coordinates y^i in the most convenient way. We take $y^i = (z^\alpha, t^a)$, where t^a are coordinates along the orbits. Locally, this is always possible. In these coordinates, the vectors X^i_a have components

$$X^i_a = (0, \delta^b_a), \tag{2.A.3a}$$

and σ_{ij} reads

$$\sigma_{ij} = \begin{pmatrix} \sigma_{\alpha\beta}(z,t) & 0 \\ & \\ 0 & 0 \end{pmatrix}, \tag{2.A.3b}$$

since $\sigma_{ij}\, X^i_a = 0$.

Because σ_{ij} is closed, $\sigma_{\alpha\beta}$ cannot depend on t^a $(\sigma_{\alpha\beta,a} = 0)$, and furthermore one must have $\sigma_{[\alpha\beta,\gamma]} = 0$. In addition, the rank of σ_{ij} is equal to $N - 2M$, so that $\sigma_{\alpha\beta}$ must be invertible. This proves the assertion that a well-defined two-form exists in the quotient space. Note that the closedness of σ_{ij} plays a key role in guaranteeing that $\sigma_{\alpha\beta}$ only depends on z^γ.

The Poisson bracket in the reduced phase space of the functions z^α is given by

$$[z^\alpha, z^\beta] = \sigma^{\alpha\beta}(z) \tag{2.A.4a}$$

with

$$\sigma^{\alpha\beta}\,\sigma_{\beta\gamma} = \delta^\alpha_\gamma. \tag{2.A.4b}$$

Now, one can extend the functions z^α off the constraint surface and compute their brackets as functions of the coordinates x^λ for the whole phase space,

$$\big[z^\alpha(x), z^\beta(x)\big] = \Big(\frac{\partial z^\alpha}{\partial x^\lambda}\,\sigma^{\lambda\mu}\,\frac{\partial z^\beta}{\partial x^\mu}\Big)(x). \tag{2.A.5}$$

Because the functions z^α are gauge invariant and the constraints are first class, the value of the bracket (2.A.5) on the constraint surface is independent of how one extends $z^\alpha(y^i)$ off $\gamma_a = 0$. Furthermore, because the bracket of two first-class quantities is also first class, (2.A.5) depends only on the z^γ themselves (on $\gamma_a = 0$). So the question is: what is the relationship between (2.A.4) and (2.A.5)?

It is straightforward to check that these two expressions are equal. This can again be done by observing that (2.A.5) does not depend on the choice of phase space coordinates. Hence, one can simply verify the equality of (2.A.5) and (2.A.4a) in any convenient set of coordinates. If one takes $x^\lambda = (z^\alpha, t^a, \gamma_a)$, one finds that the brackets $[z^\alpha(x), z^\beta(x)]$ are equal to the (α, β)-components of the tensor $\sigma^{\lambda\mu}$ in that coordinate system. It also follows from (2.A.3b) that the two-form $\sigma_{\lambda\mu}$ is given on $\gamma_a = 0$ by

$$\sigma_{\lambda\mu} = \begin{array}{c} \\ z^\alpha \\ t^a \\ \gamma_a \end{array} \begin{array}{c} \begin{array}{ccc} z^\beta & t^b & \gamma_b \end{array} \\ \begin{pmatrix} \sigma_{\alpha\beta} & 0 & A_{\alpha b} \\ 0 & 0 & D_{ab} \\ -A_{\beta a} & -D_{ab} & B_{ab} \end{pmatrix} \end{array} \tag{2.A.6}$$

with $\det D \neq 0$. The (α, β)-components of the inverse tensor $\sigma^{\lambda\mu}$ are then easily seen to be equal to $\sigma^{\alpha\beta}$, Eq. (2.A.4b). This proves the assertion.

Finally, we note that one can represent the reduced phase space by choosing a canonical gauge $C_a = 0$, which selects one point on each orbit. In any such gauge, $\sigma^{\alpha\beta} = [z^\alpha, z^\beta]^*$ because $[z^\alpha, z^\beta] = [z^\alpha, z^\beta]^*$ for gauge-invariant functions (Exercise 1.18). In the presence of a Gribov obstruction, this description of the reduced phase space by means of a gauge condition is only local, even if the reduced phase space is smooth.

A case of a smooth reduced phase space that cannot be reached by means of global gauge conditions is given by a constraint surface that is a nontrivial principal fiber bundle with fibers that are the gauge orbits. The reduced phase space is just the base space and is smooth. However, there exists no global section on the bundle, *i.e.*, no global canonical gauge condition. For example, if the constraint surface is the group manifold of $SU(2)$ (the three-sphere) and the gauge orbits are left cosets of the $U(1)$ subgroup $\exp i\phi\sigma_3$, then the reduced phase space is $SU(2)/U(1) = S^2$, the two-sphere. The corresponding principal bundle (Hopf-bundle) is nontrivial because $S^3 \neq S^2 \times S^1$. A set of constraints that realizes this example can be written down taking phase space as $\mathbb{R}^4$ and a single constraint $\gamma = (q^1)^2 + (q^2)^2 + p_1^2 + p_2^2 - 1 = 0$.

EXERCISES. CHAPTER TWO

2.1. Show that when the regularity condition is imposed, the separation of the constraints into first-class and second-class, based on the matrix of the Poisson brackets of the constraints, is well defined. That is, show that, on the constraint surface, the rank of the Poisson bracket of the constraints is invariant under redefinition of the constraints that preserve the regularity conditions.

2.2. (For §2.2.1.) The vectors $e^\lambda_i = \partial x^\lambda / \partial y^i$ tangent to the constraint surface $x^\lambda = x^\lambda(y^i)$ and the constraint functions γ_a are related by $e^\lambda_i \, \partial_\lambda \gamma_a \approx 0$. Furthermore, the equations $a_\lambda \, e^\lambda_i \approx 0$ implies $a_\lambda \approx a^a \partial_\lambda \gamma_a$.

(a) Use this information to show that any vector X^λ that is skew-orthogonal ($\sigma_{\lambda\mu} X^\lambda Y^\mu = 0$) to all vectors Y^μ tangent to the surface $\gamma_a = 0$ must be a linear combination of $X^\lambda_a = \sigma^{\lambda\mu} \, \partial_\mu \, \gamma_a$.

(b) Conclude that if rank $\sigma_{ij} = N - 2M$, the vectors X^λ_a must be tangent to $\gamma_a = 0$ and, thus, $[\gamma_a, \gamma_b] \approx 0$.

2.3. Show that if the rank of the matrix of the Poisson brackets of the constraints is constant over the constraint surface Σ, then the rank of the induced two-form σ_{ij} is also constant over Σ.

2.4. Consider the Lagrangian $L(q, \dot{q}) \equiv 0$ of §1.6.2.

(a) Evaluate the two-form induced on the constraint surface.

(b) Determine the null submanifold and compare it with the gauge orbits.

2.5. Consider the Lagrangian $L = q^2$. Show that the action obtained by inserting the primary ($p = 0$) and secondary ($q = 0$) constraints inside the canonical action $S_H[q, p, u] = \int dt(p\dot{q} + q^2 - up)$ is the same as the action obtained by (i) applying in reverse the Lagrange multiplier method (*i.e.*, eliminating u and p); and (ii) getting rid of q, which appears as an auxiliary field, by solving its own equation of motion.

2.6. Consider the action

$$S[\phi, \pi, \lambda] = \int dt\, dx [\pi\dot{\phi} - \tfrac{1}{2}(\pi^2 + \phi'^2) - \lambda(\pi - \phi')^2]$$

for a chiral scalar field in two dimensions obeying the boundary conditions $\phi(-\infty) = -\phi(+\infty)$.

(a) Observe that the Lagrange multiplier method by itself does not guarantee that one can solve the constraint $\pi - \phi' = 0$ inside the action because the regularity condition is not fulfilled by $(\pi - \phi')^2$ (the gradient of $(\pi - \phi')^2$ vanishes on $(\pi - \phi')^2 = 0$).

(b) Check that this is, however, permissible and relate this possibility to the second-class property of the constraints.

(c) What would happen if ϕ were allowed to possess a constant mode, $\phi = \mathring{\phi} + \bar{\phi}$, $\mathring{\phi}' = 0$, $\bar{\phi}(-\infty) = -\bar{\phi}(+\infty)$ (in which case there is a first-class constraint contained in $\pi - \phi' = 0$)?

2.7. Let σ_{ij} be a 2-form, and let X^i_a be a complete set of null eigenvector fields, $\sigma_{ij} X^j = 0 \Rightarrow X^j = \lambda^a X^j_a$. Show that these vector fields are integrable if σ_{ij} is closed.

2.8. Verify explicitly that the inverse matrix of (2.A.6) has components compatible with $[z^\alpha, \gamma_a] \approx 0\,,\ \ [\gamma_a, \gamma_b] \approx 0$.

2.9. Let $\chi_\alpha = 0$ be second-class constraints and $\psi_\alpha = (\chi_\alpha)^2$. The Dirac bracket associated with ψ_α is ill defined.

(a) Show that the definition $[A, B]^* = \lim_{\varepsilon \to 0}[A, B]^*_\varepsilon$, where $[A, B]^*_\varepsilon$ is the Dirac bracket associated with $\psi^\varepsilon_\alpha \equiv \varepsilon\, \chi_\alpha + (\chi_\alpha)^2$ yields a well-defined quantity that coincides with the Dirac bracket associated with any regular representation of the constraint surface.

(b) Show that no such regularization can be used to define a Dirac bracket $[A, B]^*$ associated with first-class constraints (for arbitrary A and B).

2.10. Show that there exist canonical coordinates Q^n, P_n such that the constraint surface $\phi_j = 0$ is locally equivalent to $P_\alpha = 0, Q^\alpha = 0, P_A = 0$. The number of canonical pairs (P_α, Q^α) constrained to vanish is equal to half the number of second-class constraints. The number of momenta P_A constrained to vanish is equal to the number of first-class constraints. [Hint: Take P_1 to be any function vanishing on $\phi_j = 0$ with $dP_1 \neq 0$. Construct then Q_1 solution of the differential equation $\mathcal{L}_{X_{P_1}} Q_1 \equiv [Q_1, P_1] = 1$. Show that if P_1 is second class, one can impose as initial condition $Q_1 = 0$ on $\phi_j = 0$; while if P_1 is first class, Q_1 cannot be constant on $\phi_j = 0$. Complete P_1, Q_1 to a canonical coordinate system $P_1, Q_1, P'_2, Q'_2, \ldots$ and show that the constraint surface is equivalent to $P_1 = 0, \bar{\phi}_{j'} = 0$ (P_1 first class) or $P_1 = 0, Q_1 = 0, \bar{\phi}_{j''} = 0$ (P_1 and Q_1 second class), where the functions $\bar{\phi}_{j'}$ or $\bar{\phi}_{j''}$ do not depend on either P_1 or Q_1. Repeat then the argument. Note: The proof sketched here goes along the same lines as the proof of the Darboux theorem establishing the local existence of canonical coordinates for an arbitrary symplectic 2-form.]

CHAPTER THREE

GAUGE INVARIANCE OF THE ACTION

Gauge transformations were defined in Chapter 1 as transformations that do not change the physical state of the system. It was not discussed, however, in what sense these transformations were a symmetry of the action.

This chapter provides the explicit relation between the first-class constraints and the gauge invariance of the action. This is done both for the extended Hamiltonian action and for the original action in Lagrangian form.

An interesting application of the analysis arises when the Lagrangian gauge symmetries are known in closed form. One can then use this information to determine the number of the first-class constraints directly, without having to go through all the details of the constraint algorithm. In the absence of second-class constraints, this enables one to count the degrees of freedom by mere inspection of the gauge transformations.

Conversely, if the Lagrangian gauge symmetries are not known beforehand, one can find them from the first-class constraints, which, in turn, can be systematically derived within the Dirac formalism.

3.1. STRUCTURE OF THE GAUGE SYMMETRIES

3.1.1. Notations

Our first task is to clarify the structure of the set of gauge symmetries leaving a given action invariant. This action can be the original action (1.1), the total action (1.41), the extended action (1.42a), or, actually, any action that may even involve higher order time derivatives. As the results to be derived below apply equally to all cases, we denote the variables of the variational principle by y^i $(i = 1, \ldots, n)$ without further specification. These variables can be the q's or the q's, the p's, and the u's, etc., depending on which action principle is considered.

The only restriction imposed in the analysis is that the action be the integral of a local Lagrangian,

$$S[y(t)] = \int_{t_1}^{t_2} L\,dt. \tag{3.1a}$$

A local Lagrangian is, by definition, a function of y^i and of its successive time derivatives $\dot{y}^i, \ddot{y}^i, \ldots, \overset{(k)}{y}{}^i$ up to some *finite* order k,

$$L = L(y^i, \dot{y}^i, \ldots, \overset{(k)}{y}{}^i). \tag{3.1b}$$

Here, $\overset{(k)}{y}{}^i \equiv d^k y^i / dt^k$.

The equations of motion read

$$\frac{\delta S}{\delta y^i(t)} \equiv \frac{\delta L}{\delta y^i}(y(t)) = 0, \tag{3.2a}$$

where $\delta L / \delta y^i$ are the "variational derivatives" of L,

$$\frac{\delta L}{\delta y^i} = \frac{\partial L}{\partial y^i} - \frac{d}{dt}\frac{\partial L}{\partial \dot{y}^i} + \frac{d^2}{dt^2}\frac{\partial L}{\partial \ddot{y}^i} - \cdots + (-)^k \frac{d^k}{dt^k}\frac{\partial L}{\partial \overset{(k)}{y}{}^i}. \tag{3.2b}$$

The derivatives $\delta S / \delta y^i(t)$ themselves are known as the "functional derivatives" of the action and are such that

$$\delta S = \int \frac{\delta S}{\delta y^i(t)}\, \delta y^i(t)\, dt \tag{3.2c}$$

for arbitrary variations $\delta y^i(t)$ that vanish appropriately at the boundary.

The Lagrangian can be viewed as a function of $k \times n$ independent variables $y^i, \dot{y}^i, \ldots, \overset{(k)}{y}{}^i$. The total time derivative operator is then

$$\frac{d}{dt} = \dot{y}^i \frac{\partial}{\partial y^i} + \ddot{y}^i \frac{\partial}{\partial \dot{y}^i} + \cdots + \overset{(k+1)}{y}{}^i \frac{\partial}{\partial \overset{(k)}{y}{}^i}. \tag{3.2d}$$

One can also view the Lagrangian as defining, for each t, a functional of the history $y^i(t')$ that assigns to $y^i(t')$ the value taken by the Lagrangian at t,

$$L(t)\left[y^i(t')\right] = L\left(y^i(t), \dot{y}^i(t), \ldots, \overset{(k)}{y}{}^i(t)\right). \tag{3.3a}$$

When the second point of view is adopted, one writes $L(t)$ instead of just L. As a functional of $y^i(t')$, $L(t)$ possesses functional derivatives $\delta L(t)/\delta y^i(t')$. These are *not* the same as the variational derivatives $\delta L/\delta y^i$ defined by (3.2b). For instance, for y^j, one finds

$$\frac{\delta y^j}{\delta y^i} = \delta^j_i, \tag{3.3b}$$

but

$$\frac{\delta y^j(t)}{\delta y^i(t')} = \delta^j_i\, \delta(t-t'). \tag{3.3c}$$

Similarly, one has

$$\frac{\delta \dot{y}^j}{\delta y^i} = 0, \tag{3.3d}$$

but

$$\frac{\delta \dot{y}^j(t)}{\delta y^i(t')} = \delta^j_i\, \frac{d}{dt}\, \delta(t-t'). \tag{3.3e}$$

In general,

$$\frac{\delta L(t)}{\delta y^i(t')} = \delta(t-t')\frac{\partial L}{\partial y^i}(t) + \delta'(t-t')\frac{\partial L}{\partial \dot{y}^i}(t) + \cdots + \delta^{(k)}(t-t')\frac{\partial L}{\partial \overset{(k)}{y}{}^i}(t). \tag{3.3f}$$

The notations can be easily extended to the case of field theory in any number of spacetime dimensions.

3.1.2. Gauge Transformations

A gauge transformation of the action S is a transformation of the y's that (i) can be prescribed independently at each time t; and (ii) leaves S invariant.

Gauge transformations are parametrized by arbitrary functions of time—as opposed to rigid symmetry transformations—and take the form

$$\delta_\varepsilon y^i = \bar{R}^i_{(0)\alpha}\, \varepsilon^\alpha + \bar{R}^i_{(1)\alpha}\, \dot{\varepsilon}^\alpha + \cdots + \bar{R}^i_{(s)\alpha}\, \frac{d^s \varepsilon^\alpha}{dt^s} \tag{3.4a}$$

for some s. Here, the coefficients $\bar{R}^i_{(0)\alpha}, \bar{R}^i_{(1)\alpha}, \ldots, \bar{R}^i_{(s)\alpha}$ depend on y^i and their time derivatives up to some finite order, and $\varepsilon^\alpha(t)$ are arbitrary gauge parameters. Invariance of the action under (3.4a) means that for any choice of $\varepsilon^\alpha(t)$ one has

$$\delta_\varepsilon L = \frac{dK_\varepsilon}{dt} \tag{3.4b}$$

for some function K of the y's, the ε's, and their derivatives; so

$$\delta_\varepsilon S = K_\varepsilon(t_2) - K_\varepsilon(t_1) = 0 \tag{3.4c}$$

if $\varepsilon^\alpha(t)$, as well as an appropriate number of its time derivatives, vanishes at t_1 and t_2.

3.1.3. Noether Identities

It is at this point convenient to adopt a condensed notation, where the indices i, α also include t (*i.e.*, $i \leftrightarrow (i,t), \alpha \leftrightarrow (\alpha, t)$) and a summation over i, α implies an integration over t. In these notations, y^i stands for an entire history and (3.4a) becomes

$$\delta_\varepsilon y^i = R^i_\alpha\ \varepsilon^\alpha \big(\Leftrightarrow \delta_\varepsilon y^i(t) = \int dt'\ R^i_\alpha(t,t')\, \varepsilon^\alpha(t') \big) \tag{3.5a}$$

with

$$R^i_\alpha(t,t') = \bar{R}^i_{(0)\alpha}(t)\, \delta(t-t') + \bar{R}^i_{(1)\alpha}(t)\, \delta'(t-t') + \cdots . \tag{3.5b}$$

When one employs the condensed notations, $\delta/\delta y^k$ stands for the functional derivative and not for the variational derivative.

The invariance of the action,

$$\delta_\varepsilon S = \frac{\delta S}{\delta y^i}\, \delta_\varepsilon\, y^i = \frac{\delta S}{\delta y^i}\, R^i_\alpha\ \varepsilon^\alpha = 0, \tag{3.6a}$$

for any function $\varepsilon^\alpha(t)$ yields the local relations (Noether identities)

$$\frac{\delta S}{\delta y^i} R^i_\alpha = 0, \tag{3.6b}$$

which can be rewritten more explicitly as

$$\frac{\delta L}{\delta y^i} \bar{R}^i_{(0)\alpha} - \frac{d}{dt}\Big(\frac{\delta L}{\delta y^i} \bar{R}^i_{(1)\alpha} \Big) + \cdots = 0. \tag{3.6c}$$

One consequence of the local identities (3.6b)–(3.6c) is that the field equations are not independent. The existence of a gauge symmetry implies the presence of arbitrary functions of time in the general solution of the equations of motion, since a solution is mapped on a solution by a gauge transformation [$\delta_\varepsilon(\delta S/\delta y^j)$ vanishes when $\delta S/\delta y^i = 0$, as can be seen by differentiating (3.6b)].The functions $y^i(t)$ are then not fully determined once initial conditions are given. Further consequences of the Noether identities are examined in Exercises 3.3 and 3.4

3.1.4. Gauge Group—Gauge Algebra

For a given action functional, there will be a certain number of gauge transformations. What is the structure of the set of all the gauge transformations?

Two things that can be said without having to make any calculations is that the infinitesimal gauge transformations form a Lie algebra and that the finite gauge transformations form a Lie group. There is no escape to that result because invertible transformations leaving something (here the action) invariant always obey the group axioms. The algebra of all gauge transformations is denoted by $\bar{\mathcal{G}}$ in the sequel.

The reader may easily check that if $\delta_\eta y^i$,

$$\delta_\eta \, y^i = S^i_A \, \eta^A,$$

where η^A are arbitrary functions of time, is another gauge transformation ($\delta_\eta S = 0$), then both $\lambda\delta_\varepsilon y^i + \mu\delta_\eta y^i$ and $[\delta_\varepsilon, \delta_\eta]y^i = \delta_\varepsilon(\delta_\eta y^i) - \delta_\eta(\delta_\varepsilon y^i)$ also leave the action invariant. Here, λ and μ are arbitrary real numbers. Furthermore, $[\lambda\delta_\varepsilon, \delta_\eta]y^i = \lambda[\delta_\varepsilon, \delta_\eta]y^i$. Accordingly, the infinitesimal gauge transformations form a Lie algebra.

Now, it turns out that the gauge algebra $\bar{\mathcal{G}}$ of all gauge transformations is huge and contains a lot of irrelevant transformations. To develop the formalism, it is necessary to describe the gauge transformations more economically in terms of a much smaller set. The key requirement on this smaller set is that it should contain all the information about the Noether identities.

3.1.5. Trivial Gauge Transformations

Consider the following transformations,

$$\delta_\mu \, y^i = \mu^{ij} \frac{\delta S}{\delta y^j}, \tag{3.7a}$$

where μ^{ij} is arbitrary and antisymmetric,

$$\mu^{ij} = -\mu^{ji}, \tag{3.7b}$$

and may involve the fields. It is easy to see that (3.7a) leaves the action S invariant, no matter what μ^{ij} is,

$$\delta S = \frac{\delta S}{\delta y^j} \frac{\delta S}{\delta y^i} \, \mu^{ij} = 0. \tag{3.8}$$

This is because the product $(\delta S/\delta y^j)(\delta S/\delta y^i)$ is symmetric in i, j while μ^{ij} is antisymmetric. So, (3.7) defines a gauge transformation.

The commutator of any gauge transformation of the type (3.7) with an arbitrary gauge transformation—actually, with any local or global symmetry transformation—is again a transformation (3.7). Therefore, the set of all the gauge transformations (3.7) forms an ideal $\mathcal{N}$ of the full gauge algebra $\bar{\mathcal{G}}$. Indeed, if $\delta S = 0$ for $\delta_\rho y^i = \rho^i$,

$$\frac{\delta S}{\delta y^i}\rho^i = 0, \tag{3.9}$$

then one finds, using (3.9),

$$[\delta_\mu, \delta_\rho]y^i = \left(\frac{\delta \rho^i}{\delta y^k}\mu^{kj} - \frac{\delta \rho^j}{\delta y^k}\mu^{ki} - \rho^k \frac{\delta \mu^{ij}}{\delta y^k}\right)\frac{\delta S}{\delta y^j}, \tag{3.10}$$

which is of the form (3.7).

The transformations (3.7) are of no physical significance because in the Hamiltonian formalism they are not generated by a constraint. Indeed, (3.7) vanishes when the equations of motion hold ("on-shell") and therefore does not lead to the presence of arbitrary functions of time in their solution [the phase space conserved quantity associated with (3.7) is the identically vanishing function and generates the identity transformation—see Exercise 3.6]. Ultimately, the triviality of the transformations (3.7) stems from the fact that they are present for any given action. These "trivial gauge transformations" or "equations of motion gauge symmetries" are therefore to be disregarded. The relevant invariance algebra of the action is given by the factor algebra $\mathcal{G} \equiv \bar{\mathcal{G}}/\mathcal{N}$ of all the gauge transformations modulo the trivial ones, a concept that is mathematically well defined as the trivial transformations form an ideal.

The trivial gauge transformations are quite clearly disregarded in theories without gauge invariance, since they are never mentioned to begin with (see, however, Exercises 3.6–3.9). The reason that they have been noticed in gauge theories is that they often appear as commutators of nontrivial gauge transformations or in redefinitions of gauge transformations.

Before closing this section, we mention the following useful theorem.

Theorem 3.1. *Under suitable regularity assumptions on the functions $\delta S/\delta y^i$, to be precised below, any gauge transformation that vanishes on-shell can be written as in (3.7),*

$$\delta y^i \approx 0 \quad \text{and} \quad \delta y^i \frac{\delta S}{\delta y^i} = 0 \Rightarrow \delta y^i = \varepsilon^{ij}\frac{\delta S}{\delta y^j} \tag{3.11}$$

for some $\varepsilon^{ij} = -\varepsilon^{ji}$.

This theorem will be proved in §10.2.2 and §17.2.2.

3.1.6. Independent Noether Identities

The trivial gauge transformations were discarded because they imply no degeneracy of the equations of motion. This is not the end of the story, however; the remaining gauge transformations do not all lead to independent degeneracies of the variational principle. This can be seen as follows.

Let $\delta_\varepsilon y^i$ be gauge transformations

$$\delta_\varepsilon \, y^i = R^i_\alpha \, \varepsilon^\alpha, \tag{3.12}$$

leading to the Noether identities

$$\frac{\delta S}{\delta y^i} R^i_\alpha = 0. \tag{3.13}$$

Consider next the transformations

$$\delta_\eta y^i = (R^i_\beta \, M^\beta_A) \, \eta^A, \tag{3.14}$$

where M^β_A is some matrix that is allowed to depend on the fields. These transformations also leave the action invariant.

From the point of view of Lie algebras, the transformations $\delta_\eta y^i$ are linearly independent from the transformations R^i_α whenever the M's depend on y^i because in that case $\delta_\eta y^i$ is not a combination of R^i_α with coefficients that are real numbers. However, the Noether identities that follow from $\delta_\eta S = 0$,

$$\frac{\delta S}{\delta y^i} \, R^i_\alpha \, M^\alpha_A = 0, \tag{3.15}$$

are not independent from the Noether identities (3.13) in the sense that (3.15), viewed as an equation restricting the functional form of S, is a direct consequence of (3.13). Hence, there is no new information contained in the invariance of S under (3.14).

3.1.7. Generating Sets

This leads one to the concept of a generating set. A set G of gauge transformations

$$\delta_\varepsilon \, y^i = R^i_\alpha \, \varepsilon^\alpha \tag{3.16a}$$

is a "generating set," or a "complete set," if it contains all the information about the Noether identities. More precisely, G is a generating set if any gauge transformation δy^i can be written in terms of the elements of G as

$$\delta y^i \frac{\delta S}{\delta y^i} = 0 \Rightarrow \delta y^i = \mu^\alpha R^i_\alpha + M^{ij} \frac{\delta S}{\delta y^j}, \qquad M^{ij} = -M^{ji} \tag{3.16b}$$

with coefficients μ^α and M^{ij} *that may involve the fields.* Given a generating set, the Lie algebra of all the gauge transformations is spanned by (3.16b). Note that a generating set is, in general, *not* a basis in the Lie algebraic sense.

As the commutator of two elements of the generating set is also a gauge transformation, it must be expressible as in (3.16b). So, one finds

$$R^j_\alpha \frac{\delta R^i_\beta}{\delta y^j} - R^j_\beta \frac{\delta R^i_\alpha}{\delta y^j} = C_{\alpha\beta}{}^\gamma R^i_\gamma + M^{ij}_{\alpha\beta} \frac{\delta S}{\delta y^j}, \qquad M^{ij}_{\alpha\beta} = -M^{ji}_{\alpha\beta}, \tag{3.17}$$

where, in general, $C_{\alpha\beta}{}^\gamma$ and $M^{ij}_{\alpha\beta}$ involve the dynamical variables.

To analyze the dynamics of a gauge system, it is enough to consider only generating sets. This is because generating sets contain all the information about the Noether identities, about the degeneracy of the action principle, and about the number of required gauge conditions.

3.1.8. "Open Algebras"

When the generating set chosen to describe the gauge symmetry is such that $M^{ij}_{\alpha\beta} \neq 0$, one says that the generating set defines an "open algebra." It defines a "closed algebra" when $M^{ij}_{\alpha\beta} = 0$, even though $C_{\alpha\beta}{}^\gamma$ may involve the dynamical variables; the "closed algebra" is a true Lie algebra only when the $C_{\alpha\beta}{}^\gamma$'s are constant.

It should be stressed again that the terminology applies to the generating set and not to the set of all the gauge transformations, which is always a Lie algebra.

Gauge theories with a generating set G that is a Lie algebra are very special in that one can think of the transformations of G "abstractly," *i.e.*, independently of what the dynamics or the field content are. In that case, one can construct the generating set before writing down the action. This is a very lucky instance, however, and many interesting gauge theories are characterized by generating sets that do not form a Lie algebra. It is, therefore, necessary to develop the formalism in the general case.

3.1.9. Reducible Generating Sets

Although generating sets should be complete, they can contain some redundancy. This occurs when there are some relations among the generators, *i.e.*, when there exist some nontrivial μ^α such that the following identities hold,

$$\mu^\alpha R^i_\alpha = N^{ij} \frac{\delta S}{\delta y^j}. \tag{3.18}$$

By "nontrivial," it is meant that μ^α does not vanish on-shell. The coefficients N^{ij} are antisymmetric in i, j, as the righthand side of (3.18) is a gauge transformation.

So, if the only solution of (3.18) reads

$$\mu^\alpha = M^{\alpha i} \frac{\delta S}{\delta y^i}, \tag{3.19}$$

one says that the generating set is irreducible. Otherwise, it is reducible. The terminology "(ir)reducible" is justified because, as we shall see, generating sets that are (ir)reducible are associated with (ir)reducible systems of constraints as defined in Chapter 1.

A simple example of a reducible generating set can be exhibited in the theory described by the Lagrangian

$$L = \tfrac{1}{2} \dot{q}^T R^T R \dot{q}, \tag{3.20a}$$

where R is the 2-by-2 matrix

$$R = \begin{pmatrix} 0 & 1 \\ 0 & 0 \end{pmatrix} \tag{3.20b}$$

and where q is a two-dimensional column vector,

$$q(t) = \begin{pmatrix} q_1(t) \\ q_2(t) \end{pmatrix}. \tag{3.20c}$$

A complete set of gauge transformations is given by

$$\delta_\varepsilon q = R\varepsilon, \tag{3.20d}$$

where ε is an arbitrary time-dependent column-vector. The invariance of the Lagrangian (3.20a) follows from $R^2 = 0$. The transformations (3.20d) are reducible for the same reason: if $\varepsilon = R\eta$ for some η, the variation $\delta_\varepsilon q$ vanishes.

3.1.10. Relation between Different Generating Sets

Although the gauge algebras $\mathcal{G}$, $\mathcal{N}$, and $\bar{\mathcal{G}}$ are entirely determined by the action itself, there is an enormous freedom in the choice of the generating sets.

Two generating sets R^i_α and R^i_A ($\alpha = 1, \ldots, m$; $A = 1, \ldots, M$) are related as

$$R^i_\alpha = t^A_\alpha R^i_A + M^{ij}_\alpha \frac{\delta S}{\delta y^j}, \qquad M^{ij}_\alpha = -M^{ji}_\alpha, \tag{3.21a}$$

$$R^i_A = \bar{t}^\alpha_A R^i_\alpha + M^{ij}_A \frac{\delta S}{\delta y^j}, \qquad M^{ij}_A = -M^{ji}_A. \tag{3.21b}$$

This follows from (3.16b).

We will always take generating sets that are parametrized by a finite number of arbitrary functions of time (or of spacetime, in the case of field theory). As shown below, such a choice is always possible. This excludes taking the full algebra $\bar{\mathcal{G}}$ itself as a generating set. Further requirements (covariance, locality) may also be imposed to narrow down the choice. Yet, this still leaves a huge freedom. A basic condition on the theory is that its predictions do not depend on which particular generating set is adopted. As will be seen, this, indeed, turns out to be the case.

3.1.11. Generating Sets and Gauge Orbits

The gauge transformations map the stationary surface $\delta S/\delta y^i = 0$ on itself. Furthermore, on that surface the transformations of a generating set are integrable, *i.e.*, obey the Frobenius integrability condition: the extra term proportional to the equations of motion in (3.17) vanishes. Therefore, these transformations generate well-defined surfaces, the "gauge orbits." The gauge orbits do not depend on the choice of a generating set because of (3.21).

The number of elements in an irreducible generating set is equal to the dimension of the gauge orbits on the stationary surface. This gives a geometrical explanation of why generating sets play a central role. By contrast, the dimension of the Lie algebra $\bar{\mathcal{G}}$ containing all the gauge transformations is much greater: $\bar{\mathcal{G}}$ is far from having an irreducible action on the gauge orbits. (The word "dimension" is used here in a loose sense, since the orbits, the generating sets, and $\bar{\mathcal{G}}$ are all infinite-dimensional. In the present general discussion of the gauge symmetries of the action, we deliberately use the suggestive terminology of finite-dimensional manifold theory, as our aim is to convey informally the general ideas.)

The above observation yields a criterion more practical than (3.16b) for deciding whether a set of gauge transformations is complete. The set $\delta_\varepsilon \phi^i = R^i_\alpha \, \varepsilon^\alpha$ is complete if it accounts for all the arbitrary functions present in the general solution of the equations of motion. In that instance, the transformations $\delta_\varepsilon \phi^i = R^i_\alpha \, \varepsilon^\alpha$ completely generate the gauge orbits on the stationary surface.

One may express the requirement of completeness of the generating set by means of a condition on the second functional derivatives of the action: all the null eigenvectors of the matrix $\delta^2 S / \delta y^i \, \delta y^j$ should be spanned on-shell by R^i_α, *i.e.*,

$$\frac{\delta^2 S}{\delta y^j \, \delta y^i} \xi^j = 0 \Rightarrow \xi^i = \lambda^\alpha R^i_\alpha \qquad \left(\text{on } \frac{\delta S}{\delta y^i} = 0\right). \tag{3.22}$$

Indeed, if δy^i is a gauge transformation, one finds, upon differentiation of $(\delta S / \delta y^i) \delta y^i = 0$ and use of (3.22), that δy^i is equal to $\lambda^\alpha R^i_\alpha$ on the stationary surface. Hence, $\delta y^i - \lambda^\alpha R^i_\alpha$ is a gauge transformation that vanishes on shell and, thus, it is a trivial gauge transformation by Theorem 3.1. This implies that $\delta y^i = \lambda^\alpha R^i_\alpha + M^{ij} \, \delta S / \delta y^j$, as required.

3.2. GAUGE TRANSFORMATIONS OF THE EXTENDED ACTION

3.2.1. Algebra of the Constraints

We now turn to the problem of determining a complete set of gauge symmetries of the extended action (1.42a),

$$S_E \left[q^n(t), p_n(t), u^a(t), u^\alpha(t) \right] = \int (p_n \dot{q}^n - H - u^a \gamma_a - u^\alpha \chi_\alpha) \, dt. \tag{3.23}$$

We recall that the γ's are the first-class constraints, while the χ's are second class. Furthermore, we take H to be the first-class Hamiltonian.

Because γ_a and H are first class, one has

$$[\gamma_a, \gamma_b] = C_{ab}{}^c \gamma_c + T^{\alpha\beta}_{ab} \chi_\alpha \chi_\beta, \tag{3.24a}$$
$$[\gamma_a, \chi_\alpha] = C^b_{a\alpha} \gamma_b + C^\beta_{a\alpha} \chi_\beta, \tag{3.24b}$$
$$[H, \gamma_a] = V_a{}^b \gamma_b + V_a^{\alpha\beta} \chi_\alpha \chi_\beta, \tag{3.24c}$$
$$[H, \chi_\alpha] = V^b_\alpha \gamma_b + V^\beta_\alpha \chi_\beta. \tag{3.24d}$$

Only squares of second-class constraints can appear in (3.24a) and (3.24c) because the bracket of two first-class quantities is also first class. The functions $T^{\alpha\beta}_{ab}$ and $V^{\alpha\beta}_a$ can be assumed to be symmetric in α, β. Furthermore, the bracket $[\chi_\alpha, \chi_\beta]$ is given by

$$[\chi_\alpha, \chi_\beta] = C_{\alpha\beta}, \quad \det C_{\alpha\beta} \neq 0. \tag{3.24e}$$

For the sake of comparing (3.24) with the formalism in which the second-class constraints are eliminated, it is useful to rewrite the constraint algebra in terms of the Dirac bracket. One gets

$$[\gamma_a, \gamma_b]^* = C'^{c}_{ab}\gamma_c + T'^{\alpha\beta}_{ab}\chi_\alpha\chi_\beta, \tag{3.25a}$$

$$[\gamma_a, \chi_\alpha]^* = 0, \tag{3.25b}$$

$$[H, \gamma_a]^* = V'^{b}_{a}\gamma_b + V'^{\alpha\beta}_{a}\chi_\alpha\chi_\beta, \tag{3.25c}$$

$$[H, \chi_\alpha]^* = 0. \tag{3.25d}$$

We leave it to the reader to determine the precise relation between $C_{ab}{}^{c}$, $T^{\alpha\beta}_{ab}, \ldots$ on the one hand and $C'^{c}_{ab}, T'^{\alpha\beta}_{ab}, \ldots$ on the other hand. That relation will not be explicitly needed in the sequel.

3.2.2. Gauge Transformations

The general solution $q^n(t), p_n(t), u^a(t), u^\alpha(t)$ of the equations of motion contain as many arbitrary functions of time as there are first-class constraints. These arbitrary functions can be taken to be the multipliers $u^a(t)$. A complete set of gauge transformations should thus contain as many gauge parameters as there are first-class constraints.

It is easy to exhibit such a complete set. Indeed, two gauge-related states of the system are obtained by integration of the infinitesimal transformations generated by the first-class constraints. So, the gauge transformations should read

$$\delta_\varepsilon F = \varepsilon^a [F, \gamma_a] \tag{3.26a}$$

where the arbitrary gauge parameters ε^a can be taken to depend on time only. Here, F is any function of the q's and the p's.

The variation δS_E vanishes if one transforms simultaneously the multipliers u^a, u^α as

$$\delta_\varepsilon u^a = \dot{\varepsilon}^a + u^c \varepsilon^b C_{bc}{}^{a} + u^\alpha \varepsilon^b C^a_{b\alpha} - \varepsilon^b V_b{}^{a}, \tag{3.26b}$$

$$\delta_\varepsilon u^\alpha = u^c \varepsilon^b T^{\alpha\beta}_{bc}\chi_\beta - \varepsilon^b V_b^{\alpha\beta}\chi_\beta + u^\beta \varepsilon^b C^\alpha_{b\beta}. \tag{3.26c}$$

That this is so can be seen as follows. First, (3.26a) is a time-dependent canonical transformation. Therefore, the variation of the kinetic term $p\dot{q}$ is a boundary term, plus the integral of the explicit time derivative of the canonical generator,

$$\delta(\dot{q}^n p_n) = \frac{dM}{dt} + \frac{\partial}{\partial t}(\varepsilon^a \gamma_a) \tag{3.27a}$$

$$= \frac{dM}{dt} + \dot{\varepsilon}^a \gamma_a, \tag{3.27b}$$

with M given by

$$M = \varepsilon^a \left(\frac{\partial \gamma_a}{\partial p_n} p_n - \gamma_a \right). \tag{3.27c}$$

Next, one observes that the variations of H, γ_a, and χ_α in δS_E are cancelled by the terms containing $C_{bc}{}^a, C^a_{b\alpha}, V_b{}^a, T^{\alpha\beta}_{bc}, V_b^{\alpha\beta}, C^\alpha_{b\beta}$ in (3.26b) and (3.26c), thanks to the relations (3.24). Hence, the variation of the extended action S_E reduces to the boundary term,

$$\delta_\varepsilon S_E = M(t_2) - M(t_1), \tag{3.28}$$

which is zero if $\varepsilon^a(t_2) = \varepsilon^a(t_1) = 0$.

Equations (3.26b) and (3.26c) are very useful in practice to determine the structure functions $C_{bc}{}^a$, $C^a_{b\alpha}$, $V_b{}^a$, $T^{\alpha\beta}_{bc}$, $V_b^{\alpha\beta}$, and $C^\alpha_{d\beta}$ (which depend in general on the q's and the p's) of the algebra of the gauge generators γ_a. In fact, in many cases one knows beforehand, on geometrical grounds or otherwise, how the Lagrangian multipliers must transform so as to leave the action invariant. One can then use (3.26b) and (3.26c) to find $C_{bc}{}^a, C^a_{b\alpha}, \ldots$. This procedure avoids direct computation of the brackets $[\gamma_a, \gamma_b]$, $[\gamma_a, H]$, and $[\gamma_a, \chi_\alpha]$, which may be quite laborious in complicated theories. (In some cases, the geometrical argument yields the structure functions as functions of the coordinates q^n and their velocities $\dot{q}^n$ rather than as phase space functions. Yet this is very useful information, since the structure functions are then determined as functions of q^n and p_n on the surface of the primary constraints.)

3.2.3. Another Generating Set

The transformations (3.26a)–(3.26c), where $\varepsilon^a(t)$ are arbitrary functions of time, account for all the degeneracies of the action. Hence, they form a complete set: any gauge transformation can be written in terms of them as in (3.16b). Another generating set, which differs from (3.26a)–(3.26c) by a redefinition of the gauge parameters and the addition of trivial gauge transformations, is obtained by replacing in (3.26a) the Poisson bracket by the Dirac bracket. One then finds as gauge transformations (Exercise 3.15)

$$\delta'_\varepsilon F(q,p) = \varepsilon^a(t)[F, \gamma_a]^*, \tag{3.29a}$$

$$\delta'_\varepsilon u^a = \dot{\varepsilon}^a + u^c \varepsilon^b C'^a_{bc} - \varepsilon^b V'^a_b, \tag{3.29b}$$

$$\delta'_\varepsilon u^\alpha = u^c \varepsilon^b T'^{\alpha\beta}_{bc} \chi_\beta - \varepsilon^b V'^{\alpha\beta}_b \chi_\beta - \frac{d}{dt}\left(\varepsilon^a C^{\alpha\beta}[\chi_\beta, \gamma_a]\right), \tag{3.29c}$$

which imply that

$$\delta'_\varepsilon \chi_\alpha = 0. \tag{3.29d}$$

The interest of the representation (3.29) of the gauge symmetry is that it can be compared directly with the gauge symmetries of the reduced theory in which the second-class constraints are eliminated (see Exercises 3.16 and 3.17).

3.2.4. Gauge Transformations as Canonical Transformations

According to (3.16b), the most general gauge transformation is given by

$$\delta y^i = R^i_{\ \alpha}\, \mu^\alpha + M^{ij} \frac{\delta S}{\delta y^j}, \tag{3.30}$$

where μ^α and M^{ij} are arbitrary functions that may involve time as well as the dynamical variables and their derivatives. If one writes down explicitly the term $R^i_{\ \alpha}\, \mu^\alpha$ for the generating set (3.26), one finds that $\delta_\mu F$ takes the form (3.26a) with ε^a replaced by μ^a, provided one keeps μ^a outside the Poisson bracket (Exercise 3.18).

When μ^a depends on q^n, p_n but not on their time derivatives, it is convenient to push μ^a inside the bracket

$$\delta_\mu F = \mu^a [F, \gamma_a] \to \bar{\delta}_\mu F = [F, \mu^a \gamma_a].$$

This is because the gauge transformation $\bar{\delta}_\mu$ is then a canonical transformation in the phase space of the q's and the p's (depending possibly on $t, u^a, \dot{u}^a, \ldots, u^\alpha, \dot{u}^\alpha, \ldots$).

The replacement of δ_μ by $\bar{\delta}_\mu$ can be achieved by adjusting appropriately the equations-of-motion symmetries in (3.30). One finds (Exercise 3.19)

$$\bar{\delta}_\mu F = [F, \mu^a \gamma_a], \tag{3.31a}$$

$$\bar{\delta}_\mu u^a = \frac{D\mu^a}{Dt} + [\mu^a, H_E] + u^c \mu^b C_{bc}{}^a + u^\alpha \mu^b C^a_{b\alpha} - \mu^b V_b{}^a, \tag{3.31b}$$

$$\bar{\delta}_\mu u^\alpha = u^c \mu^b T^{\alpha\beta}_{bc} \chi_\beta - \mu^b V^{\alpha\beta}_b \chi_\beta + u^\beta \mu^b C^\alpha_{b\beta}. \tag{3.31c}$$

Here, D/Dt measures both the explicit time dependence and the implicit time dependence occurring through the Lagrange multipliers and their derivatives,

$$\frac{D}{Dt} = \frac{\partial}{\partial t} + \dot{u}^a \frac{\partial}{\partial u^a} + \ddot{u}^a \frac{\partial}{\partial \dot{u}^a} + \cdots + \dot{u}^\alpha \frac{\partial}{\partial u^\alpha} + \ddot{u}^\alpha \frac{\partial}{\partial \dot{u}^\alpha} + \cdots. \tag{3.32a}$$

One has

$$\frac{d}{dt} = \frac{D}{Dt} + \dot{q}^n \frac{\partial}{\partial q^n} + \dot{p}_n \frac{\partial}{\partial p_n}. \tag{3.32b}$$

A useful application of (3.31) is obtained by taking $\mu^a = a^a_b(q,p)\,\bar{\varepsilon}^b(t)$ and setting $u^a = a^a_b \bar{u}^b$ in (3.31). One then finds that the gauge transformations (3.31a)–(3.31c) become those of the generating set associated with the equivalent representation $\bar{\gamma}_a = a^a_b \gamma_b$ of the first-class constraints. This provides an explicit verification that two equivalent choices of first-class constraints lead to gauge symmetries that are related according to (3.21).

The same considerations apply, of course, to the representation (3.29) of the gauge symmetries, in which the Dirac bracket is used instead of the Poisson bracket.

3.2.5. Open and Closed Algebras

If one computes the commutator of two gauge symmetries (3.26), one finds a transformation of the type (3.30),

$$[\delta_\varepsilon, \delta_\eta] y^i = \delta_\nu y^i + \delta_M y^i, \tag{3.33a}$$

$$\delta_\nu y^i = R^i_\alpha\, \nu^\alpha, \quad \delta_M y^i = M^{ij} \frac{\delta S}{\delta y^j}, \tag{3.33b}$$

where δ_ν and δ_M are given by

$$\nu^a = -C_{bc}{}^a \varepsilon^b \eta^c, \tag{3.33c}$$

$$\delta_M F = [F, C_{bc}{}^a]\, \frac{\delta S_E}{\delta u^a} \varepsilon^b \eta^c, \tag{3.33d}$$

$$\begin{aligned} \delta_M u^a = & -[q^n, C_{bc}{}^a]\, \frac{\delta S_E}{\delta q^n} \varepsilon^b \eta^c - [p_n, C_{bc}{}^a]\, \frac{\delta S_E}{\delta p_n} \varepsilon^b \eta^c \\ & + 12 \overset{(2)}{U}{}^{ae}_{bcd}\, \varepsilon^b \eta^c u^d \frac{\delta S_E}{\delta u^e} + 4 \overset{(2)}{H}{}^{ae}_{bc}\, \varepsilon^b \eta^c \frac{\delta S_E}{\delta u^e}. \end{aligned} \tag{3.33e}$$

For simplicity, we have assumed no second-class constraints. The inclusion of the second-class constraints in the algebra of the gauge transformations is direct. It only makes the formulas more cumbersome (Exercise 3.20).

In (3.33e), $\overset{(2)}{U}{}^{ae}_{bcd}$ and $\overset{(2)}{H}{}^{ae}_{bc}$ are the "second-order structure functions" implicitly defined by

$$2 \overset{(2)}{U}{}^{ae}_{bcd}\, \gamma_e = \overset{(1)}{D}{}^a_{[bcd]}, \quad \overset{(2)}{U}{}^{ae}_{bcd} = -\overset{(2)}{U}{}^{ea}_{bcd}, \tag{3.34a}$$

$$\overset{(1)}{D}{}^a_{bcd} = -\tfrac{1}{2}[C_{bc}{}^a, \gamma_d] + \tfrac{1}{2} C_{bc}{}^e C_{de}{}^a, \tag{3.34b}$$

$$2\overset{(2)}{H}{}^{ae}_{bc}\gamma_e = \overset{(1)}{E}{}^{a}_{[bc]}, \quad \overset{(2)}{H}{}^{ae}_{bc} = -\overset{(2)}{H}{}^{ea}_{bc}, \tag{3.34c}$$

$$\overset{(1)}{E}{}^{a}_{bc} = \tfrac{1}{2}[H, C_{bc}{}^{a}] - [V_b{}^a, \gamma_c] + \tfrac{1}{2}V^a_e C_{bc}{}^e - V^e_b C_{ec}{}^a. \tag{3.34d}$$

That the equations (3.34a) and (3.34c) possess a solution for $\overset{(2)}{U}{}^{ae}_{bcd}$ and $\overset{(2)}{H}{}^{ae}_{bc}$ follows from the Jacobi identity and will be analyzed in more detail in the discussion of the BRST formalism, where the higher-order structure functions turn out to play a central role (see §9.4.3 and Exercises 9.3 and 11.2).

It follows from (3.33) that the algebra of the gauge transformations (3.26) is open whenever the structure functions $C_{bc}{}^a$ in $[\gamma_b, \gamma_c] = C_{bc}{}^a\gamma_a$ are not constant. Conversely, when the C's are constant, $\delta_M F$ in (3.33d) vanishes and $\delta_M u^a$ reduces to the terms containing the functional derivatives $\delta S_E/\delta u^e$. But the coefficients of these terms are determined by (3.34) up to $\mu^{aef}\delta S_E/\delta u^f$, where μ^{aef} is completely antisymmetric. Because $\overset{(1)}{D}{}^{a}_{[bcd]} = 0$ (Exercise 3.21), one can use this freedom to set $\overset{(2)}{U}{}^{ae}_{bcd}$ equal to zero. With that choice, $\delta_M u^a$ becomes

$$\delta_M u^a = 4\,\overset{(2)}{H}{}^{ae}_{bc}\,\varepsilon^b\eta^c\frac{\delta S_E}{\delta u^e}.$$

The algebra is thus open (in the Lagrange multiplier sector) if the structure functions $\overset{(2)}{H}{}^{ae}_{bc}$ differ from zero and are not of the form $\mu^{aef}_{bc}\delta S_E/\delta u^f$, that is, $\overset{(1)}{E}{}^{a}_{bc} \neq 0$. Actually, since the generators γ_a define a group action in phase space, one can adjust H so that $[H, \gamma_a] = 0$ (Exercises 1.24 and 3.22). With a strongly gauge-invariant Hamiltonian, $\overset{(1)}{E}{}^{a}_{bc}$ vanishes and the gauge transformations form a Lie algebra.

It should be recalled again that the concept of being "open" or "closed" applies to a definite generating set, *i.e.*, to a definite choice of the constraint functions and of the Hamiltonian. It does not apply to the set of all gauge transformations, which is always closed.

3.2.6. Reducible First-Class Constraints

When the first-class constraints are irreducible, the gauge transformations (3.26) are independent. This is because the only combinations of the gauge transformations that are zero on-shell are themselves characterized by on-shell vanishing gauge parameters: the equation $\bar{\delta}_\mu F \equiv [F, \mu^a\gamma_a] \approx 0$ for all F implies $\mu^a \approx 0$.

This is no longer the case when the first-class constraints are reducible,

$$Z^a_A \gamma_a = 0. \tag{3.35}$$

By taking

$$\mu^a = \mu^A Z^a_A, \tag{3.36a}$$

one finds

$$\bar{\delta}_\mu F = 0 \tag{3.36b}$$

for all F. So, the gauge transformations (3.36a) possess a redundant action on the q's and the p's. Furthermore, if one takes the functions μ^A solution of

$$\frac{D\mu^A}{Dt} + [\mu^A, H_E] = 0, \tag{3.36c}$$

one also obtains

$$\bar{\delta}_\mu u^a \approx 0, \qquad \bar{\delta}_\mu u^\alpha \approx 0. \tag{3.36d}$$

This follows from (3.31) and the identities $[H, Z^a_A \gamma_a]=0$, $[\gamma_b, Z^a_A \gamma_a]=0$, $[\chi_\alpha, Z^a_A \gamma_a]=0$, as well as from the bracket relations (3.24). The general solution of (3.36c) is characterized by "initial data" that can be taken to be arbitrary functions of q^n, p_n, u^a, u^α and their time derivatives. The gauge transformations (3.26) are therefore reducible.

The reducibility conditions (3.36a) and (3.36c) determine gauge functions μ^a, whose time dependence is fixed by a differential equation. In that sense, the μ^a are not local in time. To maintain a formalism that is local in time, it is necessary to enlarge the set (3.26) of gauge symmetries by adding the new transformation

$$\delta_\rho F = 0, \quad \delta_\rho u^a = -\rho^A Z^a_A, \quad \delta_\rho u^\alpha = 0, \tag{3.37a}$$

where ρ^A is arbitrary.

The addition of the gauge transformations (3.37) is permissible and only increases the reducibility. The new reducibility relations read

$$\mu^a = \mu^A Z^a_A \tag{3.37b}$$

as before, but (3.36c) is replaced by

$$\rho^A = \frac{D\mu^A}{Dt} + [\mu^A, H_E]. \tag{3.37c}$$

The relations (3.37b) and (3.37c) can be solved by choosing μ^A arbitrarily. The equation (3.37c) determines then ρ^A as a local function of the dynamical variables.

3.2.7. Conclusions

In this section, we have investigated the gauge symmetries of the extended action. We have found that at any given time, they reduce to the gauge transformations defined in Chapter 1 as those transformations that do not change the physical state, completed to include a transformation law for the Lagrange multipliers.

It is also of interest to investigate the global symmetries of the extended action. One finds again that the concept of "first class" plays a central role. Indeed, the global symmetries of S_E turn out to be generated by the first-class functions that are constant of the motion (Exercise 3.24).

3.3. GAUGE TRANSFORMATIONS OF THE ORIGINAL LAGRANGIAN ACTION

3.3.1. Gauge Symmetries of S_T and S_L

Our next problem is to determine the relationship between the first-class constraints and the gauge symmetries of the original action S_L (Eq. (1.1)). To that end, it is necessary to find a complete set of gauge transformations for S_L.

This can be done by constructing a complete set of gauge symmetries for the nonextended Hamiltonian action S_T (1.41), which incorporates explicitly only the primary constraints. As any symmetry of S_T defines a symmetry of S_L upon elimination of the conjugate momenta and of the Lagrange multipliers associated with the primary constraints, one automatically gets a complete set of gauge symmetries for S_L.

The converse—namely, that any symmetry of S_L can be extended to a symmetry of S_T—is also true. Because this result will not be needed explicitly here, we will not provide a proof. (A proof is sketched in Exercise 3.17.)

3.3.2. Proof of the Dirac Conjecture under Simplifying Assumptions

The problem of determining all the gauge symmetries of S_L is of interest only for systems for which the property conjectured by Dirac holds. Indeed, when that property does not hold, one has to postulate extra gauge symmetries besides those exhibited by the general solution of the Lagrangian equations of motion. So, a complete set of gauge symmetries of S_L would still be physically incomplete in that case.

Now, there exist counterexamples to the Dirac conjecture (§1.2.2 and 1.6.3). Our first task, therefore, is to impose a series of extra conditions on the constraints that guarantee the validity of the conjecture.

The conditions adopted here are not the weakest possible ones but provide a good illustration of the standard situation found in practice. These conditions are:

(*i*) The consistency algorithm leading to the secondary, tertiary, $\ldots$, constraints never "bifurcates," which implies that the generation to which a constraint belongs is well defined and excludes, for instance, the third and fourth examples analyzed in §1.6.3. This hypothesis can be lifted to some extent (Exercise 3.29), but this will not be done in the text.

(*ii*) The first- and second-class constraints are not mixed in the consistency algorithm. That is, the bracket $[\gamma_a, H + u^{a_1}\gamma_{a_1} + u^{\alpha_1}\chi_{\alpha_1}]$ of γ_a with H_T involves only first-class constraints and primary second-class constraints, *i.e.*, $V_a^{\alpha\beta} = 0, T_{ab_1}^{\alpha\beta} = 0, C_{a\beta_1}^{\alpha} = 0$ for $\alpha \neq \alpha_1$ (where the indices a_1, α_1 and β_1 refer to the primary constraints); and, similarly, the consistency conditions $\dot{\chi}_\alpha = 0$ for the second-class constraints do not generate first-class constraints. This excludes the first two examples of §1.6.3.

(*iii*) The functions ${V_a}^b$ appearing in $[H, \gamma_a] = {V_a}^b\gamma_b$ obey appropriate rank conditions on the constraint surface that exclude the counterexample of §1.2.2 to the Dirac conjecture. These conditions will be made more precise below.

(*iv*) The first-class constraints are irreducible. This last condition is imposed just for the sake of keeping the formulas simple. It can be lifted without changing the conclusions (Exercise 3.26).

When conditions (*i*)—(*iii*) hold, the Dirac conjecture is correct. This can be seen by analyzing in more detail the constraint algorithm. For that purpose, one can assume for simplicity that there are no second-class constraints, since these do not mix with the first-class constraints and are irrelevant to the Dirac conjecture. We denote the primary, secondary, $\ldots$, constraints by $\phi_{m_1}, \phi_{m_2}, \ldots$, *i.e.*,$\gamma_a \equiv (\phi_{m_1}, \phi_{m_2}, \ldots)$.

Now, if the primary constraints $\phi_{m_1} \approx 0$ are first class, one must have

$$[\phi_{m_1}, \phi_{m'_1}] = C^{m''_1}_{m_1 m'_1}\, \phi_{m''_1} \tag{3.38}$$

with only primary constraints in the right-hand side of (3.38) ($m_1, m'_1, m''_1 = 1, \ldots, M_1$). In that case, the consistency equations following from the nonextended action S_T

$$\dot{\phi}_{m_1} = [\phi_{m_1}, H] + u^{m'_1}[\phi_{m_1}, \phi_{m'_1}] = 0 \tag{3.39a}$$

reduce to

$$[\phi_{m_1}, H] = 0 \quad \text{on} \quad \phi_{m_1} \approx 0 \tag{3.39b}$$

and manifestly impose no restriction on the Lagrange multipliers u^{m_1}. If there were secondary constraints in the righthand side of (3.38), the consistency algorithm would bifurcate as in the Example 3 of §1.6.3, a possibility that we are excluding.

The secondary constraints are contained in $[\phi_{m_1}, H] = 0$ on $\phi_{m_1} \approx 0$. If one denotes by ϕ_{m_2} $(m_2 = 1, \ldots, M_2)$ a complete irreducible set of secondary constraints, one then has

$$[H, \phi_{m_1}] = V_{m_1}^{m_1'} \phi_{m_1'} + V_{m_1}^{m_2} \phi_{m_2}. \tag{3.40}$$

We will assume that $M_2 \leq M_1$ and that the rank of the matrix $V_{m_1}^{m_2}$ is "maximum," *i.e.*, equal to M_2 everywhere on $\phi_{m_1} \approx 0$. This excludes the counterexample of §1.2.2, since $V_{m_1}^{m_2}$ vanishes on $\phi_{m_2} \approx 0$ in that case.

The next step is to demand the preservation in time of the secondary constraints. By the same reasoning as the one that led to (3.38) and (3.40), one finds

$$[\phi_{m_1}, \phi_{m_2}] = C_{m_1 m_2}^{m_1'} \phi_{m_1'} + C_{m_1 m_2}^{m_2'} \phi_{m_2'} \tag{3.41a}$$

and

$$[H, \phi_{m_2}] = V_{m_2}^{m_1} \phi_{m_1} + V_{m_2}^{m_2'} \phi_{m_2'} + V_{m_2}^{m_3} \phi_{m_3} \tag{3.41b}$$

if no restriction is to arise on the multipliers u^{m_1}. Here, ϕ_{m_3} $(m_3 = 1, \ldots, M_3 \leq M_2)$ is a complete set of independent "tertiary constraints" and the matrix $V_{m_2}^{m_3}$ is assumed to be of maximum rank M_3 on the constraint surface. Because of (3.41), the conditions $\dot{\phi}_{m_2} = 0$ do not restrict the Lagrange multipliers u^{m_1} and imply instead that $\phi_{m_3} = 0$.

The consistency algorithm leads accordingly to a chain of constraints $\phi_{m_s} \approx 0$ $(m_s = 1, \ldots, M_s \leq M_{s-1},\ s = 1, \ldots, L)$, with

$$[\phi_{m_1}, \phi_{m_s}] = \sum_{i \leq s} C_{m_1 m_s}^{m_i} \phi_{m_i} \tag{3.42a}$$

$$[H, \phi_{m_s}] = \sum_{i \leq s+1} V_{m_s}^{m_i} \phi_{m_i}, \tag{3.42b}$$

and, if the consistency algorithm ends at order L,

$$[H, \phi_{m_L}] = \sum_{i \leq L} V_{m_L}^{m_i} \phi_{m_i}. \tag{3.42c}$$

The matrices $V_{m_s}^{m_{s+1}}$ are of maximum rank M_{s+1} for each $s \leq L-1$.

We have shown in §1.2.1 that the Poisson bracket $[\phi_{m_1}, H]$ of any first-class primary constraint ϕ_{m_1} with the Hamiltonian generates a gauge transformation that does not change the physical state. It follows from the consistency algorithm and the manner in which the secondary constraint ϕ_{m_2} arises that all the secondary (first-class) constraints generate gauge transformations in that sense. This is because the Poisson bracket $[\phi_{m_1}, H]$ contains all the secondary constraints ϕ_{m_2} as $V_{m_1}^{m_2}$ is of maximal rank on the constraint surface.

Similarly, the Poisson brackets $[\phi_{m_2}, H]$ contain all the tertiary constraints, etc., ..., so that all the first-class constraints, primary as well as secondary of any generation, generate gauge transformations that do not change the physical state of the system at a given time. This proves the Dirac conjecture under the above assumptions.

Lastly, we observe that the constraints of order s can be redefined by adding combinations of the previous constraints as

$$\phi_{m_s} \rightarrow \bar{\phi}_{m_s} = \sum_{i \leq s} A_{m_s}^{m_i} \phi_{m_i}, \tag{3.43a}$$

where the matrix $A_{m_s}^{m'_s}$ is invertible. Similarly, the Hamiltonian H is only well defined on the surface of the primary constraints,

$$H \rightarrow H + c^{m_1} \phi_{m_1}. \tag{3.43b}$$

The formulas (3.42) are form-invariant under such redefinitions.

The above redefinitions (3.43) exhaust the ambiguity of the unextended Hamiltonian formalism. In the extended formalism, in which the distinction between primary and secondary constraints is given up, the redefinition freedom is much greater. One can then mix the constraints without restriction on i in (3.43a) or add to H secondary constraints as well.

3.3.3. Lagrangian Form of the Gauge Transformations—Basic Equations

When the constraint algebra is given by (3.42), it is straightforward to derive a complete set of gauge symmetries for S_L—or, what is the same, for S_T. This is because the action S_T,

$$S_T[q^n, p_n, u^{m_1}] = \int (p_n \dot{q}^n - H - u^{m_1} \phi_{m_1})\, dt, \tag{3.44a}$$

is obtained from the extended action S_E,

$$S_E[q^n, p_n, u^{m_i}] = \int (p_n \dot{q}^n - H - u^{m_i} \phi_{m_i})\, dt, \tag{3.44b}$$

by imposing the gauge conditions

$$u^{m_i} = 0, \qquad i \geq 2, \tag{3.45}$$

on the Lagrange multipliers $u^{m_2}, u^{m_3}, \ldots$, associated with the secondary constraints.

These gauge conditions can be reached by gauge transformation (3.26) or (3.31); furthermore, it is permissible to inject them inside the action, since one loses no equation by doing so. The constraints $\phi_{m_2} = 0$, $\phi_{m_3} = 0, \ldots$, obtained by extremizing S_E with respect to the Lagrange multipliers that are set equal to zero in (3.44) reappear through the consistency algorithm.

The gauge symmetries of S_T (and S_L) are therefore just the residual gauge symmetries of the extended action S_E in the gauge (3.45). So, they read

$$\delta F = [F, G], \qquad G = \mu^a \gamma_a, \tag{3.46a}$$

with the condition that $\mu^a(q^n, p_n, u^{m_1}, \dot{u}^{m_1}, \ldots)$ should preserve the gauge condition (3.45). That is, using (3.31),

$$\begin{aligned} &\frac{D\mu^{m_i}}{Dt} + [\mu^{m_i}, H] + u^{m_1}[\mu^{m_i}, \phi_{m_1}] \\ &\quad - \sum_{j \geq i-1} \mu^{m_j} V^{m_i}_{m_j} - u^{m_1} \sum_{j \geq i} \mu^{m_j} C^{m_i}_{m_1 m_j} = 0, \end{aligned} \tag{3.46b}$$

for $i \geq 2$.

The variation of u^{m_1} is, on the other hand, given by

$$\begin{aligned} \delta u^{m_1} &= \frac{D\mu^{m_1}}{Dt} + [\mu^{m_1}, H] + u^{m_1'} [\mu^{m_1}, \phi_{m_1'}] \\ &\quad - \sum_j \mu^{m_j} V^{m_1}_{m_j} - u^{m_1'} \sum_j \mu^{m_j} C^{m_1}_{m_1' m_j}. \end{aligned} \tag{3.46c}$$

One may verify directly (although this is not necessary) that (3.46) leaves the action S_T invariant up to a boundary term. It is also clear from the form of the gauge condition ($u^{m_i} = 0$, $i \geq 2$) why only the structure functions $C^{m_i}_{m_1 m_j}$ appear in Eqs. (3.46).

3.3.4. Solution of the Basic Equations

With (3.42), the general solution of (3.46b) regarded as a system of equations for the μ^{m_i} can be constructed step by step, starting from the last equation with $i = L$ and working one's way down to $i = 1$.

Equation (3.46b) with $i = L$ can be rewritten as

$$\mu^{m_{L-1}} V^{m_L}_{m_{L-1}} = \frac{D\mu^{m_L}}{Dt} + [\mu^{m_L}, H] + u^{m_1}[\mu^{m_L}, \phi_{m_1}] \\ - u^{m_1} \mu^{m'_L} C^{m_L}_{m_1 m'_L} - \mu^{m'_L} V^{m_L}_{m'_L}. \quad (3.47)$$

The general solution of (3.47) contains m_{L-1} arbitrary functions of time. This can be seen as follows. First, the μ^{m_L} are not determined by (3.47). Without loss of generality, one can take them to be arbitrary functions depending on t only.

Because the rank of $V^{m_L}_{m_{L-1}}$ is maximal, (3.47) determines then M_L of the parameters $\mu^{m_{L-1}}$ in terms of μ^{m_L}, $\partial\mu^{m_L}/\partial t$, q, p, and u^{m_1}. The remaining $M_{L-1} - M_L$ parameters can be taken to be any solution of $\mu^{m_{L-1}} V^{m_L}_{m_{L-1}} = 0$ depending on t, and possibly on q and p as well if $V^{m_L}_{m_{L-1}}$ is a function of q and p. For instance, one can take $\mu^{m_{L-1}} = k^\rho(t)\, A^{m_{L-1}}_\rho(q,p)$ with $A^{m_{L-1}}_\rho V^{m_L}_{m_{L-1}} = 0$ and $k^\rho(t)$ arbitrary.

Thus, at this stage, there are $M_L + (M_{L-1} - M_L) = M_{L-1}$ arbitrary functions in the solution of (3.46), and M_L of them appear together with their first time derivatives.

One pursues the analysis of the remaining equations (3.46b) in the same fashion: eq. (3.46b) with $i = k$ takes the form

$$\mu^{m_{k-1}} V^{m_k}_{m_{k-1}} = \frac{D\mu^{m_k}}{Dt} + [\mu^{m_k}, H] + u^{m_1}[\mu^{m_k}, \phi_{m_1}] \\ - \sum_{j \geq k} \mu^{m_j} V^{m_k}_{m_j} - u^{m_1} \sum_{j \geq k} \mu^{m_j} C^{m_k}_{m_1 m_j}. \quad (3.48)$$

The righthand side of (3.48) only depends on q, p; on u^{m_1} and its time derivatives up to order $L - k$; and on the arbitrary functions μ^{m_s} of order greater than $k-1$ introduced in the analysis of Eqs. (3.46b) with $i > k$, and their time derivatives. A key term in (3.48) is $D\mu^{m_k}/Dt$ because this term increases the order of the time derivatives by one unit.

Since $V^{m_k}_{m_{k-1}}$ is of maximal rank M_k, Eq. (3.48) determines M_k of the M_{k-1} functions $\mu^{m_{k-1}}$. The remaining ones, $M_{k-1} - M_k$ in number, are arbitrary. Thus, at each stage, one increases the order of the time derivatives of the already introduced gauge parameters by one unit, and one introduces $M_{k-1} - M_k$ new arbitrary functions. Finally, the last equation (3.46c) determines δu^{m_1} in terms of the rest. The total number of arbitrary functions is equal to M_1. This number is equal to the number of arbitrary functions in the general solution of the equations of motion following from S_T. Hence, our construction yields a complete set of gauge transformations for S_T.

An equivalent complete set of gauge symmetries is obtained by starting from (3.26a), (3.26b), and (3.26c), where ε^a is a function of time and of q, p, u and their derivatives. One gets equations for ε^a similar to (3.46b). Their analysis proceeds along exactly the same lines (see Exercise 3.27).

3.3.5. Lagrange Multiplier Dependence of Gauge Transformations

We stress that the general solution μ^{m_k} of (3.46) depends, in the generic case, not only on the arbitrary gauge functions but also on q, p, as well as on the Lagrange multipliers u^{m_1} and their time derivatives. Thus, it is essential in the analysis to allow from the beginning for that general dependence. Had we required μ^{m_k} to depend on time only, $\mu^{m_k} = \mu^{m_k}(t)$, we would have found the system (3.46) to possess much fewer solutions (if any at all).

A complete set of solutions $\mu^{m_k}(t)$ depending on t only explicitly would exist when the following additional conditions are fulfilled by the constraints and the Hamiltonian:

$$(i) \qquad [\phi_{m_i}, \phi_{m_1}] \text{ contains primary constraints only} \qquad (3.49a)$$

(so as to eliminate the u^{m_1} term in (3.46b); and

$$(ii) \qquad V^{m_i}_{m_j} \text{ is independent of } q, p \qquad (\text{for } i \geq 2) \qquad (3.49b)$$

(to get rid of q and p in (3.46b).

Now, it is possible, at least in principle, to achieve (3.49) locally in phase space by means of appropriate redefinitions of the constraints and of the Hamiltonian H (see Exercise 5.3 of Chapter 5). However, this observation is not of great help, neither practically, since the necessary redefinitions may be quite complicated, nor theoretically, since the independence of μ^{m_k} on $p, q, u^{m_1}, \dot{u}^{m_1}, \ddot{u}^{m_1}, \ldots$ is *not* a statement invariant under allowed redefinitions of the constraints.

Indeed, Eqs. (3.49) are not invariant under (3.43), so a dependence of the coefficients μ^{m_k} on u^{m_1} and its time derivatives may be induced under redefinitions of the constraints. Accordingly, to be able to write the gauge transformations in an arbitrary basis for the constraints, one must start with a gauge generator G that contains $u^{m_1}, \dot{u}^{m_1}, \ddot{u}^{m_1}, \ldots$, etc. Only in this case does one get a formalism that is manifestly invariant under the allowed redefinitions in the theory.

As an example of a gauge transformation that involves explicitly the Lagrange multiplier u^{m_1}, consider the following Lagrangian, $L = (\frac{1}{2})[(\dot{q}_2 - e^{q_1})^2 + (\dot{q}_3 - q_2)^2]$. The primary constraint is $p_1 = 0$, and the Hamiltonian reads $H = e^{q_1} p_2 + q_2 p_3 + \frac{1}{2}(p_2)^2 + \frac{1}{2}(p_3)^2$. The secondary and tertiary constraints can be, respectively, taken to be $\phi_2 = e^{q_1} p_2$

and $\phi_3 = e^{q_1}p_3$. With this choice, the structure functions are constant, namely, $V_1^2 = V_2^3 = 1$, $C_{12}^2 = C_{13}^3 = -1$, whereas the other structure functions vanish.

The gauge generator G given by the above algorithm reads $G = [\ddot{\varepsilon} + 2\dot{\varepsilon}u^1 + \varepsilon\dot{u}^1 + \varepsilon(u^1)^2]p_1 + (\dot{\varepsilon} + \varepsilon u^1)p_2 e^{q_1} + \varepsilon p_3 e^{q_1}$ and involves both u^1 and $\dot{u}^1$. Here $\varepsilon \equiv \mu^3$ is an arbitrary function of time. The Lagrangian gauge transformations are obtained by eliminating p_i and $u^1 = \dot{q}^1$ from the Hamiltonian transformations. One finds $\delta q^1 = \ddot{\varepsilon} + 2\dot{\varepsilon}\dot{q}^1 + \varepsilon\ddot{q}^1 + (\dot{q}^1)^2\varepsilon$, $\delta q^2 = e^{q_1}(\dot{\varepsilon} + \varepsilon\dot{q}^1)$ and $\delta q^3 = \varepsilon e^{q_1}$. The Lagrangian transformations involve the velocity $\dot{q}^1$ that cannot be expressed in terms of q and p only and the acceleration $\ddot{q}^1$ that is not determined by the equations of motion. The formalism is capable of handling such cases without difficulty.

In this example, the redefinition of the gauge parameter $\varepsilon e^{q_1} \to \eta$ simplifies the form of the gauge transformations, which become $\delta q^1 = e^{-q_1}\ddot{\eta}$, $\delta q^2 = \dot{\eta}$, and $\delta q^3 = \eta$. This amounts to redefining the constraints as $\phi_1' = e^{-q_1}p_1$, $\phi_2' = p_2$, and $\phi_3' = p_3$. The new constraints obey (3.49), and the new generator G' reads $G' = \ddot{\varepsilon}\phi_1' + \dot{\varepsilon}\phi_2' + \varepsilon\phi_3'$. For more complicated models, however, finding the analogous simplifying step may be quite an involved task.

3.3.6. Gauge Invariance and Degree of Freedom Count

The total number of independent arbitrary functions of time appearing in the gauge generator $G = \sum \mu^{m_k}\phi_{m_k}$ is equal to

$$M_L + (M_{L-1} - M_L) + \cdots + (M_1 - M_2) = M_1; \tag{3.50}$$

i.e., is equal to the number of (first-class) primary constraints. Of these gauge parameters, $M_1 - M_2$ appear undifferentiated, $M_2 - M_3$ occur together with their first time derivatives, $M_3 - M_4$ appear together with their first and second time derivatives, etc. In other words, the number of generations of constraints associated with a given gauge transformation is equal to the order of the highest time derivative of the gauge parameter plus one. Furthermore, the gauge parameters and their time derivatives all appear in the transformation laws

$$\delta F = [F, G] \tag{3.51}$$

because the constraints are assumed to be irreducible.

At any given time, and in particular at the time at which the initial data are given, one can choose arbitrarily any function and its successive time derivatives. So, if one counts independently the gauge parameters and their time derivatives, one finds a total number of arbitrary gauge parameters at $t = t_0$ equal to the total number of first-class constraints (primary and secondary, tertiary, ...). Indeed, one has

$$M_1 - M_2 + 2(M_2 - M_3) + 3(M_3 - M_4) + \cdots + k(M_k - M_{k+1}) + \cdots + LM_L \\ = M_1 + M_2 + \cdots + M_L = M. \qquad (3.52)$$

The order of the time derivatives appearing in the gauge transformations can be used in reverse to determine the number of first-class constraints by mere inspection. For that purpose, one can go to the Lagrangian form of the transformation laws, which is usually what is known, and analyze the variation of the q^i's and of those velocities that can be expressed in terms of the momenta only. This latter step (analysis of the velocity variations) may actually even be unnecessary if all the gauge parameters and their relevant derivatives can be shown to already act on the q's.

We have thus established the following criterion: the number of first-class constraints is equal to the number M of independent gauge parameters effectively appearing in the gauge transformation laws at any given time (the gauge parameters and their derivatives are independent at any given time). Hence, *in the absence of second-class constraints,* the number of degrees of freedom is equal to the number of conjugate pairs minus the number of independent gauge parameters acting effectively at a given time.

The fact that the number of generations of constraints is determined by the order of the time derivatives of the gauge parameters in the transformation laws of the coordinates is also a consequence of the so-called "Noether's second theorem" (see Exercise 3.28). However, the precise relationship between the consistency algorithm of the Hamiltonian constraints and the form of the gauge transformations is not determined by that theorem.

This criterion for counting the number of degrees of freedom can be easily checked on the simple examples of §1.6.1 and §1.6.2.

3.3.7. Total and Extended Hamiltonians Compared and Contrasted

The previous discussion enables one to clarify further the relationship between the total and extended Hamiltonians.

As we have seen, the gauge parameters and their time derivatives are independent at a given time. Therefore, the data characterizing the system at a given time (Cauchy data) can be modified by the gauge transformations as if the multipliers μ^{m_i} for all the first-class constraints were independent gauge parameters. From that point of view, there is no difference between the total and the extended Hamiltonian. However, it is only the extended Hamiltonian that makes the property manifest.

Even though we have not yet discussed the quantum aspects of constrained systems, it is illuminating to indicate now the quantum mechanical analog of these remarks. In the quantum theory, the physical states must be gauge invariant, *i.e.*, must be invariant under the transformations generated by the gauge generator G of Eq. (3.46a). This means that they must be annihilated by G,

$$G\,|\psi\rangle = 0 \Leftrightarrow \sum_i \mu^{m_i}\phi_{m_i}|\psi\rangle = 0. \tag{3.53}$$

This condition should hold at each instant of time and for any permissible choice of the gauge parameters. This implies, since μ^{m_i} are independent at a given time, that each term in (3.53) should vanish,

$$\phi_{m_i}|\psi\rangle = 0. \tag{3.54}$$

But this equation leads, in turn, to

$$\sum_i \varepsilon^{m_i}\phi_{m_i}\,|\,\psi\rangle = 0, \tag{3.55}$$

where there is now no restriction on the function ε^{m_i}. The physical states are thus invariant under the wider set of gauge transformations of the extended action.

3.4. NONCANONICAL GAUGES

3.4.1. Derivative Gauges

It was shown in Chapter 1 that canonical gauges were appropriate for eliminating the redundancy in the description of the physical state at a given time.

When one deals with histories instead of configurations at one time, the problem of gauge fixation takes the following form. Extremization of the action S_E in a class of paths obeying definite boundary conditions at t_1 and t_2 does not yield a unique history. Rather, the extremals differ from each other by a gauge transformation (3.26). *That gauge transformation must vanish at the endpoints to preserve the boundary conditions,*

$$\varepsilon^a(t_1) = \varepsilon^a(t_2) = 0. \tag{3.56}$$

The gauge transformations obeying (3.56) exhaust all the ambiguity of the variational principle except for "unfortunate" choices of the boundary data. The question is then: can one eliminate the redundancy in the description of the classical solution by means of gauge conditions? To that end, one cannot impose an arbitrary canonical gauge $C_a(q,p) = 0$ for all times, since to reach such a gauge would, in general, violate (3.56) and, hence, modify the given class of paths. A different kind of gauge-fixing conditions is needed.

Because there are two boundary conditions on $\varepsilon^a(t)$, the searched-for gauge conditions $C_a(q,p,\dot{q},\dot{p},\ldots,u^a,\dot{u}^a,\ldots) = 0$ should lead to a second-order differential system for $\varepsilon^a(t)$. As $u^a(t)$ transforms with the first-time derivative of ε^a, one possibility is to impose

$$\dot{u}^a = f^a(q,p,u). \tag{3.57}$$

To check whether (3.57) is admissible, let us assume that we found a history that obeys (3.57) and let us make a gauge transformation. The gauge condition transforms as

$$\delta(\dot{u}^a - f^a) = \ddot{\varepsilon}^a + M^a{}_b(q,p,u)\dot{\varepsilon}^b + L^a{}_b(q,p,u)\varepsilon^b. \tag{3.58}$$

This will be zero only if $\varepsilon^a(t) = 0$ for all t [use $\varepsilon^a(t_1) = \varepsilon^a(t_2) = 0$], except for some "unfortunate" choices of the time interval. The gauge conditions (3.57) are thus good gauge conditions, at least locally in trajectory space (see Exercise 3.30).

Alternative ways to fix the gauge freedom (3.26) may be devised by considering gauge conditions involving the second time derivatives of the canonical variables, or higher order derivatives of the multipliers together with prescribed values of $u^a(t)$ at the endpoints. In practice, such gauge conditions are, however, awkward.

Yet another possibility is useful for the comparison of the extended and total Hamiltonian formalism along the lines of Sec. 3.3. It is obtained by imposing $u^{m_i} = 0$ for $i \geq 2$, as in (3.35), and by restricting the L-th derivative of the multiplier u^{m_1}. For instance, for a system containing one primary and one secondary constraint, one may take $u_2 = 0$ and $\ddot{u}_1 = 0$.

Because the gauge conditions that are compatible with the boundary conditions (3.56) contain the time derivatives of the multipliers or of the canonical variables, they are called "derivative gauges."

The preceding discussion can be directly extended to off-shell histories when the gauge transformations of the generating set (3.26) close off-shell. When the transformations do not close off-shell, a naive extension of the above reasoning would seem to indicate that additional gauge conditions are necessary to fix the extra off-shell freedom coming in through commutators. However, this is not so, and one may admit in the path integral only histories obeying (3.57) and no further restriction. This assertion cannot be justified as before in terms of gauge orbits but can be shown to be correct through BRST theory.

3.4.2. Multiplier Gauges

The "multiplier gauges" lie between canonical gauges and derivative gauges and are defined by conditions that fix the Lagrange multipliers in terms of the canonical variables,

$$u^a + \chi^a(q,p) = 0. \tag{3.59}$$

The simplest case is just $u^a = 0$.

To reach (3.59), one must combine the gauge freedom vanishing at the endpoints with a transformation generated by the first-class constraints at either t_1 or t_2. Such a transformation generically modifies the boundary conditions; that is, it modifies the class of paths under consideration.

3.4.3. Reducible Gauge Transformations—Redundant Gauge Conditions

The previous analysis assumed that the gauge transformations were irreducible. In the reducible case, there is an extra term in the variation of u^a involving the extra gauge parameter ρ^A of Eq. (3.37a). Furthermore, since some of the gauge transformations do not act on the q's, the p's, and the u's, good gauge conditions should fix the gauge parameters ε^a and ρ^A only up to ineffective gauge transformations of the form (3.37b)–(3.37c).

The derivative gauge conditions (3.57) no longer meet this requirement. Indeed, Eq. (3.58) is replaced by

$$\ddot{\varepsilon}^a + M^a{}_b(q,p,u)\dot{\varepsilon}^b + L^a{}_b(q,p,u)\varepsilon^b - (\rho^A Z_A{}^a)^{\cdot} = 0, \tag{3.60a}$$

where the ε's are now required to fulfill

$$\varepsilon^a(t_2) = Z_A{}^a\mu_1^A, \quad \varepsilon^a(t_2) = Z_A{}^a\mu_2^A \tag{3.60b}$$

for some arbitrary μ_1^A and μ_2^A in order to leave the boundary conditions on the q's and the p's unchanged [see (3.36b)]. Up to an ineffective gauge transformation, the general solution of Eqs. (3.60a)–(3.60b) is

$$\varepsilon^a = 0, \qquad \rho^A = C^A \tag{3.61a}$$

with

$$(C^A Z_A{}^a)^{\cdot} = 0. \tag{3.61b}$$

The transformations (3.61) involve an arbitrary integration "constant" $C^A(q,p,u,\dot{q},\dot{p},\ldots,t=t_0)$ and act nontrivially on the u^a's associated with the nonindependent constraints. To fix the leftover transformation (3.61), it is necessary to impose a further condition, which can be a multiplier gauge condition

$$T^A{}_a(q,p)u^a = 0 \tag{3.62}$$

at a single time $t = t_0$, where $T^A{}_a$ is such that the equations $T^A{}_a u^a = 0$ and $u^a = Z^a{}_B\, \rho^B = 0$ imply $u^a = 0$.

It turns out to be more convenient to impose the conditions (3.62) at all times and to choose the functions f^a in the derivative gauge conditions in such a way that $\dot{u}^a - f^a = 0$ is compatible with (3.62). The gauge conditions are then not independent, *i.e.*, *redundant.* The standard example of redundant gauge conditions arises in the context of an abelian 2-form gauge field $A_{\mu\nu} = -A_{\nu\mu}$, for which the gauge transformations read (see Chapter 19)

$$\delta A_{\mu\nu} = \partial_\mu \varepsilon_\nu - \partial_\nu \varepsilon_\mu. \tag{3.63}$$

The q's and the p's are then the spatial components A_{ij} and their conjugate momenta π^{ij}, while the mixed components A_{0i} play the role of Lagrange multipliers for the redundant constraints $-2\,\partial_j \pi^{ji} \approx 0$ (see Sec. 19.2). The ε_i's are the analog of the above ε^a's, whereas ε_0 stand for the gauge parameter ρ^A. The Lorentz gauge condition is

$$\partial^\mu A_{\mu\nu} = 0 \Leftrightarrow \begin{cases} \partial^0 A_{0i} + \partial^j A_{ji} = 0, & \text{(3.64a)} \\ \partial^i A_{i0} = 0. & \text{(3.64b)} \end{cases}$$

The conditions (3.64a) are derivative gauge conditions; the condition (3.64b) is a multiplier gauge. Equation (3.64b) is preserved in time by (3.64a), since $\partial^\nu \partial^\mu A_{\mu\nu}$ identically vanishes.

EXERCISES. CHAPTER THREE

3.1. Let $f(y^i, \dot{y}^i, \ldots)$ be the conserved charge ("Noether charge") of a global symmetry. Show that f is on-shell gauge invariant. [Hint: Knowing that the solutions of the equations of motion are determined up to a gauge transformation, could f be conserved if $\delta_\varepsilon f$ were not zero on-shell?]

3.2. Show that the commutator of a rigid (= global) symmetry with a gauge transformation is a gauge transformation.

3.3. (a) Show that (3.4b) implies that

$$\frac{\delta L}{\delta y^i} \delta_\varepsilon y^i = \frac{dQ}{dt}, \tag{i}$$

where the "Noether charge" Q takes the form (up to an irrelevant additive constant)

$$Q = \lambda^i_{(0)} \frac{\delta L}{\delta y^i} + \lambda^i_{(1)} \frac{d}{dt} \frac{\delta L}{\delta y^i} + \cdots + \lambda^i_{(r)} \frac{d^r}{dt^r} \frac{\delta L}{\delta y^i}. \tag{ii}$$

Hence, the Noether charge vanishes when the equations of motion hold. [Hint: Expand the Noether charge as $Q = \varepsilon^\alpha Q_{(0)\alpha} + \dot{\varepsilon}^\alpha Q_{(1)\alpha} + \cdots$ and use the fact that (i) is an identity valid for any ε^α. This enables one to determine successively $Q_{(S)\alpha}, Q_{(S-1)\alpha}, \dots, Q_{(0)\alpha}$, which are found to be of the required form.]

(b) Conversely, let $\bar{\delta} y^i$ be a transformation leaving S invariant up to a boundary term and such that the corresponding Noether charge vanishes on-shell, as in (ii). Prove then that $\bar{\delta} y^i$ can be extended to a gauge transformation

$$\delta_\varepsilon y^i = (\bar{\delta} y^i)\varepsilon + \lambda^i_{(0)}\dot{\varepsilon} - (\lambda^i_{(1)}\dot{\varepsilon})^{\cdot} + \cdots, \tag{iii}$$

where $\varepsilon(t)$ is an arbitrary function of time. This gauge transformation reduces to the given $\bar{\delta} y^i$ for $\varepsilon = 1$.

3.4. Repeat the analysis of Exercise 3.3 for field theories. [Hint: The analog of (i) is

$$\frac{\delta \mathcal{L}}{\delta \phi^i} \delta_\varepsilon \phi^i = \partial_\mu j^\mu$$

where j^μ is the "Noether current"; the analog of (ii) is

$$j^\mu = \lambda^{\mu i} \frac{\delta \mathcal{L}}{\delta \phi^i} + \lambda^{\mu i \nu} \partial_\nu \frac{\delta \mathcal{L}}{\delta \phi^i} + \cdots + \lambda^{\mu i \nu_1 \dots \nu_s} \partial_{\nu_1} \dots \partial_{\nu_s} \frac{\delta \mathcal{L}}{\delta \phi^i} + \partial_\nu S^{\mu\nu}$$

where $S^{\mu\nu} = -S^{\nu\mu}$ is identically conserved, $\partial_\mu S^{\mu\nu} \equiv 0$; the analog of (iii) is

$$\delta_\varepsilon \phi^i = \bar{\delta}\phi^i \varepsilon + \lambda^{\mu i} \partial_\mu \varepsilon - \partial_\nu (\partial_\mu \varepsilon \lambda^{\mu i \nu}) + \cdots. \quad]$$

3.5. Write explicitly (3.7a) and (3.7b) for

$$\mu^{ij}(t,t') = k_0^{ij}(t)\,\delta(t-t') + k_1^{ij}(t)\,\frac{d}{dt}\delta(t-t') + \cdots,$$

where the arbitrary functions $k_0^{ij}, k_1^{ij}, \dots$ may involve the variables y^i and their derivatives.

3.6. The purpose of Exercises 3.6 through 3.9 is to show how equation-of-motion symmetries can arise already in the case of systems without gauge freedom.

(a) Let $L = \frac{1}{2}\dot{q}^2$. Show that $S = \int L\,dt$ is invariant under

$$\delta q = 2\,\varepsilon\,\ddot{q} + \dot{\varepsilon}\,\ddot{q}$$

where $\varepsilon(t)$ is an arbitrary function of time. Compute the Noether charge and show that it vanishes identically when expressed as a phase space function (using the dynamical equation of motion $\ddot{q} = 0$).

(b) Write $\delta q(t)$ explicitly as $\delta q(t) = \int dt'\, \mu(t,t')[\delta S/\delta q(t')]$ with some appropriate $\mu(t,t') = -\mu(t',t)$.

3.7. (a) Let $L = \frac{1}{2}(\dot{q}^1)^2 + \frac{1}{2}(\dot{q}^2)^2$. Show that $\delta q^1 = \dot{q}^2, \delta q^2 = \dot{q}^1$ and $\bar{\delta} q^1 = \dot{q}^1$, $\bar{\delta} q^2 = 0$ are both global symmetry transformations of the action.

(b) Compute the commutator $[\delta, \bar{\delta}]$ and prove that it is a nonvanishing, equation-of-motion gauge transformation. So, such transformations can be generated through the commutator of global symmetries, even for systems without true gauge freedom and even when the global symmetries do not involve the accelerations.

(c) Observe that although the commutator $[\delta, \bar{\delta}]$ involves the equations of motion, there is no problem with the representation of the symmetries δ and $\bar{\delta}$ in quantum mechanics. This is because, as is always the case, these symmetries, when rewritten in Hamiltonian form ($\delta q^1 = p^2, \delta q^2 = p^1, \delta p^1 = 0, \delta p^2 = 0, \bar{\delta} q^1 = p^1, \bar{\delta} q^2 = \bar{\delta} p^1 = \bar{\delta} p^2 = 0$) commute everywhere in phase space without having to invoke equations of motion: the trivial symmetries are factored out. The canonical generators for δ and $\bar{\delta}$ are, respectively, $p^1 p^2$ and $\frac{1}{2}(p^1)^2$.

3.8. Let $L(q^n, \dot{q}^n)$ be a regular Lagrangian and let $\delta q^n = f^n(q, \dot{q}, \ddot{q}, \dddot{q}, \dots, t)$ be a symmetry transformation of L, $\delta L = dK/dt$. The purpose of this exercise is to show that one can redefine the symmetry transformation $\delta q^n \rightarrow \delta' q^n = f'^n$ by adding trivial transformations so as to get rid of the accelerations and the higher order time derivatives in δq^n.

(a) Show that f^n can be written as

$$f^n = f'^n(q, \dot{q}, t) + M^n$$

where M^n is a linear combination of the equations of motion,

$$M^n(q, \dot{q}, \dots, t) = \int dt' \, \rho^{nn'}(t, t') \, \frac{\delta L}{\delta q^{n'}}(t')$$

Here, $\rho^{nn'}$ may also involve the dynamical variables. [Use the fact that any function that vanishes on-shell is a combination of $\delta L/\delta q^n$ and its time derivatives; see Chapter 17.]

(b) Prove that f'^n and M^n separately define symmetry transformations of L and that the transformation $\delta' q^n = f'^n$ is such that

$$\delta' L = \frac{dK'}{dt},$$

where $K'(q, \dot{q}, t)$ is the on-shell value of K. [Hint: Analyze the identity $I \equiv A_n f^n + dF/dt = 0$ in terms of the variables $q^n, \dot{q}^n, A_n, \dot{A}_n, \ddot{A}_n, \dots$. Here, $A_n \equiv -\delta L/\delta q^n$. Show that the relation $A_n f'^n + dF'/dt = 0$ holds, where $F'(q, \dot{q}, A_n = 0, \dot{A}_n = 0, \ddot{A}_n = 0, \dots)$ is the on-shell value of F.]

(c) Use Theorem 3.1 to infer that $\rho^{nn'}(t, t')$ in (i) can be taken to be antisymmetric, $\rho^{nn'}(t, t') = -\rho^{n'n}(t', t)$ and, hence, defines a trivial gauge transformation.

Conclusion: In the case of regular second-order Lagrangians, one can assume without loss of generality that the symmetry transformations do not involve the accelerations and their derivatives. Note, however, that the set of such symmetry transformations does not form a subgroup (see Exercise 3.7).

3.9. Let $S_H[q^n, p_n] = \int (p_n \dot{q}^n - H) dt$ be a first-order action in canonical form.

(a) Repeat the analysis of Exercise 3.8 in that case, *i.e.*, prove that any symmetry transformation of S can be assumed, without loss of generality, to depend on q and p only and not on their derivatives,

$$\delta_H q^n = \delta_H q^n(q,p,t) \quad (= [q^n, Q(q,p,t)])$$
$$\delta_H p_n = \delta_H p_n(q,p,t) \quad (= [p_n, Q(q,p,t)]).$$

Here, Q is the conserved charge.

(b) Show that this condition is now preserved under the commutator.

(c) In the case when S_H has been obtained from a regular second-order action principle, prove that the comparison between the symmetry transformations in Hamiltonian form and the corresponding symmetry transformations in Lagrangian form actually involves equation-of-motion symmetries, except in the particular case of "point transformations" for which δq^n does not contain p_n (or $\dot{q}^n$). [Hint: Observe that the Lagrangian transformation $\delta_L q^n(q,\dot{q},t)$ is related to $\delta_H q^n(q,p,t)$ as $\delta_H q^n(q,p,t) = \delta_L q^n(q,\dot{q},t) + \rho^n_{n'}\, \delta S_H/\delta p_{n'}$ with $\rho_{n'}{}^n = 0$ if and only if $\delta_L q^n$ does not involve the velocities. When $\rho^n{}_{n'} \neq 0$, it follows that $\delta_H p_n$ differs from the variation of p_n viewed as a function of the coordinates and the velocities by q-equations of motion. See Exercise 3.17 in this context.]

3.10. Consider the action $S[A_\mu] = \int d^3x\, \varepsilon^{\mu\nu\rho} F_{\mu\nu} A_\rho$ for pure abelian Chern-Simons theory in three dimensions. Here, $F_{\mu\nu} = \partial_\mu A_\nu - \partial_\nu A_\mu$. This action is invariant under ordinary gauge transformations $\delta A_\mu = \partial_\mu \Lambda$ and diffeomorphisms, $\delta A_\mu = \xi^\rho \partial_\rho A_\mu + \partial_\mu \xi^\rho A_\rho$. Show that the diffeomorphisms differ from ordinary gauge transformations by trivial gauge symmetries.

3.11. Consider the action $S[q,p] = \int (p\dot{q} - H)dt$. Show that it is invariant under $\delta q = \varepsilon(\dot{q} - \partial H/\partial p), \delta p = \varepsilon(\dot{p} + \partial H/\partial q)$. Check that the algebra of these trivial gauge transformations is isomorphic with the algebra of diffeomorphisms in one dimension. Observe that when $H = 0$, these transformations actually reduce to standard reparametrizations "along the world line."

3.12. Check that $\delta y^i = \lambda^\alpha R^i_\alpha$ defines a trivial gauge transformation when $\lambda^\alpha = M^{\alpha i}\delta S/\delta y^i$.

3.13. Check explicitly that the transformations (3.26a)–(3.26c)) map a solution of the equations of motion on another solution. How important is the fact that the brackets $[\gamma_a, \gamma_b]$ and $[H, \gamma_a]$ both involve the second-class constraints at least squared?

3.14. Let $G \equiv D^{\alpha\beta}\chi_\alpha\chi_\beta$. The function G is first-class but vanishes quadratically with the constraints. Show that the transformations $\delta q^i = [q^i, G]$ and $\delta p_i = [p_i, G]$ leave the extended action invariant provided one defines δu^a and δu^α appropriately, but they are trivial gauge transformations.

3.15. Check that (3.26) and (3.29) differ by equations-of-motion symmetries.

3.16. In the extended action principle, perform the change of variables $q^n, p_n \to y^i, \chi_\alpha$, where y^i are coordinates on $\chi_\alpha = 0$ and obey $[y^i, \chi_\alpha] = 0$ on $\chi_\alpha = 0$ (see §2.3.2). Show that the variables (u^α, χ_α) can be viewed as auxiliary fields, *i.e.*, that the equations $\delta S_E/\delta u^\alpha = 0$, $\delta S_E/\delta \chi_\alpha = 0$ can be solved for u^α and χ_α and yield $u^\alpha = 0, \chi_\alpha = 0$. Observe that when u^α and χ_α are eliminated by means of their own equations of motion, one gets the action (2.26) in which there is no second-class constraint left.

3.17. Let $S[y^i(t), z^A(t)]$ be an action depending on auxiliary fields $z^A(t)$. That is, the equations of motion $\delta S/\delta z^A = 0$ can be solved to yield z^A as a function of y^i and its derivatives (Exercises 1.4 and 1.5). We denote the solution of $\delta S/\delta z^A = 0$ by $z(y)$. Let $S_R[y^i(t)]$ be the action obtained by elimination of the auxiliary fields, $S_R[y^i(t)] \equiv S[y^i(t), z^A(y^i)(t)]$. Assume S to be gauge invariant under $\delta y^i = Y^i, \delta z^A = Z^A$.

(a) Show that S_R is invariant under $\bar{\delta} y^i = Y^i(y, z(y))$ and that one can redefine the gauge symmetries of S by adding on-shell trivial transformations so that they coincide with those of S_R (*i.e.*, S is invariant under the same gauge transformation $\bar{\delta} y^i$ as S_R is, provided one transforms z^A appropriately).

(b) Prove that actually any symmetry (global or local) of S_R can be extended to z^A so as to be a symmetry of S. [Hint: (*i*) Write $S = S_R + T$ and show that $\delta T/\delta y^i$ vanishes when $\delta T/\delta z^A = 0$, so that $\delta T/\delta y^i = \lambda^A{}_i(\delta T/\delta z^A)$ for some $\lambda^A{}_i$. (*ii*) Observe that $\delta T = 0$ for any δy^i and $\delta z^A = -\lambda^A{}_i\, \delta y^i$.]

(c) Prove that if $(\delta y^i, \delta z^A)$ is a complete set of gauge symmetries for S, then $\bar{\delta} y^i$ is a complete set for S_R. Conversely, any complete set of gauge symmetries for S_R can be extended to a complete set for S as in (b).

3.18. Write down explicitly the functions R^i_α defined by the generating set (3.26). Expand out $R^i_\alpha\, \mu^\alpha + M^{ij}(\delta S/\delta y^j)$ for arbitrary functions μ^α, M^{ij} of time and of the dynamical variables.

3.19. Check that (3.31) differs from (3.26) by a trivial gauge transformation.

3.20. Compute the algebra of the gauge transformations (3.26) when second-class constraints are present.

3.21. Assume $C_{ab}{}^c$ constant in $[\gamma_a, \gamma_b] = C_{ab}{}^c \gamma_c$.

(a) Show that the Jacobi identity for the Poisson bracket implies then the Jacobi identity for the structure constants $C_{ab}{}^c$.

(b) Verify that $D^e_{[abc]} = 0$, with D^e_{abc} given by (3.34b).

3.22. When the $C_{ab}{}^c$ are constant, the transformations $\delta_\varepsilon F(q,p) = [F, \gamma_a]\varepsilon^a$ form a Lie algebra.

(a) Show that one can define gauge orbits everywhere in phase space and not just on the constraint surface.

(b) Reanalyze Exercise 1.24 in the light of this result.

3.23. A set Z^α_Γ of reducibility coefficients is complete if any solution λ^α of $\lambda^\alpha R^i_\alpha = N^{ij}(\delta S/\delta y^j)$ can be written as

$$\lambda^\alpha = \lambda^\Gamma Z^\alpha_\Gamma + \rho^{\alpha j} \frac{\delta S}{\delta y^j}.$$

Consider the transformation

$$\delta y(t) = \dot{\varepsilon} = \int dt'\, R(t,t')\, \varepsilon(t'), \quad R(t,t') = \frac{d}{dt}\delta(t-t'),$$

which leaves the action $S[y(t)] \equiv 0$ strictly invariant for any choice of $\varepsilon(t)$. This gauge transformation is reducible, since $\varepsilon =$ constant leads to $\delta y = 0$.

(a) Show that one cannot find a local, complete set $Z^\alpha_\Gamma = Z(t', t'')$ of reducibility coefficients (to be local, $Z(t', t'')$ should contain $\delta(t' - t'')$ and a finite number of its derivatives).

(b) Write instead $\delta y(t) = \dot{\varepsilon}^1 + \varepsilon^2 = \int dt' [R_1(t,t')\,\varepsilon^1(t) + R_2(t,t')\,\varepsilon^2(t)]$, $R_1(t,t') = (d/dt)\delta(t-t'), R_2(t,t') = \delta(t-t')$. Show that now one can find a complete, local, reducibility set. [In the Lagrangian BRST formalism, the first description of the gauge symmetries would require ghost of ghosts that are global in time, which is awkward. The second description requires ghosts of ghosts that are ordinary dynamical variables depending on time.]

3.24. This exercise investigates the global symmetries of S_E.

(a) Consider the transformations

$$\delta q^n = Q^n, \quad \delta p_n = P_n, \quad \delta u^a = U^a, \quad \delta u^\alpha = U^\alpha,$$

where Q^n, P_n, U^a, and U^α are functions of $q, p, u, \dot{u}, \ddot{u}, \ldots, \overset{(k)}{u}$ and t. Show that these transformations leave S_E invariant up to a boundary term if and only if

$$Q^n = \frac{\partial F}{\partial p_n}, \qquad P_n = -\frac{\partial F}{\partial q_n} \tag{i}$$

with

$$\frac{DF}{Dt} + [F, H] + u^a[F, \gamma_a] + u^\alpha[F, \chi_\alpha] = U^a\gamma_a + U^\alpha\chi_\alpha. \tag{ii}$$

[D/Dt is defined by (3.32a).]

(b) In view of the results of Exercises 3.16 and 3.17, it is enough to investigate Eqs. (i) and (ii) in the presence of only first-class constraints. Show that the general solution of (ii) is given by

$$F = f^a(q, p, u, \dot{u}, \ldots, \overset{(k-1)}{u}, t)\gamma_a + \bar{F}(q, p, t)$$

where $\bar{F}$ is a first-class function that is a constant of the motion,

$$[\bar{F}, \gamma_a] \approx 0, \qquad \frac{\partial \bar{F}}{\partial t} + [\bar{F}, H] \approx 0.$$

The term $f^a\gamma_a$ defines a gauge transformation, the term $\bar{F}$ defines a global symmetry (when nonweakly vanishing).

(c) Prove that the allowance of a possible dependence of Q^n, P_n, U^a and U^α on the time derivatives of q and p does not bring in new transformations. That is, this dependence can be absorbed in equations-of-motion symmetries or gauge transformations generated by $f^a(q, p, \dot{q}, \dot{p}, \ldots)\gamma_a$.

(d) Show that any global symmetry of S_E can be improved by the addition of a gauge transformation so as to define a global symmetry of S_T and S_L.

3.25. (a) Let $F(q, \dot{q}, \ddot{q}, \ldots)$ be a function that is on-shell invariant under the gauge transformations of the Lagrangian formalism, $\delta_\varepsilon F = 0$ when $\delta L/\delta q^i = 0$. Show that when rewritten in phase space, $F(q,p)$ is weakly invariant under the gauge transformations of the extended formalism, $[F, \gamma_a] \approx 0$. [Hint: Use the results of Exercise 3.17 to go to the total formalism and express F as a function of q, p by means of the equation $\delta S_T = 0$ (show that u^{m_1}, which transforms with the highest order time derivative of the gauge parameter, can be eliminated from F if F is to be gauge invariant, while u^{α_1} vanishes on-shell).] Note that the modification $F \to F + \lambda^\alpha\chi_\alpha$ enables one to achieve also $[F, \chi_\alpha] \approx 0$.

(b) Let $\delta_\varepsilon q^i = Q^i(q, \dot{q}, \ldots)$ be a symmetry of S_L. Show that its conserved charge Q, when rewritten in phase space, generates a symmetry of S_E. [Hint: Use (a) and Exercises 3.1 and 3.24.] Prove, furthermore, that $Q^i(q, \dot{q}, \ldots)$ and $[q^i, Q(q,p)]$ differ on-shell at most by a gauge transformation. [Hint: The transformations $Q^i(q, \dot{q}, \ldots)$ and $[q^i, Q(q,p)]$ are both symmetries of S_L and have the same on-shell Noether charge; use then Exercise 3.3(b).]

3.26. Repeat the analysis of the gauge symmetries of S_L when the constraints are reducible. Recall that the gauge transformations for the Lagrange multipliers now include (3.37a) in addition to (3.31). Discuss separately the cases when the primary constraints are irreducible and when they are reducible.

3.27. Repeat the analysis of the gauge symmetries of S_L with the parametrization (3.26) of the gauge transformations, where ε is outside the bracket.

(a) Derive the basic equations replacing (3.46).

(b) Show that the numbers of arbitrary functions and of their time derivatives occurring in the new form of the gauge transformations are the same as those found in the text for the parametrization (3.31).

3.28. Let $L(q, \dot{q})$ be a Lagrangian invariant up to a total time derivative under

$$\delta_\varepsilon q^i = Q^i_{(0)\alpha}\varepsilon^\alpha + Q^i_{(1)\alpha}\dot{\varepsilon}^\alpha + \cdots + Q^i_{(k)\alpha}\overset{(k)}{\varepsilon}{}^\alpha,$$

where $Q^i_{(j)\alpha}$ depends on q^i and $\dot{q}^i$ but not on the higher order time derivatives.

(a) Show that the Noether charge Q appearing in the Noether identity

$$\frac{\delta L}{\delta q^i}\delta_\varepsilon q^i = \frac{dQ}{dt}$$

cannot depend on the accelerations or the higher order time derivatives of q^i and involves $\varepsilon^\alpha, \dot{\varepsilon}^\alpha, \ldots$ up to $\overset{(k-1)}{\varepsilon}{}^\alpha$. By expanding Q as

$$Q = \varepsilon^\alpha Q_{(0)\alpha} + \dot{\varepsilon}^\alpha Q_{(1)\alpha} + \cdots + \overset{(k-1)}{\varepsilon}{}^\alpha Q_{(k-1)\alpha},$$

show that the on-shell conservation of Q for any choice of ε^α successively implies the constraints $Q_{(k-1)\alpha} = 0, Q_{(k-2)\alpha} = 0, \ldots, Q_{(0)\alpha} = 0$ on q^i and $\dot{q}^i$.

(b) Compare with the Hamiltonian analysis of Sec. 3.3. [Note: The primary constraints of the Hamiltonian analysis are not seen in the Lagrangian formalism because they become identically zero when expressed in terms of q^i and $\dot{q}^i$. This is why there are only $k-1$ generations of constraints in the Lagrangian formalism.]

3.29. Consider a system that fulfills (3.42b)–(3.42c) but not (3.42a). That is, the brackets $[\phi_{m_1}, \phi_{m_s}]$ may contain constraints ϕ_{m_i} of order $i > s$. Assume that, as in Example 3 of §1.6.3, the consistency algorithm leads to the sole possibility $\phi_{m_k} \approx 0$. Show that the analysis of the residual symmetries in the gauge (3.45) reducing S_E to S_T can be carried along the lines of §3.3.4 and leads to the same conclusions. [Hint: Expand the unknown functions μ^{m_i} in powers of $u^{m_1}, \mu = \mu_0^{m_i} + \mu_1^{m_i} + \mu_2^{m_i} + \cdots$ where $\mu_k^{m_i}$ is homogeneous of degree k in u^{m_1} and its derivatives. Determine successively $\mu_0^{m_i}, \mu_1^{m_i}, \ldots$ by proceeding exactly as in §3.3.4.]

3.30. Consider a system with a single first-class constraint and Lagrange multiplier u. Show that $U = \int_{t_1}^{t_2} u(t)\, dt$ is invariant under gauge transformations that vanish at the endpoints. Study U as function of $u_1 \equiv u(t_1)$ for solutions of $\dot{u} = f(u)$. Show that when $u(t)$ remains finite for finite t, then as u_1 runs from $-\infty$ to $+\infty$, so does U. Conversely, show that $U(u_1)$ may not cover the entire real line when the solution of $\dot{u} = f(u)$ becomes infinite in a finite time (and is ill defined later on). In that case, the derivative gauge condition $\dot{u} = f(u)$ cannot be reached by a gauge transformation that vanishes at the endpoints for those values of U out of the range of $U(u_1)$.

3.31. Inject the multiplier gauge conditions $u^a = \chi^a(q,p)$ inside S_E. Show that the new variational principle is equivalent to the original one only if one imposes by hand $\gamma_a = 0$ at one time.

3.32. Show that the Poisson bracket relations (3.42) imply $[\phi_{m_k}, \phi_{m_{k'}}] \approx 0$ for arbitrary k and k' and not just $[\phi_{m_1}, \phi_{m_{k'}}] \approx 0$. [Hint: Use the Jacobi identity $[[H, \phi_{m_k}], \phi_{m_s}] + [[\phi_{m_k}, \phi_{m_s}], H] + [[\phi_{m_s}, H], \phi_{m_k}] = 0$ with $s = 1, 2, \ldots,$ as well as the definition of the secondary, tertiary, $\ldots$, constraints.]

CHAPTER FOUR

GENERALLY COVARIANT SYSTEMS

4.1. INTRODUCTION

One normally describes the motion of the system by giving the canonical variables as a function of time. The time is assumed to have a direct physical significance but is not itself a dynamical variable.

There exists a different formulation of the dynamics in which the physical time and the dynamical variables of the system are treated more symmetrically. This formulation includes the time among the canonical variables and describes the relations ("correlations") between the original dynamical variables and the physical time by giving the enlarged set of canonical variables in terms of an arbitrary parameter. The arbitrary parameter does not possess any physical significance, and the formalism is therefore invariant under reparametrizations of it, or, as one says, the formalism is "generally covariant." In field theory, one also introduces arbitrary labels for the spatial coordinates, and the theory becomes then invariant under arbitrary changes of the spacetime coordinates.

In practice, generally covariant systems arise in two different ways. One may have a system in which originally the physical time was not included as a canonical variable and proceed to "parametrize" the theory to achieve general covariance. This can *always* be done. Or the system may be "already generally covariant." The already generally covariant system "per excellence" is the gravitational field in general relativity.

Attempts at "deparametrizing" already generally covariant systems have not been quite successful. It seems thus preferable to aim at both formulating and answering questions while treating all variables on the same footing. This coincides with the point of view about gauge invariance that has been taken throughout this book. It is not accidental that this should be so, since, as we shall see below, in a generally covariant theory the motion is just the unfolding of a gauge transformation.

4.2. TIME AS A CANONICAL VARIABLE—ZERO HAMILTONIAN

4.2.1. Parametrized Systems

Consider a system with canonical variables q^i, p_i, Hamiltonian $H_0(q,p)$ and, for simplicity, no constraint. The action reads

$$S[q^i(t), p_i(t)] = \int_{t_1}^{t_2} \left(p_i \frac{dq^i}{dt} - H_0\right) dt. \tag{4.1}$$

Introduce now the time $t \equiv q^0$ and its conjugate p_0 as canonical variables by replacing (4.1) by

$$S[q^0(\tau), q^i(\tau), p_0(\tau), p_i(\tau), u^0(\tau)] = \int_{\tau_1}^{\tau_2} [p_0\dot{q}^0 + p_i\dot{q}^i - u^0(p_0 + H_0)]\, d\tau, \tag{4.2}$$

where the dot denotes derivative with respect to the parameter τ.

The motion obtained by demanding that (4.2) be stationary is equivalent to that following from extremizing (4.1). This is seen by extremizing first (4.2) with respect to u^0 and p_0, which yields

$$\gamma_0 \equiv p_0 + H_0 = 0 \tag{4.3}$$

and

$$\dot{t} - u^0 = 0. \tag{4.4}$$

Equations (4.3) and (4.4) may be solved to express those variables that were varied in terms of the others. It is then legitimate to replace in (4.2) p_0 by $-H_0$ (and u^0 by $\dot{t}$, which is not seen) to obtain a reduced action for the remaining variables. That reduced action depends only on $q^\mu(\tau)$ $(\mu = 0, i)$ and $p_i(\tau)$ and reads

$$\int_{\tau_1}^{\tau_2} (p_i\dot{q}^i - H_0\dot{t})\, d\tau = \int_{t_1}^{t_2} \left(p_i \frac{dq^i}{dt} - H_0\right) dt. \tag{4.5}$$

For Eq. (4.5) to hold, t must be a monotonous function of τ so that its inverse exists. However, the generally covariant version does not need that assumption and is, therefore, more flexible. For example, in the path integral for generally covariant systems (Chapter 16), one admits trajectories that go back in the time t, *even in nonrelativistic particle mechanics.* The resulting path integral coincides, nevertheless, in that case with that of the reduced action (4.1).

4.2.2. Zero Hamiltonian

The action (4.2) contains one extra canonical pair over (4.1) but also contains the constraint $\gamma_0 \approx 0$. This constraint—being the only one—is first class. Thus, according to the counting of §1.4.2, the number of independent degrees of freedom is the same for (4.1) and (4.2), in agreement with the discussion leading to (4.5).

An important property of (4.2) is that there is no first-class Hamiltonian H' in it. *Thus, the extended Hamiltonian contains only the constraint* γ_0. Therefore, the motion is the unfolding of a gauge transformation.

If the original theory has other gauge generators $\gamma_{a'}$ $(a' = 1, \ldots, m)$ and second-class constraints $\chi_\alpha = 0$ to start with, the action replacing (4.2) reads

$$S = \int_{\tau_1}^{\tau_2} (p_\mu \dot{q}^\mu - H_E)\, d\tau, \tag{4.6a}$$

where the extended Hamiltonian is now a combination of all the constraints

$$H_E = u^a \gamma_a + u^\alpha \chi_\alpha \qquad (a = 0, \ldots, m). \tag{4.6b}$$

4.2.3. Parametrization and Explicit Time Dependence

Incorporating time as a canonical variable in a theory originally not written in generally covariant form does not only enable one to reach a more symmetrical formulation of the motion. It is also useful in practice to deal with problems in which there is an explicit time dependence in the constraints. Such a dependence complicates the formalism because equations expressing the preservation in time of the constraints involve "explicit time derivatives" and are not formulated just in terms of brackets.

This difficulty disappears when time is introduced as a canonical variable because after that step is taken, no explicit time derivative is left. The analysis proceeds then as in the ordinary case but in the enlarged phase space containing t and p_0. The notion of first class is, in particular, understood to include also $p_0 + H(p, q, t)$ among the constraints.

4.3. TIME REPARAMETRIZATION INVARIANCE

4.3.1. Form of Gauge Transformations

The action (4.6) is invariant under the transformation

$$\delta q = \dot{q}\varepsilon, \tag{4.7a}$$

$$\delta p = \dot{p}\varepsilon, \tag{4.7b}$$

$$\delta u^a = (u^a \varepsilon)^{\cdot} \tag{4.7c}$$

$$\delta u^\alpha = (u^\alpha \varepsilon)^{\cdot}, \tag{4.7d}$$

with

$$\varepsilon(\tau_1) = \varepsilon(\tau_2) = 0. \tag{4.7e}$$

The transformation (4.7) is an infinitesimal reparametrization of amount $\varepsilon(\tau)$, *i.e.*, is obtained from $\tau \to \bar{\tau} = \tau - \varepsilon(\tau)$. Equations (4.7a)–(4.7b) state that q and p transform as scalars, whereas (4.7c)–(4.7d) state that the multipliers u^a, u^α transform as scalar densities. The condition (4.7e) states that the endpoints τ_1 and τ_2 are left invariant.

The reparametrization (4.7) differs from the gauge symmetries generated by the γ_a by an "equation-of-motion symmetry" and, therefore, is not an independent symmetry. Indeed, (4.7) is equal on-shell to a gauge invariance (3.26) of the action with parameters

$$\varepsilon^a = u^a \varepsilon \tag{4.8}$$

(Exercises 4.3 and 4.4).

4.3.2. Must the Hamiltonian Be Zero for a Generally Covariant System?

We have just shown that the extended action S_E with zero Hamiltonian is invariant under arbitrary reparametrizations of the τ-variable. One may then ask the converse question, namely, whether general covariance implies a zero Hamiltonian.

If the q's and p's transform as scalars under reparametrizations, the $p\dot{q}$-term in the action transforms as a scalar density, and its time integral is therefore invariant by itself. Furthermore, since the γ_a and the χ_α are functions of p and q, they also transform as scalars. One therefore adjusts the transformation law for the multipliers u^a and u^α to be that of a scalar density so as to make the $u^a\gamma_a$ and $u^\alpha\chi_\alpha$ terms also scalar densities. Thus, once it is assumed that the p's and q's are scalars, it is crucial to have an independent variable in front of each constraint function to be able to achieve invariance. If one had in the action a nonzero first-class Hamiltonian $H'(q,p)$ besides the $u^a\gamma_a + u^\alpha\chi_\alpha$ term, one could not achieve reparametrization invariance, since $H'(q,p)$ would transform as a scalar and not as a density.

Thus, if q and p transform as scalars under time reparametrizations, the Hamiltonian is (weakly) zero for a generally covariant system. This is the case found so far in practice because one normally formulates theories in terms of variables with a simple geometrical meaning. For that reason, one usually says that a zero Hamiltonian is the distinguishing mark of a generally covariant system.

However, it is important to realize that one may also have general covariance with a nonzero Hamiltonian. Indeed, one knows that a time-dependent canonical transformation changes the value of the Hamiltonian. Thus, by performing a canonical transformation that depends explicitly on the parameter τ, one may bring in a nonzero Hamiltonian while keeping time reparametrization invariance of the action. Since τ itself does not transform as a scalar, the new canonical variables will not be scalars in agreement with the discussion above.

4.3.3. Simple Example of a Generally Covariant System with a Nonzero Hamiltonian

A simple and interesting example of this possibility is to consider the action (4.2) and define the variable

$$T(\tau) = t(\tau) - \tau, \tag{4.9}$$

while keeping all the other variables unchanged. The action becomes

$$S = \int_{\tau_1}^{\tau_2} \{p_0\dot{T} + p_i\dot{q}^i - H_0 - \lambda(p_0 + H_0)\}\, d\tau \tag{4.10}$$

with $\lambda = u^0 - 1$. This action is generally covariant and possesses a nonzero Hamiltonian H_0. The canonical coordinate T does not transform as a scalar under reparametrizations but rather inhomogeneously as a "connection"

$$\delta T = \dot{T}\varepsilon + \varepsilon. \tag{4.11}$$

The example (4.10) shows clearly that there is really no distinguishing technical feature of general covariance over other gauge transformations.

Indeed, once one has a gauge generator the action can always be made time reparametrization invariant by appropriately defining the transformations of the variables: one first makes a time-dependent canonical transformation to go to a zero Hamiltonian, and one defines the new canonical variables to transform as scalars. This can be achieved without introducing an explicit time dependence in the constraints (see Exercise 5.7)

4.4. "TRUE DYNAMICS" VERSUS GAUGE TRANSFORMATIONS

4.4.1. Interpretation of the Formalism

Once there is an arbitrary function of time in the general solution of the equations of motion, one cannot tell from the equations themselves whether this is because some of the canonical variables are not observable or whether the time (*i.e.*, the variable over which the Lagrangian is integrated to form the action) is not observable.

Normally, the choice is made based on information provided from outside. For example, in Maxwell's equations in Minkowski space one assumes that the time variable has an objective meaning and interprets the gauge invariance as indicating that not all components of the vector potential are observable. Ultimately, this interpretation is based on the fact that the Maxwell field is just a subsystem of the universe and that there are other systems interacting with it that are used to provide a time standard.

Conversely, when one deals, for example, with a parametrized nonrelativistic particle, one normally assumes that the variables t and q^i have a physical meaning and that τ is unobservable. Again, this is based on information not contained in the action itself.

However, when one tries to apply the formalism to the universe as a whole—as one does in general relativity—one has no "outside" available, and it is desirable to have an interpretation of the formalism that does not need foreign ingredients. For that purpose, the most symmetric form of the theory is that in which the Hamiltonian is weakly zero, *i.e.*, in which it is a linear combination of the constraints. One then says that all physical questions should be formulated in terms of functions that have zero brackets with the constraints. By doing so, one treats on the same footing "ordinary gauge-invariant quantities" (such as the magnetic field in electromagnetism) and constants of the motion (such as the initial position or the energy for a free particle). This is because, when the Hamiltonian weakly vanishes, the first-class functions with no explicit τ-dependence are constants of the motion.

At first sight this approach would seem to exclude asking questions such as "what is the position q at time t_0" for the particle. However, this is not so. Indeed, there is a constant of the motion $q_{t_0}(\tau)$ that is equal to the position at time t_0. For a free nonrelativistic particle with mass m and momentum p, it is simply

$$q_{t_0}(\tau) = q(\tau) - \frac{p(\tau)}{m}\big(t(\tau) - t_0\big) \tag{4.12}$$

and the above question is equivalent to "what is the value of the first-class function $q_{t_0}(\tau)$?"

4.4.2. Reduced Phase Space

According to the terminology introduced in Chapters 1 and 2, the functions having vanishing brackets with the constraints are the "classical observables" and are defined over the reduced phase space. So when the extended Hamiltonian is a linear combination of the constraints, an observable takes the same value on an entire classical history and on the gauge-related ones. It may thus be thought of as a function in the space of classical solutions of the equations of motion.

The set of observables is easily characterized in the case of a parametrized system. The constraint $p_0 + H_0 = 0$ can be solved for p_0. Accordingly, functions on the constraint surface can be viewed as functions of q^i, p_i, and t. the condition $[A(q^i, p_i, t), p_0 + H_0] \approx 0$ can then be integrated, at least in principle. This is equivalent to solving the equations of motion and completely determines the time dependence of A.

The set of the observables is thus isomorphic to the set of functions $A(\dot{q}, \dot{p})$ of the initial data. The space of the initial data $\dot{q}, \dot{p}$ is itself isomorphic to the space of the q's and p's at any fixed time. Therefore, the observables are in bijective correspondence with the dynamical variables of the original, unparametrized theory.

Going to the reduced phase space does not amount to freezing the dynamics at a given instant of time. Rather, it is equivalent to describing the system in terms of constants of the motion. In these canonical coordinates, the Hamiltonian vanishes and, allowing for an explicit τ-dependence, the equations of motion reduce to $dA/d\tau = \partial A/\partial\tau$.

One may reach the reduced phase space for a parametrized system by imposing the canonical gauge condition $t = t_0$. This condition defines a slice that cuts all the histories once and only once. In the gauge $t = t_0$, q and p become equal to the constants of the motion q_{t_0} and p_{t_0}; *i.e.*, these constants of the motion are the gauge-invariant extensions of q and p off the canonical gauge $t = t_0$.

In practice, the explicit construction of the reduced phase space in terms of the given canonical variables may not be possible, since it requires solving the equations of motion for arbitrary initial data. This is a hopeless task for systems with a complicated Hamiltonian, not to mention general relativity. Therefore, it appears necessary to develop the theory without reducing the system, a point of view that has already been put forward in Chapter 2.

EXERCISES. CHAPTER FOUR

4.1. Consider the action

$$S[x^\mu(\tau)] = -m \int_{\tau_1}^{\tau_2} \left(- \frac{dx^\mu}{d\tau} \frac{dx_\mu}{d\tau} \right)^{1/2} d\tau$$

describing the motion of a relativistic particle in Minkowski space ($\mu = 0, 1, \ldots, n-1$). This action is generally covariant under reparametrizations of the worldline. Perform the canonical analysis. Show that there is one constraint, the "mass-shell condition" $\mathcal{H} = p_\mu p^\mu + m^2 \approx 0$, and that $H_E = u\mathcal{H}$.

4.2. Consider the action

$$S[x^\mu, e] = \int [e^{-1} \dot{x}_\mu \dot{x}^\mu - em^2]\, d\tau.$$

(a) Prove that it is equivalent to the spinless relativistic particle action of Exercise 1 but that the mass-shell condition arises now as a *secondary* constraint.

(b) Show that the action $S[x^\mu, e]$ can be interpreted as the Klein–Gordon action for n scalar fields $x^\mu(\tau)$ in one spacetime dimension, coupled to the background metric $g_{00}(\tau) = e^2(\tau)$. The mass term appears as a one-dimensional "cosmological constant" term.

(c) Relate the "einbein" $e(\tau)$ to $u(\tau)$.

(d) Conclusion: By adding new variables, primary constraints may become secondary (and vice versa), and the Lagrange multipliers may become part of the canonical variables.

4.3. Show that in the absence of second-class constraints, the transformation (4.7) can be rewritten as

$$\delta q^\mu = \frac{\delta S_E}{\delta p_\mu}\, \varepsilon + \varepsilon^a [q^\mu, \gamma_a], \tag{i}$$

$$\delta p_\mu = -\frac{\delta S_E}{\delta q^\mu}\, \varepsilon + \varepsilon^a [p_\mu, \gamma_a], \tag{ii}$$

$$\delta u^a = \dot{\varepsilon}^a, \tag{iii}$$

and hence differs from (3.26) with $\varepsilon^a = u^a \varepsilon$ by a trivial gauge transformation.

4.4. Show that when second-class constraints are present, the equalities of Exercise 4.3 hold only if one sets $u^\alpha = 0$. As $u^\alpha = 0$ follows from the equations of motion, this means that there must be further equation-of-motion symmetries in (i)–(iii) (see

Theorem 3.1). Write explicitly these equation-of-motion symmetries for the simple action $S_E = \int (p\dot{q} - u^1 p - u^2 q)\, dt$.

4.5. Write explicitly all the equations of motion following from the action (4.2). Show that for solutions of the equations of motion, a change of sign of $u^0(\tau)$ can be interpreted as a switchback of the trajectories on themselves. Show that these switchbacks can be eliminated by going to a new parametrization (this does not modify the action), so that one can assume that t is a monotonous function of τ.

4.6. In (nonparametrized) classical mechanics, it is useful to define the unequal time bracket $[A(t), B(t')]$ of phase space functions. This is done by expressing $q(t), p(t)$ in A and $q(t'), p(t')$ in B in terms of $q(\bar{t}), p(\bar{t})$ by means of the equations of motion (where $\bar{t}$ is arbitrary) and by using $[q(\bar{t}), p(\bar{t})] = 1$, $[q(\bar{t}), q(\bar{t})] = [p(\bar{t}), p(\bar{t})] = 0$.

(a) Show that this leads to a well-defined unequal bracket (*i.e.*, the answer does not depend on which $\bar{t}$ one takes).

(b) Show that $[A(t), B(t')]$ depends, in general, on t and t'.

(c) Show that the unequal time bracket coincides with the reduced phase space bracket $[A_t, B_{t'}]$ of the parametrized theory.

4.7. Relate, for a parametrized system, the gauge-invariant extensions q_{t_0}, p_{t_0} of q and p in the gauge $t = t_0$ to the gauge-invariant extensions q_{t_1}, p_{t_1} of q and p in the gauge $t = t_1$. Observe, in particular, that the relationship is a canonical transformation with no explicit τ-dependence (so that H remains zero).

4.8. It is sometimes of interest to consider τ-dependent gauge conditions. The analysis of such gauge conditions was not covered in the discussion of Chapter 1 and is carried out in this exercise.

(a) Show that in the canonical gauge $C_a(q, p, \tau) = 0$, the motion is not generated by the Hamiltonian in the Dirac bracket. (Hint: Is $dC_a/d\tau = [C_a, H_E]^* + \partial C_a/\partial\tau$ zero?]

(b) Show that the correct time evolution for q and p contains an extra term, besides $[q, H_E]^*$ and $[p, H_E]^*$, which is a gauge transformation with an appropriate coefficient. Determine that coefficient.

4.9. (a) Prove that it is permissible to insert time-dependent canonical gauge conditions inside the action.

(b) After this is done, the action reads, in terms of unconstrained variables y^i,

$$S[y^i(\tau)] = \int [a_i(y, t)\dot{y}^i - H(y, \tau)]\, d\tau.$$

Write explicitly the equations of motion following from this action. Show that the extra term in the equations of motion (besides $[y^i, H]^*$) stems from the time dependence of the one-form a_i from which the bracket derives.

(c) Using the Darboux theorem, show that one can redefine new variables $\bar{y}^i = y^i(y, \tau)$ that have time-independent brackets and for which the equations of motion take the standard form.

(d) Conclusion: The presence of an extra term in $\dot{y}^i = [y^i, H]^* +$ "extra" has nothing mysterious. It is simply related to the fact that the y's are time-dependent coordinates in the sense that they have time-dependent (Dirac) brackets.

4.10. Solve for the constraint $p_t + H = 0$ and the gauge condition $t - \tau = 0$ inside the canonical action for a parametrized system, and check that one gets the original Hamiltonian action. Show that this gauge condition can also be reached by first making the τ-dependent canonical transformation (4.9) and imposing $T = 0$, using the standard theory for time-independent gauge conditions.

CHAPTER FIVE

FIRST-CLASS CONSTRAINTS: FURTHER DEVELOPMENTS

5.1. PRELIMINARIES AND NOTATIONS

To develop the subsequent formalism in as concise and clear a way as possible, we will assume from now on that all the second-class constraints have been eliminated and that one has been able to find an explicit, complete set of canonical variables subject only to the first-class constraints. The elimination of the second-class constraints may have been achieved by means of the Dirac bracket method or by means of the alternative procedure described in §1.4.3, which consists of adding new variables and increasing the gauge symmetry. In the presence of the second-class constraints, the equalities derived in the sequel should be understood as equalities up to second-class constraints and the bracket $[\ ,\ ]$ should be rewritten as $[\ ,\ ]^*$.

The canonical variables that describe the system after the second-class constraints have been eliminatedwill again by denoted by q^i, p_i $(i = 1, \ldots, n)$ and are assumed for simplicity to form conjugate pairs. The collective notation $z^A \equiv (q^i, p_i)$ $(A = 1, \ldots, 2n)$ will also be used. If the variables do not split into canonically conjugate pairs, it is only necessary to rewrite the formulas below in a form covariant under changes of coordinates in phase space by inserting explicitly the symplectic two-form when this is necessary. The first-class canonical Hamiltonian will be denoted by $H_0(q,p)$, while $G_a \approx 0\,(a = 1,\dots,m)$ stand for all the first-class constraints, including those which may have been added in the process of eliminating the second-class ones. It will be assumed, as always, that the first-class constraints obey the regularity assumption. They might be reducible,

$$Z^a_\alpha G_a = 0 \qquad (\alpha = 1,\dots,\bar{m}). \tag{5.1}$$

A more detailed analysis of the reducibility conditions (5.1) will be provided in Chapter 10.

In the absence of second-class constraints, the first-class conditions read

$$[H_0, G_a] = V_a{}^b G_b, \tag{5.2a}$$
$$[G_a, G_b] = C_{ab}{}^c G_c. \tag{5.2b}$$

The motion is generated by the extended Hamiltonian H_E containing the new constraints (if any) multiplied by their Lagrange multipliers,

$$H_E = H_0 + \lambda^a G_a, \tag{5.3}$$

and the corresponding action and its gauge symmetries reduce to

$$S_E\left[q^i(t), p_i(t), \lambda^a(t)\right] = \int (\dot{q}^i p_i - H_0 - \lambda^a G_a)\, dt, \tag{5.4}$$
$$\delta_\varepsilon F = \varepsilon^a [F, G_a], \tag{5.5a}$$
$$\delta_\varepsilon \lambda^a = \dot{\varepsilon^a} + \lambda^c \varepsilon^b C_{bc}{}^a - \varepsilon^b V_b{}^a. \tag{5.5b}$$

Here F is any function of the q's and the p's.

5.2. ABELIANIZATION OF CONSTRAINTS

5.2.1. Ambiguity in the Description of the Constraint Surface

The constraint functions G_a are not uniquely determined. They can always be replaced by an equivalent set,

$$\bar{G}_b(q,p) = M_b{}^a(q,p)\, G_a(q,p), \qquad \det M \neq 0 \tag{5.6}$$

because the equations $G_a = 0$ are equivalent to $\bar{G}_a = 0$. The constraint surfaces $G_a = 0$ and $\bar{G}_a = 0$ are therefore identical.

That the condition $\det M_a{}^b \neq 0$ is necessary for (5.6) to be acceptable is not entirely obvious. What is clear is that the equivalence of $G_a = 0$ with $\bar{G}_a = 0$ implies

$$\bar{G}_b = M_b{}^a G_a, \qquad G_a = N_a{}^c \bar{G}_c,$$

for some matrices $M_b{}^a(q,p)$ and $N_a{}^c(q,p)$. Substitution of one of the equations into the other yields successively

$$(M_b{}^a N_a{}^c - \delta_b{}^c)\bar{G}_c = 0$$

and

$$(N_a{}^c M_c{}^b - \delta_a{}^b)G_b = 0.$$

In the irreducible case, this leads to the equations (see Sec. 9.1 below and Theorem 9.1)

$$\begin{aligned} M_b{}^a N_a{}^c - \delta_b{}^c &= \rho_b{}^{cd}\bar{G}_d, & \rho_b{}^{cd} &= -\rho_b{}^{dc}, \\ N_a{}^c M_c{}^b - \delta_a{}^b &= \sigma_a{}^{bd}G_d, & \sigma_a{}^{bd} &= -\sigma_a{}^{db}. \end{aligned}$$

It follows that the matrices M and N are invertible on the constraint surface ($MN = NM = I$) and hence also in its vicinity by continuity. But far from $G_a = 0$, $M_b{}^a$ and $N_a{}^c$ may fail to be invertible.

However, the matrices M and N are not uniquely determined off $G_a = 0$, since one can replace $M_b{}^a$ and $N_a{}^c$, respectively, by $M_b{}^a + m_b{}^{ac}G_c$, $m_b{}^{ac} = -m_b{}^{ca}$ and $N_a{}^c + n_a{}^{cd}\bar{G}_d$, $n_a{}^{cd} = -n_a{}^{dc}$, without changing G_a and $\bar{G}_a$. This freedom can be used to make $M_b{}^a$ and $N_b{}^c$ invertible at any point off $G_a = 0$. We will assume that this can be achieved in a smooth way.

Similar considerations apply to the reducible case, for which there is the even greater freedom of adding to $M_b{}^a$ an arbitrary combination of the null vectors Z^a_α.

The change (5.6) in the generators implies a change in the description of the gauge transformations and, thus also, in the structure functions. The change is such that if $(\partial M_b{}^a/\partial q^i) \neq 0$, $(\partial M_b{}^a/\partial p_i) \neq 0$, the gauge transformations associated with the second set $\bar{G}_b$,

$$\bar{\delta}_\varepsilon F = \varepsilon^a[F, \bar{G}_a], \tag{5.7a}$$

are combinations of the gauge transformations associated with the first set G_a,

$$\delta_\varepsilon F = \varepsilon^a[F, G_a], \tag{5.7b}$$

only on the surface where G_a (and $\bar{G}_a$) vanish.

One expects, of course, that the two descriptions will be equivalent, but that equivalence is not transparent. This is because the change (5.6) of constraint generators cannot, in general, be implemented by a symmetry transformation of the theory, *i.e.*, by a canonical transformation. For instance, if the original G_a were abelian, $[G_a, G_b] = 0$, one could choose $M_b{}^a$ such that $[\bar{G}_a, \bar{G}_b] \neq 0$. The replacement $G_a \rightarrow \bar{G}_a$ of the constraint functions then clearly cannot be induced by a canonical transformation, which leaves by definition the bracket invariant. It is one virtue of the BRST formalism and the introduction of the ghost variables that the equivalence becomes manifest.

5.2.2. Abelianization Theorem

Since one admits coefficients that depend on the canonical variables in the linear combination (5.6), the flexibility of the structure functions is enormously greater than the one available when changing the basis of the generators of a Lie algebra (see Exercise 5.1).

Indeed, one can even achieve that the constraints become abelian, namely, that the structure functions all vanish. This can always be done in principle—at least locally, that is—in a region of phase space. By a canonical transformation, one can then take the new constraint functions to be some of the new momenta. [A necessary and sufficient condition for a set of functions to be some of the momenta of a new (local) canonical coordinate system is that the Poisson brackets of these functions among themselves all vanish.]

This possibility is extremely useful for proving general properties. The equivalent abelian constraints play a role similar to normal coordinates in Riemannian geometry, or of symplectic coordinates in symplectic geometry. However, the abelianization is not of much use in practice because simplicity requirements in the functional form of the G_a's in terms of the original canonical variables, which are especially important in the passage to quantum mechanics (such as polynomial structure or locality in field theory), are usually absent in the abelianized constraints.

The problem of abelianization is the following. Given a set of constraints G_a obeying

$$[G_a, G_b] = C_{ab}{}^c G_c, \tag{5.8}$$

one wants to find an invertible matrix $M_a{}^b$ such that

$$F_a = M_a{}^b G_b \tag{5.9a}$$

obeys

$$[F_a, F_b] = 0. \tag{5.9b}$$

The fact that (5.9b) can always be achieved by redefinitions of the constraints is a consequence of the canonical representation of the constraint surface discussed in Exercises 1.22 and 2.10. For pure first-class systems, however, there exists a more direct proof, which is reproduced here.

One constructs the F_a by first splitting the G_a into an irreducible subset $G_{a'}$ ($a' = 1, \ldots, m'$) such that $dG_{a'}$ is of "maximal rank" and the remaining constraints $G_{a''}$ are given by $X_{a''}{}^{a'} G_{a'}$. Since $dG_{a'}$ is of maximal rank, one can solve the equations $G_{a'} = 0$ to express m' canonical variables $z_{a'}$ in terms of the others, z_Δ ($\Delta = 1, \ldots, 2n - m'$),

$$G_{a'} = 0 \Leftrightarrow z_{a'} = z_{a'}(z_\Delta). \tag{5.10a}$$

The searched-for functions $F_{a'}$ are just given by

$$F_{a'} = z_{a'} - z_{a'}(z_\Delta). \tag{5.10b}$$

Indeed, one has

$$[z_{a'}, z_{b'}] = 0 \text{ or } \pm 1, \tag{5.11a}$$

$$[z_{a'}, z_\Delta] = 0 \text{ or } \pm 1, \tag{5.11b}$$

$$[z_\Delta, z_\Gamma] = 0 \text{ or } \pm 1. \tag{5.11c}$$

Therefore, the bracket

$$[F_{a'}, F_{b'}] = [z_{a'}, z_{b'}] - [z_{a'}, z_{b'}(z_\Delta)] - [z_{a'}(z_\Delta), z_{b'}] + [z_{a'}(z_\Delta), z_{b'}(z_\Gamma)]$$

only depends on z_Δ and does not involve $z_{a'}$. Accordingly, $[F_{a'}, F_{b'}]$ is independent of any particular value taken by $z_{a'}$. But we know that $[F_{a'}, F_{b'}] = 0$ when the constraints hold (first-class property), *i.e.*, when $z_{a'} = z_{a'}(z_\Delta)$. Therefore, $[F_{a'}, F_{b'}]$ must vanish identically,

$$[F_{a'}, F_{b'}] = 0, \tag{5.12a}$$

as stated. We have thus achieved part of our goal, namely, we have replaced the irreducible subset $G_{a'}$ by equivalent constraints

$$F_{a'} = M_{a'}{}^{b'} G_{b'} \tag{5.12b}$$

identically in involution. (Note that $M_{a'}{}^{b'}$ is invertible and that $M_{a'}{}^{b''}$ is zero.)

We now turn to the construction of the reducible abelian constraints $F_{a''}$. We may take

$$M_{a''}{}^{b''} = \delta_{a''}{}^{b''}, \qquad M_{a''}{}^{b'} = -X_{a''}{}^{b'}. \tag{5.13a}$$

This yields

$$\begin{aligned} F_{a''} &= M_{a''}{}^{b'} G_{b'} + M_{a''}{}^{b''} G_{b''} \\ &= 0, \end{aligned} \tag{5.13b}$$

i.e., $F_{a''}$ identically vanishes and thus clearly possesses vanishing bracket with everything else. Furthermore, the matrix $M_a{}^b$,

$$M_a{}^b = \begin{pmatrix} M_{a'}{}^{b'} & 0 \\ X_{a''}{}^{b'} & \delta_{a''}{}^{b''} \end{pmatrix} \tag{5.13c}$$

is invertible, so that the choice (5.13a) is permissible.

This achieves the proof of local abelianization. As a by-product of our proof, we have also established that the new reducibility coefficients $Z_{a''}{}^a$ in the reducibility conditions

$$Z_{a''}{}^a F_a = 0 \tag{5.14a}$$

can be taken to be constant, namely,

$$Z_{a''}{}^{a'} = 0, \qquad Z_{a''}{}^{b''} = \delta_{a''}{}^{b''}. \tag{5.14b}$$

It cannot be stressed enough that the abelianization of the constraints can, in general, only be achieved locally. Furthermore, this locality is usually of a very restricted nature in the sense that it is, in general, impossible to abelianize the constraints in a single neighborhood of the entire constraint surface. Rather, abelianization by a definite set of functions F^a will generically only hold in a region of phase space that does not cover the whole constraint surface but only part of it.

5.3. EXTERIOR DERIVATIVE OPERATOR ALONG THE GAUGE ORBITS ("LONGITUDINAL DERIVATIVE")

5.3.1. Definition of Longitudinal Derivative

The concept of gauge invariance can be rephrased more geometrically by introducing a differential operator whose importance turns out to be fundamental in the BRST formulation of gauge theories.

Through each point on the constraint surface, there is a gauge orbit defining "longitudinal" (or "vertical") directions. A vector field on the constraint surface is said to be longitudinal if it is everywhere tangent to the gauge orbits. Thus, in particular, the vector fields associated with the constraints and defining the infinitesimal gauge transformations,

$$\partial_{X_a} F \equiv \partial_a F = [F, G_a], \tag{5.15}$$

are, according to our discussion of Sec. 2.2, longitudinal vector fields.

One defines longitudinal p-forms to be antisymmetric multilinear objects acting on longitudinal vectors. That is, if α is a longitudinal

p-form, the number $\alpha(X_1, \ldots, X_p)$ is defined only when the vector X's are all tangent to the gauge orbits, and one has

$$\alpha(X_1, \ldots, X_i, X_{i+1}, \ldots, X_p) \\ = -\alpha(X_1, \ldots, X_{i+1}, X_i, \ldots, X_p), \tag{5.16a}$$

$$\alpha(\lambda X_1 + \mu Y_1, \ldots, X_p) \\ = \lambda\alpha(X_1, \ldots, X_p) + \mu\alpha(Y_1, \ldots, X_p). \tag{5.16b}$$

The exterior product of longitudinal forms is defined in the standard manner. We shall, however, slightly depart from the usual conventions by including an extra sign factor. If α is a p-form and β a q-form, we define

$$\alpha\beta = \alpha \wedge \beta(-)^{qp}, \tag{5.17a}$$

where $\alpha \wedge \beta$ would be the standard definition. With this convention, an odd left derivation of degree one for the product $\alpha \wedge \beta$ becomes an odd right derivation of same degree for the product $\alpha\beta$. This right action for derivative operators makes easier the comparison with the Hamiltonian formulas in which the generators are put to the right in the Poisson bracket [see, *e.g.*, (5.15)].

So, in particular, if α and β are 1-forms and X, Y longitudinal vectors, one finds

$$(\alpha \wedge \beta)(X, Y) = \alpha(X)\beta(Y) - \alpha(Y)\beta(X), \tag{5.17b}$$

while

$$(\alpha\beta)(X, Y) = -\alpha(X)\beta(Y) + \alpha(Y)\beta(X). \tag{5.17c}$$

The extra minus sign in (5.17c) may be thought of as arising from the commutation of the 1-form β with the 1-vector X. Namely, in order to write $\alpha(X)\beta(Y)$ from $(\alpha\beta)(X, Y)$, one needs to pass X through β. This brings the extra minus sign if one regards 1-vectors as odd objects (like 1-forms).

Because the Lie bracket of two longitudinal vector fields is again a longitudinal vector field, one can define an exterior derivative d acting on longitudinal forms and taking only antisymmetrized derivatives along the gauge orbits. This longitudinal d is characterized by the following defining properties,

$$\text{(i)} \qquad (dF)(X) = \partial_X F, \tag{5.18a}$$

where F is a function on the constraint surface and X a longitudinal vector;

$$\text{(ii)} \qquad d^2 = 0; \tag{5.18b}$$

$$\text{(iii)} \qquad d(\alpha\beta) = \alpha\, d\beta + (-)^q\, (d\alpha)\beta, \tag{5.18c}$$

for arbitrary longitudinal p-form α and q-form β (see Figure 3). The explicit form of d is given by the Eilenberg–Cartan formula,

$$\begin{aligned}(d\alpha)(X_0, \ldots, X_p) &= \sum_{j=0}^{p} (-)^j\, \partial_{X_j} \alpha(X_0, \ldots \hat{X}_j, \ldots, X_p) \\ &+ \sum_{0 \le i < j \le p} (-)^{p-i-j} \alpha(X_0, \ldots, \hat{X}_i, \ldots, \hat{X}_j, \ldots, X_p, [X_i, X_j]),\end{aligned} \tag{5.19}$$

where the caret ˆ denotes omission. (Recall that according to the conventions explained in Chapter 2, the Lie bracket used here differs from the standard Lie bracket by a minus sign.)

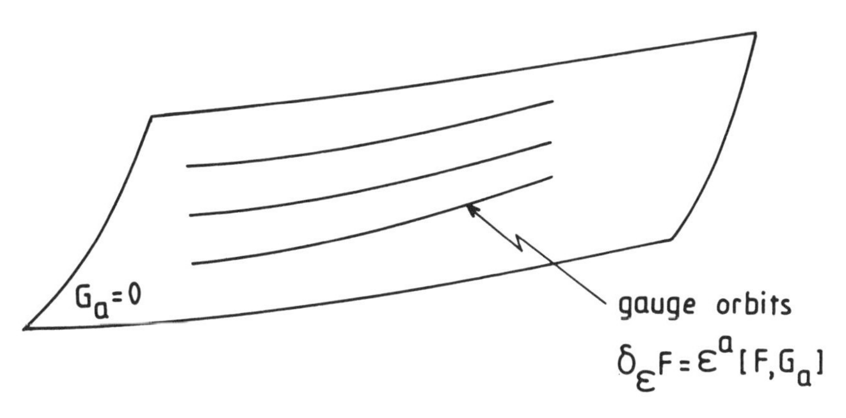

Figure 3: The first-class constraint functions G_a define not only the constraint surface Σ through $G_a = 0$ but also the gauge orbits through $\delta_\varepsilon F = \varepsilon^a (X_a F)$ with $X_a F \equiv [F, G_a]$. The gauge orbits define the longitudinal directions. The longitudinal derivative d takes derivatives only along the longitudinal directions.

In local coordinates (x, y) such that the equations of the gauge orbits are $x =$ const., one can identify longitudinal forms with the polynomials in dy, the coefficients of which depend on both x and y. For instance,

$$\alpha = \alpha_i(x, y)\, dy^i \tag{5.20a}$$

is the most general longitudinal 1-form. When computing $d\alpha$, the variables x are treated as external parameters, *i.e.*, one only differentiates with respect to y,

$$d\alpha = -\frac{\partial \alpha_i}{\partial y^j}\, dy^j\, dy^i. \tag{5.20b}$$

This representation, however, is usually only local.

An alternative definition of longitudinal forms and of the longitudinal d is worked out in Exercise 5.4.

5.3.2. Longitudinal Cohomology

Of special interest are the closed longitudinal p-forms obeying

$$d\alpha = 0. \tag{5.21a}$$

Since $d^2 = 0$, exact p-forms

$$\alpha = d\beta \tag{5.21b}$$

are closed. The longtitudinal cohomology is defined as the quotient space of the closed p-forms modulo the exact ones. The longitudinal cohomological spaces are denoted by

$$\left(\frac{\operatorname{Ker} d}{\operatorname{Im} d}\right)^p \qquad \text{or} \qquad H^p(d). \tag{5.22}$$

For $p = 0$, (5.22) yields the gauge-invariant functions. So, the "observables" can be identified with the cohomological class of d with degree zero. The higher cohomology groups $H^p(d)$ with $p > 0$ depend on the cohomology of the orbits and on the way they are stacked ("twisted") to fill the constraint surface. Their physical role is not well understood.

5.3.3. Representation of Longitudinal Derivative in the Irreducible Case

In the irreducible case, the vector fields X_a associated with the constraints are independent. One can thus define everywhere on the constraint surface a basis ω^a of longitudinal 1-forms by the conditions

$$\omega^a(X_b) = \delta_b{}^a, \qquad a, b = 1, \ldots, m. \tag{5.23}$$

This enables one to view longitudinal forms as polynomials in the ω^a's with coefficients that are functions on the constraint surface,

$$\alpha = \frac{1}{p!}\, \alpha_{a_1 \ldots a_p}\, \omega^{a_1} \ldots \omega^{a_p}. \tag{5.24}$$

In that representation of longitudinal forms, d is completely characterized by the rules

$$dF = (\partial_a F)\,\omega^a, \tag{5.25a}$$
$$d\omega^a = \tfrac{1}{2}\omega^b\omega^c C_{cb}{}^a, \tag{5.25b}$$

as well as the derivation property (5.18c). The equation (5.25b) follows from the fact that the basis $\{X_a\}$ is not holonomic but, rather, obeys

$$[X_a, X_b] = C_{ab}{}^c X_c \quad (\text{on } G_a = 0) \tag{5.26}$$

[see (2.17b)].

5.3.4. Representation of Longitudinal Derivative in the Reducible Case

The difference between the reducible and the irreducible cases is that in the reducible case, the vectors X_a associated with the constraints, even though completely spanning the tangent space to the orbits, are no longer linearly independent. As a result, the "components" $\alpha_{a_1\ldots a_q}$ of α,

$$\alpha_{a_1\ldots a_q} \equiv \alpha(X_{a_1}, \ldots, X_{a_q}) \tag{5.27a}$$

are not independent. One finds, instead,

$$\alpha_{a_1\ldots a_q} Z_\alpha^{a_q} = 0 \qquad \text{on} \qquad G_a = 0\,, \tag{5.27b}$$

since one has

$$Z_\alpha^{a_q} X_{a_q} \approx 0. \tag{5.27c}$$

Conversely, any set of completely antisymmetric functions $\alpha_{a_1\ldots a_q}$ obeying the algebraic condition (5.27b) can be viewed as the components of a longitudinal form α in the overcomplete set of tangent vectors X_a, *i.e.*,

$$\alpha_{a_1\ldots a_q} = \alpha(X_{a_1}, \ldots, X_{a_q}).$$

It will be useful in the sequel to describe a longitudinal form by means of its overcomplete set of components $\alpha_{a_1\ldots a_q}$, subject to (5.27b). One can then formally introduce 1-forms ω^a "dual to X_a" and identify longitudinal forms with polynomials in the ω^a, which identically vanish when one ω^a is replaced by any of the reducibility functions Z_α^a,

$$\alpha = \frac{1}{p!}\alpha_{a_1\ldots a_p}\omega^{a_1}\ldots\omega^{a_p}, \tag{5.28a}$$

$$\alpha_{a_1\ldots a_p} Z_\alpha^{a_1}\,\omega^{a_2}\ldots\omega^{a_p} = 0. \tag{5.28b}$$

It should be stressed that in order for α to be a longitudinal form, the coefficients $\alpha_{a_1\ldots a_p}$ in (5.28a), which are functions on the constraint surface, must obey the condition (5.28b). This means that the ω^a themselves are not longitudinal one-forms, and as a matter of fact they do not appear to have a direct geometrical interpretation.

The action of the longitudinal exterior derivative operator d in terms of the overcomplete components (5.27a) is just the same as in the irreducible case once the p-forms are identified with polynomials in ω^a of order p, *i.e.*,

$$d\omega^a = \tfrac{1}{2}\omega^b\omega^c C_{cb}{}^a, \tag{5.29a}$$

$$dF = \partial_a F \omega^a \qquad (\partial_a F \equiv [F, G_a]). \tag{5.29b}$$

Here, the $C_{cb}{}^a$ are again the structure functions of the set of vectors X_a,

$$[X_a, X_b] = C_{ab}{}^c X_c. \tag{5.30}$$

The rules (5.29) can be used to define the exterior derivative of arbitrary polynomials in the ω^a's. However, the nilpotency of the longitudinal exterior derivative is guaranteed only for true longitudinal forms. It does not hold for arbitrary polynomials in ω^a, and, for example, one finds that $d^2\omega^a$ is, in general, different from zero. This is because one allows in (5.29a) for an arbitrary set of $C_{bc}{}^a$ compatible with the first-class property; *i.e.*, one can freely add to $C_{bc}{}^a$ a term of the form $M_{bc}{}^{ad} G_d + N^{\alpha}_{bc} Z^a_{\alpha}$, so that even if a choice of $C_{bc}{}^a$ is such that $d^2 = 0$ on arbitrary polynomials, this may not be true for $C_{bc}{}^a + M_{bc}{}^{ad} G_d + N^{\alpha}_{bc} Z^a_{\alpha}$.

One can define an extended exterior derivative that is nilpotent on arbitrary polynomials. This is done by introducing additional formal objects at higher degree. This question will be examined in Chapter 10.

5.3.5. Phase Space Characterization of principle. Longitudinal Forms

The longitudinal forms are defined on the constraint surface Σ. It is, however, useful to give their components as functions over phase space because in practice it is hard to display a coordinate system on the constraint surface. Such a characterization relies on observing that the algebra $C^\infty(\Sigma)$ of functions on the constraint surface Σ can be viewed as the quotient algebra $C^\infty(P)/\mathcal{N}$ of phase space functions modulo functions that vanish on Σ (see §1.5.1). Thus, one can alternatively regard the longitudinal forms as polynomials in ω^a with coefficients that belong to $C^\infty(P)$,

$$\alpha = \frac{1}{p!}\alpha_{a_1\ldots a_p}(q,p)\,\omega^{a_1}\ldots\omega^{a_p}, \tag{5.31}$$

provided one identifies two forms that coincide on Σ,

$$\alpha \sim \alpha' \Leftrightarrow \alpha'_{a_1\ldots a_p}(q,p) \approx \alpha_{a_1\ldots a_p}(q,p) \tag{5.32}$$

In the reducible case, one must demand, in addition, that $\alpha_{a_1\ldots a_p} Z^{a_p}_{\alpha}$ should be weakly zero,

$$\alpha_{a_1\ldots a_p} Z^{a_p}_{\alpha} \approx 0. \tag{5.33}$$

Because the gauge transformations and the structure functions $C_{ab}{}^{c}(q,p)$ are defined everywhere in phase space and not just on Σ, one can define the exterior derivative of the polynomials (5.31) by exactly the same rules (5.25) and (5.29) as before. However, one finds, in general, that d^2 [acting on polynomials obeying (5.33)] is zero only on the constraint surface

$$d^2 \approx 0. \tag{5.34}$$

This is because the vector fields X_q in general fail to close off Σ.

In terms of the phase space description, the longitudinal cohomology is obtained by identifying two weakly closed polynomials (5.31) that differ by a weakly exact one.

5.4. HAMILTON–JACOBI THEORY

5.4.1. Unconstrained Systems—Complete Integrals

In the case of unconstrained systems, the Hamilton–Jacobi theory provides a bridge between classical and quantum mechanics. The Hamilton–Jacobi theory for constrained systems plays a similar role. In order to understand its main features, we first briefly review how the Hamilton–Jacobi equation arises in the standard classical mechanics of unconstrained systems.

The equations of motion define a canonical transformation between the coordinates q^i and the momenta p_i at time t on the one hand and the initial coordinates q^i_0 and momenta p^0_i at time t_0 on the other hand. Let (α^i, β_i) be canonical coordinates obtained from (q^i_0, p^0_i) by a time-independent canonical transformation. The transformation $(q^i, p_i) \rightarrow (\alpha^i, \beta_i)$ is also canonical, and we assume that its generating function $S(q^i, \alpha^i, t)$ can be expressed in terms of q^i and α^i. One has

$$p_i = \frac{\partial S}{\partial q^i}, \tag{5.35a}$$

$$\beta_j = -\frac{\partial S}{\partial \alpha^j}, \tag{5.35b}$$

and

$$\det \frac{\partial^2 S}{\partial \alpha^j \, \partial q^i} \neq 0. \tag{5.35c}$$

The variables α^i and β_i are constants of the motion. Hence, the Hamiltonian $\bar{H}(\alpha^i, \beta_i) = H + \partial S/\partial t$ describing the evolution of α^i and β_i can be taken to vanish. This leads to the Hamiltonian-Jacobi equation

$$\frac{\partial S}{\partial t} + H\left(q^i, \frac{\partial S}{\partial q^i}\right) = 0. \tag{5.36}$$

A solution $S(q^i, \alpha^i)$ of (5.36) depending on n variables α^i such that (5.35c) holds is called a complete integral. Knowledge of a complete integral enables one to construct the general solution of the equations of motion from (5.35b) by mere substitution,

$$\beta_j = -\frac{\partial S}{\partial \alpha^j} \Leftrightarrow q^i = q^i(\alpha^j, \beta_k, t). \tag{5.37}$$

The solution of (5.36) for which the time-independent canonical variables (α^i, β_j) coincide with the initial data (q_0^i, p_j^0) is denoted by $W(q^i, q_0^i, t)$ and is called the Hamilton principal function. One can show that it is equal to the "classical action," *i.e.*, that it is given by

$$W(q^i, q_0^i, t) = \int_{q_0^i, t_0}^{q^i, t} du \, L(q, \dot{q}), \tag{5.38}$$

where the integral is evaluated along the solution of the equations of motion joining $q_0^i(t_0)$ to $q^i(t)$.

5.4.2. Unconstrained Systems—Incomplete Integrals

The complete solutions of the Hamilton–Jacobi equations are not the only ones of physical interest. Solutions depending on fewer integration constants are also important. Such solutions are called "incomplete integrals" and are obtained by setting m of the α's equal to definite values in a complete integral.

We denote by α^A the α's that are unspecified, $A = 1, \ldots, n-m$, and by α_a those that are set equal to definite values, $a = 1, \ldots, m$. Without loss of generality, one can assume those definite values to be zero, so the incomplete integral $S(q^i, \alpha^A, t)$ is related to the complete integral $S(q^i, \alpha^A, \alpha_a, t)$ by

$$S(q^i, \alpha^A, t) = S(q^i, \alpha^A, \alpha_a = 0, t). \tag{5.39}$$

Once α_a is set equal to zero, the dependence of S on α_a is lost. Therefore, the conjugate variable $\beta^a = -\partial S/\partial \alpha_a$ is unknown. This means that through the equations

$$p_i = \frac{\partial S}{\partial q^i}, \tag{5.40a}$$

$$\beta_A = -\frac{\partial S}{\partial \alpha^A}, \tag{5.40b}$$

$$\text{rank} \; \frac{\partial^2 S}{\partial \alpha^A \, \partial q^i} = n - m, \tag{5.40c}$$

with α^A and β_A given, a complete integral $S(q^i, \alpha^A, t)$ can no longer determine a unique solution $q^i(t), p_i(t)$ of the equations of motion. Indeed, the constants of the motion α^A, β_A and $\alpha_a = 0$ do not characterize a single classical trajectory. Rather, they characterize all the trajectories that have the same values of α^A, β_A and $\alpha_a = 0$ but that differ in the value of the unknown conjugate β^a to α_a.

So, if (q^i, p_i) is a solution of (5.40) at time t, and $(q^i + \delta q^i, p_i + \delta p_i)$ a solution of (5.40) at time $t + \delta t$, one would have

$$\delta q^i = \frac{\partial H}{\partial p_i}(q,p)\,\delta t, \qquad \delta p_i = -\frac{\partial H}{\partial q^i}(q,p)\,\delta t,$$

in the case when $(q^i + \delta q^i, p_i + \delta p_i)$ and (q^i, p_i) happen to give the same values of the conjugate momenta β^a, *i.e.*, lie on the same classical trajectory. This follows from the Hamilton–Jacobi theory for a complete integral. However, $(q^i + \delta q^i, p_i + \delta p_i)$ and (q^i, p_i) may yield different values of β^a. Since the shifts in β^a are generated by the conjugate variables α_a, one finds in general

$$\delta q^i = \left[\frac{\partial H}{\partial p_i}(q,p) + \lambda^a \frac{\partial \alpha_a}{\partial p_i}(q,p)\right] \delta t, \tag{5.41a}$$

$$\delta p_i = \left[-\frac{\partial H}{\partial q^i}(q,p) - \lambda^a \frac{\partial \alpha_a}{\partial q^i}(q,p)\right] \delta t, \tag{5.41b}$$

for some λ^a. In addition, the α_a's vanish,

$$\alpha_a(q,p) = 0. \tag{5.41c}$$

If λ^a is equal to zero in (5.41a)–(5.41b), then (q^i, p_i) and $(q^i + \delta q^i, p_i + \delta p_i)$ are on the same classical trajectory. If $\lambda^a \neq 0$, then (q^i, p_i) and $(q^i + \delta q^i, p_i + \delta p_i)$ are on two different trajectories characterized by different values of β^a.

The extreme case is of direct interest in quantum mechanics and is the one for which the solution $S(q^i, t)$ involves no integration constant at all ($m = n$). In that instance, any two solutions contained in $S(q^i, t)$ have the same values of a complete set of commuting conserved quantities but differ in their conjugates. Intermediate cases with only some integration constants α^A in $S(q^i,\alpha^A,t)$ are relevant to the study of first-class constrained systems.

5.4.3. Constrained Systems

The Hamilton–Jacobi theory for constrained systems is almost identical with the theory of incomplete solutions developed in the previous section. This appears most transparently if one identifies the α_a with an abelian representation of the constraint surface $G_a = 0$. The conjugate variables β^a are then pure gauge, whereas the variables α^A and their conjugates β_A, which commute with α_a, form a complete set of gauge-invariant functions. They can be regarded as canonical coordinates on the reduced phase space.

With this identification, the generating function $S(q^i, \alpha^A, \alpha_a, t)$ defines a canonical transformation

$$(q^i, p_i) \rightarrow \alpha^A, \beta_A, \alpha_a, \beta^a \tag{5.42}$$

such that the constraints become just $\alpha_a = 0$. So, if one sets $\alpha_a = 0$ in $S(q^i, \alpha^A, \alpha_a, t)$, one gets a function $S(q^i, \alpha^A, t)$ that fulfills

$$G_a\left(q^i, \frac{\partial S}{\partial q^i}\right) = 0, \tag{5.43a}$$

$$\frac{\partial S}{\partial t} + H_0\left(q^i, \frac{\partial S}{\partial q^i}\right) = 0, \tag{5.43b}$$

$$\text{rank}\ \frac{\partial^2 S}{\partial \alpha^A\, \partial q^i} = n - m. \tag{5.43c}$$

At the same time, the information on $\partial S/\partial \alpha_a = -\beta^a$ is lost, and the conjugate variables β^a are arbitrary. The equations (5.43a)–(5.43b) are the Hamilton–Jacobi equations for the constrained system under consideration.

If $\left(q^i(t), p_i(t)\right)$ is, for each t, a solution of

$$p_i = \frac{\partial S}{\partial q^i}, \tag{5.44a}$$

$$\beta_A = -\frac{\partial S}{\partial \alpha^A}, \tag{5.44b}$$

5.4.5. Hamilton Principal Function

An important solution of the Hamilton–Jacobi equations is given by the "Hamilton principal function" $W(q_2^i, t_2; q_1^i, t_1)$. This function is well defined when the boundary problem of finding the solution of the equations of motion that takes arbitrarily prescribed values q_2^i at t_2 and q_1^i at t_1 possesses a unique solution up to gauge transformations.

This condition will be assumed throughout the section. This implies that the constraints $G_a(q,p) = 0$ can be solved for some of the p's. The matrix $\partial G_a / \partial p_i$ is thus of maximum rank. One can then devise gauge conditions that involve only the q's because there is no gauge transformation that leaves a given set of q's invariant.

The integral

$$W(q_2^i, t_2; q_1^i, t_1) = \int_{q_1, t_1}^{q_2, t_2} (p_i \, dq^i - H_0 \, dt - \lambda^a G_a \, dt), \tag{5.53}$$

evaluated on a classical history joining $q^i(t_1) = q_1^i$ to $q^i(t_2) = q_2^i$, does not depend on which particular history is chosen among the gauge-related histories that are solutions of the boundary problem. This is because the action is invariant under gauge transformations, leaving the given arguments q_2^i and q_1^i of W invariant, *i.e.*, vanishing at the time boundaries. (The gauge is fixed at the boundaries by q_1^i and q_2^i, as in §3.4.1.) Accordingly, the Hamilton principal function W is well defined by (5.53).

Straightforward calculations yield

$$p_i^2 \equiv p_i(t_2) = \frac{\partial W}{\partial q_2^i} \tag{5.54a}$$

and

$$\frac{\partial W}{\partial t_2} = -(H_0 + \lambda^a G_a)\Big(q_2, \frac{\partial W}{\partial q_2}\Big). \tag{5.54b}$$

Since q_2^i, p_i^2 are solutions of the equations of motion, one finds

$$G_a\Big(q_2^i, \frac{\partial W}{\partial q_2^i}\Big) = 0, \tag{5.55a}$$

and hence also from (5.54b),

$$\frac{\partial W}{\partial t_2} + H_0\Big(q_2^i, \frac{\partial W}{\partial q_2^i}\Big) = 0. \tag{5.55b}$$

This shows that W is a solution of the Hamilton–Jacobi equations (5.43).

Similarly, one gets at t_1 the time-reversed equations,

$$G_a\left(q_1^i, -\frac{\partial W}{\partial q_1^i}\right) = 0, \tag{5.56a}$$

$$-\frac{\partial W}{\partial t_1} + H_0\left(q_1^i, -\frac{\partial W}{\partial q_1^i}\right) = 0. \tag{5.56b}$$

For the example of the pure gauge theory with as many inhomogeneous constraints (5.51) as coordinates and with Hamiltonian $H_0 \equiv 0$, the integral (5.53) becomes

$$\begin{aligned} W(q_2^i, t_2; q_1^i, t_1) &= \int_{q_1,t_1}^{q_2,t_2} \frac{\partial V}{\partial q^i}\,\dot{q}^i\,dt \\ &= V(q_2^i) - V(q_1^i). \end{aligned} \tag{5.57}$$

The fact that q_2^i and q_1^i are not mixed in W reflects the absence of true physical degres of freedom: p_i^2 and p_i^1 are, respectively, completely determined by q_2^i or q_1^i and do not depend on the q's at the other time boundary.

EXERCISES. CHAPTER FIVE

5.1. Let $G_1 = p_1$, $G_2 = e^{x^1} p_2$ be the constraint functions of a system with canonical pairs (x^1, p_1) and (x^2, p_2).

(a) Check that G_1 and G_2 form a nonabelian group.

(b) Show that one can define new equivalent constraint functions $\bar{G}_1$ and $\bar{G}_2$ that form an abelian group.

Conclusion: The freedom (5.6) can even change the group structure (when there is one).

5.2. Prove that one can use the redefinition freedom of the unextended formalism [Eq. (3.43a)] to abelianize the constraints ϕ_{m_k} when they obey (3.42a). In other words, one can abelianize the primary constraints ϕ_{m_1} by a redefinition $\phi_{m_1} \to \bar{\phi}_{m_1} = M_{m_1}{}^{m_1'}\,\phi_{m_1'}$ involving only the primary constraints, the secondary constraints ϕ_{m_2} by a redefinition $\phi_{m_2} \to \bar{\phi}_{m_2} = M_{m_2}{}^{m_2'}\,\phi'_{m_2} + M_{m_2}{}^{m_1}\,\phi_{m_1}$ involving only the primary and secondary constraints, ..., etc.

5.3. The previous exercise indicates that one can locally assume the constraints to be "pure momenta" even in the unextended formalism,

$$\phi_{m_1} \equiv p_{m_1}, \phi_{m_2} \equiv p_{m_2}, \ldots, \phi_{m_L} \equiv p_{m_L}.$$

(a) Show from (3.42b) that H is given by

$$\begin{aligned} H = {} & f^{m_2}(q^{m_1}, \ldots, q^{m_L}; p_{m_2}, \ldots, p_{m_L})p_{m_2} \\ & + f^{m_3}(q^{m_2}, \ldots, q^{m_L}; p_{m_3}, \ldots, p_{m_L})p_{m_3} \\ & + \cdots + f^{m_k}(q^{m_{k-1}}, \ldots, q^{m_L}; p_{m_k}, \ldots, p_{m_L})p_{m_k} \\ & + \cdots + f^{m_L}(q^{m_{L-1}}, q^{m_L}; p_{m_L})p_{m_L} + f_0 \end{aligned}$$

up to primary constraints, with $\partial f^{m_i}/\partial q^{m_{i-1}}$ of maximum rank. (The extra, gauge-invariant canonical pairs, which may be present in f^{m_i} or f_0, are not explicitly written.)

(b) Use this information to show that one can redefine the constraints so as to fulfill the conditions (3.49). Hence, there exists (locally) a representation of the constraints and of H so that the gauge parameters μ^{m_i} of the unextended formalism depend only on t. (Hint: Start the analysis from the last generation of constraints.)

5.4. Let $\beta, \beta', \ldots$ be ordinary p-forms on the constraint surface with both longitudinal and transverse components. One defines an equivalence relation as follows: $\beta \sim \beta'$ iff $\beta(X_1, \ldots, X_p) = \beta'(X_1, \ldots, X_p)$ whenever the vectors $X_1, \ldots, X_p$ are all tangent to the gauge orbits.

(a) Show that $\beta \sim \beta'$ is indeed an equivalence relation and that it is compatible with the exterior product ($\beta \sim \beta'$ and $\gamma \sim \gamma' \Rightarrow \beta\gamma \sim \beta'\gamma'$). Check that longitudinal p-forms as defined in the text can be identified with equivalence classes of $\sim$.

(b) Prove that if $\beta \sim \beta'$, then $d\beta \sim d\beta'$ where d is the ordinary exterior derivative on the constraint surface. Use this property to provide an alternative definition of the longitudinal exterior derivative.

(c) A p-form β is said to be co-longitudinal if $\beta(X_1, \ldots, X_p)$ vanishes when *all* the vectors $X_1, \ldots, X_p$ are tangent to the gauge orbits. Observe that $\beta \sim \beta'$ iff $\beta - \beta'$ is co-longitudinal. Prove that co-longitudinal forms define an ideal for the exterior product, *i.e.*, if β is co-longitudinal and γ arbitrary, $\beta\gamma$ is co-longitudinal.

5.5. Assume that the constraint surface $G_a = 0$ has a product structure $B \times O$, where B is the "reduced phase space" and O is a generic gauge orbit (this is not always the case). Show that the cohomology of the longitudinal d is given by the product

$$\left(\frac{\text{Ker } d}{\text{Im } d}\right)^p \simeq C^\infty(B) \otimes \left(\frac{\text{Ker } d}{\text{Im } d}\right)^p_{DR/O},$$

where (i) $C^\infty(B)$ denotes the space of C^∞–functions on B; and (ii) $(\text{Ker } d/\text{Im } d)^p_{DR/O}$ is the standard De Rham cohomology of O.

5.6. Let G_a be given first-class constraints, $[G_a, G_b] = C_{ab}{}^c(q,p)G_c$. The abelianized constraints $F_a = M_a{}^b(q,p)G_b$ can be derived by trying to construct explicitly $M_a{}^b$ along the following lines:

(a) Write down the differential equations on the coefficients $M_a{}^b$ that follow from the requirement $[F_a, F_b] = 0$.

(b) Prove explicitly that the integrability conditions for the equations on $M_a{}^b$ are fulfilled everywhere on the constraint surface. (Note: Of course, one knows that these integrability conditions are fulfilled everywhere as a result of the abelianization theorem.)

5.7. Consider a constrained system with first-class Hamiltonian $H(q,p)$ and constraints $G_a(q,p) = 0$. Use the abelianization theorem to show that the time-dependent canonical transformation that makes $H = 0$ can be chosen so as to induce no explicit time dependence in the constraints. (Hint: The constraints may be assumed to be $P_k = 0$ for some momenta and H can be taken not to depend on either P_k or Q_k.)

5.8. Gauge-invariant integrals from elements of $H^p(d)$.Let α be a longitudinal p-form along the gauge orbits and let S be an arbitrary closed p-surface lying on a definite orbit.

(a) Prove that $\int_S \alpha = \int_{S'} \alpha$ for all S' obtained from S by a gauge transformation if and only if $d\alpha = 0$, where d is the longitudinal exterior derivative.

(b) Show that $\int_S \alpha$ vanishes if α is exact or if S is the boundary of a $p+1$ surface.

5.9. Verify that the integrability conditions for the Hamilton–Jacobi equations (5.43a)–(5.43b) hold.

5.10. Let $p_t + H(q^i, p_i) \approx 0$ be the constraint for a "parametrized" system. Verify that the Hamilton–Jacobi theory for this system, viewed as a constrained Hamiltonian system, is identical to the standard Hamilton–Jacobi theory for the original unconstrained system (q^i, p_i) with Hamiltonian $H(q^i, p_i)$ and time t.

5.11. Let σ be the group of canonical transformations that map the constraint surface on itself, and let $\mathcal{C}$ be the subgroup of σ that maps the gauge orbits on themselves (that is, a point in an orbit is mapped on a point in the same orbit by $\mathcal{C}$). Show that $\mathcal{C}$ is a normal subgroup of σ and that the Lie algebra of the infinitesimal generators of $\sigma/\mathcal{C}$ is isomorphic with the Lie algebra of the infinitesimal generators of the canonical transformations of the reduced phase space.

5.12. Let Σ and $\bar{\Sigma}$ be two first-class constraint surfaces of respective equations

$$G_a = 0, \quad \bar{G}_a = 0 \qquad (a = 1, \dots, m).$$

Assume that the matrices $\partial G_a/\partial p_i$ and $\partial \bar{G}_a/\partial p_i$ are of maximum rank m (so that $\delta q^i = [q^i, G_a]\varepsilon^a = 0 \Rightarrow \varepsilon^a = 0$ and $\delta \bar{q}^i = [\bar{q}^i, \bar{G}_a]\bar{\varepsilon}^a = 0 \Rightarrow \bar{\varepsilon}^a = 0$).

Let $S(q^i, \bar{q}^i)$ be a solution of

$$G_a\left(q^i, \frac{\partial S}{\partial q^i}\right) = 0, \quad \bar{G}_a\left(\bar{q}^i, -\frac{\partial S}{\partial \bar{q}^i}\right) = 0,$$

with

$$\text{rank}\left(\frac{\partial^2 S}{\partial \bar{q}^i\, \partial q^i}\right) = n - m.$$

(a) Show that the equations

$$p_i = \frac{\partial S}{\partial q^i}, \quad \bar{p}_i = -\frac{\partial S}{\partial \bar{q}^i},$$

define implicitly a mapping from Σ to $\bar{\Sigma}$ that associates with each point of Σ a gauge orbit of $\bar{\Sigma}$ and vice versa. [Hint: By differentiating the equations $G_a(q^i, \partial S/\partial q^i) = 0$, $G_a(\bar{q}^i, -\partial S/\partial \bar{q}^i) = 0$ with respect to q^i and $\bar{q}^i$, show that $\delta p_i = [p_i, G_a]\varepsilon^a$, $\delta \bar{p}_i = 0$ for $\delta q^i = [q^i, G_a]\varepsilon^a$, $\delta \bar{q}^i = 0$, and vice versa.]

(b) Show that the induced transformation of the reduced phase spaces associated with Σ and $\bar{\Sigma}$ is canonical.

(c) Let $\Sigma' : G'_a = 0$ be a third first-class constraint surface and let $\bar{S}(\bar{q}^i, q'^i)$ be a generating function relating $\bar{\Sigma}$ and Σ' as above. Show that the combined transformation $\Sigma \to \Sigma'$ is generated by $S'(q^i, q'^i) = S + \bar{S}$, where the "intermediate" $\bar{q}^i$ are eliminated through $\partial(S + \bar{S})/\partial \bar{q}^i = 0$.

5.13. Let $S(q^i, \alpha^A, t)$ be a complete solution of the Hamilton–Jacobi equations. Show that $W(q_2^i, t_2; q_1^i, t_1)$ is given by $S(q_2^i, \alpha^A, t_2) - S(q_1^i, \alpha^A, t_1)$ where the variables α^A are eliminated by means of

$$\frac{\partial(S_2 - S_1)}{\partial \alpha^A} = 0.$$

(Hint: This is a special case of Exercise 5.12c. Why is there a minus sign?)

5.14. Let $S(q^i, \alpha_A, t)$ be a solution of the Hamilton–Jacobi equation (5.36) depending on $n-m$ constants α_A. Show that there exists a complete integral $S(q^i, \alpha_A, \alpha_a, t)$ such that $S(q^i, \alpha_A, t) = S(q^i, \alpha_A, \alpha_a = 0, t)$. {Hints: Construct the canonical variables $\alpha_A(q,p,t)$, $\alpha_a(q,p,t)$ and their conjugates $\beta^A(q,p,t)$, $\beta^a(q,p,t)$ by observing that the equations $p_i = \partial S/\partial q^i(q^j, \alpha_A, t)$ not only imply m relations $\alpha_a(q,p,t) = 0$ but also enable one to express, on $\alpha_a = 0$, the α_A as functions of q and p, $\alpha_A = \alpha_A(q,p,t)$. One then finds $\beta_A = -\partial S/\partial \alpha_A$ as a function of q, p, and t on $\alpha_a = 0$. Show that one has on $\alpha_a = 0$ $[\alpha_A, \alpha_B] = 0$, $[\alpha_A, \alpha_a] = 0$, $[\alpha_a, \alpha_b] = 0$, $[\alpha_A, \beta^B] = \delta_A{}^B$, $[\alpha_a, \beta^B] = 0$. [To that end, differentiate $\alpha_a(q, \partial S/\partial q(q,\alpha)) = 0$, $\alpha_A = \alpha_A(q, \partial S/\partial q(q,\alpha))$, $(-\partial S/\partial \alpha^A) = \beta_A(q^i, \partial S/\partial q(q,\alpha))$ with respect to q^i and α_A.] Accordingly, one can redefine $\alpha_a \to \alpha'_a = M_a{}^b \alpha_b$ and extend α_A and β_A off $\alpha_a = 0$ so as to preserve these bracket relations.}

5.15. Let G_a be first-class constraints that are linear in the momenta, but not necessarily homogeneous,

$$G_a = a_a{}^i(q) p_i + f_a(q) \approx 0.$$

(a) Show that the gauge transformations are well defined in configuration space even when $f_a \neq 0$, *i.e.*, $\delta_\varepsilon q^i$ only involves q^i.

(b) Prove that the functions $a_a{}^i(q)$ define vector fields that close in the Lie bracket in configuration space. Do the structure functions $C_{ab}{}^c$ in $[\gamma_a, \gamma_b] = C_{ab}{}^c \gamma_c$ depend on the momenta?

(c) Show that $f_a(q) = a_a{}^i(q) \partial_i F$ (locally, at least), so that G_a can be written as

$$G_a = a_a{}^i(q) p_i$$

by redefining the momenta as $p_i \to p_i - \partial_i F$ (canonical "phase" transformation). What are the canonical transformations that leave G_a linear and homogeneous in p_i? Similar question for the redefinitions $G_a \to \bar{G}_a = M_a{}^b G_b$ of the constraints.

(d) Show that the variation (3.28) of the extended action S_E vanishes for the constraints of (c), no matter what value ε^a takes at the endpoints.

5.16. Consider a system with Hamiltonian H and constraints $G_a = 0$. Let $q^{*\alpha}, p^*_\alpha$ be a complete set of gauge-invariant functions of q^i and p_i ("observables") with canonical brackets.

(a) Show that the action $\int (p^*_\alpha \dot{q}^{*\alpha} - H(q^*, p^*))\, dt$ induced in the reduced phase space weakly differs from the (q^i, p_i)-action by a boundary term that may not be invariant under the transformations generated by G_a.

(b) Relate the boundary term to the generating function of the canonical transformation from (q^i, p_i) to $(q^*, p^*, \beta, \gamma)$, where β, γ are pure gauge pairs necessary to complete the coordinate system based on q^*, p^*.

CHAPTER SIX

FERMI DEGREES OF FREEDOM: CLASSICAL MECHANICS OVER A GRASSMANN ALGEBRA

Our discussion in the previous chapters applies to systems described by dynamical variables that are real or complex functions of time and, hence, that belong to a commutative algebra. These variables are quantized by means of commutators and describe, therefore, bosonic degrees of freedom.

The purpose of this chapter and the next is to extend the formalism to include fermionic degrees of freedom. These turn out to be described classically by anticommuting variables belonging to a Grassmann algebra. The classical equations of motion for the fermionic variables can be derived from an action that belongs to the commuting part of the algebra, and a Hamiltonian formalism having much in common with the usual canonical formalism can be consistently constructed.

Even for purely bosonic systems, the consideration of Grassmann mechanics is a necessary step when there is gauge freedom. This is because the study of those systems leads to the introduction of additional "ghost" variables, which obey Fermi statistics.

This chapter develops the canonical formalism for Fermi degrees of freedom in the absence of constraints. The constrained case comes next (Chapter 7).

6.1. FERMIONS AND ANTICOMMUTING c-NUMBERS

The canonical quantization procedure is based on the correspondence

$$\text{(Poisson bracket)} \rightarrow (i\hbar)^{-1} \text{ (commutator)}. \tag{6.1}$$

This prescription is a consistent one in that the objects being associated have the same algebraic properties:

$$[A, B]_- = -[B, A]_-, \tag{6.2a}$$

$$[A, BC]_- = [A, B]_- C + B[A, C]_-, \tag{6.2b}$$

$$\big[A, [B, C]_-\big]_- + \big[C, [A, B]_-\big]_- + \big[B, [C, A]_-\big]_- = 0. \tag{6.2c}$$

Insofar as we treat integer spin systems, the correspondence (6.1) may be safely applied, and with some luck we may end up with a consistent quantum theory. However, when we deal with half-integer spin we are bound to find serious difficulty in applying (6.1). The difficulty stems from the fact that half-integer spin systems obey Fermi statistics and are quantized with anticommutation rules such as

$$\big[\psi_a(\mathbf{x}), \psi^\dagger_\beta(\mathbf{x}')\big]_+ = \hbar\, \delta_{\alpha\beta}\, \delta(\mathbf{x}, \mathbf{x}') \tag{6.3}$$

for the electron field. Thus the Poisson bracket is turned into an anticommutator in the quantum theory.

Now, the association of an ordinary Poisson bracket with an anticommutator is not one that can be uniformly applied to all the dynamical variables in the theory because the ordinary Poisson bracket does not have the same algebraic properties as the anticommutator. This is a serious obstacle.

One might argue at this point that there is a good reason for the difficulty. Fermions disappear in the classical limit, and hence we are not supposed to have a classical formalism to describe them. However, if this argument were taken seriously, it would mean that we have to be so clever as to guess the complete quantum theory for a Fermi system at the very start. This is the way it should be, of course, but regretfully it is not the way it is.

In fact, it is most valuable to have the possibility of attacking the problem of constructing the quantum theory of a Fermi system by steps, without having to solve all problems at once. This is what is achieved by describing Fermions classically by means of anticommuting c-numbers. Pragmatically speaking, what we construct is a formalism that enables one to learn from a Fermi quantum system as much as an ordinary classical theory permits one to learn from a Bose quantum system. Thus we can, for example, write down Hamiltonians and study invariance properties without having to worry about operator ordering ambiguities or infinities.

However, in the Fermi case, the classical theory itself does not describe directly the physical world in an approximative sense in the way in which classical mechanics describes macroscopic nature. This is due to the fact that the classical limit is obtained by letting $\hbar$ go to zero and at the same time allowing the quantum numbers to become very large. In this limit, classical mechanics may be applied directly to the physical world, and its variables acquire measurable meaning. But, for Fermi systems the quantum numbers typically have an upper bound on account of the exclusion principle, and hence it is not possible to let them grow indefinitely. On the other hand, the limit $\hbar \rightarrow 0$, with the quantum numbers held fixed, is not empty and provides a good starting point to build a quantum theory on the basis of "anticommuting c-numbers."

The anticommutativity of our new c-numbers follows from letting $\hbar \rightarrow 0$ in equations such as (6.3). This does not mean that the anticommutator structure is lost in that limit, since it appears in the next order in $\hbar$. In fact, we will see below that there exists a nontrivial generalization of the ordinary Poisson bracket, which is the classical analog of an anticommutator. The generalized Poisson bracket will permit the canonical quantization prescription to be applied uniformly to all the dynamical variables (up to factor ordering problems, of course). This property is important in supersymmetric theories, where Bosons and Fermions are mixed by transformations that leave the theory invariant.

6.2. FORMAL PROPERTIES OF ANTICOMMUTING c-NUMBERS

6.2.1. Grassmann Algebra

A general Grassmann algebra with n generators $\mathcal{G}_n$ is defined as follows: (*i*) $\mathcal{G}_n$ is a vector space over the complex numbers; (*ii*) a product is defined over $\mathcal{G}_n$, which is associative and bilinear with respect to addition and multiplication by scalars; (*iii*) $\mathcal{G}_n$ contains the unit element for this product; (*iv*) $\mathcal{G}_n$ is generated by n elements ξ^A, $A = 1, \ldots, n$, which obey the relation,

$$\xi^A \xi^B + \xi^B \xi^A = 0.$$

Furthermore, there is no other independent relations among the generators.

It follows from the anticommutation of the ξ's that $\mathcal{G}_n$ is 2^n-dimensional as a vector space. A basis of $\mathcal{G}_n$ is given by the monomials $1, \xi^A, \xi^A \xi^B$ $(A < B), \ldots, \xi^1 \xi^2 \ldots \xi^n$. A general element g of $\mathcal{G}_n$ reads

$$g = g_0 + g_A \xi^A + g_{AB} \xi^A \xi^B + \cdots + g_{A_1 \ldots A_n} \xi^{A_1} \ldots \xi^{A_n},$$

where the coefficients $g_{AB}, \ldots, g_{A_1 \ldots A_n}$ can be assumed to be completely antisymmetric. The coefficient g_0 is the component of g along unity.

It is easily seen that (*i*) two Grassmann algebras with same number n of generators are isomorphic and (*ii*) one can give a matrix representation of the generators ξ^A, as 2^n-by-2^n matrices (see Exercise 6.18). In practice, however, the matrix representation of the generators is not useful and is even cumbersome. The reader should thus get familiar with the "abstract" generators ξ^A and their defining properties and should become able to manipulate them without thinking in terms of a matrix representation.

Infinite-dimensional Grassmann algebras are defined along similar lines. They are infinite-dimensional vector spaces with an infinite number of anticommuting generators ξ^A.

We will distinguish in what follows between even dynamical variables ("commuting c-numbers") and odd dynamical variables ("anticommuting c-numbers"). The following properties are dictated by the idea that these variables are the $\hbar \to 0$ limit of Bosonic and Fermionic operators, respectively:

$$\theta^\alpha \theta^\beta + \theta^\beta \theta^\alpha = 0, \qquad \text{for } \theta^\alpha, \theta^\beta \text{ odd,} \tag{6.4a}$$

$$\theta^\alpha q^i - q^i \theta^\alpha = 0, \qquad \text{for } \theta^\alpha \text{ odd, } q^i \text{ even,} \tag{6.4b}$$

$$q^i q^j - q^j q^i = 0, \qquad \text{for } q^i, q^j \text{ even.} \tag{6.4c}$$

We shall realize the dynamical variables as even (q^i) or odd (θ^α) elements of a Grassmann algebra with time-independent generators ξ^A,

$$q^i(t) = q^i_0(t) + q^i_{AB}(t) \xi^B \xi^A + \cdots, \tag{6.5a}$$

$$\theta^\alpha(t) = \theta^\alpha_A(t) \xi^A + \theta^\alpha_{ABC}(t) \xi^C \xi^B \xi^A + \cdots. \tag{6.5b}$$

The coefficients $q^i_0, q^i_{AB}, \theta^\alpha_A, \ldots$ are complex numbers, and it is easy to check that (6.4) follows from (6.5). The series (6.5a) and (6.5b) terminate in the case of a finite dimensional Grassmann algebra because any product with a repeated generator ξ^A vanishes. Otherwise, one gets infinite formal series. The ξ-independent component q^i_0 of q^i is called its component along unity (or "body").

As will be explained below (§6.5.3), the ξ^A are just an auxiliary device and play no role either classically or quantum-mechanically. They are introduced simply to make the presentation similar to familiar presentations of ordinary vector analysis.

6.2.2. Superfunctions

A Grassman-valued function of the dynamical variables (q^i, θ^α) is an element of the Grassmann algebra, to which q^i and θ^α belong, which depends on q^i and θ^α. In terms of components, a function $f(q^i, \theta^\alpha)$ is equivalent to a set of functions $f_0, f_A, f_{AB}, \ldots$ of the components $q^i_0, q^i_{AB}, \ldots, \theta^\alpha_A, \ldots$ of q^i and θ^α, such that

$$f = f_0(q^i_0, q^i_{AB}, \ldots, \theta^\alpha_A, \ldots) + f_A(q^i_0, q^i_{AB}, \ldots, \theta^\alpha_A, \ldots)\,\xi^A + \cdots. \tag{6.6}$$

Of particular importance are the so-called "superfunctions." These depend on the individual components $q^i_0, q^i_{AB}, \ldots, \theta^\alpha_A, \ldots$ only through the entire series (6.5a)–(6.5b), *i.e.*, only through q^i and θ^α, and have no explicit dependence on ξ^A. A general superfunction can be expanded in powers of the odd variables θ^α as

$$f(q, \theta) = f_0(q) + f_\alpha(q)\theta^\alpha + f_{\alpha\beta}(q)\,\theta^\beta\theta^\alpha + \cdots, \tag{6.7}$$

where the functions $f_{\alpha_1 \ldots \alpha_k}(q)$ of the commuting q's are fully antisymmetric in $\alpha_1, \ldots, \alpha_k$.

It will be seen that the Poisson bracket is defined only for the q^i and the θ^α as a whole and not for their individual components (Sec. 6.5 below). For this reason, these individual components are of no interest classically or quantum-mechanically. Only superfunctions are relevant.

Consequently, one should only focus at the classical level on those "generic" properties that hold solely as consequences of the basic relation (6.4) and do not depend on the expansion of the basic variables in terms of a basis. In particular, one should not demand the algebra to have a particular number of generators ξ^A, which is the same as saying that its dimension should be taken as infinite. Because only superfunctions will be dealt with in the sequel, we shall usually call them more simply "functions."

Left derivatives are defined by

$$\delta f = \delta\theta^\alpha \frac{\partial^L f}{\partial\theta^\alpha} \tag{6.8}$$

with $\delta\theta^\alpha$ on the left. If we were dealing with a finite-dimensional Grassmann algebra, this relationship would not completely define $\partial^L f/\partial\theta^\alpha$, since $\partial^L f/\partial\theta^\alpha$ could have an arbitrary component along the product $\Pi_A \xi^A$ of all the ξ's without changing (6.8). So, the component of $\partial^L f/\partial\theta^\alpha$ along $\Pi_A\xi^A$ would really be undefined by (6.8). This awkward feature does not appear in the case of an infinite-dimensional Grassmann algebra, thus providing a confirmation of the need for working in a Grassmann algebra with an unspecified number of generators. In that case, the left derivative of f with respect to θ^α is simply obtained by pulling θ^α to the left in the expansion (6.7) and dropping it.

By using the chain rule, it is easy to check that the extremum equations $\partial f/\partial\theta^\alpha = 0$, $\partial f/\partial q^i = 0$, in which one first differentiates with respect to θ^α and q^i and then expresses θ^α and q^i in terms of their components, are equivalent to the equations $\partial f/\partial$ (components of θ) $= 0$, $\partial f/\partial$ (components of q) $= 0$, in which one performs the two operations in reversed order.

Some care is needed to keep track of minus signs when applying familiar differentiation rules. For example,

$$\frac{\partial^L(OA)}{\partial\theta^\alpha} = -O\,\frac{\partial^L A}{\partial\theta^\alpha} + \frac{\partial^L O}{\partial\theta^\alpha}\,A \tag{6.9}$$

if O is odd and A is arbitrary. Right derivatives are defined similarly by

$$\delta f = \frac{\partial^R f}{\partial\theta^\alpha}\,\delta\theta^\alpha. \tag{6.10}$$

6.2.3. Grassmann parity

Any function (6.7) can be decomposed into "even" and "odd" components,

$$f = f_{\mathrm{E}} + f_{\mathrm{O}}. \tag{6.11}$$

The even (odd) component contains only the even (odd) powers of θ^α in the expansion of f.

The Grassmann parity ε_f of a function f of definite parity is defined to be equal to 0 (mod 2) if f is even and to 1 (mod 2) if f is odd. One has

$$fg = (-)^{\varepsilon_g\varepsilon_f}\,gf \tag{6.12}$$

for any pair of functions f and g of respective definite parity ε_f and ε_g.

6.2.4. Complex Conjugation

Another useful concept is the one of complex conjugation (involution), which will be the classical analog of Hermitian conjugation for operators. It has the properties

$$(AB)^* = B^* A^*, \tag{6.13a}$$

$$(A^*)^* = A, \tag{6.13b}$$

$$(\alpha A)^* = \alpha^* A^*, \tag{6.13c}$$

where α is a complex number.

A variable is called real if $A^* = A$ and imaginary if $A^* = -A$. Note that for odd variables, $\theta_\alpha \theta_\beta$ is imaginary if θ_α and θ_β are real.

The basic dynamical variables q^i and θ^α (as well as the Grassmann generators ξ^A) usually will be taken to be real.

6.3. CHANGES OF VARIABLES

The variables (q^i, θ^α) will be regarded in the sequel as coordinates of a "supermanifold." It will be often necessary to perform changes of variables $(q^i, \theta^\alpha) \to (q'^i, \theta'^\alpha)$ to new coordinates. We examine here the conditions under which these changes of variables can be inverted.

6.3.1. Invertible Matrices

Lemma 6.1. *A matrix $C(q^i, \theta^\alpha)$ that is a function of the variables q^i, θ^α,*

$$C(q^i, \theta^\alpha) = C_0(q^i) + C_\alpha(q^i)\theta^\alpha + C_{\alpha\beta}(q^i)\theta^\beta\theta^\alpha + \dots \tag{6.14a}$$

(where C_0, C_α, and $C_{\alpha\beta}\dots$ are matrices) has an inverse if and only if its component $C_0(q^i)$ has an inverse.

Proof. Let B be the searched-for inverse matrix to C,

$$B(q^i, \theta^\alpha) = B_0(q^i) + B_\alpha(q^i)\theta^\alpha + B_{\alpha\beta}(q^i)\theta^\beta\theta^\alpha + \cdots . \tag{6.14b}$$

The condition

$$CB = I \tag{6.14c}$$

implies

$$C_0 B_0 = I, \tag{6.14d}$$

$$C_0 B_\alpha + C_\alpha B_0 = 0, \tag{6.14e}$$

$$C_0 B_{\alpha\beta} - \tfrac{1}{2}(C_\alpha B_\beta - C_\beta B_\alpha) + C_{\alpha\beta} B_0 = 0, \tag{6.14f}$$

$$\vdots$$

This system of equations can be solved if and only if C_0 has an inverse. In that case, the solution is unique and has the form

$$B_0 = C_0^{-1}, \tag{6.14g}$$

$$B_\alpha = -C_0^{-1} C_\alpha C_0^{-1}, \tag{6.14h}$$

$$B_{\alpha\beta} = [\tfrac{1}{2}\, C_0^{-1}(C_\beta C_0^{-1} C_\alpha C_0^{-1} - C_\alpha C_0^{-1} C_\beta C_0^{-1}) - C_0^{-1} C_{\alpha\beta} C_0^{-1}], \tag{6.14i}$$

$$\vdots$$

The left inverse is also a right one, *i.e.*, we have,

$$CB = BC = I,$$

anticommuting c-numbers notwithstanding. This completes the proof.

Remark. The matrix elements of $C_0(q^i)$ belong to the Grassmann algebra and can, in turn, be expanded in powers of the generators ξ^A,

$$C_0(q^i) = \overset{(0)}{C}_0 + \overset{(1)}{C}_{0A}\, \xi^A + \cdots . \tag{6.15}$$

By the same argument as the one used above, one shows that C_0 can be inverted if and only if $\overset{(0)}{C}_0$—which is now an ordinary complex matrix depending on q_0^i—is invertible, *i.e.*, if and only if $\det \overset{(0)}{C}_0 \neq 0$.

6.3.2. Invertible Changes of Variables

Let $q'^i(q^i, \theta^\alpha)$ and $\theta'^\alpha(q^i, \theta^\alpha)$ be functions of q^i and θ^α with the same respective Grassmann parity $[\varepsilon(q'^i) = \varepsilon(q^i) = 0\,,\ \ \varepsilon(\theta'^\alpha) = \varepsilon(\theta^\alpha) = 1]$.

Just as in standard analysis, one defines the super-Jacobian matrix by

$$J = \begin{pmatrix} A & B \\ C & D \end{pmatrix}, \tag{6.16a}$$

$$A = \frac{\partial q'^i}{\partial q^j}, \qquad B = \frac{\partial^R q'^i}{\partial \theta^\beta},$$
$$C = \frac{\partial \theta'^\alpha}{\partial q^j}, \qquad D = \frac{\partial \theta'^\alpha}{\partial \theta^\beta}. \tag{6.16b}$$

The even-even part A and the odd-odd part D are square matrices, which are both even. The even-odd part B and the odd-even part C are both odd. We are taking a right derivative for $\partial^R q'^i/\partial\theta^\beta$ $(= -\partial^L q'^i/\partial\theta^\beta)$ in (6.16b) because this choice is more convenient when the index i is regarded as a row index and the index β as a column index, as is conventional. Indeed, the chain rule for right derivatives takes then the form of the standard multiplication of the super-Jacobian matrices. Had we taken left derivatives in the definition of the super-Jacobian matrix, it would then have been necessary to use transpose-matrix notations.

Because the matrices B and C are odd, they possess no component along unity. Hence, the super-Jacobian matrix J is invertible if and only if both A and D are invertible. In that case, one can write the unique inverse matrix as

$$J^{-1} = \begin{pmatrix} (A - BD^{-1}C)^{-1} & 0 \\ -D^{-1}C(A - BD^{-1}C)^{-1} & I \end{pmatrix} \begin{pmatrix} I & -BD^{-1} \\ 0 & D^{-1} \end{pmatrix}, \tag{6.16c}$$

and also as

$$J^{-1} = \begin{pmatrix} I & -A^{-1}B(D - CA^{-1}B)^{-1} \\ 0 & (D - CA^{-1}B)^{-1} \end{pmatrix} \begin{pmatrix} A^{-1} & 0 \\ -CA^{-1} & I \end{pmatrix}, \tag{6.16d}$$

and one easily checks that $J^{-1}J = I$. [Note that the components along unity of $A - BD^{-1}C$ and $D - CA^{-1}B$, respectively, are equal to the components along unity of A and D, since B and C are odd].

Theorem 6.1. *The change of variables* $q'^i = q'^i(q^i, \theta^\alpha)$, $\theta'^\alpha = \theta'^\alpha(q^i, \theta^\alpha)$ *can locally be inverted if and only if the super-Jacobian matrix J can be inverted.*

Proof. (i) if $q'^i(q^i, \theta^\alpha)$, $\theta'^\alpha(q^i, \theta^\alpha)$ can be inverted, then the super-Jacobian matrix J possesses an inverse by the chain rule,

$$J^{-1} = \frac{\partial^R(q^i, \theta^\alpha)}{\partial(q'^j, \theta'^\beta)}.$$

(ii) The proof of the converse statement is straightforward but tedious. It goes as follows. First, using power expansions in θ^α and the fact that A is invertible, one shows that q^i can be expressed as a function of q'^i and θ^α, *i.e.*, that one can invert the transformation in the Bose sector (with θ^α viewed as parameters). Hence, one can express q^i and θ'^α as functions of q'^i and θ^α. Next, one observes that $\theta'^\alpha = \theta'^\alpha(q'^j, \theta^\beta)$ can also be inverted to yield θ^α as a function of q'^j, θ'^α (use power expansions in θ^α again, and the fact that D is invertible). The inversion of the transformation in the Fermi sector, $\theta^\alpha = \theta^\alpha(q'^j, \theta'^\alpha)$, completes the inversion of the full transformation, since one then gets q^i as a function of q'^j and θ'^β by mere substitution of $\theta^\alpha(q'^i, \theta'^\alpha)$ in the already obtained $q^i(q'^j, \theta^\alpha)$.

Remark. As it never was necessary to introduce the ξ's explicitly in the calculation, the inverse functions $q^i(q'^i, \theta'^\alpha)$ and $\theta^\alpha(q'^i, \theta'^\alpha)$ are superfunctions. Accordingly, the change of variables associates with any superfunction $F(q^i, \theta^\alpha)$ of the original variables, a superfunction $F\big(q^i(q', \theta'), \theta^\alpha(q', \theta')\big)$ of the new ones and vice-versa.

6.4. CANONICAL FORMALISM IN THE PRESENCE OF ODD VARIABLES

We will assume that the classical equations of motion for our system follow from extremizing an action functional of the form

$$S = \int_{t_1}^{t_2} L(q^i, \dot{q}^i, \theta^\alpha, \dot{\theta}^\alpha)\, dt, \tag{6.17}$$

with q^i and θ^α given at the endpoints. The Lagrangian L—and hence the action—will be taken to be even and real, as this is the case encountered in all known physical applications. We will also assume in this section that the Lagrangian is not degenerate. The degenerate case will be treated at length later on.

The canonical momenta are defined as usual by

$$p_i = \frac{\partial L}{\partial \dot{q}^i}, \tag{6.18a}$$

$$\pi_\alpha = \frac{\partial^L L}{\partial \dot{\theta}^\alpha}. \tag{6.18b}$$

The function p_i is even and real, whereas π_α is odd and imaginary. The derivatives of the Lagrangian with respect to the velocities are assumed in this chapter to be invertible functions of the velocities, and we conventionally take left derivatives in the definition of the conjugate momenta (6.18b).

The canonical Hamiltonian is

$$H = \dot{q}^i p_i + \dot{\theta}^\alpha \pi_\alpha - L \tag{6.19}$$

and is real and even. Using the inverse transformation to (6.18), it can be expressed as a function of the coordinates and the momenta. The $\dot{\theta}^\alpha$ have been put on the left on the π_α in (6.19) because of our convention of using left derivatives in (6.18b). With this definition of H, one finds from (6.19) and the chain rule that the inverse transformation to (6.18) reads

$$\dot{q}^i = \frac{\partial H}{\partial p_i}, \tag{6.20a}$$

$$\dot{\theta}^\alpha = \frac{\partial^R H}{\partial \pi_\alpha} = -\frac{\partial^L H}{\partial \pi_\alpha}. \tag{6.20b}$$

From this point on, most of the usual canonical analysis goes through if we are careful with the ordering of the anticommuting factors. For example, the equations of motion in Hamiltonian form follow by formally extremizing the action

$$S = \int (\dot{q}^i p_i + \dot{\theta}^\alpha \pi_\alpha - H)\, dt \tag{6.21}$$

as a functional of the coordinates and the momenta, keeping the coordinates q^i and θ^α fixed at the endpoints.

There is, however, one important new feature that appears when odd variables are introduced, namely, the generalization of the Poisson bracket. This problem is examined in the next section.

6.5. GENERALIZED POISSON BRACKET

6.5.1. Definition

If we carry out the extremization of (6.21), we find the following equations of motion:

$$q^i = \frac{\partial H}{\partial p_i}, \qquad \dot{p}_i = -\frac{\partial H}{\partial q^i}, \tag{6.22a}$$

$$\dot{\theta}^\alpha = -\frac{\partial^L H}{\partial \pi^\alpha}, \qquad \dot{\pi}_\alpha = -\frac{\partial^L H}{\partial \theta^\alpha}. \tag{6.22b}$$

Note that the sign in the $\dot{\theta}^\alpha$ equation differs from the corresponding one for the even case. We can now use (6.22) to determine the change in time of a general function F of the canonical variables

$$\frac{d}{dt}F = \left[\frac{\partial H}{\partial p_i}\frac{\partial F}{\partial q^i} - \frac{\partial H}{\partial q^i}\frac{\partial F}{\partial p_i}\right] - \left[\frac{\partial^L H}{\partial \pi_\alpha}\frac{\partial^L F}{\partial \theta^\alpha} + \frac{\partial^L H}{\partial \theta^\alpha}\frac{\partial^L F}{\partial \pi_\alpha}\right]. \tag{6.23}$$

Demanding

$$\dot{F} = [F, H], \tag{6.24}$$

we can infer from (6.23) a definition for the Poisson bracket of any function F with an even function E

$$[F, E] = \left[\frac{\partial E}{\partial p_i}\frac{\partial F}{\partial q^i} - \frac{\partial E}{\partial q^i}\frac{\partial F}{\partial p_i}\right] - \left[\frac{\partial^L E}{\partial \pi_\alpha}\frac{\partial^L F}{\partial \theta^\alpha} + \frac{\partial^L E}{\partial \theta^\alpha}\frac{\partial^L F}{\partial \pi_\alpha}\right]. \tag{6.25}$$

Now, we want our Poisson brackets to have the same algebraic properties as their quantum counterparts. Therefore, we must require that

$$[F, E] = -[E, F] \tag{6.26}$$

because the right structure to use in the quantum-mechanical case when at least one of the variables is even is a commutator, which is antisymmetric. If F is even, (6.26) is automatically satisfied by (6.25). If F is odd, we obtain from (6.26) a definition for the odd-even case.

It remains to define the bracket $[0_1, 0_2]$ of two odd functions. To this end we return once more to our basic requirement and demand that that bracket should have the same properties as an anticommutator $[0_1, 0_2]_+$. In particular, we must have

$$[0_1, 0_2] = [0_2, 0_1] \tag{6.27a}$$

and

$$[0_1 0_2, 0_3] = 0_1[0_2, 0_3] - [0_1, 0_3]0_2. \tag{6.27b}$$

From (6.27a) and the general structure of the bracket (6.25) we make the ansatz

$$[0_1, 0_2] = \alpha\left[\frac{\partial 0_1}{\partial q^i}\frac{\partial 0_2}{\partial p_i} + \frac{\partial 0_2}{\partial q^i}\frac{\partial 0_1}{\partial p_i}\right] + \beta\left[\frac{\partial 0_1}{\partial \theta_\alpha}\frac{\partial 0_2}{\partial \pi_\alpha} + \frac{\partial 0_2}{\partial \theta^\alpha}\frac{\partial 0_1}{\partial \pi_\alpha}\right],$$

which satisfies (6.27b) if and only if $\alpha = 1 = -\beta$. So, we finally have

$$[0_1, 0_2] = \left[\frac{\partial 0_1}{\partial q^i}\frac{\partial 0_2}{\partial p_i} + \frac{\partial 0_2}{\partial q^i}\frac{\partial 0_1}{\partial p_i}\right] - \left[\frac{\partial 0_1}{\partial \theta^\alpha}\frac{\partial 0_2}{\partial \pi_\alpha} + \frac{\partial 0_2}{\partial \theta^\alpha}\frac{\partial 0_1}{\partial \pi_\alpha}\right]. \tag{6.28}$$

Equations (6.25) and (6.28) can be summarized as

$$[F,G] = \left[\frac{\partial F}{\partial q^i}\frac{\partial G}{\partial p_i} - \frac{\partial F}{\partial p_i}\frac{\partial G}{\partial q^i}\right] + (-)^{\varepsilon_F}\left[\frac{\partial^L F}{\partial \theta^\alpha}\frac{\partial^L G}{\partial \pi_\alpha} + \frac{\partial^L F}{\partial \pi_\alpha}\frac{\partial^L G}{\partial \theta^\alpha}\right], \tag{6.29}$$

where ε_F is the Grassmann parity of F.

Formula (6.29) can be succinctly rewritten provided one admits both left and right derivatives. It reads

$$[F,G] = \frac{\partial^R F}{\partial z^A} C^{AB} \frac{\partial^L G}{\partial z^B}, \tag{6.30}$$

where z^A stands for all the canonical variables q^i, p_i, θ^α, and π_α, and where C^{AB} is defined through

$$[z^A, z^B] = C^{AB}. \tag{6.31a}$$

The form (6.30) may be taken as the more fundmental one, since there is no longer any mention of F or G in C^{AB}—unlike (6.29). The basic non-vanishing Poisson brackets between the coordinates and the momenta are explicitly given by

$$[q^i, p_j] = -[p_j, q^i] = \delta^i_j, \tag{6.31b}$$

$$[\theta^\alpha, \pi_\beta] = [\pi_\beta, \theta^\alpha] = -\delta^\alpha_\beta. \tag{6.31c}$$

Note that the bracket $[F,G]$ is defined only for superfunctions F, G, for which the derivatives $\partial^R F/\partial z^A$ and $\partial^L G/\partial z^B$ are defined.

6.5.2. Properties of the Generalized Poisson Brackets

The algebraic properties which we used above to arrive at the Poisson bracket (6.30) can be written as

$$[F,G] = -(-)^{\varepsilon_F \varepsilon_G}[G,F], \tag{6.32}$$

$$[F, G_1 G_2] = [F, G_1]G_2 + (-)^{\varepsilon_F \varepsilon_{G_1}} G_1 [F, G_2]. \tag{6.33}$$

One also easily finds from (6.29) that

$$[F,G]^* = -[G^*, F^*] \tag{6.34}$$

and

$$\varepsilon([F,G]) = \varepsilon_F + \varepsilon_G. \tag{6.35}$$

When the brackets are multiplied by i, the relation (6.34) yields the appropriate hermitian properties of the commutators and anticommutators. It expresses the antihermiticity of the Poisson bracket. This property appears more transparently if one uses a real variable $\pi'_\alpha = i\pi_\alpha$ instead of π_α. In that case one finds then $(C^{AB})^* = -C^{BA}$.

Besides (6.32)–(6.35), the bracket (6.30) obeys a generalized version of the Jacobi identity, which reads

$$\begin{aligned}\big[[F_1, F_2], F_3\big] + (-)^{\varepsilon_{F_1}(\varepsilon_{F_2}+\varepsilon_{F_3})}\big[[F_2, F_3], F_1\big] \\ + (-)^{\varepsilon_{F_3}(\varepsilon_{F_1}+\varepsilon_{F_2})}\big[[F_3, F_1], F_2\big] = 0.\end{aligned} \tag{6.36}$$

This identity takes exactly the same form as in the pure bosonic case, except when two variables are odd, F_1 and F_2, say, and the other is even, in which case the phase of the middle term of (6.36) is equal to minus one.

The identity (6.36) and other properties of the generalized Poisson bracket are most easily obtained with the help of a formal device, the so-called odd constants. An odd constant is an odd number that has zero brackets with all the canonical variables.

An odd dynamical variable becomes an even one upon multiplication by an odd constant. It is then a helpful technique in calculations to accompany odd functions with a constant odd factor. In this way one needs to use only the properties of brackets of even functions, with which one is more familiar. At the end of the calculation the odd constants may be dispensed with and the answer stated in terms of the original odd functions.

6.5.3. Algebra of Superfunctions over Phase Space as the Central Object in Grassmann Mechanics

The introduction of the generators ξ^A was motivated by the desire to view the dynamical evolution as defining a trajectory in a definite space. The coordinates of that space are the components of the z's $[z \equiv (q, p, \theta, \pi)]$ in the Grassmann algebra generated by the ξ's. This description, however, is extremely uneconomical because nowhere in the analysis do the ξ's explicitly enter. So, the ξ's are really unnecessary.

For this reason, it appears preferable to shift the emphasis from the points to the functions, *i.e.*, to view the dynamical evolution as defining a map from the algebra of phase space superfunctions onto itself. This algebra is generated by just q, p, θ, and π, which are finite in number when there is a finite number of degrees of freedom. The equations of motion (6.22) are then regarded as mapping the functions q, p, θ, and π on new functions, rather than defining a point transformation. In concrete terms, the motion permits the expression of $z(t)$ in terms of $z(t_0)$. One may then take the $z(t_0)$ for any t_0 as generating the algebra. What appears naturally in mechanics is the expansion in terms of the $z(t_0)$ and not the expansion in terms of the ξ's.

6.6. PHYSICAL FERMIONS NEED FIRST-ORDER EQUATIONS

6.6.1. A Simple Model System

We now study a simple model system that contains important features found in gauge-fixed systems of actual physical interest.

The model contains just two real odd variables, C and $\bar{C}$, together with an even variable A. (To have a nondegenerate Lagrangian with Fermi degrees of freedom, one must bring in at least two odd variables, since it is impossible to form a nonlinear kinetic expression with a single odd velocity. The commuting variable A is brought in to make possible the discussion in §6.6.3 below. In this section, which focuses on C and $\bar{C}$, A is just carried along.)

The action is

$$S[A, C, \bar{C}] = \int dt \, (\tfrac{1}{2} \dot{A}^2 + i\dot{\bar{C}}\dot{C}). \tag{6.37}$$

The conjugate momenta are

$$\pi = \frac{\partial L}{\partial \dot{A}} = \dot{A}, \tag{6.38a}$$

$$\bar{\mathcal{P}} = \frac{\partial^L L}{\partial \dot{C}} = -i\dot{\bar{C}}, \tag{6.38b}$$

$$\mathcal{P} = \frac{\partial^L L}{\partial \dot{\bar{C}}} = i\dot{C}, \tag{6.38c}$$

so that the canonical Hamiltonian is given by

$$H = \tfrac{1}{2} \pi^2 + i\bar{\mathcal{P}}\mathcal{P}. \tag{6.39}$$

It is easy to check that the equations of motion generated by (6.39) are equivalent to the variational equations of (6.37).

Upon quantization, A, π, C, $\bar{C}$, $\bar{\mathcal{P}}$, and $\mathcal{P}$ become operators obeying ($\hbar = 1$)

$$A\pi - \pi A = \quad i, \tag{6.40a}$$

$$C\bar{\mathcal{P}} + \bar{\mathcal{P}}C = -i, \tag{6.40b}$$

$$\bar{C}\mathcal{P} + \mathcal{P}\bar{C} = -i, \tag{6.40c}$$

as it follows from the classical Poisson brackets. The other commutators/anticommutators vanish. The operators $A, \pi, C, \bar{C}$ should be hermitian, since their classical counterparts are real. The momenta $\bar{\mathcal{P}}, \mathcal{P}$ should be antihermitian.

An elementary consequence of the basic anticommutators is that the space of quantum states must include negative norm states if it is to yield a representation of (6.40). This may be seen as follows. Define new hermitian operators

$$\gamma_0 = C - i\bar{\mathcal{P}}, \tag{6.41a}$$

$$\gamma_1 = C + i\bar{\mathcal{P}}. \tag{6.41b}$$

They obey

$$\gamma_0^2 = -1, \tag{6.42a}$$

$$\gamma_1^2 = +1, \tag{6.42b}$$

$$\gamma_1\gamma_2 + \gamma_2\gamma_1 = 0. \tag{6.42c}$$

The operator γ_0 has a square equal to -1. Therefore, the Hilbert space must contain nonvanishing states with negative norms in the inner product that makes γ_0 hermitian.

6.6.2. Negative Norm States Generic for Nondegenerate Fermionic Lagrangians

The presence of negative norm states is not peculiar to the particular example considered here but is rather a generic feature characteristic of any nondegenerate Lagrangian containing odd variables.

Although leading to negative norm states, the Lagrangian (6.37) is physically relevant to the quantization of gauge models. The odd variables C and $\bar{C}$ correspond then to the Faddeev-Popov ghosts. The negative norm states are eliminated in that case by means of a subsidiary condition, the "BRST condition."

When one deals with physical fermions, rather than ghosts, negative norms are avoided in a different way, namely, by using a degenerate Lagrangian. Hence, constrained systems are unavoidable in the description of physical fermions. It turns out that the simplest system then has only one anticommuting *phase space* variable, which leads to a "half-degree of freedom." (See §7.1.1.) This feature is the basic building block of more complicated models.

6.6.3. Supersymmetry

The Lagrangian (6.37) is invariant under the following transformation

$$\delta A = \dot{C}\varepsilon, \tag{6.43a}$$

$$\delta C = 0, \tag{6.43b}$$

$$\delta \bar{C} = i\dot{A}\varepsilon, \tag{6.43c}$$

where ε is an odd imaginary parameter, independent of time.

The transformation (6.43) possesses one remarkable feature: it mixes variables quantized according to Bose statistics (bosons) with variables quantized according to Fermi statistics (fermions). Such a transformation is called a "supersymmetry transformation." Just as for usual symmetries, one can distinguish between two kinds of supersymmetries: global ones ("rigid supersymmetry"), like (6.43), in which the infinitesimal parameter ε is a constant, and local ones ("gauge supersymmetry"), in which the infinitesimal parameter ε is an arbitrary function of time.

6.7. GEOMETRY OF PHASE SPACE IN THE ANTICOMMUTING CASE

6.7.1. Phase Space

The space of the q's, the p's, the θ's, and the π's is called phase space, just as in the pure bosonic case. It is denoted by P. Thus, a point in phase space is a set of given q^i, p_i, θ^α, and π_α, which are, respectively, even and odd elements of the underlying Grassmann algebra generated by the ξ's. The dimensions of phase space is said to be equal to $2(N+M)$ if N is the number of q's and M is the number of θ's. (One can define the concept of supermanifold without introducing the ξ's, but this will not be done here.)

The real manifold $\tilde{P}$ obtained by identifying the points whose coordinates have the same components along unity is called the "underlying manifold" or "body" of P. It is the bosonic phase space with coordinates q, p viewed as ordinary real variables. If $\tilde{O}$ is an open set of $\tilde{P}$, the superdomain O over $\tilde{O}$ is the set of all point that "project" on a point of $\tilde{O}$. To an open covering of $\tilde{P}$ corresponds a covering of P by superdomains. A property is said to hold locally on P if it holds on a superdomain.

6.7.2. Supersymplectic Geometry

Because the phase space has a nondegenerate Poisson bracket obeying the Jacobi identity, one can define, just as in the bosonic case, an associated invertible two-form that is closed. This two-form is named the symplectic two-form. Its new feature is that it is now symmetric in the fermionic indices, since the corresponding bracket is also symmetric. The study of the geometry of phase space is called "supersymplectic geometry," or, more simply, just "symplectic geometry."

The exterior calculus of p-forms on a supermanifold can be developed along the same lines as for ordinary real manifolds. One important difference is that one can construct forms of arbitrary high degree, since the exterior forms now possess a symmetric part. Also, it has not been possible so far to develop the integration theory of p-forms on p-manifolds in a satisfactory manner because a p-form does not transform as a density, *i.e.*, it does not get divided by the super-Jacobian. Rather, the integration theory must be developed in terms of superdensities. These points are discussed in Chapter 20 ("Complementary Material"), which contains the main formulas of the exterior calculus, as well as a brief survey of the integration theory on a supermanifold.

An important theorem of supersymplectic geometry is Darboux theorem. It states, as in the bosonic case, that there is locally only one symplectic structure. More precisely, given an arbitrary invertible two-form that is closed, one can find coordinates that "diagonalize" it, *i.e.*, which have brackets equal to 0, ± 1, or $\pm i$. These canonical coordinates are defined up to a canonical transformation. The proof of Darboux theorem in the pure fermionic case turns out to be algebraic and is sketched in Exercise 6.13. More information on noncanonical coordinates may be found in Exercise 6.17.

EXERCISES. CHAPTER SIX

6.1. Show that the right and left derivatives obey the following commutation relations: $(\partial^L/\partial\theta^\alpha)(\partial^L/\partial\theta^\beta) = -(\partial^L/\partial\theta^\beta)(\partial^L/\partial\theta^\alpha)$ and that $(\partial^L/\partial\theta^\alpha)(\partial^R/\partial\theta^\beta) = (\partial^R/\partial\theta^\beta)(\partial^L/\partial\theta^\alpha)$.

6.2. Show that any variable A can be separated into real and imaginary parts, just as with ordinary complex conjugation.

6.3. Let F be an even, real function of θ^α. Show that $(\partial F/\partial\theta^\alpha)$ is imaginary and odd. What happens if F is odd and real?

6.4. Show that

$$\frac{\partial^R F}{\partial\theta^\alpha} = -(-)^{\varepsilon_F}\,\frac{\partial^L F}{\partial\theta^\alpha}$$

and

$$\frac{\partial^R F}{\partial q^i} = \frac{\partial^L F}{\partial q^i},$$

so that one has

$$\frac{\partial^R F}{\partial y^A} = (-)^{\varepsilon_F \varepsilon_A + \varepsilon_A} \frac{\partial^L F}{\partial y^A}$$

with $y^A \equiv (q^i, \theta^\alpha)$, $\varepsilon_A \equiv \varepsilon(y^A)$.

6.5. "Cauchy–Riemann" conditions. Let f be a superfunction of q^i and θ^α, with components f_0, f_A, f_{AB}, etc.,... . Show that the derivatives of f with respect to the components of q^i and θ^α are related as follows,

$$\frac{\partial f}{\partial q^i_{AB}} = \xi^B \xi^A \frac{\partial f}{\partial q^i_0},$$

$$\frac{\partial f}{\partial q^i{}_{ABCD}} = \xi^D \xi^C \xi^B \xi^A \frac{\partial f}{\partial q^i_0},$$

$$\vdots$$

$$\frac{\partial f}{\partial \theta^\alpha_{ABC}} = \xi^{[C} \xi^B \frac{\partial f}{\partial \theta^\alpha_{A]}},$$

$$\vdots$$

etc.

By expanding f in terms of its Grassmann components, infer the analog of the "Cauchy–Riemann" conditions for the derivatives with respect to the components of q^i and θ^α, *e.g.*,

$$\frac{\partial f_0}{\partial q^i_{AB}} = 0, \qquad \frac{\partial f_{AB}}{\partial q^i{}_{CD}} = \frac{1}{2}(\delta^C_A \delta^D_B - \delta^D_A \delta^C_B)\frac{\partial f_0}{\partial q^i_0} \quad \text{etc.}, \ldots .$$

6.6. The transformation $q'^i = q'^i(q^i, \theta^\alpha)$, $\theta'^\alpha = \theta'^\alpha(q^i, \theta^\alpha)$ can be reexpressed as an ordinary change of variables among the Grassmann components of q'^i, θ'^α and q^i, θ^α, *i.e.*,

$$q'^i_0 = q'^i_0(q^j_0)$$

$$q'^i_{AB} = q'^i_{AB}(q^j_0, q^j{}_{AB}, \theta^\alpha_A)$$

$$\vdots$$

Relate the Jacobian of this ordinary change of variables to the super-Jacobian (6.16a), and check explicitly that the Jacobian is invertible if and only if the super-Jacobian is.

6.7. Show that (6.20) would not hold if one had put the $\dot\theta^\alpha$ to the right of π_α in (6.19), with π_α still defined with left derivatives as in (6.18b).

6.8. Derive the Jacobi identity (6.36) (i) directly from the expression (6.30); and (ii) using the odd constant device.

6.9. Let y^i be functions of the canonical variables and F be a function of y^i. Establish the "derivation property" $[F,G] = (\partial^R F/\partial y^i)\,[y^i,G]$ and $[G,F] = [G,y^i]\,(\partial^L F)(\partial y^i)$.

6.10. Let F be an odd function obeying $[F,F]=0$.
(a) Observe that $[F,F]=0$ is a nontrivial statement; exhibit an odd function G such that $[G,G]\neq 0$.
(b) Show that $[[K,F],F]$ is equal to zero for any K.

6.11. Let F be a function of definite (even or odd) parity. Prove that $[[F,F],F]=0$ (identically).

6.12. (a) Show that the transformation (6.43) is nilpotent ($\delta_\eta\delta_\varepsilon = 0$) when one uses the equations of motion.
(b) Compute the canonical generator Ω of (6.43). (Note that the Lagrangian is only invariant modulo a total time derivative.) Show that $[\Omega,\Omega]=0$.
The transformation (6.43) is very similar to the BRST symmetry of electromagnetism.

6.13. Darboux theorem in the fermionic case.
(a) Let y^A be real odd variables with nondegenerate Poisson brackets $[y^A,y^B]=\sigma^{AB}(y)$, $\det \overset{(0)}{\sigma}{}^{AB}\neq 0$, where $\overset{(0)}{\sigma}{}^{AB}$ is the component of σ^{AB} along the identity ($A=1,\ldots,N$). From (6.34) and (6.32), $(\sigma^{AB})^* = -\sigma^{AB} = -\sigma^{BA}$, and σ^{AB} is even. Let σ_{AB} be the inverse matrix to $\sigma^{AB}(y)$,

$$\sigma_{AB}\sigma^{BC} = \delta^C_A.$$

Show that the Jacobi identity for $\sigma^{AB}(y)$ is equivalent to the statement that $\sigma_{AB}(y)$ is closed,

$$\frac{\partial\sigma_{AB}}{\partial y^C}+\frac{\partial\sigma_{BC}}{\partial y^A}+\frac{\partial\sigma_{CA}}{\partial y^B}=0.$$

(b) Infer by using power expansions in y^C that σ_{AB} is exact,

$$\sigma_{AB}=\frac{\partial a_B}{\partial y^A}+\frac{\partial a_A}{\partial y_B}.$$

Note that this is a purely algebraic statement. The potential a_B is imaginary and odd.
(c) Show that one can find new real variables $y'^A = y'^A(y)$ such that the components a'^A of the one-form a_A in the new coordinate system read

$$a'_A(y')=\frac{\partial y^C}{\partial y'^A}\,a_C(y)=\pm\frac{i}{2}\,y'^A.$$

(Use power expansions again.) The new variables y'^A possess canonical brackets,

$$[y'^A,y'^B]=iv_A\delta^{AB}\,,\quad v_A=\pm 1.$$

(d) If the y^C are even in number (which is not implied by $\det\sigma^{AB}\neq 0$ and does not necessarily hold), show the existence of new variables that form canonically conjugate pairs. If there are as many v_A equal to $+1$ as there are v_A equal to -1, the new coordinates can be chosen real and the new momenta pure imaginary.

6.14. Same notations as for Exercise 6.13. Let F be an odd function such that $[F, F]$ has a nonzero component along unity. Show that one can find an even function g and odd variables $\bar{y}^a$, $(a = 2, \ldots, N)$ such that the functions $gF, \bar{y}^a$ define a canonical coordinate system,

$$[gF, gF] = \pm i, \qquad [gF, \bar{y}^a] = 0, \qquad [\bar{y}^a, \bar{y}^b] = \pm i\delta^{ab}.$$

(Hint: Use power expansions.)

6.15. Let $*$ be a product obeying a commutation relation

$$a * b = (-)^{\sum_i \varepsilon_a^i \varepsilon_b^i}\, b * a$$

involving many gradings ε_a^i $(i = 1, \ldots, M)$. Show that one can redefine the product by including an appropriate phase

$$ab = (-)^{f(\varepsilon_a^i, \varepsilon_b^i)}\, a * b$$

in such a way that only the total grading $\varepsilon = \sum_i \varepsilon^i$ occurs in the new commutation relations,

$$ab = (-)^{\varepsilon_a \varepsilon_b}\, ba.$$

6.16. Let G be an even, "infinitesimal" function. Consider the infinitesimal transformation, $A' = A + [A, G]$. Verify $[z'^A, z'^B] = [z^A, z^B]$ for the canonical variables, *i.e.*, that the transformation is canonical.

6.17. Consider the first-order action principle

$$S[(z)^A(t)] = \int dt(\dot{z}^A a_A(z) - H). \tag{i}$$

Show that the Euler-Lagrange equations of motion are equivalent to the Hamiltonian equations

$$\dot{z}^A = [z^B, H], \tag{ii}$$

where the bracket is defined by

$$[z^A, z^B] = \sigma^{AB}, \tag{iii}$$

$$\sigma^{AB}\sigma_{BC} = \delta^A_C, \tag{iv}$$

$$\sigma_{AB} = -\frac{\partial^R a_A}{\partial z^B} - (-)^{(\varepsilon_A+1)(\varepsilon_B+1)}\frac{\partial^R a_B}{\partial z^A}. \tag{v}$$

The symplectic 2-form

$$\sigma = \tfrac{1}{2}\sigma_{AB}\, dz^B\, dz^A \tag{vi}$$

(with $\sigma_{AB} = (-)^{(\varepsilon_A+1)(\varepsilon_B+1)}\sigma_{BA}$) is closed,

$$d\sigma = 0. \tag{vii}$$

[We have assumed σ_{AB} invertible throughout.]

6.18. (a) Show that any Grassmann algebra (more generally, any associative algebra) possesses a matrix representation. [Hint: The generators ξ^A define a linear operator in $\mathcal{G}_n$, the operator of multiplying by ξ^A. The searched-for matrices can be taken to be the matrices of the linear operators $\xi^1, \xi^2, \ldots, \xi^n$ in the basis $1, \xi^A, \xi^A \xi^B$ $(A < B)$, etc. ...]

(b) Construct explicitly the matrix representation for a Grassmann algebra with two generators.

CHAPTER SEVEN

CONSTRAINED SYSTEMS WITH FERMI VARIABLES

The extension of the formalism of Chapter 1 to the case with Fermi degrees of freedom is straightforward. When the Legendre transformation from the velocities to the momenta cannot be inverted, one finds again constraints in the Hamiltonian formalism. The constraints can either be associated with a gauge symmetry and then be first class or they can be second-class, in which case they lead to a modification of the Poisson bracket. The new bracket—still named "Dirac bracket"—is defined as before and enables one to set consistently to zero the second-class constraints.

The geometric distinction between first-class and second-class constraints, based on the rank of the two-form induced on the constraint surface, also remains valid. All the considerations of Chapter 2 and Chapter 3 still apply (see Exercises 7.1–7.4).

As in the bosonic case, the formalism can be satisfactorily developed only when the regularity conditions are fulfilled and when the two-form induced on the constraint surface can be brought to the form

$$\begin{pmatrix} 0 & 0 \\ 0 & M \end{pmatrix}$$

where M is weakly invertible. Examples that violate the conditions are given in Exercises 7.5–7.7. These examples are constructed by using the property that there exist nonzero dynamical variables whose product is nevertheless zero. They describe models that are not equivalent to an unconstrained Hamiltonian reduced system.

The following differences with respect to the pure bosonic case are, however, to be noted:

(*i*) The phase space obtained by elimination of the second-class constraints may be odd dimensional.
(*ii*) Appropriate sign factors need to be included in some of the previous formulas to account for the anticommutativity of the Fermi degrees of freedom.

These two points are examined by Sec. 7.1 and Sec. 7.2, respectively.

7.1. ODD-DIMENSIONAL PHASE SPACE

7.1.1. Example

An interesting feature of constrained Hamiltonian systems with Fermi degrees of freedom is that the number of second-class constraints need not be even. This is because the matrix of the bracket of the constraints $[\chi_\alpha, \chi_\beta]$ is symmetric in the indices associated with the fermionic constraints. As a result, the dimension of the second-class constraint surface may itself be odd.

To illustrate this fact, we will consider an example in which one has three second-class constraints. This example is interesting in that it also indicates how physical fermions avoid the negative norm states problem encountered in §6.6.1.

Consider a system with three real anticommuting variables θ^i ($i = 1, 2, 3$) described by the first-order Lagrangian

$$L = -\frac{i}{2}\dot{\theta}^i \theta^j \delta_{ij}. \tag{7.1}$$

The primary constraints are

$$p_i + \frac{i}{2}\delta_{ij}\theta^j = 0, \tag{7.2a}$$

and the Hamiltonian vanishes, since L is homogeneous of first degree in the velocities,

$$H = 0. \tag{7.2b}$$

There is no further constraint.

The constraints (7.2a) are all second class,

$$\left[p_i + \frac{i}{2}\,\delta_{ij}\,\theta^j\,,\ p_k + \frac{i}{2}\,\delta_{k\ell}\,\theta^\ell\right] = -i\delta_{ik}, \tag{7.3a}$$

and can be used to eliminate the momenta in terms of the θ's. The basic Dirac brackets, from which all others can be inferred upon use of (7.2a), read

$$[\theta_i\,,\theta_j]^* = -i\delta_{ij}. \tag{7.3b}$$

The phase space of the independent variables θ_i is clearly odd dimensional. Because the θ's are odd in number, one cannot split them into canonically conjugate pairs of "coordinates" and "momenta." If one inserts (7.2a) into the standard canonical action $\int(\dot{\theta}p - H)\,dt$, one recovers the original action (7.1), which was, in that sense, already in canonical form.

In the quantum theory, the θ's become operators $\hat{\theta}^i$ obeying anticommutation relations that follow from the basic Dirac brackets (7.3b) according to the rule "anticommutator $= i$ times Dirac bracket,"

$$\hat{\theta}^i\hat{\theta}^j + \hat{\theta}^j\hat{\theta}^i = \delta^{ij}. \tag{7.4a}$$

As is well known, these anticommutation relations can be realized in a positive definite two-dimensional Hilbert space, provided one takes

$$\hat{\theta}^i = \frac{1}{\sqrt{2}}\,\sigma^i, \tag{7.4b}$$

where σ^i are the standard hermitian Pauli matrices.

Accordingly, we see that by allowing first-order Lagrangians with appropriate kinetic terms in the fermionic sector, one avoids the negative norm problem, since the resulting Dirac brackets admit realizations in a positive definite Hilbert space.

Also to be emphasized again is the fact that the infinite number of Grassmann components of θ^i are *not* realized as operators in (7.4) and that the Grassmann generators ξ^A have thus no significance in the quantum theory (the quantum states belong to an ordinary complex vector space).

7.1.2. Boundary Term in Action Principle

Finally, one additional word concerning the classical variational principle $\delta S = 0$ from which the equations of motion of the example

$$\dot{\theta}^i = 0 \tag{7.5}$$

derive. These equations of motion should be obtained by extremizing the action under small deformations of the history $\theta^i(t)$ of the system. The allowed deformations must obey boundary conditions, the number of which is equal to the number of integration constants in the general solution of the equations of motion. For Bose variables that obey second-order equations, one usually demands that the coordinates be fixed at initial and final times. It is not possible, however, to impose a similar requirement on the θ's, as this would imply two boundary conditions for a first-order differential equation. Another way to say the same thing is that the θ's are self-conjugate, so that giving the θ's amounts to giving simultaneously the coordinates and the momenta. As a consequence, the action for our system cannot be taken to be just the integral of the Lagrangian (7.1) but must be supplemented by a boundary term.

Instead of following an inductive procedure, we will simply exhibit the correct action principle and show that it satisfies all the necessary requirements.

We write the action as

$$S = \int_{t_1}^{t_2} L\, dt - \frac{i}{2}\, \theta^i(t_1)\, \theta^j(t_2)\, \delta_{ij} \tag{7.6}$$

and state that the solutions of the equations of motion are those histories that yield no variation of S under the conditions

$$\delta[\theta(t_1) + \theta(t_2)] = 0. \tag{7.7}$$

To verify that this action principle is a suitable one, we must verify two properties:

(*i*) Extremization of S under condition (7.7) should yield just the equations of motion without additional restrictions.

(*ii*) Those equations of motions should have a unique solution consistent with arbitrary given values of $\theta(t_1) + \theta(t_2)$ (with the possible exception of some "unfortunate" choices of the boundary conditions).

If we vary $\theta(t)$, we find

$$\begin{aligned} \delta S &= \int_{t_2}^{t_1} dt\Big(- \frac{i}{2}\, \dot\theta^i \delta\theta_i - \frac{i}{2}\, \delta\dot\theta^i \theta_i\Big) - \frac{i}{2}\, \delta\theta^i(t_1)\, \theta_i(t_2) - \frac{i}{2}\, \theta^i(t_1)\, \delta\theta_i(t_2) \\ &= \int_{t_2}^{t_1} dt(-i\dot\theta^i \delta\theta_i) + \frac{i}{2}\big[\delta\theta(t_1) + \delta\theta(t_2)\big]\big[\theta(t_1) - \theta(t_2)\big]. \end{aligned}$$

The boundary term vanishes because of condition (7.7), and extremization of S yields just $\dot\theta^i = 0$, as needed.

Let us suppose that $\theta(t_1) + \theta(t_2)$ is given as 2ξ, say. In that case, there is a unique solution to equation (7.5) with that boundary condition, namely, $\theta(t) = \xi$ for all t so that requirement (*ii*) is also fulfilled and the action (7.6) meets all necessary conditions.

It should be mentioned that had we attempted to fix $\theta(t_1) - \theta(t_2)$, instead of $\theta(t_1) + \theta(t_2)$, we could have also satisfied requirement (*i*) by properly adjoining the boundary term to the action, but we would not have been able to satisfy (*ii*), since, in that case, the boundary condition would not have fixed the solution of the equation of motion.

7.1.3. Alternative Boundary Conditions in the Hamiltonian Variational Principle for Bosonic Variables

The variational principle (7.6)–(7.7) clearly treats the coordinates and the momenta on the same footing, since the θ's are self-conjugate. It suggests considering an alternative variational principle even when the phase space coordinates can be split into coordinates (q) and momenta (p). This new variational principle treats the coordinates and the momenta more symmetrically and is appropriate to both the fermionic and bosonic cases. It corresponds to boundary data of the form

$$q^i(t_1) + q^i(t_2) = 2Q^i, \tag{7.8a}$$

$$p_i(t_1) + p_i(t_2) = 2P_i. \tag{7.8b}$$

The variational principle suitable to (7.8) is given by

$$\bar{S} = \int (\tfrac{1}{2}\,\dot{q}^i p_i - \tfrac{1}{2}\,q^i \dot{p}_i - H)\,dt + \tfrac{1}{2}[q^i(t_1)\,p_i\,(t_2) - q^i(t_2)\,p_i\,(t_1)]. \tag{7.8c}$$

It differs from the standard variational principle in which the coordinates are fixed at both endpoints by a surface term,

$$\bar{S} = S - \tfrac{1}{2}[q^i(t_2) - q^i(t_1)]\,[p_i(t_2) + p_i(t_1)], \tag{7.9a}$$

$$S = \int (\dot{q}^i p_i - H)\,dt. \tag{7.9b}$$

It is the analog of the variational principle (7.6) and (7.7) for self-conjugate fermions.

If one denotes (q^i, p_i) by z^A and defines

$$z_1 \circ z_2 = \sigma_{AB}\, z_1^A z_2^B = q_1^i\, p_{2i} - q_2^i\, p_{1i} = -z_2 \circ z_1, \tag{7.10a}$$

one can rewrite (7.8c) as

$$\bar{S} = \int (\tfrac{1}{2}\dot{z} \circ z - H)\,dt + \tfrac{1}{2}z(t_1) \circ z(t_2). \tag{7.10b}$$

7.2. INCORPORATION OF APPROPRIATE SIGN FACTORS

We now turn to the discussion of how the formulas of the bosonic case get modified by phase factors in the presence of Grassmann-odd variables. We will only reproduce here the final formulas, as the calculations themselves are direct and not very illuminating.

7.2.1. Gauge Transformations

The first-class constraints are taken to be real and of definite Grassmann parity ε_a,

$$\varepsilon(G_a) = \varepsilon_a, \qquad G_a^* = G_a. \tag{7.11}$$

Their "algebra" is still written as

$$[G_a, G_b] = C_{ab}{}^c \, G_c, \tag{7.12a}$$

$$[H_0, G_a] = V_a{}^b \, G_b, \tag{7.12b}$$

withthe G's to the right of the structure functions in the right-hand side of (7.12). The structure functions obey

$$C_{bc}{}^a = -(-)^{\varepsilon_b \varepsilon_c} \, C_{cb}{}^a, \qquad \varepsilon(C_{bc}{}^a) = \varepsilon_a + \varepsilon_b + \varepsilon_c, \tag{7.13a}$$

$$(C_{bc}{}^a)^* = C_{bc}{}^a \, (-)^{\varepsilon_a + \varepsilon_a(\varepsilon_b + \varepsilon_c) + \varepsilon_b \varepsilon_c}, \tag{7.13b}$$

$$(V_b{}^a)^* = V_b{}^a \, (-)^{\varepsilon_a + \varepsilon_a \varepsilon_b}, \qquad \varepsilon(V_b{}^a) = \varepsilon_a + \varepsilon_b. \tag{7.13c}$$

The gauge transformations are given by

$$\delta F = [F, \varepsilon^a G_a] \tag{7.14a}$$

$$= (-)^{\varepsilon_F \varepsilon_a} \, \varepsilon^a \, [F, G_a], \qquad \varepsilon^a = \varepsilon^a(t). \tag{7.14b}$$

The extended action

$$S_E = \int [\dot{z}^A \, a_A(z) - H_0 - \lambda^a \, G_a] \, dt \tag{7.15a}$$

is invariant under (7.14) if one transforms the multipliers as

$$\delta_\varepsilon \, \lambda^a = \dot{\varepsilon}^a + \lambda^c \varepsilon^b C_{bc}{}^a - \varepsilon^b \, V_b{}^a, \tag{7.15b}$$

as in (5.5b).

The Lagrange multiplier λ^a and the gauge parameter ε^a possess the same parity as the constraints G_a and are real or imaginary according to whether G_a is even or odd,

$$\varepsilon(\lambda^a) = \varepsilon(\varepsilon^a) = \varepsilon_a, \tag{7.16a}$$

$$(\lambda^a)^* = (-)^{\varepsilon_a} \, \lambda^a, \qquad (\varepsilon^a)^* = (-)^{\varepsilon_a} \, \varepsilon^a. \tag{7.16b}$$

The reality and parity conditions on λ^a and ε^a are dictated by the fact that $\lambda^a G_a$ and $\varepsilon^a G_a$ should both be real and even.

The reducibility conditions (if any) will still be taken to be of the form

$$Z^a_\alpha G_a = 0 \tag{7.17a}$$

(without sign factors). The action is then also invariant under

$$\delta\lambda^a = \mu^\alpha Z^a_\alpha. \tag{7.17b}$$

Each term in the left-hand side (7.17a) can be assumed to be of definite Grassmann parity ε_α, and so

$$\varepsilon(Z^a_\alpha) = \varepsilon_\alpha + \varepsilon_a. \tag{7.17c}$$

Similarly, one may assume that the function $Z^a_\alpha G_a \equiv Z_\alpha$ is such that $Z^*_\alpha = -(-)^{\varepsilon_\alpha} Z_\alpha$, *i.e.*,

$$(Z^a_\alpha)^* = -(-)^{\varepsilon_\alpha+\varepsilon_a+\varepsilon_\alpha\varepsilon_a} Z^a_\alpha. \tag{7.17d}$$

This somewhat unnatural choice will actually turn out to be convenient in subsequent formulas.

Finally, we note that the on-shell trivial gauge transformations (3.7) become

$$\delta_\mu y^i = \mu^{ij} \frac{\delta^R S}{\delta y^j}, \tag{7.18a}$$

where μ^{ij} is now such that

$$\mu^{ij} = (-)^{(\varepsilon_i+1)(\varepsilon_j+1)} \mu^{ji}, \qquad \varepsilon(\mu^{ij}) = \varepsilon_i + \varepsilon_j. \tag{7.18b}$$

7.2.2. Gauge Orbits—Exterior Derivative along the Gauge Orbits

The constraint functions G_a define (right) vector fields X_a according to

$$f\overleftarrow{\partial}_{X_a} \equiv f\overleftarrow{X}_a = [f, G_a]. \tag{7.19a}$$

One has, with the conventions explained in Sec. 20.1,

$$\varepsilon(X_a) = \varepsilon_a + 1, \tag{7.19b}$$

as $\varepsilon(f\overleftarrow{X}_a) = \varepsilon_f + \varepsilon(X_a) + 1$ and $\varepsilon\big([f, G_a]\big) = \varepsilon_f + \varepsilon_a$. In the sequel, we will often write $X_a f$ instead of $f\overleftarrow{X}_a$. It should be kept in mind, however, that X_a has a right action.

If the constraints are reducible, the vectors X_a are not linearly independent on $G_a = 0$. Equation (7.17a) on the constraint functions leads to

$$(X_a) Z^a_\alpha (-)^{\varepsilon_a(\varepsilon_\alpha+1)} \approx 0 \tag{7.19c}$$

for the corresponding vector fields.

One easily finds

$$[X_a, X_b] \approx X_c \, C_{ab}{}^c \, (-)^{\varepsilon_c+\varepsilon_c(\varepsilon_a+\varepsilon_b)}, \tag{7.20}$$

and the integral submanifolds tangent to X_a on $G_a = 0$ are again the gauge orbits.

The forms ω^a dual to X_a $[\omega^a(X_b) = \delta^a{}_b]$ fulfill (see Sec. 20.1 for sign and symmetry conventions)

$$d\omega^a = \tfrac{1}{2}(-)^{\varepsilon_a+\varepsilon_b} \, \omega^b \, \omega^c \, C_{cb}{}^a, \tag{7.21a}$$

$$\varepsilon(\omega^a) = \varepsilon_a + 1, \qquad \omega^a \omega^b = (-)^{(\varepsilon_a+1)(\varepsilon_b+1)} \, \omega^b \, \omega^a, \tag{7.21b}$$

and one has

$$dF = [F, G_a] \, \omega^a \tag{7.21c}$$

for the longitudinal derivative of a function F.

If one expands an arbitrary longitudinal p-form α as

$$\alpha = \frac{1}{p!} \omega^a \omega^b \ldots \omega^f \, \alpha_{f \ldots ba}, \tag{7.22a}$$

one gets

$$\alpha_{ab \ldots c} = (-)^{(\varepsilon_a+1)(\varepsilon_a+\cdots+\varepsilon_c+p)} \, \alpha(X_a, \ldots, X_c) \tag{7.22b}$$

and

$$\alpha_{ab \ldots cd \ldots f} = (-)^{(\varepsilon_c+1)(\varepsilon_d+1)} \alpha_{ab \ldots dc \ldots f}. \tag{7.22c}$$

Furthermore, for reducible constraints, the components of α in the overcomplete basis $\{X_a\}$ are subject to

$$Z^a_\alpha \, \omega^b \ldots \omega^c \, \alpha_{c \ldots ba} \approx 0. \tag{7.23}$$

EXERCISES. CHAPTER SEVEN

7.1. Check that the proof of the abelianization theorem for first-class constraint goes through, even when fermionic constraints are included.

7.2. More generally, show that any set of constraints, fermionic and bosonic, obeying the regularity assumptions discussed in the text can be replaced by equivalent ones (a) obeying also the regularity assumptions and (b) having Poisson brackets among each other equal to 0, ± 1, or $\pm i$. (Hint: See Exercises 1.22, 2.10, and 6.14.)

7.3. Let ϕ_α be fermionic constraints obeying the regularity conditions.
(a) Show that $F \approx 0$ imply $F = \lambda^\alpha \phi_\alpha$, just as in the bosonic case.
(b) Show that if the constraints are irreducible, the equations $F^\alpha \phi_\alpha = 0$ imply $F^\alpha = F^{\alpha\beta}\phi_\beta$ with $F^{\alpha\beta} = F^{\beta\alpha}$.
(c) Extend the analysis to mixed constraints. How important are the regularity conditions?

7.4. Consider the second-class constraint $\chi \equiv \pi + \frac{i}{2}\theta = 0$, with θ and π anticommuting. Show that by adding a single, self-conjugate fermionic variable ψ,

$$[\psi, \psi] = -i,$$

one can replace $\chi = 0$ by an equivalent constraint that is first class.

7.5. Consider the Lagrangian $L = \theta^1\theta^2\dot{\theta}^1\dot{\theta}^2$, with θ^1 and θ^2 anticommuting. Show that the constraints between the θ's and the π's do not define a submanifold in the sense of Sec. 20.1 and that the Hamiltonian cannot be a superfunction of the θ's and the π's. (Observe that any superfunction of θ and π is linear in the velocities $\dot{\theta}$ when the momenta are expressed in terms of $\dot{\theta}$.)

7.6. Consider the bosonic constraint $\phi \equiv p_1 p_2 = 0$ where p_1 and p_2 are fermionic momenta. Verify that the constraint is first class but that there is no reduced phase space. (There is no way to fix the gauge by canonical methods because there is no function χ such that $[\phi, \chi]$ can be inverted.)

7.7. Consider the fermionic constraint $\phi \equiv p_1 + \theta_1\theta_2\theta_3$. This constraint is not first class. Show that one cannot, nevertheless, define a Dirac bracket that would enable one to set ϕ strongly equal to zero. (This model is excluded by the regularity condition on the two-form induced on the constraint surface.)

7.8. Check that the transformations (7.18) leave the action $S[y^i]$ invariant.

7.9. Check that the coefficient of $\omega^b\omega^c$ in (7.21a) acquires the sign factor $(-)^{(\varepsilon_b+1)(\varepsilon_c+1)}$ when one permutes b and c.

CHAPTER EIGHT

GRADED DIFFERENTIAL ALGEBRAS—ALGEBRAIC STRUCTURE OF THE BRST SYMMETRY

8.1. INTRODUCTION—GHOSTS

The concept of ghost has steadily emerged as one of central importance in the study of the dynamics of physical systems endowed with a local geometric structure.

Ghosts were first encountered in quantum field theory as fields with the "wrong" relation between spin and statistics that were necessary in addition to those that appeared in the correspondent classical problem. The ghost fields enabled one to maintain a local description of quantum gauge theories in terms of elementary processes involving free propagation and local vertices. They also ensured that the theory would be unitary and independent of the gauge choice. This they did through contributions to virtual process only, thus avoiding conflict with the spin-statistics theorem.

After the advent of the Faddeev-Popov measure for summing over equivalence classes of gauge field histories, the ghosts were regarded as just an artifact leading to a useful representation of that measure.

They were restored to a somewhat more respectable role when it was realized that in more complicated theories it was necessary to include self-interactions of the ghosts in order to achieve unitarity, and this could not be fitted in the Faddeev-Popov scheme.

However, it was the discovery of the Becchi-Rouet-Stora-Tyutin (BRST) symmetry that raised the ghost to a prominent role. Since this symmetry mixes the ghosts with the other fields, it became clear that one should regard all the fields, including the ghosts, as different components of a single geometrical object. This point of view emerged as the logical development of the idea of gauge invariance.

The need for ghosts and the symmetry that reveals their profound importance were first established in quantum mechanics. It was only afterwards realized that they have a natural and necessary place within classical mechanics as well. Indeed, the BRST symmetry could have been discovered long ago within a strictly classical context by mathematicians dealing with the geometry of phase space had they only been willing to extend their analysis to Grassmann variables.

Having said this, it should be emphasized that we are not advocating that ghosts should have direct physical meaning within classical mechanics. It is nevertheless extremely useful to be able to discuss the BRST symmetry classically. One can then bring in this concept as a powerful tool in the actual construction of the quantum theory, rather than having to wait for the quantum theory to discuss it.

The central idea of the BRST theory is to substitute for the original gauge symmetry a fermionic rigid symmetry acting on an appropriately extended phase space. That symmetry—the BRST symmetry—completely captures the original gauge invariance and leads to a simpler formulation of the theory.

To develop the BRST theory in a crisp manner, it is helpful to use some concepts from homological algebra. This chapter is devoted to those concepts and to an overview of how they underlie the BRST theory.

8.2. GRADED DIFFERENTIAL ALGEBRAS

We first recall a few definitions.

8.2.1. Supercommutative Algebras

An associative algebra over the complex numbers is a set A with the following properties:

(i) A is a vector space over $\mathbb{C}$;

(ii) for every pair (x, y) of elements of A, there is a product $xy \in A$ that satisfies the bilinearity conditions and that is associative. These conditions read explicitly

$$(x_1 + x_2)y = x_1 y + x_2 y, \quad x(y_1 + y_2) = xy_1 + xy_2, \tag{8.1a}$$

$$\alpha(xy) = (\alpha x)y = x(\alpha y), \quad \alpha \in \mathbb{C}, \tag{8.1b}$$

$$(xy)z = x(yz). \tag{8.2}$$

The algebra is said to be $\mathbb{Z}_2$-graded-commutative or, in an abbreviated fashion, *supercommutative,* if A is the direct sum

$$A = A_0 \oplus A_1 \tag{8.3a}$$

of two complementary subspaces such that

$$xy = (-)^{\varepsilon_x \varepsilon_y}\, yx \tag{8.3b}$$

with $\varepsilon_u = 0$ when $u \in A_0$ and $\varepsilon_u = 1$ when $u \in A_1$, which implies

$$A_0 A_0 \subset A_0, \qquad A_0 A_1 \subset A_1, \qquad A_1 A_1 \subset A_0. \tag{8.4}$$

From now on, we will only consider "homogeneous" elements $x, y, \ldots \in A$ of definite parity.

The direct sum (8.3) defines a "Z_2-grading." The parity ε_x of a homogenenous element x is called its "Grassmann parity" in analogy with the theory of Grassmann algebras.

Unless specified otherwise, the (associative) algebras considered in this book are not only supercommutative, but they also have a unit, that is, an element 1 such that $1x = x1 = x$ for any element x of A. The complex multiples of the unit form a subalgebra of A_0, which is isomorphic to $\mathbb{C}$. A complex conjugation ("involution") obeying the properties (6.13) is also assumed to be defined. The subset of real elements contains the unit.

8.2.2. Examples

The simplest algebras are the *finitely generated supercommutative free algebras.* These algebras are polynomial algebras with a finite number of generators. The generators can be either commuting (x_i) or anticommuting (y_α)

$$x_i x_j = x_j x_i, \qquad i = 1, \ldots, m, \tag{8.5a}$$

$$x_i y_\alpha = y_\alpha x_i, \qquad \alpha = 1, \ldots, n, \tag{8.5b}$$

$$y_\alpha y_\beta = -y_\beta y_\alpha. \tag{8.5c}$$

They obey no further relation besides those implied by the associativity (they are "free"). The generators can be taken to be real.

A general element of the algebra generated by x_i and y_α is an arbitrary polynomial in x_i and y_α with complex coefficients. This algebra is denoted by $\mathbb{C}[x_1, \ldots, x_m; y_1, \ldots, y_n]$ and is the tensor product of the symmetric algebra with m generators and the antisymmetric algebra with n generators. This latter is just a Grassmann algebra and is sometimes denoted by $\Lambda[y_1, \ldots, y_m]$.

Other important examples of algebras are the algebras $C^\infty(M)$ of complex-valued smooth functions on a real manifold. The functions are smooth in the real sense, *i.e.*, both their real and imaginary parts are smooth. Similarly, the algebra of smooth functions (superfunctions) on a supermanifold M is denoted by $C^\infty(M)$.

Finally, the exterior algebra on a supermanifold M is the algebra of p-forms on M; the product is the exterior product. In the case of the longitudinal forms along the gauge orbits generated by irreducible constraints, the exterior longitudinal algebra is isomorphic to the tensor product algebra $C^\infty(\Sigma) \otimes \mathbb{C}[\omega^a]$ of polynomials in ω^a with coefficients that belong to $C^\infty(\Sigma)$ (see §5.3.3).

8.2.3. Graded Lie Algebra of Graded Derivations

Let V be a Z_2-graded vector space,

$$V = V_0 \oplus V_1. \tag{8.6}$$

The set End (V) of linear transformations $V \rightarrow V$ is an associative algebra with unit. It is not supercommutative. The grading (8.6) of V endows End (V) with a natural grading: a linear transformation M is said to have definite parity ε_M $(= 0$ or $1)$ if Mx has definite parity $\varepsilon_M + \varepsilon_x$ for any homogeneous element $x \in V$,

$$\varepsilon(Mx) = \varepsilon_M + \varepsilon_x. \tag{8.7}$$

An arbitrary element of End (V) can be decomposed into an even and an odd linear transformation,

$$\text{End}\,(V) = \text{End}\,_0(V) \oplus \text{End}\,_1(V). \tag{8.8}$$

The ordinary multiplication of linear transformations preserves the grading,

$$\varepsilon(M_1 M_2) = \varepsilon_{M_1} + \varepsilon_{M_2}. \tag{8.9}$$

Given two linear transformations M_1 and M_2, their graded commutator $[M_1, M_2]$ is defined by

$$[M_1, M_2] = M_1 M_2 - (-)^{\varepsilon_{M_1} \varepsilon_{M_2}} M_2 M_1. \tag{8.10}$$

The graded commutator is antisymmetric, except when M_1 and M_2 are both odd, in which case it is symmetric.

The graded commutator—simply referred to as the "commutator" in the sequel—obeys a graded version of the Jacobi identity,

$$\begin{aligned} \big[[M_1, M_2], M_3\big] + (-)^{\varepsilon_{M_1}(\varepsilon_{M_2}+\varepsilon_{M_3})} \big[[M_2, M_3], M_1\big] \\ + (-)^{\varepsilon_{M_3}(\varepsilon_{M_1}+\varepsilon_{M_2})} \big[[M_3, M_1], M_2\big] = 0. \end{aligned} \tag{8.11}$$

One says that the vector space End (V), with the commutator structure (8.10), is a *(graded) Lie algebra.*

Because any supercommutative algebra A is a Z_2-graded vector space, the set End (A) of linear transformations of A is a graded Lie algebra. Of particular interest are the linear transformations that obey the Leibnitz rule,

$$D(xy) = x\, Dy + (-)^{\varepsilon_D \varepsilon_y} (Dx) y. \tag{8.12}$$

These are called (graded) *derivations.* We adopt the convention of a right action for D.

The set Der (A) of all the derivations of A is a graded Lie subalgebra of End (A). This is because the graded commutator of two derivations is again a derivation (Exercise 8.3).

Examples of derivations are the vector fields on a supermanifold M, which are derivations of $C^\infty(M)$. Another example is given by the exterior derivative operator d, which is a derivation of the exterior algebra on M.

8.2.4. Gradings

The algebras occurring below possess another grading besides the Z_2-grading (8.3a). That is, these algebras split as

$$A = \bigoplus_n A_n \tag{8.13a}$$

where n is either a positive or negative integer (Z-grading) or a nonnegative integer (N-grading). Furthermore, the unit belongs to A_0 and the multiplication is such that

$$A_n A_m \subset A_{n+m}. \tag{8.13b}$$

If $x \in A_n$, one says that the degree of x is equal to n,

$$\deg x = n \Leftrightarrow x \in A_n. \tag{8.13c}$$

The grading (8.13) may have no relationship with the Z_2-grading, namely, A_n may contain both commuting and anticommuting elements. Note that the subspace A_0 is a subalgebra of A by (8.13b).

The grading of A induces again a grading in the spaces End (A) and Der (A). A linear transformation M has degree $\deg M$ iff

$$\deg(Mx) = \deg M + \deg x. \tag{8.14}$$

One easily gets

$$\deg(M_1 M_2) = \deg M_1 + \deg M_2, \tag{8.15a}$$

$$\deg[M_1, M_2] = \deg M_1 + \deg M_2. \tag{8.15b}$$

The degree $\deg M$ of a linear transformation M may be negative even if the grading of A is ≥ 0 because M may decrease the degree in (8.14).

In the case of the exterior algebra over P, the form degree defines a N-grading. Moreover, $\deg d = 1$. Given a grading, one can define a "counting operator" N by

$$Nx = (\deg x)x \tag{8.16}$$

for elements $x \in A$ of definite degree. One extends N to all of A by linearity. The counting operator is even and of degree zero,

$$\varepsilon(N) = 0, \qquad \deg(N) = 0. \tag{8.17a}$$

Because of (8.13b), it is also a derivation,

$$N(xy) = (Nx)y + x(Ny). \tag{8.17b}$$

8.2.5. Ideals

An ideal B is a vector subspace of A such that $AB \subset B$, *i.e.*, $xy \in B$ whenever y (or x) $\in B$. An ideal is, thus, also a subalgebra.

Given an ideal B, the product in A induces a well-defined product in the space A/B of equivalence classes of elements of A that differ by an element of B. This product is supercommutative and associative if the product in A is. All the algebraic properties of A pass on to A/B (see Exercise 8.5).

8.2.6. Differentials—Cohomology Algebras

A differential D is an odd derivation that is nilpotent of order two,

$$D^2 \equiv \tfrac{1}{2}[D, D] = 0, \tag{8.18a}$$

$$\varepsilon(D) = 1. \tag{8.18b}$$

From now on, the terms "nilpotent of order two" will be replaced by just the word "nilpotent" (we will not encounter higher order nilpotency). We will assume that there is a N-grading such that

$$\deg D = \pm 1. \tag{8.19}$$

If $\deg D = -1$, the differential D decreases the grading and behaves like a boundary operator. If $\deg D = +1$, the differential D increases the grading and behaves like a coboundary operator. We will also consider differentials in an algebra with a Z-grading. Then, the sign of $\deg D$ is a matter of convention and can be taken to be positive.

One has $D(A_n) \subset A_{n\pm 1}$ depending on $\deg D = \pm 1$, *i.e.*,

$$\cdots \xrightarrow{D} A_n \xrightarrow{D} A_{n+1} \xrightarrow{D} A_{n+2} \longrightarrow \cdots \quad (\deg D = 1), \tag{8.20a}$$

$$\cdots \xrightarrow{D} A_n \xrightarrow{D} A_{n-1} \xrightarrow{D} A_{n-2} \longrightarrow \cdots \quad (\deg D = -1). \tag{8.20b}$$

The graded algebra A with the differential D is called a *graded differential algebra.* A *differential complex* is by definition a N- (or Z-)graded vector space $V = \bigoplus_n V_n$ with a nilpotent linear operator of degree ± 1. Accordingly, a graded differential algebra is a differential complex in which a multiplication is defined, for which the differential D is a derivation.

Given D, one defines $\operatorname{Ker} D$ as the set of elements of A that are annihilated by D,

$$x \in \operatorname{Ker} D \Leftrightarrow Dx = 0. \tag{8.21a}$$

The subspace $\operatorname{Ker} D$ is a subalgebra of A because D is a derivation. The elements of $\operatorname{Ker} D$ are said to be *D-closed.* The subspace $\operatorname{Im} D$ is by definition the image of A under D, *i.e.*,

$$x \in \operatorname{Im} D \Leftrightarrow \; \exists\, y \in A : Dy = x. \tag{8.21b}$$

The elements of $\operatorname{Im} D$ are said to be *D-exact.*

The subspace $\operatorname{Im} D$ is an ideal of $\operatorname{Ker} D$. Therefore, one can define the quotient algebra

$$\frac{\operatorname{Ker} D}{\operatorname{Im} D}. \tag{8.22a}$$

That quotient algebra is denoted by $H^*(D)$ and called the cohomology algebra—or just cohomology—of D when $\deg D = 1$; it is denoted by

$H_*(D)$ and called the homology algebra—or just homology—of D when $\deg D = -1$. Both $H^*(D)$ and $H_*(D)$ inherit the gradings of A, *i.e.*,

$$H^*(D) = \bigoplus_n H^n(D), \tag{8.22b}$$

$$H_*(D) = \bigoplus_n H_n(D). \tag{8.22c}$$

An element of $\operatorname{Ker} D$ that is not in $\operatorname{Im} D$ is called a *nontrivial (co)cycle.* If $H_k(D)$ [or $H^k(D)$] vanishes, the differential D is said to be *acyclic in degree* k.

8.2.7. Contracting Homotopy

Let Λ be a linear operator that can be diagonalized, *i.e.*, $A = \oplus \bar{A}_\lambda$ with $\Lambda x = \lambda x$ for $x \in \bar{A}_\lambda$. Assume also that the subspace $\bar{A}_0$ associated with the eigenvalue zero is contained in the subspace A_0 of degree zero, $\bar{A}_0 \subset A_0$. A linear operator σ whose anticommutator with D yields Λ,

$$\sigma D + D\sigma = \Lambda \tag{8.23}$$

is called a *contracting homotopy for* Λ with respect to D.

The interest of contracting homotopies resides in the following theorem.

Theorem 8.1. *If there exists a contracting homotopy, the differential* D *is acyclic in degree* k $\neq 0$,

$$H^k(D) = 0, \quad k \neq 0 \qquad [\text{or } H_k(D) = 0, \quad k \neq 0]. \tag{8.24}$$

Proof. Equation (8.23) implies $[\Lambda, D] = 0$. Consequently, the components of any D-closed element along different eigenspaces of Λ must separately be D-closed. One can thus analyze the cohomology of D at a fixed eigenvalue λ of Λ. Let x be such that $\Lambda x = \lambda x,\ Dx = 0,\ \lambda \neq 0$. One finds

$$x = \frac{1}{\lambda}\Lambda x = \frac{1}{\lambda}\Big(\sigma D + D\sigma\Big)x = D\Big(\frac{1}{\lambda}\sigma x\Big),$$

which shows that x is D-exact. Hence, the cohomology of D can only lie at eigenvalue $\lambda = 0$. As the eigenspace $\bar{A}_0$ with $\lambda = 0$ is contained in A_0, this proves the theorem.

8.2.8. Cohomology for the Lie Algebra of Derivations

The differential D also defines a cohomology for the Lie algebra of the derivations of A.

A derivation Δ is *D-invariant*, or *D-closed*, if it commutes with D (in the graded sense), *i.e.*,

$$[\Delta, D] = 0 \Leftrightarrow \Delta \text{ is } D\text{-closed}. \tag{8.25a}$$

The set of D-closed derivations is a Lie subalgebra of the set Der (A) of the derivations of A. This follows from the Jacobi identity.

A derivation Δ is D-exact if it can be written as the commutator of another derivation with D,

$$\Delta = [\bar{\Delta}, D] \Leftrightarrow \Delta \text{ is } D\text{-exact}. \tag{8.25b}$$

D-exact derivations are D-closed. They actually form an ideal (for the commutator) of the Lie algebra of D-closed derivations. Hence, one can define the quotient graded Lie algebra $\mathcal{H}^*(D)$ [or $\mathcal{H}_*(D)$] of equivalence classes of D-closed derivations modulo D-exact ones.

8.2.9. Differential modulo δ

Let δ be a differential and d be an odd derivation. One says that d is a *differential modulo* δ when the following two conditions hold: (i) d commutes with δ (in the graded sense),

$$d\delta + \delta d = 0; \tag{8.26a}$$

and (ii) d^2 is δ-exact, *i.e.*,

$$d^2 = -[\delta, \Delta] \qquad (\equiv -\delta\Delta - \Delta\delta) \tag{8.26b}$$

for some derivation Δ.

The first condition implies that d induces a derivation in $H_*(\delta)$. This derivation is still denoted by d. The second condition expresses that the induced derivation is a differential, namely, $d^2 = 0$ in $H_*(\delta)$. Hence, one can consider the cohomology of d in $H_*(\delta)$, which is denoted by $H^k(d\,|H_*(\delta))$ or just $H^k(d)$ when no confusion can arise. A cohomological class of d is thus determined by a δ-closed element x that is d-closed modulo δy for some y,

$$\delta x = 0, \tag{8.27a}$$

$$dx = \delta y; \tag{8.27b}$$

and one must identify x with $x' = x + dz + \delta z'$ where $\delta z = 0$,

$$x \sim x + dz + \delta z', \qquad \delta z = 0. \tag{8.27c}$$

An interesting example of a differential modulo δ arises as follows. Let s be a differential and let us assume that there is a grading of the algebra such that the expansion of s according to that grading starts with δ. Let d be the next term in the expansion. That is,

$$s = \delta + d + \text{"higher orders,"} \tag{8.28a}$$

$$\deg(d) = 1 + \deg(\delta), \tag{8.28b}$$

$$\deg(\text{higher orders}) > \deg(d). \tag{8.28c}$$

Then, the equation $s^2 = 0$ implies not only that δ is a differential, $\delta^2 = 0$, but also that d is a differential modulo δ. Hence, differentials modulo δ automatically appear in the expansions of true differentials according to appropriate auxiliary degrees.

Conversely, one may ask the question: given a differential δ and a differential d modulo δ, can one find a differential s that combines δ with d as in (8.28)? The answer to this question is, in general, negative. It is, however, affirmative in the cases of interest to the BRST theory. Furthermore, in those cases, $H^k(s) = H^k(d\,|H_*(\delta))$.

What makes possible the merging of δ with d in the BRST context is that δ possesses then very special properties. Namely, it defines a *resolution,* a key concept to which we now turn.

8.3. RESOLUTION

8.3.1. Definition

Let A be an algebra. A *(homological) resolution* of A is a N-graded differential algebra $\bar{A}$ with differential δ of degree minus one such that

$$H_k(\delta) = 0, \qquad k \neq 0, \tag{8.29a}$$

$$H_0(\delta) = A. \tag{8.29b}$$

So, a resolution represents A as the quotient algebra $H_*(\delta) \equiv H_0(\delta)$ of the N-graded differential algebra $(\bar{A}, \delta)$. The grading of $\bar{A}$ is called the resolution degree. The elements of A are obtained by identifying elements of $\bar{A}$. When passing to $\bar{A}$, one undoes the identification, *i.e.*, one "resolves" A. A *cohomological resolution* of A is defined in a similar manner.

In practice, one often wishes the resolution to have nice additional properties. For instance, $\bar{A}$ may be desired to be a free supercommutative algebra, so that its structure is simpler than the one of A.

8.3.2. Example

An important example of resolution is provided by the following result. Let $\mathbb{C}[x_i]$ be the polynomial algebra in variables x's that can be either even or odd. Let $\mathbb{C}[x_i, z_\alpha, \mathcal{P}_\alpha]$ be the polynomial algebra in the x's, as well as in other variables z_α and $\mathcal{P}_\alpha$, whose respective parities are related as

$$\varepsilon(\mathcal{P}_\alpha) = \varepsilon(z_\alpha) + 1. \tag{8.30a}$$

Define a grading r and a derivation δ by

$$r(x_i) = 0, \qquad r(z_\alpha) = 0, \qquad r(\mathcal{P}_\alpha) = 1, \tag{8.30b}$$
$$\delta(x_i) = 0, \qquad \delta(z_\alpha) = 0, \qquad \delta(\mathcal{P}_\alpha) = z_\alpha. \tag{8.30c}$$

One extends r and δ to arbitrary polynomials in the x's, the z's, and the $\mathcal{P}$'s so that they are, respectively, even and odd derivations.

One easily checks that δ is a differential of r-degree minus one,

$$\delta^2 = 0, \tag{8.30d}$$

$$r(\delta) = -1. \tag{8.30e}$$

Theorem 8.2. *The differential complex* $\big(\mathbb{C}[x_i, z_\alpha, \mathcal{P}_\alpha], \delta\big)$ *provides a resolution of the algebra* $\mathbb{C}[x_i]$,

$$\begin{cases} H_0(\delta) = \mathbb{C}[x_i], & (8.31a) \\ H_k(\delta) = 0, \qquad k \neq 0. & (8.31b) \end{cases}$$

Furthermore, the homology of δ *in the Lie algebra of derivations is given by*

$$\begin{cases} \mathcal{H}_0(\delta) = \textit{Lie algebra of derivations of } \mathbb{C}[x_i], & (8.31c) \\ \mathcal{H}_k(\delta) = 0, \qquad k \neq 0. & (8.31d) \end{cases}$$

Proof. (i) The subalgebra of $\mathbb{C}[x_i, z_\alpha, \mathcal{P}_\alpha]$ of resolution degree zero is isomorphic to $\mathbb{C}[x_i, z_\alpha]$. It coincides with $(\operatorname{Ker}\delta)_0$ because x_i and z_α are both annihilated by δ. As the most general element of degree one is given by $m^\alpha \mathcal{P}_\alpha$ with $m^\alpha \in \mathbb{C}[x_i, z_\alpha]$, one finds that $(\operatorname{Im}\delta)_0$ is given by the polynomials of $\mathbb{C}[x_i, z_\alpha]$ of degree at least one in z_α, *i.e.*,

$$p \in (\operatorname{Im}\delta)_0 \Leftrightarrow p = m^\alpha z_\alpha, \qquad m^\alpha \in \mathbb{C}[x_i, z_\alpha].$$

The quotient algebra $(\operatorname{Ker}\delta)_0/(\operatorname{Im}\delta)_0$ is thus isomorphic to the algebra of the polynomials in the x's. This proves (8.31a).

(ii) Let us introduce a grading $\bar{n}$ that counts the number of z's and $\mathcal{P}$'s, *i.e.*,

$$\bar{n}(x_i) = 0, \quad \bar{n}(z_\alpha) = 1, \quad \bar{n}(\mathcal{P}_\alpha) = 1. \tag{8.32}$$

The corresponding counting operator $\bar{N}$ has nonnegative eigenvalues. Its eigenspace with eigenvalue zero coincides with $\mathbb{C}[x_i]$ and, thus, is contained in the subalgebra $\mathbb{C}[x_i, z_\alpha]$ of resolution degree zero.

An odd derivation σ can be defined by

$$\sigma(x_i) = 0, \quad \sigma(z_\alpha) = \mathcal{P}_\alpha, \quad \sigma(\mathcal{P}_\alpha) = 0 \tag{8.33a}$$

(and is extended to an arbitrary element of $\mathbb{C}[x_i, z_\alpha, \mathcal{P}_\alpha]$ so as to be indeed an odd derivation). One has

$$\sigma\delta + \delta\sigma = \bar{N}. \tag{8.33b}$$

Therefore, σ is a contracting homotopy for $\bar{N}$ with respect to δ. By using the Theorem 8.1, one gets (8.31b).

(*iii*) Let Δ be a δ-closed derivation. One can decompose Δ according to its components of definite $\bar{N}$-degree, which map by definition an element of degree k on an element of degree $k + \ell$,

$$\Delta = \Sigma \Delta_\ell, \quad [\bar{N}, \Delta_\ell] = \ell \Delta_\ell.$$

If Δ is δ-closed, each of its components Δ_ℓ must be separately δ-closed because $\bar{N}$ commutes with δ. Furthermore, one finds for $\ell \neq 0$,

$$\begin{aligned}\Delta_\ell &= \tfrac{1}{\ell}[\bar{N}, \Delta_\ell] \\ &= \tfrac{1}{\ell}\big[[\sigma, \delta], \Delta_\ell\big] \\ &= -\tfrac{1}{\ell}\big[[\Delta_\ell, \sigma], \delta\big],\end{aligned}$$

and so, Δ_ℓ ($\ell \neq 0$) is δ-exact in the space of derivations ($[\Delta_\ell, \sigma]$ is a derivation because the contracting homotopy σ itself is a derivation). So, only Δ_0 can be nontrivial, as stated in (8.31d).

Now, the most general derivation is determined by its action on the generators x_i, z_α, and $\mathcal{P}_\alpha$. If Δ_0 is of $\bar{N}$-degree zero, one must have

$$\begin{aligned}\Delta_0 x_i &= \mu_i, \\ \Delta_0 z_\alpha &= \nu_\alpha{}^\beta z_\beta + \rho_\alpha{}^\beta \mathcal{P}_\beta, \\ \Delta_0 \mathcal{P}_\alpha &= \lambda_\alpha{}^\beta z_\beta + \sigma_\alpha{}^\beta \mathcal{P}_\beta,\end{aligned}$$

with μ_i, $\nu_\alpha{}^\beta$, $\rho_\alpha{}^\beta$, $\lambda_\alpha{}^\beta$, and $\sigma_\alpha{}^\beta \in \mathbb{C}[x_i]$. The condition $[\Delta_0, \delta] = 0$ implies $\rho_\alpha{}^\beta = 0$ and $\sigma_\alpha{}^\beta = (-)^{\varepsilon(\Delta_0)} \nu_\alpha{}^\beta$. It then follows that when μ_i vanishes, Δ_0 is δ-exact,

$$\Delta_0 = [\kappa, \delta] \qquad (\mu_i = 0).$$

Here, κ is, for instance, defined by

$$\kappa x_i = 0, \qquad \kappa z_\alpha = \sigma_\alpha{}^\beta \mathcal{P}_\beta, \qquad \kappa \mathcal{P}_\alpha = (-)^{\varepsilon(\Delta_0)} \lambda_\alpha{}^\beta \mathcal{P}_\beta,$$

and extended to $\mathbb{C}[x_i, z_\alpha, \mathcal{P}_\beta]$ as a derivation of degree $\varepsilon(\Delta_0) + 1$. This demonstrates (8.31c) and ends the proof of the theorem.

The Theorem 8.2 gives an example of a phenomenon that will be encountered many times later on: the variables $(z_\alpha, \mathcal{P}_\alpha)$ cancel in pairs when one passes to the homology because $\mathcal{P}_\alpha$ is not δ-closed, while z_α is δ-exact and, hence, also drops out. One says that "z_α is killed by $\mathcal{P}_\alpha$."

Also to be emphasized is that the existence of a contracting homotopy that is a derivation, as here, implies the acyclicity (in degree $k \neq 0$) of δ in both the original algebra and in the algebra of the derivations. This feature will be found to hold for the resolutions that occur in BRST theory.

8.4. ELEMENTS OF HOMOLOGICAL PERTURBATION THEORY

8.4.1. Main Theorem

We now come back to the analysis of differentials modulo δ and assume throughout this section that the differential δ provides a homological resolution of $H_0(\delta)$. The central result that will be established here is that under this condition, any differential modulo δ can be combined with δ to yield a true differential s with same cohomology.

Because δ defines a homological resolution of $H_0(\delta)$, one has explicitly

$$\delta^2 = 0, \tag{8.34a}$$
$$r(\delta) = -1 \Leftrightarrow r(\delta x) = r(x) - 1, \quad \text{for } r(x) \geq 1, \tag{8.34b}$$
$$\delta x = 0, \quad \text{for } r(x) = 0, \tag{8.34c}$$
$$H_k(\delta) = 0, \quad k \neq 0 \quad (k = \text{resolution degree}). \tag{8.34d}$$

Let d be a derivation modulo δ of resolution degree zero, *i.e.*,

$$\delta d + d\delta = 0, \qquad d^2 = -[\delta, \overset{(1)}{s}], \qquad r(d) = 0, \tag{8.34e}$$

where $\overset{(1)}{s}$ is a derivation of resolution degree one. The grading associated with d is denoted by deg. We assume that it is left unchanged by δ, and we take for definiteness d to be a coboundary operator, so that one has

$$\deg(\delta) = 0, \qquad \deg(d) = 1. \tag{8.34f}$$

The total grading is defined by

$$\mathrm{gh}(x) = \deg(x) - r(x) \tag{8.35a}$$

and is called the (total) ghost number for reasons that will become clear later on. One finds from (8.34b), (8.34e), and (8.34f) that

$$\mathrm{gh}\,\delta = \mathrm{gh}\,d = 1. \tag{8.35b}$$

The "main theorem" of homological perturbation theory combines d with δ into a single differential.

Theorem 8.3. *(a) If $\mathcal{H}_k(\delta) = 0$ for all $k \neq 0$, there exists a differential s of total ghost number one that combines d with δ,*

$$s = \delta + d + \overset{(1)}{s} + \overset{(2)}{s} + \cdots, \tag{8.36a}$$

$$r(\overset{(k)}{s}) = k, \qquad \text{gh}\,(\overset{(k)}{s}) = 1, \tag{8.36b}$$

$$s^2 = 0. \tag{8.36c}$$

(b) Any differential s that combines d with δ as in (8.36a)–(8.36c) fulfills

$$H^k(s) = H^k(d), \tag{8.36d}$$

where the cohomology of d is computed in $H_0(\delta)$, i.e., is defined by the equations

$$dx = \delta y, \qquad x \sim x + dz + \delta z', \tag{8.37a}$$

with

$$r(x) = 0 = r(z), \qquad r(y) = 1 = r(z'). \tag{8.37b}$$

8.4.2. Proof of the Main Theorem: (*i*) Existence of s

Theorem 8.3 is a straightforward consequence of the Jacobi identity and of $\mathcal{H}_k(\delta) = 0$ for $k \neq 0$. It is proven "perturbatively" as follows.

The first terms δ, d, and $\overset{(1)}{s}$ in s are known and are such that the components of s^2 of resolution degrees -2, -1, and 0 vanish. It is thus only necessary to show that one can find $\overset{(k)}{s}$ with $k \geq 2$ such that the terms of order $n \geq 1$ in s^2 vanish as well. Assume thus that s has been constructed up to order $n \geq 1$,

$$s_n = \delta + d + \cdots + \overset{(n)}{s}, \tag{8.38a}$$

with

$$2s_n^2 \equiv [s_n, s_n] = \overset{(n)}{\rho} + \overset{(n+1)}{\rho} + \cdots, \tag{8.38b}$$

where $\overset{(k)}{\rho}$ is of resolution degree k. We want to show that it is possible to add to s_n a term $\overset{(n+1)}{s}$ of resolution degree $n+1$ such that $s_{n+1}^2 = (s_n + \overset{(n+1)}{s})^2$ starts at order $n+1$, i.e., that one can eliminate $\overset{(n)}{\rho}$ by a suitable choice of $\overset{(n+1)}{s}$.

A direct calculation yields

$$\big[s_n + \overset{(n+1)}{s}, s_n + \overset{(n+1)}{s}\big] = 2\big[\delta, \overset{(n+1)}{s}\big] + \overset{(n)}{\rho} + \text{“higher order.”} \tag{8.38c}$$

To remove $\overset{(n)}{\rho}$, the searched-for $\overset{(n+1)}{s}$ must be a solution of the equation

$$2\big[\delta, \overset{(n+1)}{s}\big] + \overset{(n)}{\rho} = 0. \tag{8.38d}$$

This equation possesses a solution if and only if $\big[\delta, \overset{(n)}{\rho}\big] = 0$. This condition is clearly necessary ($\delta^2 = 0$). It is also sufficient because $\mathcal{H}_n(\delta) = 0$ for $n \neq 0$.

To check that $\big[\delta, \overset{(n)}{\rho}\big]$ indeed vanishes, one observes that $\big[\delta, \overset{(n)}{\rho}\big]$ is just the component of resolution degree $n-1$ of $\big[[s_n, s_n], s_n\big]$, and so it vanishes by the Jacobi identity. This establishes the existence of s.

8.4.3. Proof of the Main Theorem: (*ii*) Evaluation of $H^k(s)$

Once the existence of s is demonstrated, it remains to show that $H^k(s) = H^k(d)$. To that end, it is convenient to expand any element of ghost number k according to the resolution degree,

$$x = \sum_{n \geq 0} \overset{(n)}{x}, \qquad r\big(\overset{(n)}{x}\big) = n, \qquad \text{gh}\, \overset{(n)}{x} = k. \tag{8.39}$$

Consider the map π that projects x on its component $\overset{(0)}{x}$ of zero resolution degree,

$$\pi(x) = \overset{(0)}{x}. \tag{8.40}$$

Because sx is given by

$$sx = d\overset{(0)}{x} + \delta\overset{(1)}{x} + \text{ higher order,} \tag{8.41}$$

one finds

$$\pi s \simeq d\pi, \tag{8.42}$$

where $\simeq$ means “equal modulo δ-exact terms.” Therefore, π induces a map of $H^k(s)$ to $H^k(d)$. One must show that this map is an isomorphism. The linearity of π and the fact that it preserves the product, $\pi(xy) = \pi(x)\pi(y)$, are obvious, so one must only prove that π is surjective and injective.

(a) π is surjective: let $\overset{(0)}{x}$ be an element of $H^k(d)$, *i.e.*, a solution of $d\overset{(0)}{x} = -\delta\overset{(1)}{x}$. The sum $\overset{(0)}{x} + \overset{(1)}{x}$ is s-closed up to terms of resolution degree one. Assume then that one has constructed

$$\overset{(n)}{y} = \overset{(0)}{x} + \cdots + \overset{(n)}{x} \tag{8.43a}$$

so that $s\overset{(n)}{y}$ starts at resolution degree $n > 0$,

$$s\overset{(n)}{y} = \overset{(n)}{t} + \overset{(n+1)}{t} + \cdots, \quad r(\overset{(k)}{t}) = k. \tag{8.43b}$$

The nilpotency of s implies that $\overset{(n)}{t}$ is δ-closed. Hence, since $H^n(\delta) = 0$, $\overset{(n)}{t}$ is also δ-exact, $\overset{(n)}{t} = -\delta\overset{(n+1)}{x}$. This implies that $\overset{(n)}{y} + \overset{(n+1)}{x}$ is s-closed up to terms of resolution degree greater than or equal to $n + 1$. Going on recursively in the same fashion, one proves the existence of x,

$$x = \overset{(0)}{x} + \sum_{n\geq 1} \overset{(n)}{x} \tag{8.43c}$$

such that $sx = 0$. By construction, $\pi(x) = \overset{(0)}{x}$. This shows that π is surjective.

(b) π is injective: let x be a s-closed element of A that is mapped on a d-exact element of $H_0(\delta)$, *i.e.*,

$$x = \sum_{n\geq 0} \overset{(n)}{x}, \qquad sx = 0, \tag{8.44a}$$

$$\overset{(0)}{x} = d\overset{(0)}{z} + \delta\overset{(1)}{z}. \tag{8.44b}$$

Set $x' = x - s(\overset{(0)}{z} + \overset{(1)}{z})$. One has $\overset{(0)}{x}{}' = 0$. Furthermore, because $sx' = sx = 0$, one finds that $\overset{(1)}{x}{}'$ obeys

$$\delta\overset{(1)}{x}{}' = 0. \tag{8.44c}$$

This implies that $\overset{(1)}{x}{}' = \delta\overset{(2)}{z}$, and so the s-closed element $x'' = x - s(\overset{(0)}{z} + \overset{(1)}{z} + \overset{(2)}{z})$, which is in the s-equivalence class of x, starts only at resolution degree two. Going on recursively along the same lines, one finally gets

$$0 = x - s\Big(\sum_{n\geq 0} \overset{(n)}{z}\Big), \tag{8.44d}$$

which shows that x is s-exact. This demonstrates that the map π is injective and ends the proof of the theorem.

8.4.4. Comments

(i) The proof of the theorem explicitly shows that the cohomological class of a s-closed element is completely determined by its component at zero resolution degree. From the point of view of the s-cohomology, only δ and d are relevant. The higher order terms $\overset{(1)}{s}, \overset{(2)}{s}, \ldots$ are important for securing the nilpotency of s but play no role at the cohomological level.

(ii) There is no *a priori* reason that the sums (8.36a), (8.43c), or (8.44d) should be finite or can be made finite by choosing approriately the available freedom at each value of the resolution degree. Accordingly, one should in principle allow for infinite formal sums $\sum_{n\geq 0} x_n$, with $r(x_n) = n$. Dealing with formal sums not involving any concept of convergence is not a problem at this stage because the inductive procedure leads to well-defined equations for x_n at each value of n. It may become a problem later when trying to interpret $\sum x_n$ as an operator in quantum mechanics. In most cases, however, the sums in question turn out to be actually finite because the resolution degree is bounded from above or for other reasons. Accordingly, we will not deepen this issue further.

(iii) For the reader familiar with algebraic topology, we note that the proof of $H^k(s) = H^k(d)$ (which is short for $H^k\big(d|H_0(\delta)\big)$ is just a rephrasing of standard spectral sequence arguments in the special case where the spectral sequence collapses after the second step.

8.5. GEOMETRIC APPLICATION: THE BRST CONSTRUCTION IN BRIEF

8.5.1. Introduction

As we indicated above, the central idea of the BRST theory is to replace the original gauge symmetry by a fermionic rigid symmetry s acting on an appropriately extended space containing new variables, the "ghosts." The key property of that symmetry—the "BRST symmetry"—is that it is a nilpotent derivation. Therefore, one can construct its cohomological groups $H^k(s)$. The original gauge invariance is recovered when one passes to the cohomology, *i.e.*, the BRST differential is constructed so as to fulfill the important requirement

$$H^0(s) = \{\text{gauge invariant functions}\}. \tag{8.45}$$

The nilpotency of s and the equality (8.45) are the fundamental equations of the BRST theory.

The fact that any gauge symmetry allows for the construction of a nilpotent derivation obeying (8.45) is already remarkable in itself. It is, however, more than just an aesthetic result for the following reasons.

(*i*) The replacement of a gauge symmetry by a rigid symmetry enables one to substitute for the original gauge invariant action one that depends on all the variables and that can be used directly in the path integral. By contrast, the original action cannot be used as such in the path integral, since the integration over the gauge degrees of freedom yields infinity (the integrand does not depend on these variables). This infinity is compensated for by the ghosts. Compensations between the ghosts and original degrees of freedom may also happen in other instances, such as in the avoidance of anomalies.

(*ii*) The use of the BRST symmetry bypasses the need to construct explicitly the reduced phase space. It allows one to work consistently with functions defined in the original phase space, thus preserving manifest locality and covariance.

The purpose of the remaining subsections is to sketch the general ideas behind the BRST symmetry without giving the explicit proofs. The details will be developed in the next two chapters.

8.5.2. Geometric Ingredients

It has been shown in Chapter 1 that the gauge invariant functions ("observables") are obtained from the phase space functions through a two-step reduction procedure:

(*i*) First, one identifies the functions that coincide on the constraint surface, *i.e.*, one replaces the algebra $C^\infty(P)$ of smooth phase space functions by the algebra $C^\infty(\Sigma)$ of smooth functions on the constraint surface.

(*ii*) Second, one imposes the gauge invariance condition on the elements of $C^\infty(\Sigma)$, *i.e.*, one requires that they be annihilated by the longitudinal exterior derivative d on Σ.

Whenever these two ingredients are present—namely, a surface Σ embedded in a manifold P and a set of vector fields tangent to Σ that close on Σ and, hence, define gauge orbits, one can construct a nilpotent derivation s, the BRST differential. This differential acts on an appropriate algebra containing $C^\infty(P)$ and is such that $H^0(s) = \{$functions of $C^\infty(\Sigma)$ constant along the gauge orbits$\}$.

8.5.3. BRST Differential

To the two steps in the construction of the functions of $C^\infty(\Sigma)$ that are constant along the gauge orbits correspond two auxiliary derivations δ, d in the construction of the BRST differential s.

(i) The first derivation δ yields a *Koszul–Tate resolution* of the algebra $C^\infty(\Sigma)$ of smooth functions on the surface Σ. It acts on polynomials in some generators $\mathcal{P}_\alpha$ to be introduced below with coefficients that are smooth functions on P, *i.e.*, on $\mathbb{C}[\mathcal{P}_\alpha] \otimes C^\infty(P)$. The differential of the Koszul–Tate complex is denoted by δ. The purpose of the generators $\mathcal{P}_\alpha$ is to kill off the functions of $C^\infty(P)$ that vanish on Σ, in much the same way as the $\mathcal{P}$'s of Theorem 8.2 kill off the polynomials that are of degree at least one in the z's. When these generators are introduced, one finds that $H_0(\delta) = C^\infty(\Sigma)$ and $H_k(\delta) = 0$ for $k \neq 0$, so that δ indeed implements the restriction to Σ through its homology.

(ii) The second differential is the longitudinal exterior derivative d, which is such that $H^0(d) = \{$functions on Σ that are constant along the gauge orbits$\}$. This differential d is defined on Σ but can be "lifted" to P (§5.3.5). The lifted derivation, still denoted by d, can furthermore be extended to the generators $\mathcal{P}_\alpha$ of the Koszul complex so as to anticommute with δ. One then finds that d^2 vanishes up to δ-exact terms, so that *d is a differential modulo δ.*

All the ingredients of homological perturbation are thus met. Theorem 8.3 guarantees accordingly the existence of a differential s,

$$s = \delta + d + \text{ "more,"} \tag{8.46}$$

whose cohomology is equal to the cohomology of the longitudinal exterior derivative d. *This differential, which combines the Koszul–Tate differential implementing the restriction to the constraint surface and the longitudinal derivative d, is the BRST differential.*

8.5.4. Canonical Action of s

There is a huge freedom in the construction of s. To begin with, there is some flexibility in the actual realizations of δ and d. In addition, even for definite δ and d, the higher order derivations $\overset{(1)}{s}, \overset{(2)}{s}, \ldots$ are determined only up to δ-exact terms.

One can partially use this freedom in such a way that the generators of the Koszul–Tate complex are in equal number as the generators of the longitudinal exterior complex. They can then be decreed to be canonically conjugate to each other, thus introducing a symplectic structure in the extended space of the original variables and of the new generators associated with δ and d. *In terms of this symplectic structure, the BRST transformation is a canonical transformation, i.e.,**

$$sx = [x, \Omega] \tag{8.47}$$

for any x in the BRST complex. Here, the bracket stands for the Poisson bracket in the extended phase space. Equation (8.47) is what makes it possible to implement the BRST symmetry in quantum mechanics.

Once it is required that s should be a canonical transformation, the BRST symmetry is unique up to a canonical change of variables, at least in the co-called "minimal sector" to be defined below. The fermionic function Ω in (8.47) is known as the BRST generator. By the Jacobi identity for the Poisson bracket, the nilpotency of s is equivalent to

$$[\Omega, \Omega] = 0, \tag{8.48}$$

which is a cornerstone of the BRST theory.

The fact that s can be chosen to have a canonical action is not guaranteed by the main theorem. One needs further results that are derived below. We will develop the BRST theory by relying on the appropriate Poisson bracket structure from the very beginning. This task will keep us busy for the next two chapters.

EXERCISES. CHAPTER EIGHT

8.1. Show that the product of x_i and y_α in (8.5) can be redefined by including a sign factor so as to fulfill $x_i * y_\alpha = -y_\alpha * x_i$ (with $x_i * y_\alpha = x_i y_\alpha (-)^{\sigma_{i\alpha}}$).

8.2. Let A and B be two supercommutative and associative algebras with a unit and a complex conjugation.

(a) Show that $(a \otimes b)(\bar{a} \otimes \bar{b}) = (a\bar{a}) \otimes (b\bar{b})(-)^{\varepsilon_{\bar{a}}\varepsilon_b}$ $(a, \bar{a} \in A,\ b, \bar{b} \in B)$ defines a supercommutative, associative product on the tensor product $A \otimes B$. The parity of $a \otimes b$ is given by $\varepsilon_a + \varepsilon_b$.

(b) Check that there is a well-defined unit and a well-defined complex conjugation. Hence, $A \otimes B$ has a natural structure of associative and supercommutative algebra with a unit and an involution.

Remark: If one identifies $a \in A$ with $a \otimes 1_B$ and $b \in B$ with $1_A \otimes b$, then the product ab is well defined and equal to $a \otimes b$. Furthermore, $ba \equiv (1_A \otimes b)(a \otimes 1_B) = (-)^{\varepsilon_a \varepsilon_b} a \otimes b$.

8.3. (a) Check explicitly that the graded commutator of two derivations is again a derivation.

(b) Prove that $D(1) = 0$ for any derivation D.

---

* An infinitesimal canonical transformation takes the form $\delta_S F = [F, S]$ with S infinitesimal and even. So for (8.47) to be a canonical transformation, one should insert an anticommuting infinitesimal parameter ε in (8.47), *i.e.*, replace $[F, \Omega]$ by $[F, \Omega\varepsilon] = [F, \Omega]\varepsilon$. For simplicity of language and notations, however, we shall refer in the sequel to (8.47) as the actual canonical transformation.

(c) Show that zD defined by$(zD)x = (Dx)z$ is a derivation of parity $\varepsilon_D + \varepsilon_z$ if D is a derivation of parity ε_D.

8.4. Show that a derivation in the free algebra $\mathbb{C}[x_1, x_2, \ldots, x_m; y_1, \ldots, y_n]$ is completely determined by its action on the generators.

8.5. Let B be an ideal of the algebra A with grading $A = \oplus A_n$.

(a) Prove that A/B is isomorphic to $\oplus(A_n/B_n)$, where $B_n = A_n \cap B$ (we assume $B = \oplus B_n$).

(b) Verify that the multiplication in A/B is supercommutative and associative if the multiplication in A is.

8.6. Show that a necessary and sufficient condition for $H^*(D) = 0$ is that $1 = Dy$ for some y.

8.7. Assume that there exists a linear operator ρ such that

$$\rho D + D\rho = I,$$

where I is the unit operator.

(a) Show that $H^k(D) = 0$ for all k's.

(b) Can ρ be a derivation? [Hint: Is I a derivation?]

8.8. Prove that the elements of the quotient graded Lie algebra $\mathcal{H}^*(D)$ define derivations of the quotient algebra $H^*(D)$.

8.9. An automorphism f of an algebra is an invertible linear transformation that preserves the product, $f(\alpha x + \beta y) = \alpha f(x) + \beta f(y)$, $f(xy) = f(x)\, f(y)$. Show that the transformation defined by

$$f(x) = (-)^{\varepsilon_x} x$$

for homogeneous elements and extended linearly, is an automorphism (known as the *parity automorphism*).

8.10. The purpose of this exercise and the next two is to derive the Frölicher and Nijenhuis (1956) classification of the derivations of the exterior algebra A on a smooth manifold M. One has $A = \oplus A_p$, where A_p is the subspace of exterior forms of degree p.

(a) Show that any derivation D of A is a "local" operator, *i.e.*, if ω_1 and ω_2 are two differential forms that coincide on an open set U, then $D\omega_1 = D\omega_2$ on U. [Hint: Let x be a point of U, and let V be an open set such that $x \in V \subset U$. Let ρ be a C^∞-function vanishing on V and equal to 1 on M/U. If χ is a differential form vanishing on U, one has $\chi = \rho\chi$. (On U, this equality reads $0 = \rho \cdot 0$, since $\chi = 0$; on $M\backslash U$, this equality reads $\chi = 1 \cdot \chi$.) The derivation property implies then $D\chi = 0$ at x and, hence, by a similar reasoning, at any point of U.]

(b) Prove that any derivation D of A is completely determined by its action on the 0-forms (*i.e.*, on A_0) and on the 1-forms (*i.e.*, on A_1). [Hint: Use (a) and the fact that any point x possesses an open neighborhood U on which any form can be expressed as a polynomial in a set of basis 1-forms, with coefficients that belong to $C^\infty(U)$.]

(c) Show that, conversely, any linear mapping $A_0 \oplus A_1 \to A$ satisfying the derivation property $D(\alpha_1\alpha_2) = \alpha_1\, D\alpha_2 + (-)^{\varepsilon_{\alpha_2}\varepsilon_D}(D\alpha_1)\alpha_2$ when $\alpha_1\alpha_2 \in A_0$ or A_1, can be uniquely extended to A so as to be a derivation.

[Hint: Show that $D\phi$ is uniquely defined on the open sets U of (b) and satisfy the compatibility conditions on the intersection of two such open sets.

(d) Show that there is no derivation of A of degree ≤ -2 (except the zero derivation).

8.11. (Inner Contraction Derivations).

(a) Let L be a vector-valued p-form, *i.e.*, L is a $\binom{1}{p}$-tensor that is completely antisymmetric in its lower indices. Verify that

$$i_L F = 0, \qquad (i_L \alpha)_{\lambda_1 \ldots \lambda_p} = \alpha_\lambda L^\lambda{}_{\lambda_1 \ldots \lambda_p}$$

defines a unique derivation of degree $p-1$ that vanishes on the scalars. Here, $F \in A_0$ and $\alpha \in A_1$.

(b) Prove that, conversely, any derivation that vanishes on the scalars is of the form $D = i_L$ for some L.

(c) Show that if L is the $\binom{1}{1}$-Kronecker tensor, then D is just the counting operator, $i_D \alpha = (\deg \alpha)\alpha$.

8.12. (Decomposition Theorem for Derivations of Exterior Algebra)

(a) Show that any derivation D of degree p defines a vector-valued p-form M through $DF = i_M\, dF, F \in A_0$. [Recall that if D is a derivation, then $DF = (\partial F/\partial x^\mu)\, Dx^\mu$, since $F(x) = F(x_0) + g_\mu(x)(x^\mu - x_0^\mu), g_\mu(x_0) = (\partial F/\partial x^\mu)(x^0)$.]

(b) Prove that a derivation that commutes with the exterior derivative d (in the graded sense) is completely determined by its action on the 0-forms and is given by

$$D = [i_M, d]$$

for some M, so the cohomology of d in the algebra of derivations is trivial (even though the cohomology of d in the original exterior algebra A may be nontrivial).

(c) Prove that any derivation D of A can be decomposed as

$$D = [i_M, d] + i_L,$$

i.e., is the sum of a derivation that is d-exact and of an inner contraction derivation.

8.13. Verify that the commutator of two derivations that commute with d also commutes with d. This enables one to define a bracket among $\binom{1}{p}$-tensors as follows.

$$[[i_L, d], [i_M, d]] = [i_{[L,M]}, d].$$

Write down the Jacobi identity for $[L, M]$ that follows from the Jacobi identity for the commutator of derivations. Compute explicitly $[L, L]$ when L is a $\binom{1}{1}$-tensor. [Answer: $\frac{1}{2}[L, L] = N$ is known as the Nijenhuis tensor of L and is a vector-valued two-form explicitly given by

$$N^\mu{}_{\alpha\beta} = L^\rho{}_\alpha\, L^\mu{}_{\beta,\rho} - L^\rho{}_\beta\, L^\mu{}_{\alpha,\rho} + L^\mu{}_\rho\, L^\rho{}_{\alpha,\beta} - L^\mu{}_\rho\, L^\rho{}_{\beta,\alpha}$$

(*i.e.*, if u and v are two vector fields, $N(u, v)$ is a vector field given by $[Lu, Lv] + LL[u, v] - L[Lu, v] - L[u, Lv]$, where [,] is the Lie bracket of vector fields).]

CHAPTER NINE

BRST CONSTRUCTION IN THE IRREDUCIBLE CASE

Although the general pattern is identical in both cases, the explicit construction of the BRST differential is technically more complicated for reducible theories than it is for irreducible ones. Accordingly, we analyze separately these two situations. We start with the simpler irreducible case, where the ideas underlying the BRST construction appear quite transparently.

The first objective is to define the Koszul–Tate resolution. Next, as explained above, one needs to put together the longitudinal exterior d and the Koszul–Tate resolution along the lines of homological perturbation theory.

9.1. KOSZUL–TATE RESOLUTION

9.1.1. Definition

The Koszul–Tate resolution must implement the restriction to the constraint surface Σ

$$G_a = 0 \tag{9.1}$$

through its homology. That is, it must be such that $H_0(\delta) = C^\infty(\Sigma)$ and $H_k(\delta) = 0$ for $k \neq 0$.

Now, we have seen in Chapter 1, §1.5.1, that the algebra $C^\infty(\Sigma)$ of functions on Σ can be identified with the quotient algebra $C^\infty(P)/\mathcal{N}$ of phase space functions P that differ by an element of the ideal $\mathcal{N}$ of functions vanishing on Σ. So, to fulfill

$$H_0(\delta) \equiv \frac{(\operatorname{Ker}\delta)_0}{(\operatorname{Im}\delta)_0} = \frac{C^\infty(P)}{\mathcal{N}}, \tag{9.2}$$

it is natural to define δ so that

$$(\operatorname{Ker}\delta)_0 = C^\infty(P) \tag{9.3a}$$

and

$$(\operatorname{Im}\delta)_0 = \mathcal{N}. \tag{9.3b}$$

To achieve (9.3a), one simply sets

$$\delta z^A = 0 \tag{9.4a}$$

for any phase space variable z^A. Because δ acts as a derivation, this implies $\delta F = 0$ for any phase space function $F(z^A)$. Therefore, if the degree of z^A, called the antighost number of z^A, is taken to be zero,

$$\operatorname{antigh} z^A = 0, \tag{9.4b}$$

the definition (9.4a) implies (9.3a).

To obtain (9.3b), we recall that an arbitrary element of $\mathcal{N}$ can be written as a combination of the constraints with coefficients that are phase space functions,

$$F \approx 0 \Leftrightarrow F = F^a G_a \tag{9.5}$$

(Theorem 1.1 with ϕ_m replaced by G_a). Hence, if one introduces as many generators $\mathcal{P}_a$ as there are constraints G_a, and sets

$$\delta \mathcal{P}_a = -G_a, \tag{9.6a}$$

one gets

$$F \approx 0 \Leftrightarrow F = \delta(-F^a \mathcal{P}_a) \in (\operatorname{Im}\delta)_0. \tag{9.6b}$$

The inclusion of a minus sign in (9.6a) is a matter of convention and is chosen for later convenience.

To satisfy the grading properties of δ, one takes

$$\operatorname{antigh} \mathcal{P}_a = 1, \tag{9.6c}$$

$$\varepsilon(\mathcal{P}_a) = \varepsilon_a + 1, \tag{9.6d}$$

with $\varepsilon_a \equiv \varepsilon(G_a)$. One extends δ to an arbitrary polynomial in the $\mathcal{P}$'s with coefficients that are phase space functions—*i.e.*, to any element of the tensor product algebra $\mathbb{C}[\mathcal{P}_a] \otimes C^\infty(P)$—by requiring that δ be an odd (right) derivation. Since δ^2 vanishes on all the generators, δ is nilpotent,

$$\delta^2 = 0. \tag{9.7}$$

Hence, δ is a differential called the Koszul–Tate differential.

With (9.4a) and (9.6a), the required condition $H_0(\delta) = C^\infty(P)/\mathcal{N}$ is fulfilled. The variables $\mathcal{P}_a$ are called the "ghost momenta" because they will become conjugate to the ghosts below.

9.1.2. Homology of δ

The algebra $\mathbb{C}[\mathcal{P}_a] \otimes C^\infty(P)$ contains as many independent generators in degree one as there are independent constraint functions. The new generators $\mathcal{P}_a$ "kill" the constraint functions G_a by means of Eq. (9.6a). The algebraic setting behind the Koszul–Tate differential is thus almost identical to the one of Theorem 8.2. It comes then as no surprise that the Koszul–Tate differential provides a resolution of the algebra $C^\infty(\Sigma)$, the resolution degree being the antighost number. This is the content of the next theorem.

Theorem 9.1. *The differential algebra* $[\mathbb{C}[\mathcal{P}_a] \otimes C^\infty(P), \delta]$ *provides a resolution of the algebra of functions on the constraint surface, i.e.,*

$$H_0(\delta) = C^\infty(\Sigma), \tag{9.8a}$$

$$H_k(\delta) = 0, \qquad k \neq 0. \tag{9.8b}$$

Proof. The detailed proof of Theorem 9.1 is technical and not particularly illuminating. For this reason, it is relegated to Appendix 9.A.

Historical Note: The resolution δ was introduced by Koszul. The extension to the reducible case to be discussed below follows the work of Tate. This is the reason why we already refer to δ as the "Koszul–Tate resolution."

9.2. EXTENDED PHASE SPACE

9.2.1. Ghosts and Longitudinal d

We have shown in Chapter 5 that in the irreducible case, the longitudinal forms along the gauge orbits can be represented as polynomials in the 1-forms ω^a dual to the vector fields defining the infinitesimal gauge transformations, with coefficients that are smooth functions on Σ. So, the exterior longitudinal algebra is isomorphic to $C^\infty(\Sigma) \otimes \mathbb{C}[\omega^a]$.

We will call from now on the 1-forms ω^a "the ghosts" and denote them by η^a. These are the standard terminology and notations of the BRST formalism. Furthermore, the form degree is known as "the pure ghost number" and is denoted by pure gh. One has

$$\text{pure gh}\,\eta^a = 1, \qquad \text{pure gh}\,z^A = 0. \tag{9.9}$$

We have also observed in Chapter 5 that the action of d is actually well defined on the algebra $C^\infty(P) \otimes \mathbb{C}[\eta^a]$ because the gauge transformations are defined everywhere in phase space and not just on the constraint surface Σ. But d is not a differential in $C^\infty(P) \otimes \mathbb{C}[\eta^a]$; its square is only weakly zero.

9.2.2. Bracket Structure—Ghost Number

A remarkable feature of the Koszul–Tate complex and of the exterior longitudinal complex is that they possess exactly the same number of generators. Furthermore, the generators have the same Grassmann parity. This is not an accident because it is the same (first-class) functions G_a that determine both the constraint surface (through $G_a = 0$) and the gauge transformations (through $\delta_\varepsilon F = [F, G_a]\varepsilon^a$).

It is, therefore, natural to declare that the ghosts η^a are conjugate to the ghost momenta $\mathcal{P}_a$, *i.e.*, to extend the bracket structure of the original phase space P to η^a and $\mathcal{P}_a$ as

$$[\mathcal{P}_a, \eta^b] = -(-)^{(\varepsilon_a+1)(\varepsilon_b+1)}[\eta^b, \mathcal{P}_a] = -\delta_a{}^b. \tag{9.10}$$

The brackets of η^a and $\mathcal{P}_a$ with the original variables are taken to be zero. The brackets of the original variables among themselves are unchanged. The space obtained by adding to P the ghost variables η^a and their momenta $\mathcal{P}_a$ is referred to as the "extended phase space" P_{ext}. The algebra of superfunctions on P_{ext} that are polynomial in η^a and $\mathcal{P}_a$ is the tensor product algebra $\mathbb{C}[\mathcal{P}_a] \otimes C^\infty(P) \otimes \mathbb{C}[\eta^a]$.

Unless otherwise specified, the ghosts η^a will be taken to be real,

$$(\eta^a)^* = \eta^a. \tag{9.11a}$$

Then, because the parity of η^a and $\mathcal{P}_a$ is given by

$$\varepsilon(\eta^a) = \varepsilon(\mathcal{P}_a) = \varepsilon_a + 1 \qquad (\bmod 2), \tag{9.11b}$$

one must take $\mathcal{P}_a$ to be pure imaginary or real according to whether the constraint function G_a is even or odd. This follows from the behavior of the bracket under complex conjugation.

The antighost number and the pure ghost number are extended to $P_{\rm ext}$ by setting

$$\text{antigh}\, \eta^a = 0, \tag{9.12a}$$

$$\text{pure gh}\, \mathcal{P}_a = 0, \tag{9.12b}$$

besides (9.4b), (9.6c), and (9.9). The "total ghost number" (or just "ghost number") is defined by

$$\text{gh}\, A = \text{pure gh}\, A - \text{antigh}\, A \tag{9.13a}$$

for any variable of definite pure ghost and antighost numbers. So, one has

$$\text{gh}\, z_a = 0, \qquad \text{gh}\, \eta^a = 1 = -\text{gh}\, (\mathcal{P}_a), \tag{9.13b}$$

and

$$\text{gh}\, (AB) = \text{gh}\, A + \text{gh}\, B. \tag{9.13c}$$

The ghosts η^a count positively, while the ghost momenta count negatively.

The interest of the ghost number is that it possesses a canonical generator, contrary to the antighost number or the pure ghost number. The ghost number generator $\mathcal{G}$ is given by

$$\mathcal{G} = i\eta^a \mathcal{P}_a \tag{9.14a}$$

and verifies

$$[A, \mathcal{G}] = i(\text{gh}\, A)\, A. \tag{9.14b}$$

It is imaginary,

$$\mathcal{G}^* = -\mathcal{G}. \tag{9.14c}$$

The factor of i has been inserted in (9.14a) so that the eigenvalues of $\mathcal{G}$ are real in quantum mechanics.

9.2.3. δ and d in the Extended Phase Space

The differentials δ and d have been defined so far, respectively, on $\mathbb{C}[\mathcal{P}_a] \otimes C^\infty(P)$ and $C^\infty(P) \otimes \mathbb{C}[\eta^a]$. It is natural to extend δ to the whole of the extended phase space by setting

$$\delta \eta^a = 0 \tag{9.15a}$$

The definition (9.15a) is the only possible one if δ is to possess antighost number minus one,

$$\text{antigh}\, \delta = -1. \tag{9.15b}$$

This is because $\delta\eta^a$ should be of antighost number minus one, which is impossible unless $\delta\eta^a = 0$.

By contrast, the requirement that d should be of antighost number zero and of total ghost number one,

$$\operatorname{antigh} d = 0, \qquad \operatorname{gh} d = 1, \tag{9.16a}$$

allows the possibility of taking $d\mathcal{P}_a$ to be linear in $\mathcal{P}$ and in η, $d\mathcal{P} \sim \mathcal{P}\eta$, with an arbitrary coefficient. One could simply set $d\mathcal{P}_a = 0$, but this definition would not make d anticommute with δ except in the abelian case. For this reason, we define instead

$$d\mathcal{P}_a = (-)^{\varepsilon_a} \eta^c \, C_{ca}{}^b \, \mathcal{P}_b. \tag{9.16b}$$

This definition yields

$$[\delta, d] = 0. \tag{9.16c}$$

Furthermore, d^2 is δ-exact as a consequence of the nilpotency of d on the constraint surface (Exercise 9.14), so that d is a differential modulo δ.

Because the ghosts η^a are δ-closed but not δ-exact, the homology of δ in the algebra $\mathbb{C}[\mathcal{P}_a] \otimes C^\infty(P) \otimes \mathbb{C}[\eta^a]$ is obtained by taking the tensor product of the right-hand side of (9.8) with the ghosts. So, δ provides a resolution of the exterior longitudinal algebra $C^\infty(\Sigma) \otimes \mathbb{C}[\eta^a]$. The cohomology of d modulo δ coincides then with the cohomology of the original longitudinal exterior derivative defined on Σ.

9.3. BRINGING δ AND d TOGETHER: THE BRST SYMMETRY AS A CANONICAL TRANSFORMATION

9.3.1. BRST Generator

The main theorem of homological perturbation theory guarantees the existence of the BRST differential $s = \delta + d +$ "more" such that $s^2 = 0$. Accordingly, our only task is to show that s can be chosen to have a canonical action. For that purpose, we look directly for the BRST function that generates s through the Poisson bracket

$$sF = [F, \Omega]. \tag{9.17}$$

The grading properties of s imply for Ω

$$\varepsilon(\Omega) = 1, \tag{9.18a}$$

$$\operatorname{gh} \Omega = 1. \tag{9.18b}$$

Furthermore, as we have indicated, the nilpotency of s is equivalent to

$$[\Omega, \Omega] = 0 \tag{9.18c}$$

by virtue of the Jacobi identity. This is a nontrivial property, since Ω is fermionic and the Poisson bracket $[\Omega, \Omega]$ is accordingly symmetric.

Now, if Ω is to reproduce δ to lowest order in the ghost momenta, it must start like

$$\Omega = \eta^a G_a + \text{"more,"} \tag{9.18d}$$

where "more" contains terms of order at least one in $\mathcal{P}_a$ (and, hence, at least two in η^a). This leads, indeed, to $[\mathcal{P}_a, \Omega] = \delta\mathcal{P}_a +$ "more." The relation (9.18d) automatically implies that Ω also incorporates dz^A, as one gets

$$\begin{aligned} [z^A, \Omega] &= [z^A, G_a]\eta^a + \text{"more"} \\ &= dz^A + \text{"more."} \end{aligned} \tag{9.18e}$$

The close connection between δ and d through the Poisson bracket that we have just found is an illustration of the fact that the restriction to the constraint surface (δ) and the gauge transformations (d) are generated by the same first-class functions G_a.

For s to fully reproduce d, one should demand that the next term in Ω generate $d\mathcal{P}_a$ and $d\eta^a$ through the Poisson bracket. However, this requirement turns out to be a consequence of (9.18d) and of the nilpotency of Ω. Accordingly, it need not be imposed separately.

We will also require that the BRST generator is real,

$$\Omega = \Omega^*. \tag{9.18f}$$

This guarantees that $sF = [F, \Omega]$ has definite reality properties (*i.e.*, is real or imaginary) whenever F is real or imaginary. This would not be the case if Ω had both nonvanishing real and imaginary parts.

9.3.2. Existence of the BRST Generator

Equations (9.18) completely determine Ω as a function of the extended phase space up to a canonical transformation. This can be seen by following the same perturbative approach as in the proof of the main theorem of homological perturbation theory. Namely, one expands the BRST generator Ω as a sum of terms of definite resolution degree, *i.e.*, of definite antighost number,

$$\Omega = \sum_{p \geq 0} \overset{(p)}{\Omega}, \quad \overset{(0)}{\Omega} = \eta^a G_a, \tag{9.19a}$$

$$\text{antigh} \overset{(p)}{\Omega} = p. \tag{9.19b}$$

The function $\overset{(p)}{\Omega}$ is a polynomial of order p in the ghost momenta,

$$\overset{(p)}{\Omega} = \overset{(p)}{U}{}^{a_1 \dots a_p} \mathcal{P}_{a_p} \dots \mathcal{P}_{a_1}, \tag{9.20a}$$

$$\overset{(p)}{U}{}^{a_1 \dots a_p} = \eta^{b_1} \dots \eta^{b_p+1} \overset{(p)}{U}{}^{a_1 \dots a_p}{}_{b_{p+1} \dots b_1}. \tag{9.20b}$$

The coefficients $\overset{(p+1)}{U}{}^{a_1 \dots a_p}{}_{b_{p+1} \dots b_1}(z^A)$ are only functions of the original phase space variables and are called *the structure functions of order p.* The constraint functions G_a are thus the structure functions of order zero.

Next, one rewrites the nilpotency condition $[\Omega, \Omega] = 0$ as a set of equations on the unknown functions $\overset{(p)}{\Omega}$. One finds explicitly (see Exercise 9.3)

$$\delta \overset{(p+1)}{\Omega} + \overset{(p)}{D} = 0, \tag{9.21}$$

where $\overset{(p)}{D}$ is given by

$$\overset{(p)}{D} = \frac{1}{2} \left[\sum_{k=0}^{p} [\overset{(k)}{\Omega}, \overset{(p-k)}{\Omega}]_{\text{orig}} + \sum_{k=0}^{p-1} [\overset{(k+1)}{\Omega}, \overset{(p-k)}{\Omega}]_{\mathcal{P},\eta} \right]. \tag{9.22}$$

Here, $[\ \ ,\ \ ]_{\text{orig}}$ refers to the Poisson bracket in the original phase space, which only acts on z^A and not on the ghosts, whereas $[\ \ ,\ \ ]_{\mathcal{P},\eta}$ refers to the Poisson bracket acting only on the ghost and ghost momenta arguments and not on the original phase space variables. One has

$$[A, B] = [A, B]_{\text{orig}} + [A, B]_{\mathcal{P},\eta}. \tag{9.23}$$

The separation (9.23) is convenient because it keeps track of the number of ghost momenta. While this number is unchanged by the original Poisson bracket,

$$\text{antigh}\left([A, B]_{\text{orig}}\right) = \text{antigh}\,(A) + \text{antigh}\,(B), \tag{9.24a}$$

one finds instead

$$\text{antigh}\left([A, B]\right)_{\mathcal{P},\eta} = \text{antigh}\,(A) + \text{antigh}\,(B) - 1, \tag{9.24b}$$

since one momentum $\mathcal{P}_a$ is cancelled by one η^a in $[A, B]_{\mathcal{P},\eta}$.

Since the functions $\overset{(p)}{D}$ involve only the lower order structure functions $\overset{(k)}{\Omega}$ with $k \leq p$, the solution of the system of Eqs. (9.21) can be constructed by induction. First, $\overset{(0)}{\Omega}$ is given by (9.19a). Second, one solves Eqs. (9.21) with $p = 0$ for $\overset{(1)}{\Omega}$. One gets, using the first-class property of the constraints,

$$\overset{(1)}{\Omega} = -\tfrac{1}{2}(-)^{\varepsilon_b} \eta^b \eta^c C_{cb}{}^a \mathcal{P}_a, \tag{9.25}$$

which shows that $[\mathcal{P}_a, \overset{(1)}{\Omega}]$ and $[\eta^a, \overset{(1)}{\Omega}]$ agree with $d\mathcal{P}_a$ and $d\eta^a$ as mentioned above.

Knowing $\overset{(0)}{\Omega}$ and $\overset{(1)}{\Omega}$, one next solves Eqs. (9.21) with $p = 1$ for $\overset{(2)}{\Omega}$ and then Eqs. (9.21) with $p = 2$ for $\overset{(3)}{\Omega}$, etc., ... Using the acyclicity of δ in positive degree and the Jacobi identity for the Poisson bracket, one can show that these equations possess a solution. Therefore, the BRST generator Ω exists. We shall not give the details here because these will be developed more completely in the next chapter on reducible theories, which contain irreducible theories as a particular case (no reducibility identity). Let us simply mention that while the proof of the main theorem of Chapter 8 required explicitly the acyclicity of δ in the space of the derivations, *i.e.*, $\mathcal{H}_k(\delta) = 0$ for $k \neq 0$, the proof of the existence of Ω uses instead the acyclicity of δ in the algebra, *i.e.*, $H_k(\delta) = 0$ for $k \neq 0$. This is because one expands here the BRST generator Ω, which is an element of the algebra, rather than the BRST derivation s itself. It is for this reason that Theorem 9.1 focuses only on the homology of the Koszul–Tate differential δ in the algebra. [The property $\mathcal{H}_k(\delta) = 0$ for $k \neq 0$ can easily be shown to hold (see the appendix), but this is not necessary.]

9.3.3. The BRST Generator Is Unique up to Canonical Transformations

It is clear that Eqs. (9.21) do not determine uniquely $\overset{(p+1)}{\Omega}$ for given $\overset{(k)}{\Omega}$'s with $k \leq p$, since one has the freedom of adding an arbitrary δ-exact term to $\overset{(p+1)}{\Omega}$. Furthermore, $\overset{(0)}{\Omega}$ itself is not unique because the constraint functions G_a can be modified as in §5.2.1. What is the corresponding change in Ω? The answer is provided by the following theorem.

Theorem 9.2. *Any two BRST generators Ω and $\bar{\Omega}$ associated with the same constraint surface differ by a canonical transformation in the extended phase space. That is, the BRST generator is unique up to canonical transformations.*

Proof. Again, this result is a particular case of similar results holding for reducible constraints and proved in the next chapter. Various aspects of Theorem 9.2 are analyzed in Exercises 9.4 and 9.5. In particular, Exercise 9.4 shows how a redefinition of the constraints $G_a \to \bar{G}_a = M_a{}^b G_b$ induces a canonical transformation in the extended phase space.

Theorem 9.2 is of fundamental importance because it indicates that in the classical theory the ambiguity in the structure functions simply leads to canonically related realizations of the BRST symmetry. One can thus say that with the introduction of the ghosts, the canonical covariance of the structure of constrained systems becomes manifest. In the quantum theory, where Ω becomes an operator, different choices of the structure functions lead formally to unitarily related quantum BRST operators—modulo, of course, the factor-ordering problem.

9.4. THE BRST GENERATOR IN SIMPLE CASES—RANK

9.4.1. Abelian Constraints

The simplest possible constraints are abelian,

$$[G_a, G_b] = 0. \tag{9.26a}$$

This case is quite important, for it covers locally the general situation.

In the case of abelian constraints, the structure functions $C_{ab}{}^c$ vanish and $\overset{(0)}{\Omega}$ is nilpotent by itself. So, the BRST generator can be taken to coincide with $\overset{(0)}{\Omega}$,

$$\Omega = \overset{(0)}{\Omega} = \eta^a G_a, \tag{9.26b}$$

$$[\Omega, \Omega] = 0. \tag{9.26c}$$

It is important to realize not only that (9.26c) follows from (9.26a) but, conversely, that (9.26c)—together with the form (9.26b) of Ω—implies that the constraints are abelian.

9.4.2. Constraints that Close according to a Group

The next case in order of increasing difficulty after the abelian case (9.26a) is given by a system with gauge transformations that form a group. In that case, the Poisson brackets of the constraints do not vanish throughout phase space. Rather, one has

$$[G_a, G_b] = C_{ab}{}^c G_c, \tag{9.27a}$$

where the $C_{ab}{}^c$ are constant. The relation (9.27a) describes a graded Lie algebra. The Jacobi identity for the Poisson brackets implies the Jacobi identity for the structure constants $C_{ab}{}^c$, which reads explicitly

$$C_{ab}{}^c C_{cd}{}^e + (-)^{(\varepsilon_b+\varepsilon_d)\varepsilon_a} C_{bd}{}^c C_{ca}{}^e + (-)^{(\varepsilon_a+\varepsilon_b)\varepsilon_d} C_{da}{}^c C_{cb}{}^e = 0. \tag{9.27b}$$

If $C_{ab}{}^c \neq 0$, $\overset{(0)}{\Omega}$ is not nilpotent by itself. One needs to add $\overset{(1)}{\Omega}$. This is enough to make Ω nilpotent because of the Jacobi identity (9.27b) and because the C's are constant. So, Ω is given by

$$\Omega = \eta^a G_a - \tfrac{1}{2}(-)^{\varepsilon_b} \eta^b \eta^c C_{cb}{}^a \mathcal{P}_a. \tag{9.27c}$$

The nilpotency of Ω, $[\Omega, \Omega] = 0$, follows from (9.27a)–(9.27b). It is, in turn, completely equivalent to the structure equations (9.27a) and (9.27b).

9.4.3. Higher Order Structure Functions

In the generic case of an open algebra, the sum $\overset{(0)}{\Omega} + \overset{(1)}{\Omega}$ is not nilpotent. Higher order terms are needed in Ω. These higher order terms characterize the more complicated structure of the gauge algebra. The second-order structure functions $\overset{(2)}{U}{}^{ae}_{bcd}$ were actually already encountered in §3.2.5.

In fact, the structure functions $\overset{(2)}{U}, \overset{(3)}{U}, \ldots$ could have been constructed without ever introducing the ghosts or their momenta. This could have been done by starting from the first-class property and systematically exploring the consequences of the Jacobi identity.

However, it is only in the extended phase space that the canonical covariance of the structure of first-class constrained systems is manifest. Furthermore, it is only after the ghosts have been introduced that the existence of the higher order structure functions is seen to have an interesting algebraic content. The use of the BRST formalism for studying the structure of open gauge algebras is thus conceptually clearer and much more economical.

9.4.4. Rank

One says that a set of constraints and of associated structure functions is of rank r if all the structure functions of order strictly greater than r vanish. This means that the corresponding BRST function contains up to r ghost momenta $\mathcal{P}_a$.

Abelian theories are of rank zero. Theories based on a true gauge group are of rank one. Ranks higher than one generically occur for theories with an open algebra.

The concept of rank is not intrinsic to the constraint surface. Rather, it is defined for a given representation of the BRST symmetry. It can be changed by modifying the structure functions. Locally, it is even possible to make the rank vanish.

The following theorem has been found useful in the determination of the rank of a system of structure functions.

Theorem 9.3. *If all structure functions of order k are zero for $r < k \leq 2r+1$, then all structure functions of order strictly greater than $2r+1$ can be taken to vanish.*

The proof of this theorem is left as an exercise (Exercise 9.6).

9.5. CONCLUSIONS

We have shown in this chapter that the BRST generator for irreducible gauge theories, defined by the equations (9.18), exists and is unique in the extended phase space of the z^A, the η^a, and the $\mathcal{P}_a$.

An important feature of the BRST generator is that it is completely gauge independent. This property is manifest in our presentation because no gauge condition was ever needed. The BRST construction depends only on the structure of the first-class constraints. These constraints, in turn, are determined by the gauge-invariant action and have nothing to do with gauge conditions.

Another important property of the BRST generator is that it is a function well defined everywhere in phase space. Global obstructions may be relevant when discussing gauge-fixing conditions (Gribov problem) but do not afflict the gauge-independent BRST symmetry.

APPENDIX 9.A: PROOF OF THEOREM 9.1 (HOMOLOGY OF δ)

9.A.1. δ-Covering of Phase Space

To prove that the differential algebra $(\mathbb{C}[\mathcal{P}_a] \otimes C^\infty(P), \delta)$ yields a resolution of the algebra $C^\infty(\Sigma)$ of functions on Σ, one needs to show that $H_k(\delta) = 0$ for $k \neq 0$ [the equality $H_0(\delta) = C^\infty(\Sigma)$ has been established in the text].

To that end, one proceeds as follows: (*i*) first, one checks the required property locally, *i.e.*, on appropriate open sets covering the phase space P; (*ii*) second, one shows that the property $H_k(\delta) = 0$ for $k \neq 0$ extends without problem from the individual sets to the whole of P. This line of demonstration will be repeated in the next chapter (Appendix 10.A).

Let $\tilde{P}$ be the real manifold underlying P and let (O_i, V_α) be an open covering of P by superdomains, whose underlying open sets are denoted by $(\tilde{O}_i, \tilde{V}_\alpha)$ and cover $\tilde{P}$. Finally, let $\tilde{\Sigma}$ be the surface of $\tilde{P}$ defined by the equations $\tilde{G}_a = 0$, where the $\tilde{G}$'s are obtained from the G's by setting the fermionic coordinates equal to zero. So, only the bosonic constraint functions are relevant for determining $\tilde{\Sigma}$.

We shall say that the covering (O_i, V_α) of P is adapted to δ, or, in short, is a δ-covering, if the following properties hold:

(i) The $\tilde{O}_i$'s cover $\tilde{\Sigma}$ and there exists on O_i a regular coordinate system in which the constraint functions G_a are the first coordinates.
(ii) At least one bosonic G_a is invertible throughout a given V_α.

The existence of δ-coverings is guaranteed by the regularity condition on the constraint functions. By assumption, one can indeed take locally the G_a's as regular coordinates in the vicinity of the constraint surface. So, one can cover a neighborhood of $\tilde{\Sigma}$ by open sets of the $\tilde{O}$-type. One then covers the remaining part of $\tilde{P}$ by open sets $\tilde{V}_\alpha$ not intersecting $\tilde{\Sigma}$. If these are small enough, at least one $\tilde{G}_a$ is different from zero everywhere on $\tilde{V}_\alpha$ (if x does not belong to $\tilde{\Sigma}$, one has $\tilde{G}_a \neq 0$ at x for at least one a, and thus also $\tilde{G}_a \neq 0$ in the vicinity of x by continuity). The corresponding (bosonic) superfunction G_a can then be inverted on the corresponding superdomain V_α.

Note that the constraint functions G_a may fail to fulfill the rank condition off $G_a = 0$, and so it is not permissible to assume that the G's can be taken as first coordinates on the open sets V_α.

9.A.2. Homology of δ on O_i at Positive Antighost Number

On a given O_i, there exists a coordinate system in which the constraint functions are the first coordinates, say, $-z_a$. The remaining coordinates are denoted by x_i. One has

$$\delta x_i = 0, \qquad \delta z_a = 0, \qquad \delta \mathcal{P}_a = z_a. \tag{9.A.1}$$

This is exactly of the form (8.30c), and so by applying Theorem 8.2, one obtains

$$H_k(\delta) = 0, \qquad k \neq 0 \qquad (\text{on } O_i). \tag{9.A.2a}$$

Similarly,

$$\mathcal{H}_k(\delta) = 0, \qquad k \neq 0 \qquad (\text{on } O_i). \tag{9.A.2b}$$

Since we have made appeal to Theorem 8.2, Eqs. (9.A.2) have really been established only for polynomials in x_i, z_a, and $\mathcal{P}_a$ and not for polynomials in $\mathcal{P}_a$ with coefficients that are arbitrary smooth functions of x_i, z_a. However, the result can be easily extended to $\mathbb{C}[\mathcal{P}_a]\otimes C^\infty(O_i)$. To that end, one rewrites δ, the counting operator $\bar{N}$, and the contracting homotopy introduced in §8.3.2 in a form that makes manifest the fact that they are derivations (acting from the right),

$$\delta = \left(\frac{\partial^R}{\partial \mathcal{P}_a}\right) z_a\,, \quad \sigma = \left(\frac{\partial^R}{\partial z_a}\right) \mathcal{P}_a, \tag{9.A.3a}$$

$$\bar{N} = \left(\frac{\partial^R}{\partial \mathcal{P}_a}\right) \mathcal{P}_a + \left(\frac{\partial^R}{\partial z_a}\right) z_a. \tag{9.A.3b}$$

The relation $\delta\sigma + \sigma\delta = \bar{N}$ holds on arbitrary element of $\mathbb{C}[\mathcal{P}_a]\otimes C^\infty(O_i)$ (and not just on the elements of the polynomial subalgebra $\mathbb{C}[\mathcal{P}_a, z_a, x_i]$). Therefore, if $F(\mathcal{P}_a, z_a, x_i) \in \mathbb{C}[\mathcal{P}_a] \otimes C^\infty(O_i)$ is δ-closed, one gets

$$\bar{N}F = \delta\sigma F. \tag{9.A.4a}$$

This implies

$$t\frac{dF}{dt}(t\mathcal{P}_a, tz_a, x_i) = (\delta\sigma)\, F(t\mathcal{P}_a, tz_a, x_i), \tag{9.A.4b}$$

where t is an arbitrary real number. If, in addition, F is of antighost number $k > 0$, one finds

$$F(t\mathcal{P}_a, tz_a, x_i) = O(t).$$

Accordingly, $F(0, 0, x_i)$ vanishes and $t^{-1}F(t\,\mathcal{P}_a, t\,z_a, x_i)$ is integrable in the vicinity of $t = 0$. It follows from (9.A.4b) and the fact that δ and σ do not depend on t that

$$F(\mathcal{P}_a, z_a, x_i) = \delta\big[\sigma G(\mathcal{P}_a, z_a, x_i)\big]$$

with

$$G(\mathcal{P}_a, z_a, x_i) = \int_0^1 \frac{dt}{t} F(t\mathcal{P}_a, tz_a, x_i).$$

The function σG is clearly a polynomial of order $k + 1$ in $\mathcal{P}_a$, so that F is δ-exact in $\mathbb{C}[\mathcal{P}_a] \otimes C^\infty(O_i)$. Hence, $H_k(\delta)$ vanishes for $k \neq 0$ in the algebra $\mathbb{C}[\mathcal{P}_a] \otimes C^\infty(O_i)$ and not just in the algebra $\mathbb{C}[\mathcal{P}_a, z_a, x_i]$. Similarly, (9.A.2b) still holds (see Exercise 9.12).

9.A.3. Homology of δ on V_α at Positive Antighost Number

The homology of δ on V_α is even easier to compute. This is because one of the bosonic G's, say, G_1, can be inverted on V_α. Therefore, the function equal to one everywhere on V_α is exact on V_α,

$$1 = \delta\Big(- \frac{\mathcal{P}_1}{G_1}\Big). \tag{9.A.5a}$$

This implies

$$H_*(\delta) = 0 \qquad (\text{on } V_\alpha), \tag{9.A.5b}$$

since the equation $\delta F = 0$ leads to $F = \delta[F(-\mathcal{P}_1/G_1)]$. In particular, one gets

$$H_k(\delta) = 0, \qquad k \neq 0 \qquad (\text{on } V_\alpha). \tag{9.A.5c}$$

The equality $\mathcal{H}_k(\delta) = 0$ on V_α is also direct (see Exercise 9.13).

9.A.4. Homology of δ

To patch the results of §9.A.2 and §9.A.3 together, one introduces a partition of unity $\{f_i, f_\alpha\}$ subordinate to the δ-covering (O_i, V_α), that is, such that the support of $\tilde{f}_i$ and $\tilde{f}_\alpha$, respectively, is contained in $\tilde{O}_i, \tilde{V}_\alpha$. One then proceeds as in Appendix 1.A. If $\delta M = 0$ with antigh $M > 0$, one gets

$$\begin{aligned} M &= \delta N_i \quad \text{on } O_i \\ &= \delta N_\alpha \quad \text{on } V_\alpha. \end{aligned}$$

Let

$$N = \sum_i f_i N_i + \sum_\alpha f_\alpha N_\alpha.$$

The element N of $\mathbb{C}[\mathcal{P}_a] \otimes C^\infty(P)$ is globally defined because each term $f_i N_i$ or $f_\alpha N_\alpha$ is itself globally defined. Since δ annihilates f_i or f_α and since the f's add to one, one finds furthermore that δN is equal to M. Hence, M is δ-exact. This demonstrates that $H_k(\delta)$ vanishes for $k \neq 0$ in the algebra $\mathbb{C}[\mathcal{P}_a] \otimes C^\infty(P)$. Note that by the same reasoning, one also obtains $\mathcal{H}_k(\delta) = 0$ for $k \neq 0$.

EXERCISES. CHAPTER NINE

9.1. Show that the differential δ is intrinsic, in the sense that a change of representation of the constraint surface, $G_a \to \bar{G}_a = M_b{}^a G_b$, simply amounts to a change of basis of the generators $\mathcal{P}_a, \mathcal{P}_a \to \bar{\mathcal{P}}_a = M_b{}^a \mathcal{P}_b$ without affecting δ (use $\delta M_b{}^a = 0$).

9.2. A "tensor" $T^{ab\ldots de\ldots gh}$ is said to be completely antisymmetric in $ab\ldots de\ldots gh$ if it fulfills

$$T^{ab\ldots de\ldots gh} = (-)^{(\varepsilon_d+1)(\varepsilon_e+1)} T^{ab\ldots ed\ldots gh} \tag{i}$$

for all permutations of adjacent indices. We take here the indices to run over the constraints.

(a) Suppose that there are two constraints, one commuting (G_1), one anticommuting (G_2). Write explicitly the symmetry properties of a second-rank antisymmetric tensor in that case.
(b) Show that in $S^{ab\ldots de\ldots gh}\mathcal{P}_h\mathcal{P}_g\ldots\mathcal{P}_e\mathcal{P}_d\ldots\mathcal{P}_b\mathcal{P}_a$, only the fully antisymmetric part $(S^{ab\ldots de\ldots gh})_A$ is relevant.
(c) Write $(S^{ab\ldots de\ldots gh})_A$ in terms of $S^{ab\ldots de\ldots gh}$ as a sum over permutations of appropriately symmetrized terms.

9.3. (a) Prove Eqs. (9.21)–(9.22).
(b) Write Eq. (9.21) in terms of the components $\overset{(p+1)}{U}{}^{a_1\ldots a_{p+1}}$ of Ω. By pulling out $\mathcal{P}_{a_{p+1}}\ldots\mathcal{P}_{a_1}$, show that this equation reads explicitly

$$(p+1)\overset{(p+1)}{U}{}^{a_1\ldots a_p a_{p+1}}G_{a_{p+1}}=\left(\overset{(p)}{D}{}^{a_1\ldots a_p}\right)_A, \tag{ii.a}$$

where $(\ )_A$ denotes complete "antisymmetrization" in $a_1,\ldots,a_p$ (with the factor $1/p!$ included) to form an object with the symmetry of $\eta^{a_1}\ldots\eta^{a_p}$, and where $\overset{(p)}{D}{}^{a_1\ldots a_p}$ are the coefficients appearing in

$$\overset{(p)}{D}=(-)^{p+\Sigma_{k=1}^{p}\varepsilon_{a_k}}\overset{(p)}{D}{}^{a_1\ldots a_p}\mathcal{P}_{a_p}\ldots\mathcal{P}_{a_1} \tag{ii.b}$$

and are given by

$$\overset{(p)}{D}{}^{a_1\ldots a_p}=\frac{1}{2}\sum_{k=0}^{p}\left[\overset{(k)}{U}{}^{a_1\ldots a_k},\overset{(p-k)}{U}{}^{a_{k+1}\ldots a_p}\right](-)^{p-k+\Sigma_{i=k+1}^{p}\varepsilon_{a_i}}$$

$$-\sum_{k=o}^{p-1}(k+1)\overset{(k+1)}{U}{}^{a_1\ldots a_k a}\,\frac{\partial^L\overset{(p-k)}{U}{}^{a_{k+1}\ldots a_p}}{\partial\eta^a}(-)^{p-k+\Sigma_{i=k+1}^{p}\varepsilon_{a_i}}. \tag{ii.c}$$

Note: Equations (ii) are Eqs. (3.34a)–(3.34b) for $p=1$.

9.4. Let $\bar{G}_a=0$ and $G_a=0$ define the same constraint surface, *i.e.*,

$$\bar{G}_a=M_a{}^bG_b,$$

where the matrix $M_a{}^b$ can be inverted.
(a) Show that by making, if necessary, the ghost canonical transformation $\eta^1\to-\eta^1$, $\mathcal{P}_1\to-\mathcal{P}_1$, $\eta^a\to\eta^a$, $\mathcal{P}_a\to\mathcal{P}_a$ $(a>1)$, one can assume $M_a{}^b$ to be in the connected component of the identity (det $\overset{(0)}{M}{}_a{}^b>0$).
(b) Prove that the infinitesimal change of constraints

$$M_a{}^b=\delta_a{}^b+\varepsilon_a{}^b$$

is induced by the canonical transformation generated by $C=\eta^a\varepsilon_a{}^b\mathcal{P}_b$.
(c) Conclude from (a) and (b) that the replacement of G_a by $\bar{G}_a$ in Ω can always be accounted for by a canonical transformation in the extended phase space.

9.5. The structure functions $C_{ab}{}^c$ are not completely determined by $[G_a,G_b]=C_{ab}{}^cG_c$, as one can add to $C_{ab}{}^c$ an antisymmetric combination of the constraints without changing $C_{ab}{}^cG_c$. Check that the ambiguity can be absorbed in a canonical redefinition of the coordinates of the extended phase space.

9.6. Supply a proof of the theorem on the rank mentioned in §9.4.4. (Hint: Show that the appropriate $\overset{(k)}{D}$ vanish.)

9.7. Consider a set of first-class constraints that are linear in the momenta,

$$G_a = a_a^i(q)p_i + f_a(q).$$

(a) What are the conditions on $a_a^i(q)$ and $f_a(q)$ if the constraints G_a are to be first class (see Exercise 5.15)?
(b) Show that by appropriate choice of the structure functions, the rank of the system can be taken to be equal to one without changing the G_a, even though the algebra of the gauge transformations may be open in phase space.

9.8. Show that the rank of a system with a finite number of bosonic constraints is necessarily finite. (This may not be the case for fermionic constraints.)

9.9. Consider the following "quadratic" algebra,

$$[L_i, L_j] = C_{ij}{}^k\, L_k; \quad [L_i, T_\alpha] = C_{i\alpha}^\beta\, T_\beta;$$
$$[T_\alpha, T_\beta] = C_{\alpha\beta}^i\, L_i + D_{\alpha\beta}^{ij}\, L_i\, L_j,$$

where $C_{ij}{}^k$, $C_{i\alpha}^\beta$, $C_{\alpha\beta}^i$, and $D_{\alpha\beta}^{ij}$ are constant. The linear span of the L_i and T_α is not a Lie algebra because of the quadratic term $D_{\alpha\beta}^{ij}\, L_i\, L_j$ in $[T_\alpha, T_\beta]$.

(a) Derive the identities on $C_{ij}{}^k$, $C_{i\alpha}^\beta$, $C_{\alpha\beta}^i$, $D_{\alpha\beta}^{ij}$, which follow from the Jacobi identity for the Poisson bracket.
(b) Prove that $\overset{(2)}{\Omega}$ is zero.
(c) Show that, in general, $\overset{(3)}{\Omega} \neq 0$.
(d) Prove that $\overset{(k)}{\Omega} = 0$ for $k \geq 4$. Hence, the theory is of rank 3 unless $\overset{(3)}{\Omega}$ is zero, in which case it is of rank 1 (or even 0 if all C's and D's vanish).
(e) Construct a nontrivial example with $\overset{(3)}{\Omega} \neq 0$. (Hint: Take, *e.g.*, T_α to be scalars for the group generated by L_i, and assume that they close proportionally to the Casimir invariant $g^{ij}\, L_i\, L_j$, with g^{ij} the invariant metric.)

9.10. Consider a general "nonlinear algebra" of homogeneous degree two,

$$[G_a, G_b] = D_{ab}{}^{cd}\, G_c\, G_d$$

with $D_{ab}{}^{cd}$ constant. The G's are assumed to be bosonic.

(a) Derive the identities on $D_{ab}{}^{cd}$ that follow from the Jacobi identity for the Poisson bracket.
(b) Show that the theory may be of arbitrarily high (but finite) rank, depending on the actual value of $D_{ab}{}^{cd}$. Write down the first structure functions.

9.11. Repeat the BRST analysis in the presence of second-class constraints. Show that one can define Ω so that $[\Omega, \Omega]^* = 0$ on the surface of the *second*-class constraints.

9.12. Let Δ be a derivation acting on $\mathbb{C}[\mathcal{P}_a] \otimes C^\infty(O_i)$, where O_i is one of the open sets of §9.A.2. The differential δ is given by (9.A.1), *i.e.*, $\delta = (\partial^R/\partial \mathcal{P}_a) z_a$.

(a) Show that the antighost number of Δ is ≥ -1. [Hint: If antigh $\Delta < -1$, then $\Delta x^i = \Delta z_a = \Delta \mathcal{P}_a = 0$.]

(b) If antigh $\Delta = -1$, then $\Delta x^i = 0$, $\Delta z_a = 0$, and $\Delta \mathcal{P}_a = F_a(x, z)$. Verify that Δ is both δ-closed and δ-exact. [$\Delta = [K, \delta]$ with $K = (\partial^R/\partial z_a) F_a(x, z)$.]

(c) Consider now the case antigh $\Delta \geq 1$,

$$\Delta = \left(\frac{\partial^R}{\partial \mathcal{P}_a}\right) L_a + \left(\frac{\partial^R}{\partial z_a}\right) M_a + \left(\frac{\partial^R}{\partial x^i}\right) N_i,$$

with antigh L_a = antigh $(M_a)+1$ = antigh $(N_i)+1 \geq 2$. Using $H_k(\delta) = 0$ for $k \neq 0$, show that $[\Delta, \delta] = 0$ implies $\Delta = [K, \delta]$ for some appropriate K.

(d) What is $\mathcal{H}_0(\delta)$? [Answer: It is isomorphic to the space of derivations $(\partial^R/\partial x^i) N_i(x)$.]

9.13. Show that on the open sets V_α of §9.A.3, the Koszul–Tate differential can be written as $\delta = \partial^R/\partial \mathcal{P}'_1$, with $\mathcal{P}'_1 = -(\mathcal{P}_1/G_1)$, $\mathcal{P}'_{\bar{a}} = \mathcal{P}_{\bar{a}} - (G_{\bar{a}} \mathcal{P}_1)/G_1$, $\bar{a} \neq 1$. Conclude that any δ-exact derivation of $\mathbb{C}[\mathcal{P}_a] \otimes C^\infty(V_\alpha)$ (= vector field with polynomial coefficients in the $\mathcal{P}$'s) does not depend on $\mathcal{P}'_1$ and, hence, is δ-exact.

9.14. Show directly from (5.25) and (9.16b) that the derivation d is a differential modulo δ. [Hint: This property need only be checked on the generators. The proof relies on $d^2 \approx 0$ and (9.8). Note that the existence proof of Ω given in the text does not use the fact that d is a differential modulo δ but does, actually, imply it.]

9.15. Let $\varepsilon : C^\infty(P) \rightarrow C^\infty(\Sigma)$ be the "augmentation map" sending the phase space functions on their restriction to Σ and let $\varepsilon' : C^\infty(\Sigma) \rightarrow 0$ be the zero map. Show that in

$$\cdots \xrightarrow{\delta} C^1 \xrightarrow{\delta} C^0 \equiv C^\infty(P) \xrightarrow{\varepsilon} C^\infty(\Sigma) \xrightarrow{\varepsilon'} 0, \tag{iii}$$

one has $\varepsilon\delta = \varepsilon'\varepsilon = 0$ (besides $\delta^2 = 0$). Here, C^k is the subspace of $\mathbb{C}[\mathcal{P}_a] \otimes C^\infty(P)$ of antighost number k (homogeneous polynomials in $\mathcal{P}_a$ of degree k). Check that the sequence (iii) is exact. (A sequence is exact if the kernel of each map is equal to the image of the previous map.)

CHAPTER TEN

BRST CONSTRUCTION IN THE REDUCIBLE CASE

10.1. THE SIMPLEST EXAMPLE

Reducible theories are characterized by the fact that the gauge transformations are not mutually independent on the constraint surface. To understand the new features that this property brings in, it is useful to study a simple reducible model that illustrates the main points. That simple model is just an arbitrary unconstrained system with variables (q^i, p_i). Any unconstrained system can always be viewed as a reducible constrained one where the constraint functions are the identically vanishing functions and the gauge transformations reduce to the identity. Of course, this is not the most convenient way to treat an unconstrained system. Yet, it is of interest to see how the BRST formalism applies in that case. We shall analyze the case of a single trivial constraint, which we regard as being bosonic,

$$G = 0, \quad \varepsilon(G) = 0. \tag{10.1a}$$

This constraint is subject to the reducibility identity,

$$1 \cdot G = 0. \tag{10.1b}$$

The observables of the theory contain all the phase space functions, since the constraint surface coincides with the phase space P (there is no independent constraint) and since each phase space point is an entire gauge orbit (the gauge transformations reduce to the identity). The BRST differential therefore must be such that $H^0(s) = C^\infty(P)$.

One way to achieve this requirement is simply to take s to be the identically vanishing differential acting in the algebra $C^\infty(P)$, without ghosts or ghost momenta. This possibility is correct but is not the one that will be followed here, since we want to investigate how the reducible constraint (10.1a) can be incorporated; so we want to introduce a ghost pair $(\eta, \mathcal{P})$ for G, fulfilling

$$[\mathcal{P}, \eta] = -1, \qquad \varepsilon(\eta) = \varepsilon(\mathcal{P}) = 1, \qquad \mathrm{gh}\,(\eta) = -\mathrm{gh}\,(\mathcal{P}) = 1. \qquad (10.2)$$

If one ignores the fact that the constraint $G = 0$ is empty and proceeds as in the previous chapter, one finds the incorrect BRST generator

$$\Omega = \eta G \equiv 0 \qquad \text{(incorrect)}. \qquad (10.3a)$$

The generator (10.3a) is incorrect because the corresponding BRST transformation in the algebra $\mathbb{C}\,[\mathcal{P}] \otimes C^\infty(P) \otimes \mathbb{C}\,[\eta]$ of polynomials in $\mathcal{P}$ and η with phase space function coefficients is trivial,

$$s\,(\text{anything}) = 0 \quad \text{in} \quad \mathbb{C}\,[\mathcal{P}] \otimes C^\infty(P) \otimes \mathbb{C}\,[\eta] \qquad \text{(incorrect)}, \; (10.3b)$$

so that any nonvanishing function on the extended phase space defines a nontrivial BRST cohomological class. In particular, one gets the incorrect result $H^0(s) = \{A(q,p) + B(q,p)\,\eta\mathcal{P}\}$, where A and B are arbitrary phase space functions.

How can one recover the correct cohomology? It is clear that the origin of the difficulty lies in the fact that η and $\mathcal{P}$ are both closed but not exact and, hence, contribute to the cohomology. One should thus find a procedure that removes η and $\mathcal{P}$ from the BRST cohomology. This can be done by adding extra variables. The new variables are known as the *ghosts of ghosts* and their conjugate momenta.

There exist two ways to remove unwanted variables from the cohomology, both of which are used in the BRST construction.

(*i*) The first method follows ideas due to Tate and applies to $\mathcal{P}$. To kill $\mathcal{P}$, one introduces a new bosonic variable π at ghost number -2, and one sets

$$s\pi = -\mathcal{P}, \qquad s\mathcal{P} = 0, \qquad \varepsilon(\pi) = 0, \qquad \mathrm{gh}\,(\pi) = -2. \qquad (10.4a)$$

The minus sign in $s\pi$ is conventional. *With the introduction of π, the ghost momentum $\mathcal{P}$ becomes exact and, hence, disappears in cohomology.* The new variable π does not contribute to $H^*(s)$ either, since it is not closed.

(*ii*) The second method is dual to the first one and applies to η. To kill η, one replaces $s\eta = 0$ by $s\eta = -\phi$, where ϕ is a new bosonic variable of ghost number two,

$$s\eta = -\phi, \qquad s\phi = 0, \qquad \varepsilon(\phi) = 0, \qquad \mathrm{gh}\,(\phi) = 2. \tag{10.4b}$$

With the introduction of ϕ, the ghost η is no longer closed and, hence, disappears also in cohomology. The new variable ϕ does not contribute to $H^*(s)$ either, since it is exact.

The correct BRST differential acts therefore in the algebra $\mathbb{C}\,[\mathcal{P},\pi]\otimes C^\infty(P)\otimes\mathbb{C}\,[\eta,\phi]$ and is obtained by adjoining

$$sq^i = sp_i = 0 \tag{10.4c}$$

to (10.4a) and (10.4b). When one passes to the BRST cohomology, the variables $(\mathcal{P},\,\pi)$ and (η,ϕ) cancel in pairs (one is the BRST variation of the other) so that one gets

$$H^0(s) = C^\infty(P), \tag{10.5}$$

as required on physical grounds. If one declares that ϕ and π are conjugate to one another, which is permissible because they have conjugate degrees, one finds that s is a canonical transformation with generator

$$\Omega = \phi\mathcal{P} \qquad \text{(correct)}. \tag{10.6}$$

This simple example clearly shows that it is necessary to enlarge the ghost spectrum of reducible theories to get the correct BRST differential. It is only under this condition that the BRST cohomology is equal, at ghost number zero, to the set of observables. The introduction of the extra variables is thus mandated by cohomological considerations.

Also to be observed is that the same theory can admit many correct BRST descriptions obtained from one another by increasing the phase space. Indeed, the BRST differential can either be taken to be $sq^i = 0 = sp_i$ in $C^\infty(P)$ or $sq^i = 0 = sp_i$, $s\eta = -\phi$, $s\phi = 0$, $s\pi = -\mathcal{P}$, and $s\mathcal{P} = 0$ in $\mathbb{C}\,[\mathcal{P}\,,\pi]\otimes C^\infty(P)\otimes\mathbb{C}\,[\eta,\phi]$. This enlargement replaces $\Omega = 0$ in P by $\Omega = \phi\mathcal{P}$ in the extended phase space and does not modify the cohomology. We shall come back to this point in the next chapter when we discuss "nonminimal solutions" (Sec. 11.3).

10.2. DESCRIPTION OF REDUCIBLE THEORIES

The general case of arbitrary reducible theories is treated along the same lines as the example. The explicit expression for the BRST differential is, however, more involved because the form of the reducibility identities is, in general, more complicated, a question to which we now turn.

10.2.1. First-Order Reducibility Functions

For later convenience, we add a suffix 0 to the index a referring to the constraints, which then read

$$G_{a_0} \approx 0, \quad a_0 = 1, \dots, m_0. \tag{10.7}$$

By definition, the gauge transformations are reducible if the vector fields X_{a_0} generating the infinitesimal gauge transformations

$$X_{a_0} F \equiv [F, G_{a_0}] \tag{10.8}$$

are not linearly independent on the constraint surface. There exist then functions $Z_{a_1}{}^{a_0} \not\approx 0$ such that

$$(X_{a_0}) Z_{a_1}{}^{a_0} (-)^{\varepsilon_{a_0}(\varepsilon_{a_1}+1)} \approx 0, \qquad a_1 = 1, \dots, m_1. \tag{10.9}$$

Recall that vector fields act from the right. Thus, $(X_{a_0})\mu^{a_0}$ denotes the vector field obtained by linear combination of the vector fields X_{a_0} with coefficients μ^{a_0}; *i.e.*, X_{a_0} does not act on μ^{a_0}. The inclusion of the phase $(-)^{\varepsilon_{a_0}(\varepsilon_{a_1}+1)}$ in the definition of $Z_{a_1}{}^{a_0}$ is a matter of convention and is done to simplify Eq. (10.10) below [compare with (7.17a) and (7.19c)].

The relations (10.9) provide an on-shell definition of reducibility in terms of the gauge transformations on the constraint surface. By appropriately extending $Z_{a_1}{}^{a_0}$ off $G_{a_0} = 0$, one can replace (10.9) by the equivalent equation

$$Z_{a_1}{}^{a_0} G_{a_0} = 0, \qquad a_1 = 1, \dots, m_1, \tag{10.10}$$

which holds strongly throughout phase space, even if the vector fields X_{a_0} happen to be independent when $G_{a_0} \neq 0$. Indeed, for an arbitrary extension $Z_{a_1}{}^{a_0}$, the Taylor expansion of $Z_{a_1}{}^{a_0} G_{a_0}$ around the constraint surface must start at second order to reproduce (10.9); *i.e.*, one must have $Z_{a_1}{}^{a_0} G_{a_0} = S_{a_1}{}^{b_0 a_0} G_{b_0} G_{a_0}$, with $S_{a_1}{}^{b_0 a_0}$ some appropriate functions. By redefining $Z_{a_1}{}^{a_0} \rightarrow Z_{a_1}{}^{a_0} - S_{a_1}{}^{b_0 a_0} G_{b_0}$, one arrives at (10.10). Therefore, the constraints G_{a_0} are reducible whenever the gauge transformations fail to be independent on-shell.

Because of (10.10), the constraint surface is not of dimension $n - m_0$, where n is the dimension of phase space. Rather, it is of dimension $n - m_0'$ where $m_0' < m_0$ is the number of independent constraints G_{a_0}. At the same time, the dimension of the gauge orbits generated by the vector fields X_{a_0} on $G_{a_0} = 0$ is not of dimension m_0 but of dimension $m_0' < m_0$.

The coefficients $Z_{a_1}{}^{a_0}$ in (10.9) and (10.10) are required to exhaust all existing relations among the vector fields X_{a_0}; *i.e.*, if a combination of the vector fields X_{a_0} vanishes, it must be expressible in terms of the $Z_{a_1}{}^{a_0}$,

$$(X_{a_0})\, \lambda^{a_0} (-)^{\varepsilon_{a_0}(\varepsilon_\lambda + 1)} \approx 0 \Rightarrow \lambda^{a_0} \approx \lambda^{a_1} Z_{a_1}{}^{a_0}, \tag{10.11a}$$

$$\varepsilon(\lambda^{a_0}) = \varepsilon_{a_0} + \varepsilon_\lambda, \qquad \varepsilon_\lambda = 0 \text{ or } 1. \tag{10.11b}$$

As we have already stressed many times, we will assume throughout that the constraints G_{a_0} can be locally separated into "independent constraint functions" G_{A_0}, $A_0 = 1, \ldots, m'_0$ and "dependent constraint functions" G_{α_0}, $\alpha_0 = m'_0 + 1, \ldots, m_0$, which are expressible in terms of G_{A_0} by (10.10) as

$$G_{\alpha_0} = N_{\alpha_0}{}^{A_0} G_{A_0},$$

with coefficients $N_{\alpha_0}{}^{A_0}$ that are regular in the vicinity of $G_{a_0} = 0$. The matrix of the gradients dG_{A_0} of the independent functions should be of maximum rank on $G_{a_0} = 0$; *i.e.*, one should be able to take locally G_{A_0} as first m'_0 coordinates of a regular (possibly noncanonical) coordinate system in phase space. The separation of the constraints into two groups as above cannot always be done globally with a single set of independent constraints G_{A_0}. It may instead be necessary to cover phase space by more than one open set with a different set of independent constraints for each open set. This is because there may be global obstructions to the existence of a single set G_{A_0}. Furthermore, even if the splitting $G_{a_0} \rightarrow G_{A_0}, G_{\alpha_0}$ possesses a global validity, it may spoil manifest Lorentz invariance (or some other invariance) or even destroy locality in physical space, and so it may be undesirable.

10.2.2. Completeness in Terms of Strong Equalities

The relation (10.11a) holds on the constraint surface. Can it be replaced by an equivalent strong equation?

If $(X_{a_0})\lambda^{a_0}(-)^{\varepsilon_{a_0}(\varepsilon_\lambda+1)}$ weakly vanishes, then, as we have just seen, λ^{a_0} can be extended off $G_{a_0} = 0$ in such a way that

$$\lambda^{a_0} G_{a_0} = 0. \tag{10.12a}$$

Furthermore, the completeness of the Z's imply

$$\lambda^{a_0} = \lambda^{a_1} Z_{a_1}{}^{a_0} + \nu^{a_0 b_0} G_{b_0} (-)^{\varepsilon_{b_0}}, \tag{10.12b}$$

where, *a priori,* the coefficients $\nu^{a_0 b_0}$ do not have any particular symmetry properties in (a_0, b_0). However, if (10.12b) is to solve (10.12a), then the $\nu^{a_0 b_0}$ may be chosen to be "antisymmetric," *i.e.*,

$$\nu^{a_0 b_0} = (-)^{(\varepsilon_{a_0}+1)(\varepsilon_{b_0}+1)} \nu^{b_0 a_0}. \tag{10.12c}$$

This is a consequence of the following theorem.

Theorem 10.1. *The conditions*

$$\lambda^{a_0} G_{a_0} = 0 \quad \textit{and} \quad \lambda^{a_0} \approx 0 \tag{10.13a}$$

imply

$$\lambda^{a_0} = \nu^{a_0 b_0} G_{b_0} (-)^{\varepsilon_{b_0}} \tag{10.13b}$$

with

$$\nu^{a_0 b_0} = (-)^{(\varepsilon_{a_0}+1)(\varepsilon_{b_0}+1)} \nu^{b_0 a_0}. \tag{10.13c}$$

Proof. The proof is technical and given in the appendix.

It follows from this theorem that the completeness requirement (10.11) on the $Z_{a_1}{}^{a_0}$, which is expressed in terms of weak equalities, is equivalent to the completeness requirement that (10.12a) should imply (10.12b)–(10.12c). This second form of the completeness of the $Z_{a_1}{}^{a_0}$ involves strong equalities. While the weak form (10.11) is easier to check in practice, it turns out that the strong one is more useful for developing the BRST formalism.

10.2.3. Higher Order Reducibility Functions

It may turn out that the relations (10.10) are independent, namely, $\lambda^{a_1} Z_{a_1}{}^{a_0} \approx 0 \Rightarrow \lambda^{a_1} \approx 0$. In that case, the number m_0' of independent constraints is equal to $m_0 - m_1$. But it may be that the system $Z_{a_1}{}^{a_0}$ is overcomplete, so that there exist nonvanishing functions λ^{a_1} with the property that $\lambda^{a_1} Z_{a_1}{}^{a_0} \approx 0$. Then, the number of independent gauge generators is greater than $m_0 - m_1$ and equal to $m_0 - m_1'$ where m_1' is the number of independent relations (10.10). The consideration of an overcomplete set of $Z_{a_1}{}^{a_0}$ may again be desirable on the grounds of preserving some manifest invariance or locality in physical space.

If $Z_{a_2}{}^{a_1}$ defines a complete set of functions that weakly annihilate $Z_{a_1}{}^{a_0}$, one has

$$Z_{a_2}{}^{a_1} Z_{a_1}{}^{a_0} = (-)^{\varepsilon_{a_0}} C_{a_2}{}^{a_0 b_0} G_{b_0}, \quad a_2 = 1, \ldots, m_2, \tag{10.14a}$$

$$\lambda^{a_1} Z_{a_1}{}^{a_0} \approx 0 \Rightarrow \lambda^{a_1} = \lambda^{a_2} Z_{a_2}{}^{a_1} + \nu^{a_1 a_0} G_{a_0}. \tag{10.14b}$$

The right-hand side of (10.14a) weakly vanishes and annihilates G_{a_0}. Hence, using Theorem 10.1, one can assume

$$C_{a_2}{}^{a_0 b_0} = (-)^{(\varepsilon_{a_0}+1)(\varepsilon_{b_0}+1)} C_{a_2}{}^{b_0 a_0}. \tag{10.14c}$$

Even though complete, the set of coefficients $Z_{a_1}{}^{a_2}$ may itself be overcomplete. This leads to a tower of reducibility equations,

$$Z_{a_k}{}^{a_{k-1}} Z_{a_{k-1}}{}^{a_{k-2}} = (-)^{\varepsilon_{a_{k-2}}} C_{a_k}{}^{a_{k-2} a_0} G_{a_0}, \tag{10.15a}$$

$$k = 1, \ldots, L, \quad a_k = 1, \ldots, m_k, \tag{10.15b}$$

where we have set $Z_{a_0}{}^{a_{-1}} \equiv G_{a_0}$. We also chose $Z_{a_k}{}^{a_{k-1}}$ such that

$$(Z_{a_k}{}^{a_{k-1}})^* = (-)^{k+k(\varepsilon_{a_k}+\varepsilon_{a_{k-1}})+\varepsilon_{a_k}\varepsilon_{a_{k-1}}} Z_{a_k}{}^{a_{k-1}}. \tag{10.15c}$$

(besides $G^*_{a_0} = G_{a_0}$). This choice is permissible (see Exercise 10.2) and, although somewhat awkward at first sight, will turn out to be convenient later on.

We will allow the order of reducibility L to be arbitrarily large or even infinite. At each stage, the $Z_{a_k}{}^{a_{k-1}}$ provide a complete set of "reducibility functions" of order k, *i.e.*,

$$\lambda^{a_{k-1}} Z_{a_{k-1}}{}^{a_{k-2}} \approx 0 \Rightarrow \lambda^{a_{k-1}} = \lambda^{a_k} Z_{a_k}{}^{a_{k-1}} + \nu^{a_{k-1}a_0} G_{a_0}. \tag{10.15d}$$

For finite order of reducibility L, this equation implies

$$\lambda^{a_L} Z_{a_L}{}^{a_{L-1}} \approx 0 \Rightarrow \lambda^{a_L} = \nu^{a_L a_0} G_{a_0} \tag{10.16}$$

for the "last" coefficients $Z_{a_L}{}^{a_{L-1}}$. The case of irreducible theories corresponds to $L = 0$. The example of the introduction has one $Z_{a_1}{}^{a_0}$, which can be taken equal to unity, and no higher order reducibility functions.

Knowing the number of reducibility equations, one can express the number of independent gauge generators as

$$m = m_0 - m_1 + m_2 - m_3 + \ldots + (-)^k m'_k, \tag{10.17a}$$

where m'_k is the number of independent $Z_{a_k}{}^{a_{k-1}}$ of order k. For theories of finite order, this sum is equal to

$$m = \sum_{i=0}^{L} (-)^i m_i. \tag{10.17b}$$

Since the G_{a_0} are of definite Grassmann parity, the reducibility equations, and hence also the Z's, may be assumed to have the same property. We recursively define ε_{a_k} by

$$\varepsilon(Z_{a_k}{}^{a_{k-1}}) = \varepsilon_{a_k} + \varepsilon_{a_{k-1}} \tag{10.18}$$

with $\varepsilon_{a_0} = \varepsilon(G_{a_0})$.

It should be stressed that the coefficients $C_{a_k}{}^{a_{k-2}a_0}$ in (10.14) and (10.15) are not required to be identically zero. If they turn out to be zero, one says that one has "off-shell reducibility." Otherwise, the reducibility only holds on-shell.

The functions $Z_{a_{k-1}}{}^{a_{k-2}}$ are taken to fulfill regularity conditions analogous to those on G_{a_0}; namely, the $Z_{a_{k-1}}{}^{a_{k-2}}$ are taken so that they can locally be split on the constraint surface and in its vicinity into independent reducibility functions $Z_{A_{k-1}}{}^{a_{k-2}}$ that contain all the information about the reducibility at order k and dependent reducibility functions denoted by $Z_{\alpha_{k-1}}{}^{a_{k-2}}$. The dependent reducibility functions can be expressed in terms of the independent ones $Z_{A_{k-1}}{}^{a_{k-2}}$ as

$$Z_{\alpha_{k-1}}{}^{a_{k-2}} = \rho_{\alpha_{k-1}}{}^{A_{k-1}} Z_{A_{k-1}}{}^{a_{k-2}} + \mu_{\alpha_{k-1}}{}^{a_{k-2}a_0} G_{a_0} (-)^{\varepsilon_{\alpha_{k-2}}}. \quad (10.19)$$

The independent reducibility equations of the next order must be equivalent to (10.19), since (10.19) are the only relations among the $Z_{\alpha_{k-1}}{}^{a_{k-2}}$ (the $Z_{A_{k-1}}{}^{a_{k-2}}$ are independent). This shows that the number of independent reducibility functions $Z_{A_k}{}^{a_{k-1}}$ of order k is equal to the number of dependent functions $Z_{\alpha_{k-1}}{}^{a_{k-2}}$ of the previous order $k-1$ (range of A_k = range of α_{k-1}). Furthermore, the matrix $Z_{A_k}{}^{\alpha_{k-1}}$ is invertible because the equations

$$Z_{A_k}{}^{\alpha_{k-1}} Z_{\alpha_{k-1}}{}^{a_{k-2}} + Z_{A_k}{}^{A_{k-1}} Z_{A_{k-1}}{}^{a_{k-2}} = (-)^{\varepsilon_{a_{k-2}}} C_{A_k}{}^{a_{k-2}a_0} G_{a_0}$$

are required to be complete and, hence, must imply (10.19).

10.2.4. Ambiguity in the Reducibility Functions

Given the constraint surface, the functions G_{a_0} are only determined modulo

$$G_{a_0} \rightarrow \bar{G}_{a_0} = M_{a_0}{}^{b_0} G_{b_0}. \quad (10.20)$$

The matrix $M_{a_0}{}^{b_0}$ is assumed to be invertible (see discussion of §5.2.1).

Given G_{a_0}, the reducibility functions of the first stage $Z_{a_1}{}^{a_0}$ are in turn only determined modulo

$$Z_{a_1}{}^{a_0} \rightarrow \bar{Z}_{a_1}{}^{a_0} = M_{a_1}{}^{b_1} Z_{b_1}{}^{a_0} + \nu_{a_1}{}^{a_0 b_0} G_{b_0} (-)^{\varepsilon_{b_0}} \quad (10.21a)$$

with

$$\nu_{a_1}{}^{a_0 b_0} = (-)^{(\varepsilon_{a_0}+1)(\varepsilon_{b_0}+1)} \nu_{a_1}{}^{b_0 a_0}. \quad (10.21b)$$

The matrix $M_{a_1}^{b_1}$ carries some ambiguity, since $M_{a_1}^{b_1}$ and $M_{a_1}^{b_1} + \sigma_{a_1}^{b_2} Z_{b_2}^{b_1} + t_{a_1}{}^{b_1 b_0} G_{b_0}$ both lead to the same $\bar{Z}_{a_1}{}^{a_0}$. This enables one to arrange so that $M_{a_1}{}^{b_1}$ is invertible everywhere in phase space (just as in §5.2.1). We assume that this condition can be—and has been—achieved in a smooth way.

Similarly, given the reducibility functions of order $\leq k-1$, the reducibility functions $Z_{a_k}{}^{a_{k-1}}$ of order k are determined up to

$$Z_{a_k}{}^{a_{k-1}} \rightarrow \bar{Z}_{a_k}{}^{a_{k-1}} = M_{a_k}{}^{b_k} Z_{b_k}{}^{a_{k-1}} + \nu_{a_k}{}^{a_{k-1} b_0} G_{b_0}, \quad (10.22)$$

where $M_{a_k}{}^{b_k}$ is again taken to be invertible.

Under the redefinitions (10.20)–(10.22), the functions $C_{a_k}{}^{a_{k-2}a_0}$ appearing in (10.15) transform in a manner that can straightforwardly be computed and that will not be reproduced here. Furthermore, even for fixed $Z_{a_k}{}^{a_{k-1}}$, the functions $C_{a_k}{}^{a_{k-2}a_0}$ carry the ambiguity of adding antisymmetric combinations of the constraints.

10.2.5. Canonical Form

One can take advantage of the freedom discussed in the previous section to bring the constraints and the functions $Z_{a_k}{}^{a_{k-1}}$ to an extremely simple form in a sufficiently small open region.

Indeed, we have assumed that the constraints G_{a_0} can be split into independent and dependent ones. The dependent constraints G_{α_0} can be expressed in terms of the independent ones G_{A_0} as $G_{\alpha_0} = N_{\alpha_0}{}^{A_0} G_{A_0}$. Redefining the dependent constraints as $\bar{G}_{\alpha_0} = G_{\alpha_0} - N_{\alpha_0}{}^{A_0} G_{A_0}$, one gets $\bar{G}_{\alpha_0} = 0$ (identically). Hence, one can assume

$$G_{a_0} = (G_{A_0}, G_{\alpha_0} = 0). \tag{10.23}$$

An independent set of reducibility functions of first order is then given—up to factors of i that are needed to comply with (10.15c) but that we will not write—by

$$Z_{A_1}{}^{a_0} = (0, \delta_{A_1}{}^{\alpha_0}), \tag{10.24}$$

where A_1 ranges over the same values as α_0 does. The remaining (dependent) reducibility functions of first order can be taken to vanish,

$$Z_{\alpha_1}{}^{a_0} = 0. \tag{10.25}$$

Going on in the same fashion, one sees that at each stage, the reducibility functions can be taken to be

$$Z_{A_k}{}^{a_{k-1}} = (0, \delta_{A_k}{}^{\alpha_{k-1}}), \tag{10.26a}$$

$$Z_{\alpha_k}{}^{a_{k-1}} = 0, \tag{10.26b}$$

where the range of A_k is equal to the range of α_{k-1}.

10.3. THE KOSZUL–TATE DIFFERENTIAL

10.3.1. Nontrivial Cycles and How to Kill Them

As in the irreducible case, the first step in the BRST construction is to implement the restriction to the constraint surface Σ. To that end, one tentatively defines the differential δ as

$$\delta z^A = 0, \qquad \text{antigh}\, z^A = 0, \tag{10.27a}$$

$$\delta \mathcal{P}_{a_0} = -G_{a_0}, \qquad \text{antigh}\, \mathcal{P}_{a_0} = 1, \qquad \varepsilon(\mathcal{P}_{a_0}) = \varepsilon_{a_0} + 1, \tag{10.27b}$$

$$\delta(AB) = A\delta B + (-)^{\varepsilon_B}(\delta A)B, \tag{10.27c}$$

$$\text{antigh}\,(\delta A) = \text{antigh}\, A - 1, \qquad \varepsilon(\delta A) = \varepsilon(A) + 1, \tag{10.27d}$$

with $\mathcal{P}_{a_0}$ being the ghost momentum associated with the constraint G_{a_0} according to the same prescriptions as before. One still finds

$$\delta^2 = 0. \tag{10.28}$$

Furthermore, one easily checks that the cohomology of δ at antighost number 0 (no $\mathcal{P}_{a_0}$) is the same as in the irreducible case, $H_0(\delta) = C^\infty(\Sigma)$.

But new features emerge at higher antighost numbers, with the appearance of nontrivial cycles associated with the reducibility identities. The differential algebra $(\mathbb{C}\,[\mathcal{P}_{a_0}] \otimes C^\infty(P), \delta)$ is no longer a resolution of the algebra $C^\infty(\Sigma)$ of functions on the constraint surface. For instance, the antighost number one polynomials $Z_{a_1}{}^{a_0}\mathcal{P}_{a_0}$ are closed because of (10.10),

$$\delta(Z_{a_1}{}^{a_0}\mathcal{P}_{a_0}) = -Z_{a_1}{}^{a_0}G_{a_0} = 0, \tag{10.29a}$$

but they cannot be all exact,

$$Z_{a_1}{}^{a_0}\mathcal{P}_{a_0} \neq \delta N, \tag{10.29b}$$

for this would imply $Z_{a_1}{}^{a_0}\mathcal{P}_{a_0} \approx 0$; *i.e.*, $Z_{a_1}{}^{a_0} \approx 0$, and the theory would be irreducible.

Acyclicity of δ at higher ghost numbers is crucial for applying the main theorem of homological perturbation theory, so it should be recovered. The method by which one does so follows exactly the first pattern devised for the model of the introduction ["method (i)," Eq. (10.4a)]. Namely, one adds extra variables—one for each cycle $Z_{a_1}{}^{a_0}\mathcal{P}_{a_0}$—and extends δ to these new variables so that the original nontrivial cycles become trivial. So, one introduces as many new $\mathcal{P}_{a_1}$ as there are $Z_{a_1}{}^{a_0}$ and kills the cycles $Z_{a_1}{}^{a_0}\mathcal{P}_{a_0}$ by setting

$$\delta\mathcal{P}_{a_1} = -Z_{a_1}{}^{a_0}\mathcal{P}_{a_0}. \tag{10.30}$$

The inclusion of the minus sign is for further convenience. By definition of $Z_{a_1}{}^{a_0}$, $\delta^2\mathcal{P}_{a_1} = 0$. To preserve the grading properties of δ, one also demands

$$\varepsilon(\mathcal{P}_{a_1}) = \varepsilon_{a_1} = \varepsilon(Z_{a_1}{}^{a_0}\mathcal{P}_{a_0}) + 1, \tag{10.31a}$$

$$\text{antigh}\,\mathcal{P}_{a_1} = 2, \tag{10.31b}$$

where ε_{a_1} is given by Eq. (10.18).

With the introduction of the "ghost-of-ghost momenta," $\mathcal{P}_{a_1}$, all the homology of δ has been killed at antighost number +1: if $\delta M = 0$, antigh $(M) = 1$, then $M = \delta N$ with antigh $(N) = 2$. This simply follows from the completeness of the first-order reducibility functions $Z_{a_1}{}^{a_0}$.

Turn now to the homology at antighost number +2. Here again, one finds nontrivial cycles. These arise when the cycles $Z_{a_0}{}^{a_0}\mathcal{P}_{a_0}$ in (10.29a) are not independent. They are given by

$$Z_{a_2}{}^{a_1}\mathcal{P}_{a_1} + \tfrac{1}{2}C_{a_2}{}^{a_0b_0}\mathcal{P}_{b_0}\mathcal{P}_{a_0}. \tag{10.32}$$

The expression (10.32) is annihilated by δ as a result of (10.14) but is not δ-exact, for the coefficient of $\mathcal{P}_{a_1}$ in (10.32) should then vanish on the constraint surface. This is not the case unless the $Z_{a_1}{}^{a_0}$ are independent, in which case the $Z_{a_2}{}^{a_1}$ are superfluous.

One thus introduces further new "ghost-of-ghost momenta" $\mathcal{P}_{a_2}$, with the properties

$$-\delta\mathcal{P}_{a_2} = Z_{a_2}{}^{a_1}\mathcal{P}_{a_1} + \tfrac{1}{2}C_{a_2}{}^{a_0b_0}\mathcal{P}_{b_0}\mathcal{P}_{a_0}, \tag{10.33a}$$

$$\varepsilon(\mathcal{P}_{a_2}) = \varepsilon_{a_2} + 1 = \varepsilon(\delta\mathcal{P}_{a_2} + 1), \tag{10.33b}$$

$$\text{antigh}\,\mathcal{P}_{a_2} = 3. \tag{10.33c}$$

This extension of δ is again nilpotent. It turns out that the introduction of $\mathcal{P}_{a_2}$ is enough to completely kill the homology at antighost number 2 (Theorem 10.3 below).

The process goes along the same lines at higher antighost numbers. This follows from the next theorem.

Theorem 10.2. *For each reducibility identity, there is a δ-closed polynomial given by*

$$Z_{a_k}{}^{a_{k-1}}\mathcal{P}_{a_{k-1}} + M_{a_k}, \qquad a_k = 1, \ldots, m_k, \tag{10.34}$$

where M_{a_k} is some polynomial involving only $\mathcal{P}_{a_u}$, $u < k-1$.

Proof. The proof of this theorem is not very illuminating and is given in the appendix. Let us simply mention here that the explicit expression of M_{a_k} is tedious, except when $C_{a_s}{}^{a_{s-2}a_0} = 0$ for all s, in which case M_{a_k} can be taken to vanish.

To kill the nontrivial cycles contained among (10.34), one introduces m_k variables $\mathcal{P}_{a_k}$, such that

$$-\delta\mathcal{P}_{a_k} = Z_{a_k}{}^{a_{k-1}}\mathcal{P}_{a_{k-1}} + M_{a_k}, \tag{10.35a}$$

$$\text{antigh}\,(\mathcal{P}_{a_k}) = k+1 \tag{10.35b}$$

$$\begin{aligned}\varepsilon(\mathcal{P}_{a_k}) &= \varepsilon(\delta\mathcal{P}_{a_k}) + 1\\ &= \varepsilon_{a_k} + k + 1 \qquad (\text{mod } 2).\end{aligned} \tag{10.35c}$$

These definitions preserve nilpotency and make the expression (10.34) manifestly δ-exact.

10.3.2. Homology of δ

The definition (10.35a) does more than just remove the cycles (10.34). It completely kills the homology of δ at antighost number greater than zero. Indeed, one has

Theorem 10.3. *The Koszul–Tate complex defined by*

$$\begin{aligned} \delta z^A &= 0, \\ \delta \mathcal{P}_{a_0} &= -G_{a_0}, \\ \delta \mathcal{P}_{a_k} &= -Z_{a_k}{}^{a_{k-1}} \mathcal{P}_{a_{k-1}} - M_{a_k}, \qquad k = 1, 2, \ldots, \end{aligned} \tag{10.36}$$

yields a resolution of the algebra $C^\infty(\Sigma)$ of functions on the constraint surface,

$$H_0(\delta) = C^\infty(\Sigma), \tag{10.37a}$$

$$H_k(\delta) = 0, \qquad k \neq 0, \tag{10.37b}$$

as in the irreducible case.

Proof. The proof is again technical and given in the appendix.

Remark. Given the $Z_{a_k}{}^{a_{k-1}}$, the M_{a_k} are not unique but defined up to a δ-exact term, $M_{a_k} \to M_{a_k} + \delta N_{a_k}$ (see appendix). This change can be absorbed in a redefinition of the generators $\mathcal{P}_{a_k}$ as $\mathcal{P}_{a_k} \to \mathcal{P}_{a_k} + N_{a_k}$. Similarly, the freedom (10.22) amounts to change $\mathcal{P}_{a_k}$ as $\mathcal{P}_{a_k} \to M_{a_k}{}^{b_k} \mathcal{P}_{b_k} + \nu_{a_k}{}^{a_{k-1} b_0} \mathcal{P}_{a_{k-1}} \mathcal{P}_{b_0}$. Hence, the Koszul–Tate differential δ is unique up to a redefinition of the momenta conjugate to the ghosts of ghosts.

10.4. MORE ON THE LONGITUDINAL EXTERIOR DIFFERENTIAL

10.4.1. Problem with the Definition of the Extended Phase Space

Now that the first building block of the BRST differential has been constructed, we can turn to the second ingredient, namely, the longitudinal exterior differential d. Here, also, one finds it necessary to introduce further variables over and above the ghosts of the irreducible case.

The Koszul–Tate differential has been defined in the tensor product algebra $\mathbb{C}[\mathcal{P}_{a_0}, \mathcal{P}_{a_1}, \ldots] \otimes C^\infty(P)$. In the irreducible case, this algebra has the same number of generators as the exterior longitudinal algebra. This enables one to define the extended phase space by just taking the tensor product of $\mathbb{C}[\mathcal{P}_a] \otimes C^\infty(P)$ with $\mathbb{C}[\eta^a]$, but this cannot be done so straightforwardly in the reducible case.

Indeed, the exterior longitudinal algebra is locally of the form $C^\infty(O) \otimes \mathbb{C}[\omega^{A_0}]$, where the ω^{A_0} constitute a basis of longitudinal 1-forms. The number of ω^{A_0} is equal to the number of independent constraints. Thus, there is a mismatch between the generators of the Koszul–Tate resolution and the generators of the longitudinal complex: there are less ω^{A_0} than there are $\mathcal{P}_{a_0}$, and there is no generator in the exterior complex corresponding to the higher order ghost momenta $\mathcal{P}_{a_1}, \mathcal{P}_{a_2}, \ldots$. Moreover, the exterior longitudinal algebra may globally fail to have a tensor product structure.

The difficulty can be solved by replacing the exterior longitudinal complex by a "model" for it. This model is a graded differential algebra with differential D that has the following properties:

(*i*) It possesses a tensor product structure.
(*ii*) It contains as many generators as the Koszul–Tate complex.
(*iii*) Its cohomology coincides with the cohomology of the exterior longitudinal d,

$$H^k(D) = H^k(d). \tag{10.38}$$

Because of (*iii*), the replacement of the exterior longitudinal complex by its model is permissible. The advantage of working with the model is that one can then define an extended phase space. For this reason, it is D that takes over the role of d in the reducible case.

We will construct in this section the appropriate model for d on the constraint surface Σ. Therefore, all the equalities in the section are equalities valid on Σ and generically valid only there. The question as to how to extend the results to the whole of phase space will come next when we discuss the BRST differential.

10.4.2. The Longitudinal Differential

We have seen in Chapter 5, §5.3.4, that the longitudinal forms α can be described in terms of an overcomplete set of "components' $\alpha_{a_0 b_0 \ldots d_0}$ in the overcomplete "basis" $\{X_{a_0}\}$. These components are given by

$$\alpha_{a_0 b_0 \ldots d_0} = (-)^{(\varepsilon_\alpha + 1)(\varepsilon_{a_0} + \cdots + \varepsilon_{d_0} + p)} \alpha(X_{a_0}, \ldots, X_{d_0}) \tag{10.39}$$

[see Eq. (7.22)] and are defined everywhere on Σ, since the vector fields X_{a_0} themselves are also defined everywhere on Σ.

The components $\alpha_{a_0 \ldots d_0}$ of the p-form α are not independent because the set $\{X_{a_0}\}$ is overcomplete. Rather, one finds from (10.9)

$$Z_{a_1}{}^{d_0} \eta^{c_0} \ldots \eta^{a_0} \alpha_{a_0 \ldots c_0 d_0} = 0 \tag{10.40}$$

[see Eq. (7.23)], as well as similar relations obtained by antisymmetrizations. The variables η^{a_0} are the formal 1-forms ω^{a_0} of §5.3.4. They are called "ghosts" because they will become again the ghosts of the BRST formalism. The ghosts η^{a_0} have parity $\varepsilon_{a_0} + 1$ and pure ghost number (= form degree) one,

$$\varepsilon(\eta^{a_0}) = \varepsilon_{a_0} + 1, \quad \text{pure gh}\, \eta^{a_0} = 1. \tag{10.41}$$

The longitudinal forms can thus be represented as polynomials in the η^{a_0} with coefficients that obey (10.40). They define a subalgebra of the algebra $C^\infty(\Sigma) \otimes \mathbb{C}\,[\eta^{a_0}]$ of arbitrary polynomials in η^{a_0}.

In $C^\infty(\Sigma) \otimes \mathbb{C}\,[\eta^{a_0}]$, the longitudinal derivation is given by (7.21), that is,

$$dF = (X_{a_0} F)\eta^{a_0}, \tag{10.42a}$$

$$d\eta^{a_0} = \tfrac{1}{2}(-)^{\varepsilon_{a_0}+\varepsilon_{b_0}} \eta^{b_0} \eta^{c_0} C_{c_0 b_0}{}^{a_0}. \tag{10.42b}$$

It is a differential ($d^2 = 0$) in the subalgebra of longitudinal forms, but for arbitrary elements of $C^\infty(\Sigma) \otimes \mathbb{C}\,[\eta^{a_0}]$—and, in particular, for η^{a_0}—d fails in general to be nilpotent. The central idea of the subsequent developments is to introduce new generators η^{a_k} and to define the new differential D in such a way that (*i*) it is nilpotent on arbitrary polynomials, and (*ii*) the algebraic condition (10.40) is automatically enforced in D-cohomology and need not be imposed by hand.

10.4.3. Auxiliary Differential Δ

The fact that d is nilpotent only in the subalgebra of longitudinal forms is somewhat reminiscent of the situation found in homological perturbation theory, where differentials modulo δ were encountered. This suggests that it should be useful to define an auxiliary differential Δ whose cohomology is given by the longitudinal forms and such that d is a differential modulo Δ.

The sought-for differential Δ should be such that the polynomials in η^{a_0} that fail to fulfill (10.40) are killed in Δ-cohomology. This is done by means of the second method mentioned in the introduction. Namely, one introduces new variables—the "ghosts of ghosts"—and one kills the polynomials in η^{a_0} not fulfilling (10.40) by arranging that they are not Δ-closed.

The precise rule for adding the new variables is that there should be one ghost-of-ghost η^{a_k} for each reducibility identity of order k, with the following gradings,

$$\text{pure gh}\,\eta^{a_k} = k+1, \qquad \varepsilon(\eta^{a_k}) = \varepsilon_{a_k} + k + 1. \tag{10.43}$$

The differential Δ is then defined by

$$\Delta F = 0, \qquad \Delta\eta^{a_k} = \eta^{a_{k+1}}\, Z_{a_{k+1}}{}^{a_k}(-)^{\varepsilon_{a_k}+k+1}, \tag{10.44}$$

where F is an arbitrary function on the constraint surface. The nilpotency of Δ (on Σ) follows from the reducibility identities (10.15). Furthermore, one has

$$\text{pure gh}\,(\Delta) = 1, \qquad \varepsilon(\Delta) = 1. \tag{10.45}$$

The definition (10.44) is such that a polynomial in η^{a_0} is Δ-closed if and only if it fulfills the condition (10.40). A stronger result actually holds, namely, the new generators η^{a_k} ($k \geq 1$) carry no cohomology.

Theorem 10.4. *The cohomology of the auxiliary differential Δ is isomorphic to the set of polynomials in $C^\infty(\Sigma) \otimes \mathbb{C}\,[\eta^{a_0}]$ that obey the algebraic condition (10.40), i.e.,*

$$H^*(\Delta) = \{\textit{longitudinal forms}\}. \tag{10.46}$$

Proof. See appendix.

10.4.4. Auxiliary Grading

It is useful to introduce a new auxiliary grading such that the cohomology of Δ lies in degree zero. One defines

$$\text{aux}\,(z^A) = 0 = \text{aux}\,(\eta^{a_0}), \tag{10.47a}$$

$$\text{aux}\,(\eta^{a_k}) = k. \tag{10.47b}$$

One then gets

$$\text{aux}\,(\Delta) = 1, \tag{10.47c}$$

and

$$H^k(\Delta) = 0, \qquad k \neq 0, \tag{10.48a}$$

$$H^0(\Delta) = \{\text{longitudinal forms}\}. \tag{10.48b}$$

One also finds

$$\text{pure gh}\,(A) = \text{aux}\,(A) + \deg\,(A),$$

where deg is the polynomial degree in the η^{a_k}'s, $\deg\eta^{a_k} = 1$. The differential Δ does not modify the polynomial degree, so $\deg(\Delta) = 0$.

The auxiliary grading manifestly shows that the differential complex $C^\infty(\Sigma)\otimes\mathbb{C}\,[\eta^{a_0},\eta^{a_1},\ldots]$ with differential Δ is a cohomological resolution of the algebra of longitudinal forms on the constraint surface.

10.4.5. The Differential D

The longitudinal derivative d defined by (10.42) is nilpotent when acting on longitudinal forms. One should thus be able to extend it to the new generators $\eta^{a_1},\eta^{a_2},\ldots$ in such a way that it becomes a differential modulo Δ and, pursuing the analogy with homological perturbation theory, there should be a true differential $D=\Delta+d+$ "more" such that $H^k(D)=H^k(d)$.

The sought-for differential D on $C^\infty(\Sigma)\otimes\mathbb{C}\,[\eta^{a_0},\eta^{a_1},\ldots]$ must fulfill the following conditions:

$$(i) \qquad \text{pure gh}\,D = \text{pure gh}\,d = \text{pure gh}\,\Delta = 1; \tag{10.49a}$$

$$(ii) \qquad D^2 = 0; \tag{10.49b}$$

$$(iii) \qquad D = \Delta + \overset{(0)}{D} + \sum_{k\leq -1} \overset{(k)}{D}, \tag{10.49c}$$

$$\text{aux}\,\big(\overset{(k)}{D}\big) = k, \tag{10.49d}$$

$$\overset{(0)}{D}F = dF, \quad \overset{(0)}{D}\eta^{a_0} = d\eta^{a_0}; \tag{10.49e}$$

where F belongs to $C^\infty(\Sigma)$;

$$(iv) \qquad H^k(D) = H^k(d) \quad (k = \text{pure ghost number}). \tag{10.49f}$$

From Theorem 8.3, a differential D fulfilling (i)–(iii) exists provided $\overset{(0)}{D}$ obeys

$$\overset{(0)}{D}\Delta + \Delta\overset{(0)}{D} = 0, \quad \overset{(0)}{D}{}^2 = -[\overset{(-1)}{D},\Delta]. \tag{10.49g}$$

for some $\overset{(-1)}{D}$. Indeed, in the proof of that part of Theorem 8.3, the fact that δ was a boundary operator rather than a coboundary operator was not used. One can thus repeat the same algebraic perturbative argument leading to the existence of a nilpotent derivation starting with Δ and $\overset{(0)}{D}$ instead of δ and d.

The direct verification of (10.49g) is outlined in Exercise 10.3. An indirect proof of the existence of D is given below within the BRST formalism. So, we now take the existence of D for granted and we proceed with the analysis of its cohomology, *i.e.*, with the proof of (iv).

10.4.6. Cohomology of D

Theorem 10.5. *Let D be a differential fulfilling (i)–(iii). Then, the cohomologies of D and d are isomorphic,*

$$H^k(D) \simeq H^k(d) \qquad (k = \textit{pure ghost number}). \tag{10.50}$$

Proof. (i) Let α be a closed longitudinal p-form. It defines an element of $C^\infty(\Sigma) \otimes \mathbb{C}\,[\eta^{a_0}]$—and hence also of $C^\infty(\Sigma) \otimes \mathbb{C}\,[\eta^{a_0}, \eta^{a_1}, \ldots]$—that verifies

$$\Delta\alpha = 0, \quad \overset{(0)}{D}\alpha = 0. \tag{10.51a}$$

[$\Delta\alpha = 0$ follows from the algebraic condition (10.40), see Theorem 10.4; $\overset{(0)}{D}\alpha = 0$ follows from (10.49e) and $d\alpha = 0$.] This implies

$$D\alpha = 0 \tag{10.51b}$$

because $\overset{(k)}{D}F = 0$ and $\overset{(k)}{D}\eta^{a_0} = 0$ for $k \leq -1$.

Furthermore, if α is exact,

$$\alpha = d\beta, \tag{10.52a}$$

where β is a longitudinal $p-1$-form ($\Delta\beta = 0$), one has

$$\alpha = D\beta \tag{10.52b}$$

because $\overset{(k)}{D}\beta \equiv 0$ for $k \leq -1$. Hence, α is also D-exact. This implies the existence of a map $\rho : H^k(d) \to H^k(D)$ sending any cohomological equivalence class of d on a cohomological class of D. One must prove that this map is an isomorphism.

(ii) That the map ρ is linear and preserves the product structure is obvious. One needs to check only that it is both surjective and injective. This is done by expanding the polynomials of $C^\infty(\Sigma) \otimes \mathbb{C}\,[\eta^{a_0}, \eta^{a_1}, \ldots]$ according to the auxiliary degree. The demonstration is similar to the proof of §8.4.3, but because Δ provides a cohomological resolution rather than a homological resolution, one needs to attack the calculation from the terms of higher auxiliary degree. The details are worked out in Exercise 10.4.

10.4.7. Conclusions

We have succeeded in constructing a model for the longitudinal exterior complex. This model is the algebra $C^\infty(\Sigma) \otimes \mathbb{C}\,[\eta^{a_0}, \eta^{a_1}, \ldots]$ with differential D. The extra variables $\eta^{a_1}, \eta^{a_2}, \ldots$ of the model—called "ghosts of ghosts"—are introduced so as to enforce the algebraic condition (10.40) where one passes to the cohomology.

The algebra $C^\infty(\Sigma) \otimes \mathbb{C}[\eta^{a_0}, \eta^{a_1}, \ldots]$ contains as many generators as the algebra $\mathbb{C}[\mathcal{P}_{a_0}, \mathcal{P}_{a_1}, \ldots] \otimes C^\infty(P)$ appearing in the Koszul–Tate resolution. This is not an accident because the same reducibility functions $Z_{a_{k+1}}{}^{a_k}$ determine both the nonindependence of the constraint functions and the nonindependence of the gauge transformations.

Because the cohomology of D is isomorphic to the cohomology of d, we will sometimes loosely refer in the sequel to D itself as being the actual exterior derivative, and to the complex $(C^\infty(\Sigma) \otimes \mathbb{C}[\eta^{a_0}, \eta^{a_1}, \ldots], D)$ as being the actual longitudinal complex.

10.5. BRST TRANSFORMATION

10.5.1. Extended Phase Space

Since there is a perfect matching between the ghosts, ghosts of ghosts,..., and the ghost momenta, one can extend the original phase space by introducing a canonical structure in $\mathbb{C}[\mathcal{P}_{a_0}, \mathcal{P}_{a_1}, \ldots] \otimes C^\infty(P) \otimes \mathbb{C}[\eta^{a_0}, \eta^{a_1}, \ldots]$, defined as follows

$$[\mathcal{P}_{a_k}, \eta^{b_k}] = -\delta_{a_k}{}^{b_k}. \qquad (10.53\text{a})$$

The Poisson brackets that are not written vanish with the exception of the ones in the original phase space, which are left unchanged. The ghosts, ghosts of ghosts, ... are taken to be real,

$$(\eta^{b_k})^* = \eta^{b_k}. \qquad (10.53\text{b})$$

This forces

$$(\mathcal{P}_{a_k})^* = -(-)^{\varepsilon_{a_k}+k} \mathcal{P}_{a_k}. \qquad (10.53\text{c})$$

Furthermore, the ghost number is defined again by

$$\text{gh}\, A = \text{pure gh}\, A - \text{antigh}\, A, \qquad (10.53\text{d})$$

so that

$$\text{gh}\, \eta^{b_k} = k+1 = -\text{gh}\, \mathcal{P}_k. \qquad (10.53\text{e})$$

The ghost number generator reads

$$\mathcal{G} = i \sum_k (k+1)\, \eta^{a_k} \mathcal{P}_{a_k} \qquad (10.53\text{f})$$

and verifies

$$[A, \mathcal{G}] = i(\text{gh}\, A) A, \qquad \mathcal{G}^* = -\mathcal{G}, \qquad (10.53\text{g})$$

as in the irreducible case.

10.5.2. Combining δ with D

With δ and D at hand, we have all the necessary building blocks for defining the BRST symmetry s. To that end, one first extends δ from the algebra $\mathbb{C}[\mathcal{P}_{a_0}, \mathcal{P}_{a_1}, \ldots] \otimes C^\infty(P)$ of polynomials in the $\mathcal{P}_{a_k}$ to the algebra $\mathbb{C}[\mathcal{P}_{a_0}, \mathcal{P}_{a_1}, \ldots] \otimes C^\infty(P) \otimes \mathbb{C}[\eta^{a_0}, \eta^{a_1}, \ldots]$ of polynomials in the $\mathcal{P}_{a_k}$ and the η^{a_k} by setting

$$\delta \eta^{a_k} = 0. \tag{10.54}$$

This is the only definition compatible with the condition $\operatorname{antigh} \delta = -1$.

Next, one extends D from $C^\infty(\Sigma) \otimes \mathbb{C}[\eta^{a_0}, \eta^{a_1}, \ldots]$ to $\mathbb{C}[\mathcal{P}_{a_0}., \mathcal{P}_{a_1}, \ldots] \otimes C^\infty(P) \otimes \mathbb{C}[\eta^{a_0}, \eta^{a_1}, \ldots]$ in such a way that D becomes a differential modulo δ. One then combines δ with D along the lines of homological perturbation theory,

$$s = \delta + D + \text{"more,"} \qquad s^2 = 0. \tag{10.55}$$

Theorem 8.3 guarantees that the cohomology of s is equal to the cohomology of D in $H_0(\delta)$, *i.e.*, to the longitudinal cohomology $H^*(d)$.

That it is possible to (i) extend D so that it is a differential modulo δ and (ii) achieve (10.55) so that s defines a canonical transformation in the bracket structure (10.53a) is most expediently proven by building directly the BRST generator,

$$sA = [A, \Omega], \tag{10.56}$$

as we now show.

10.5.3. Equations Determining the BRST Generator

The BRST generator must fulfill the following requirements.

$$(i) \qquad \operatorname{gh}(\Omega) = 1, \quad \varepsilon(\Omega) = 1, \quad \Omega^* = \Omega, \tag{10.57a}$$

$$(ii) \qquad \Omega = \eta^{a_0} G_{a_0} + \eta^{a_k} Z_{a_k}{}^{a_{k-1}} \mathcal{P}_{a_{k-1}} + \text{"more"}, \tag{10.57b}$$

$$(iii) \qquad [\Omega, \Omega] = 0, \tag{10.57c}$$

where "more" now refers to terms containing at least two η's and one $\mathcal{P}$, or two $\mathcal{P}$'s and one η.

With (10.57b), Ω reproduces through the Poisson bracket the terms linear in the $\mathcal{P}$'s in $\delta \mathcal{P}_{a_k}$. Similarly, it also reproduces $\Delta \eta^{a_k}$ and dF for $F \in C^\infty(P)$. As we shall see, this fact, combined with the nilpotency condition (10.57c), automatically implies that $s \equiv [\ ,\Omega]$ contains the full δ and D, *i.e.*, $[\ ,\Omega] = \delta + D + \sum_{k \geq 1} \overset{(k)}{s}$ with $\operatorname{antigh} \overset{(k)}{s} = k$. Accordingly, the conditions (10.57a)–(10.57c) constitute a complete set of independent requirements on Ω.

To investigate the nilpotency condition (10.57c), it is convenient to expand Ω according to the resolution degree (= antighost number), as in the proof of the main theorem of homological perturbation theory,

$$\Omega = \sum_{p\geq 0} \overset{(p)}{\Omega}, \tag{10.58a}$$

$$\text{antigh} \overset{(p)}{\Omega} = p. \tag{10.58b}$$

One finds, after elementary calculations (Exercise 10.5),

$$[\Omega, \Omega] = \sum_{p\geq 0} \overset{(p)}{B}, \tag{10.59a}$$

$$\text{antigh} \left(\overset{(p)}{B}\right) = p, \tag{10.59b}$$

$$\overset{(p)}{B} = \sum_{k=0}^{p} [\overset{(p-k)}{\Omega}, \overset{(k)}{\Omega}]_{\text{orig}} + \sum_{k=0}^{p+1} \sum_{s=0}^{k} [\overset{(p-k+s+1)}{\Omega}, \overset{(k)}{\Omega}]_{\eta_{a_s}, \mathcal{P}_{a_s}}, \tag{10.59c}$$

where the bracket $[\ ,\ ]_{\text{orig}}$ denotes again the bracket in the original phase space z^A, which does not modify the antighost number, and where $[\ ,\ ]_{\eta_{a_s}, \mathcal{P}_{a_s}}$ is the bracket with respect to the pair $\eta_{a_s}, \mathcal{P}_{a_s}$ only. This latter bracket eliminates one $\mathcal{P}_{a_s}$ and thus reduces the antighost number by $s+1$.

The nilpotency condition $[\Omega, \Omega] = 0$ is thus equivalent to the set of equations

$$\overset{(p)}{B} = 0, \quad p = 0, 1, 2, \ldots, \tag{10.60}$$

where $\overset{(p)}{B}$ involves the structure functions $\overset{(i)}{\Omega}$ up to order $p+1$. The term containing $\overset{(p+1)}{\Omega}$ in $\overset{(p)}{B}$ reads explicitly

$$2 \sum_{s=0}^{p} [\overset{(p+1)}{\Omega}, \overset{(s)}{\Omega}]_{\eta_{a_s}, \mathcal{P}_{a_s}}. \tag{10.61}$$

The sum over s in (10.61) is cut at p because the term

$$[\overset{(p+1)}{\Omega}, \overset{(p+1)}{\Omega}]_{\eta_{a_{p+1}}, \mathcal{P}_{a_{p+1}}},$$

which would seem to be present in (10.59c), is actually zero: $\overset{(p+1)}{\Omega}$, being of antighost number $p+1$, cannot contain $\mathcal{P}_{a_{p+1}}$, which is of antighost number $p+2$.

10.5.4. Existence of the BRST Generator

To prove the existence of the BRST generator, we proceed by induction. Assume that the structure functions $\overset{(i)}{\Omega}$ have been constructed up to order p, *i.e.*,

$$\overset{(k)}{B} = 0, \qquad k = 0, 1, \ldots, p-1. \tag{10.62}$$

Furthermore, assume that for $i \leq L$ (and $\leq p$) the dependence of $\overset{(i)}{\Omega}$ on η^{a_i} is given by

$$\overset{(i)}{\Omega} = -\eta^{a_i}\delta\mathcal{P}_{a_i} + \text{“more,”} \qquad i = 1, \ldots, \min(p, L), \tag{10.63}$$

where "more" does not contain η^{a_i}. This incorporates the "boundary conditions" (10.57b) on Ω. If i exceeds the order of reducibility ($i > L$), there is no η^{a_i} and no assumption of the kind (10.63) is needed. The structure function $\overset{(0)}{\Omega}$ is given by $\eta^{a_0}G_{a_0}$ and obeys (10.63). We need to show that the equation $\overset{(p)}{B} = 0$ possesses a solution $\overset{(p+1)}{\Omega}$ that can be chosen to fulfill the condition (10.63) at order $p+1$.

Let R_p be the sum of the already known structure functions $\overset{(i)}{\Omega}$,

$$R_p = \sum_{s=0}^{p} \overset{(i)}{\Omega}. \tag{10.64}$$

Because of (10.62), the bracket $[R_p, R_p]$ starts at order $\geq p$. That is, one has

$$[R_p, R_p] = 2\overset{(p)}{D} + \text{“higher order,”} \tag{10.65a}$$

where $\overset{(p)}{D}$ is of antighost number p and is explicitly given by

$$\overset{(p)}{D} = \frac{1}{2}\left\{\sum_{k=0}^{p}\left[\overset{(p-k)}{\Omega}, \overset{(k)}{\Omega}\right]_{\text{orig}} + \sum_{k=1}^{p}\sum_{s=0}^{k-1}\left[\overset{(p-k+s+1)}{\Omega}, \overset{(k)}{\Omega}\right]_{\eta_{a_s}, \mathcal{P}_{a_s}}\right\}. \tag{10.65b}$$

The unwritten higher order terms in (10.65a) are of antighost number $\geq p+1$. Moreover, because of (10.63), one finds

$$[A, R_p] = \sum_{i=0}^{p} \frac{\partial^R A}{\partial \mathcal{P}_{a_i}}\delta\mathcal{P}_{a_i} + \text{“more”} = \delta A + \text{“more”} \tag{10.65c}$$

for any function A of antighost number $\leq p+1$. "More" stands in (10.65c) for terms of antighost number $\geq$ antighost A.

Consequently, the equation $\overset{(p)}{B} = 0$, equivalent to the requirement that $[R_p + \overset{(p+1)}{\Omega}, R_p + \overset{(p+1)}{\Omega}]$ should start at order $p+1$, reads

$$\delta \overset{(p+1)}{\Omega} + \overset{(p)}{D} = 0. \tag{10.65d}$$

If $p = 0$, this equation possesses a solution $\overset{(1)}{\Omega}$ because the constraints are first class, and so $\overset{(0)}{D} \approx 0$. For $p \geq 1$, Eq. (10.65d) possesses a solution $\overset{(p+1)}{\Omega}$ if and only if $\delta \overset{(p)}{D} = 0$. This follows from the acyclicity of D at antighost number ≥ 1. But the condition $\delta \overset{(p)}{D} = 0$ is a consequence of (10.65a) and (10.65c) and of the Jacobi identity for the Poisson bracket, which implies $\big[[R_p, R_p], R_p\big] = 0$. Hence, $\overset{(p)}{D}$ is δ-exact.

Furthermore, because $\overset{(p)}{D}$ does not contain $\eta^{a_{p+1}}$, one can adjust the coefficient of $\eta^{a_{p+1}}$ in $\overset{(p+1)}{\Omega}$ to be $-\delta \mathcal{P}_{a_{p+1}}$,

$$\overset{(p+1)}{\Omega} = -\eta^{a_{p+1}} \delta \mathcal{P}_{a_{p+1}} + \text{"more,"} \tag{10.65e}$$

since $\eta^{a_{p+1}} \delta \mathcal{P}_{a_{p+1}} = \delta(\eta^{a_{p+1}} \mathcal{P}_{a_{p+1}})$ disappear from (10.65d).

Therefore, the properties (10.62) and (10.63), if true for $\overset{(i)}{\Omega}$ with $i \leq p$, pass on to $\overset{(p+1)}{\Omega}$. This proves the existence of the BRST generator Ω fulfilling all the conditions (10.57). [The reality condition $\Omega = \Omega^*$ easily follows from (10.15c) and from (10.53b) and (10.53c); see Exercise 10.6.]

The existence of Ω can be used to infer the existence of the differential D fulfilling (10.49). This differential is simply given by

$$DA = [A, \Omega]|_{\mathcal{P}=0}, \tag{10.66}$$

for any polynomial A in η^{a_i} with coefficients that are arbitrary phase space functions. It is clear that $DA \approx 0$ if $A \approx 0$, so (10.66) defines a derivation in $C^\infty(\Sigma) \otimes \mathbb{C}[\eta^{a_0}, \eta^{a_1}, \ldots]$, which is nilpotent because $[\Omega, \Omega] = 0$. The other properties (10.49) follow from (10.57b) (and from Theorem 10.5).

10.5.5. Uniqueness of the BRST Generator

It has been proven that the BRST generator exists. How unique is the solution?

To analyze this issue, consider two different, nilpotent BRST generators Ω, Ω' associated with the same constraint hypersurface but with possibly different G_{a_0}, $Z_{a_1}{}^{a_0}$, $Z_{a_2}{}^{a_1}$, etc.... We assume here that the number of reducibility functions is taken to be the same at each order for both Ω and Ω'. This guarantees that the extended phase spaces are identical. What happens when one increases the reducibility by adding trivial relations to the ones already at hand will be analyzed in the next chapter (§11.3.1).

Suppose that Ω and Ω' start differing at order $p \geq 0$. That is, the terms of antighost number $k < p$ (if any) in Ω and Ω' coincide, but $\overset{(p)}{\Omega} \neq \overset{(p)}{\Omega'}$. For $p \geq 1$, the equations determining $\overset{(p)}{\Omega}$ and $\overset{(p)}{\Omega'}$ read

$$\delta \overset{(p)}{\Omega} = \overset{(p-1)}{D} \qquad (p \geq 1), \tag{10.67a}$$

$$\delta \overset{(p)}{\Omega'} = \overset{(p-1)}{D} \qquad (p \geq 1), \tag{10.67b}$$

with the same right-hand sides and with the same δ, since δ acting on polynomials of antighost number $\leq p$ is entirely determined by $\overset{(i)}{\Omega}$, $i \leq p-1$. This implies

$$\overset{(p)}{\Omega'} = \overset{(p)}{\Omega} + \delta_\Omega \overset{(p+1)}{M}. \tag{10.68}$$

We have added the index Ω to δ in the right-hand side of (10.68) to emphasize that δ_Ω and $\delta_{\Omega'}$ may differ at order $p+1$, since the Z's and the other structure functions determining δ may differ at that order. Relation (10.68) also holds for $p = 0$ because $G'_{a_0} = M_{a_0}{}^{b_0} G_{b_0} = G_{a_0} + (M_{a_0}{}^{b_0} - \delta_{a_0}{}^{b_0}) G_{b_0}$.

If $\overset{(p)}{\Omega'} - \overset{(p)}{\Omega}$ is infinitesimally small, the change (10.68) is induced by the canonical transformation generated by $\overset{(p+1)}{M}$. Indeed, one finds from $s = \delta +$ "more,"

$$[\overset{(p+1)}{M}, \Omega] = \delta_\Omega \overset{(p+1)}{M} + \text{"higher order,"} \tag{10.69}$$

which is the desired result. The case of a finite difference $\overset{(p)}{\Omega'} - \overset{(p)}{\Omega}$ is obtained by exponentiation and discussed in Exercise 10.7. We can thus conclude that the BRST generator is unique up to a canonical transformation in the extended phase space.

The uniqueness of the BRST charge and the abelianization theorem can be used to yield a different proof of the existence of Ω (see Exercise 10.8).

10.6. CONCLUSIONS

We have shown in this chapter that the BRST differential can be constructed for arbitrary reducible gauge theories. As in the irreducible case, the BRST transformation is gauge independent, since at no stage is a gauge condition ever necessary. Furthermore, the BRST cohomology at ghost number zero is equal to the set of classical observables.

Cohomological considerations constitute the guiding thread of the entire discussion. In particular, they provide the rationale behind the enlargement of the ghost spectrum, which would otherwise appear to follow practical rules without definite theoretical foundations: the ghosts of ghosts and their momenta are introduced for the purpose of killing unwanted cohomology of the building blocks δ and D. This enables one to apply the methods of homological perturbation theory, namely, to combine δ with D into $s = \delta + D + \cdots$ in such a way that $s^2 = 0$ and $H^*(s) = H^*(D)$ [in $H_0(\delta)$] $= H^*(d)$.

The results of this chapter apply to the case of infinite order of reducibility ($L = \infty$) if the number of reducibility identities (and, hence, the number of ghosts and of ghost momenta) is finite at each order. Indeed, the recursive equations determining $\overset{(k)}{\Omega}$ involve then a finite number of terms and remain well defined. So, Ω itself is well defined as an infinite formal series. Similarly, the inductive proof that the BRST cohomology is isomorphic to $H^*(d)$ still holds.

Finally, one may wonder what happens if the regularity conditions are not fulfilled. This situation may arise in two different ways, for either of which a BRST differential with correct cohomology is not guaranteed to exist.

(*i*) The constraint functions are badly chosen [*e.g.*, $(p_1)^2 = 0$ instead of $p_1 = 0$]. Then, in all the examples known to us, the BRST cohomology is physically wrong (see Exercise 10.10).

(*ii*) The constraint equations do not define a smooth submanifold, so that no choice of the constraint functions can fulfill the regularity conditions. That case is largely unexplored at the moment of this writing.

APPENDIX 10.A. PROOFS OF THEOREMS 10.1 THROUGH 10.4

10.A.1. δ-Covering of Phase Space

An important technical tool necessary for the proofs of Theorems 10.1 through 10.4 is the concept of δ-covering of phase space.

A δ-covering in the reducible case is defined as in the irreducible case (§9.A.1), with the sole exception that it is not required that the G_{a_0} can be taken as coordinates on O_i. This is, in fact, impossible because the G_{a_0} are not independent. Rather, it is demanded that on O_i the splitting of the constraints $G_{a_0} = (G_{A_0}, G_{\alpha_0})$ into independent and dependent ones can be done and that the independent constraint functions G_{A_0} can be taken as first coordinates of a regular coordinate system. It is also demanded that on the open sets O_i, the Z's can be separated into independent and dependent reducibility functions, with a single separation on a given O_i.

10.A.2. Proof of Theorem 10.1

The strategy followed in proving that the conditions

$$\lambda^{a_0} G_{a_0} = 0, \qquad \lambda^{a_0} \approx 0, \tag{10.A.1a}$$

imply

$$\lambda^{a_0} = \nu^{a_0 b_0} G_{b_0} (-)^{\varepsilon_{b_0}}, \qquad \nu^{a_0 b_0} = (-)^{(\varepsilon_{a_0}+1)(\varepsilon_{b_0}+1)} \nu^{b_0 a_0}, \tag{10.A.1b}$$

follows the pattern adopted in Appendix 9.A. One first proves the required property on each individual open set of a δ-covering and then glues together the results obtained on the different open sets by means of a partition of unity.

That (10.A.1a) implies (10.A.1b) is obvious on the open sets V_α. One simply takes

$$\nu^{11} = 0, \quad \nu^{a_0 1} = \frac{\lambda^{a_0}}{G_1} = (-)^{(\varepsilon_{a_0}+1)} \nu^{1 a_0}, \quad \nu^{a_0 b_0} = 0, \quad a_0 \neq 1, \ b_0 \neq 1,$$

where G_1 is the invertible bosonic constraint function on V_α.

On O_i, the proof of Theorem 10.1 is also immediate. One has $G_{a_0} = (G_{A_0}, \ G_{\alpha_0} = N_{\alpha_0}{}^{A_0} G_{A_0})$ where the G_{A_0} are independent and hence such that

$$F \approx 0 \Rightarrow F = F^{A_0} G_{A_0}, \tag{10.A.2a}$$

$$\rho^{A_0} G_{A_0} = 0 \Rightarrow \rho^{A_0} = \sigma^{A_0 B_0} G_{B_0} (-)^{\varepsilon_{B_0}}, \\ \sigma^{A_0 B_0} = (-)^{(\varepsilon_{A_0}+1)(\varepsilon_{B_0}+1)} \sigma^{B_0 A_0} \tag{10.A.2b}$$

(see analysis of irreducible case). From $\lambda^{\alpha_0} \approx 0$, it follows that λ^{α_0} is a combination of the independent constraint G_{A_0}. Inserting this relation into $\lambda^{a_0} G_{a_0} = 0$ and using (10.A.2b) as well as $G_{\alpha_0} = N_{\alpha_0}{}^{A_0} G_{A_0}$, one then easily arrives at the required result.

10.A.3. Proofs of Theorems 10.2 and 10.3

Theorems 10.2 and 10.3 are proved by induction. Let K_k be the algebra $\mathbb{C}\,[\mathcal{P}_{a_0},\ldots,\mathcal{P}_{a_k}] \otimes C^\infty(P)$ of polynomials in $\mathcal{P}_{a_0},\ldots,\mathcal{P}_{a_k}$ with coefficients that are phase space functions. If $k \geq L$ where L is the order of reducibility, then $K_k = K_L$. Suppose that one has found recursively for $s = 0, 1, 2, \ldots, k$, m_s polynomials M_{a_s} belonging to K_{s-2} and such that the equations

$$\delta \mathcal{P}_{a_0} = -G_{a_0}, \tag{10.A.3a}$$

$$\delta \mathcal{P}_{a_s} = -Z_{a_s}{}^{a_{s-1}} \mathcal{P}_{a_{s-1}} - M_{a_s}, \qquad 1 \leq s \leq k, \tag{10.A.3b}$$

define a differential in K_k, *i.e.*, $\delta^2 \mathcal{P}_{a_s} = 0$. The polynomial M_{a_1} identically vanishes [Eq. (10.30)], while M_{a_2} is given by (10.33a). Theorem 10.2 states that δ can be extended from K_k to K_{k+1}, since it can be equivalently reformulated as

Theorem 10.2. *There exist m_{k+1} polynomials $M_{a_{k+1}}$ belonging to K_{k-1} such that the elements*

$$C_{a_{k+1}} = Z_{a_{k+1}}{}^{a_k} \mathcal{P}_{a_k} + M_{a_{k+1}}, \qquad a_{k+1} = 1, \ldots, m_{k+1}, \tag{10.A.4a}$$

are δ-closed,

$$\delta C_{a_{k+1}} = 0. \tag{10.A.4b}$$

Hence, the equation $\delta \mathcal{P}_{a_{k+1}} = -C_{a_{k+1}}$ defines, together with (10.A.3), a differential in K_{k+1}.

Theorem 10.2 is proved by relying on the following lemma.

Lemma 10.A.1. *For each $i < k$, any δ-closed polynomial in K_i of positive antighost number is δ-exact in K_{i+1}.*

Proof. Again, it is enough to demonstrate this property on each individual open set of a δ-covering, as in Appendix 9.A.

The differential δ is manifestly acyclic on V_α because the function 1 is δ-exact, $1 = \delta(-\mathcal{P}_1/G_1)$, $G_1 \neq 0$. Let us consider an open set O_i where one can split the constraints G_{a_0} and the reducibility functions $Z_{a_i}{}^{a_{i+1}}$ into independent and dependent ones. By elementary redefinitions of the generators $\mathcal{P}_{a_i}$, one can rewrite δ as

$$\delta \mathcal{P}_{A_0} = -G_{A_0}, \qquad \delta \mathcal{P}'_{\alpha_0} = 0, \tag{10.A.5a}$$

$$\delta \mathcal{P}'_{A_i} = Z'_{A_i}{}^{\alpha_{i-1}} \mathcal{P}'_{\alpha_{i-1}}, \qquad \delta \mathcal{P}'_{\alpha_i} = 0 \qquad (1 \leq i \leq k), \tag{10.A.5b}$$

with $Z'_{A_i}{}^{\alpha_{i-1}} = \delta_{A_i}{}^{\alpha_{i-1}}$ (see §10.2.5). The differential defined by (10.A.5) has the same form as the differential (8.30c) of Theorem 8.2: the $\mathcal{P}'_{A_i}$'s $(i = 1, 2, \ldots, k)$ kill the $\mathcal{P}'_{\alpha_{i-1}}$, while $\mathcal{P}_{A_0}$ kills G_{A_0}. Only $\mathcal{P}'_{\alpha_k}$ and the functions on the constraint surface remain in the homology of the complex (K_k, δ). The lemma then follows, since $\mathcal{P}_{\alpha_k}$ is eliminated by the condition $i < k$, while the functions on the constraint surface are eliminated by the condition of positive antighost number.

Once the lemma is proved, Theorem 10.2 is demonstrated in two lines. Indeed, the polynomial $\delta[Z_{a_{k+1}}{}^{a_k}\mathcal{P}_{a_k} - (-)^k C_{a_{k+1}}{}^{a_{k-1}b_0}\mathcal{P}_{b_0}\mathcal{P}_{a_{k-1}}]$ belongs to K_{k-2} by (10.15). Furthermore, it is manifestly δ-closed. Hence, by the lemma, there exists a polynomial $\bar{M}_{a_{k+1}}$ in K_{k-1} such that

$$\delta[Z_{a_{k+1}}{}^{a_k}\mathcal{P}_{a_k} - (-)^k C_{a_{k+1}}{}^{a_{k-1}b_0}\mathcal{P}_{b_0}\mathcal{P}_{a_{k-1}}] = -\delta\bar{M}_{a_{k+1}}, \quad \bar{M}_{a_{k+1}} \in K_{k-1} \tag{10.A.6a}$$

[note that $\bar{M}_{a_{k+1}} \neq -Z_{a_{k+1}}{}^{a_k}\mathcal{P}_{a_k} - (-)^k C_{a_{k+1}}{}^{a_{k-1}b_0}\mathcal{P}_{b_0}\mathcal{P}_{a_{k-1}}$, since this latter expression does not belong to K_{k-1}]. The searched-for M_{a_k} is

$$M_{a_{k+1}} = -(-)^k C_{a_{k+1}}{}^{a_{k-1}b_0}\mathcal{P}_{b_0}\mathcal{P}_{a_{k-1}} + \bar{M}_{a_{k+1}} \in K_{k-1}. \tag{10.A.6b}$$

The extension of δ to K_{k+1} preserves the property of Lemma 10.A.1. This leads to Theorem 10.3, since any polynomial N of antighost number g belongs to K_g. Therefore, if $\delta N = 0$ and $g \neq 0$, then $N = \delta\tilde{N}$ with $\tilde{N} \in K_{g+1}$. Hence, $H_g(\delta) = 0$ for $g \neq 0$. [The condition $H_0(\delta) = C^\infty(\Sigma)$ was proved in the text.]

10.A.4. Proof of Theorem 10.4

Theorem 10.4 is first proved on the individual open sets $O_i \cap \Sigma$ covering the constraint surface. On these open sets, one can rewrite Δ as

$$\Delta z^A = 0, \quad \Delta\eta'^{A_k} = 0, \quad \Delta\eta'^{\alpha_k} = Z'_{A_{k+1}}{}^{\alpha_k}\,\eta^{A_{k+1}}(-)^{\varepsilon_{\alpha_k}+k+1} \tag{10.A.7}$$

$(Z'_{A_{k+1}}{}^{\alpha_k} = \delta_{A_{k+1}}{}^{\alpha_k})$ by a linear redefinition of the generators η^{a_i}. This shows that Δ is of the form of the differential in Theorem 8.2: the η^{A_k} with $k \geq 1$ are killed in cohomology by $\eta^{\alpha_{k-1}}$, leaving one with only z^A and η^{A_0}. The equality (10.46) holds, therefore, in $O_i \cap \Sigma$, since the algebraic condition (10.40) defining the longitudinal forms is equivalent to η^{α_0}-independence when $Z'_{A_1}{}^{\alpha_0} = \delta_{A_1}{}^{\alpha_0}$. The theorem is then extended to the whole of the constraint surface Σ by means of a partition of unity. The glueing preserves Δ-closedness and Δ-exactness because, again, Δ annihilates the phase space functions and, in particular, the functions of the partition of unity. One also proves in the same manner that the cohomology in the space of derivations is trivial for $k \neq 0$, $\mathcal{H}_k(\Delta) = 0$ (k = auxiliary degree $\neq 0$).

EXERCISES. CHAPTER TEN

10.1. Consider two dynamical Lorentz vectors n^μ, r^μ with respective conjugate momenta π_μ, σ_μ. Let the constraints be $n^2 = 0, r^2 = 0, n \cdot r = -1, P^{\mu\nu}\pi_\nu = 0, P^{\mu\nu}\sigma_\nu = 0, \pi \cdot n = \sigma \cdot r$. Here, $P^{\mu\nu} = \eta^{\mu\nu} + n^\mu r^\nu + n^\nu r^\mu$ is the projector on the subspace orthogonal to n and r.

(a) Check that the constraints are first class.

(b) Compute the reducibility functions.

(c) Show that the system is pure gauge, *i.e.*, carries no physical degrees of freedom.

(d) Write the BRST charge.

10.2. Show that the reality condition (10.15c) on $Z_{a_k}{}^{a_{k-1}}$ is consistent, *i.e.*, yields a pure real or pure imaginary left-hand side of (10.15a). [Hint: Repeat the derivation of formula (7.17d).]

10.3. This exercise provides a direct proof of the existence of the differential D of §10.4.5.

(a) Show that the expansion of D can contain only terms of auxiliary degree ≤ 1 and that $\overset{(k)}{D} F = 0$, $\overset{(k)}{D} \eta^{a_0} = 0$ for $k \leq -1$. Here $F \in C^\infty(\Sigma)$. (Hint: Use $\text{aux}\, P = \text{pure gh}\, P - \deg P$, where $\deg P$ is the polynomial degree in the η^{a_k}'s, $\deg \eta^{a_k} = 1$.)

(b) Verify $(\overset{(0)}{D}\Delta + \Delta\overset{(0)}{D})F = 0$ and $\overset{(0)}{D}{}^2 F = -[\overset{(-1)}{D}, \Delta]F$ directly.

(c) Show next that the condition $(\overset{(0)}{D}\Delta + \Delta\overset{(0)}{D})\eta^{a_i} = 0$ and $\overset{(0)}{D}{}^2\eta^{a_i} = -[\overset{(-1)}{D}, \Delta]\eta^{a_i}$ can be used to define successively $\overset{(0)}{D}\eta^{a_{i+1}}$ and $\overset{(-1)}{D}\eta^{a_{i+1}}$. (Hint: Observe that these equations take the form $Z_{a_{i+1}}{}^{a_i}\overset{(0)}{D}\eta^{a_{i+1}} =$ given, $Z_{a_{i+1}}{}^{a_i}\overset{(-1)}{D}\eta^{a_{i+1}} =$ given. Prove that they possess solutions by differentiating the reducibility equations and using the Jacobi identity for the Lie brackets of vector fields.)

(d) Remark: The above equations for $\overset{(0)}{D}\eta^{a_{i+1}}$ and $\overset{(-1)}{D}\eta^{a_{i+1}}$ can be rewritten as $\delta(\mathcal{P}_{a_{i+1}}\overset{(0)}{D}\eta^{a_{i+1}}) =$ given, $\delta(\mathcal{P}_{a_{i+1}}\overset{(-1)}{D}\eta^{a_{i+1}}) =$ given, and so it is not surprising that combining δ and d as in the BRST formalism enables one to derive another existence proof of D.

10.4. Show that the map $\rho : H^k(d) \to H^k(D)$ defined in §10.4.6 is both surjective and injective. [Hints: (*i*) expand the polynomials of $C^\infty(\Sigma) \otimes \mathbf{C}[\eta^{a_0}, \eta^{a_1}, \ldots]$ according to the auxiliary degree. (*ii*) Show that the expansion is finite for any x of definite (finite) pure ghost number. (*iii*) Using the acyclicity of Δ in positive auxiliary degree, prove that the equation $Dx = 0$ implies $x = \overset{(0)}{x}{}' + Dy$ with $\Delta\overset{(0)}{x}{}' = 0 = d\overset{(0)}{x}{}'$, $\text{aux}\,\overset{(0)}{x}{}' = 0$ (analyze the implications of $Dx = 0$ on the component of x of higher auxiliary degree). (*iv*) Show that if, in addition, $x = Dz$, then $\overset{(0)}{x}{}' = d\overset{(0)}{z}{}'$ with $\Delta\overset{(0)}{z}{}' = 0$, $\text{aux}\,\overset{(0)}{z}{}' = 0$. Remark: In general, $\overset{(0)}{x}{}' \neq \overset{(0)}{x}$, and so the d-closed form $\overset{(0)}{x}{}'$ associated with a D-closed polynomial is not simply obtained by setting the ghosts of ghosts equal to zero in x, even though the ghosts of ghosts disappear in cohomology.]

10.5. (a) Show that $\overset{(k)}{\Omega}$ in (10.58) does not contain η^{a_s} for $s > k$ and $\mathcal{P}_{a_{s'}}$ for $s' \geq k$. (Hint: Use antigh $\overset{(k)}{\Omega} = k$, pure gh $\overset{(k)}{\Omega} = k+1$.)
(b) Verify (10.59).

10.6. Prove that $\Omega = \Omega^*$. [Hint: Show first that $(\delta \mathcal{P}_{a_k})^* = \delta \mathcal{P}_{a_k}$ and $(\delta A)^* = (-)^{\varepsilon_A} \delta A$ if A is real; verify next that $\delta \overset{(p+1)}{\Omega}$ is imaginary and Grassmann-even, so that $\overset{(p+1)}{\Omega}$ can be taken to be real and Grassmann-odd.]

10.7. Prove that the change (10.68) in $\overset{(p)}{\Omega}$ can be accounted for by a canonical transformation in the extended phase space even when $\overset{(p+1)}{M}$ is finite. (Hint: Proceed as in Exercise 9.4.)

10.8. Use the (local) abelianization theorem to supply a proof of the *global* existence of Ω. (Hint: The identities $\delta \overset{(k)}{D} = 0$ need only be checked locally.)

10.9. Perform the BRST analysis for the reducible constraints

$$G_i \equiv \varepsilon_{ijk}\, x^j p^k = 0, \quad [x^i, p^k] = \delta^{ik} \qquad (i,j,k = 1,2,3).$$

10.10. Consider the constraint $p^2 = 0$ for a single bosonic momentum p.
(a) Show that $\Omega = p^2 \eta$ is nilpotent.
(b) Show that p is BRST-closed but not BRST-exact ($p = -[\mathcal{P}/p, \Omega]$ but $\mathcal{P}/p$ is singular on the constraint surface).
(c) Show that $pq - 2\mathcal{P}\eta$ has the same properties.
(d) Compare with the correct treatment of the constrained system (q,p) with $p = 0$.

10.11. Prove that one can always construct a nilpotent function $\Omega = \eta^a G_a +$ "more" when the constraints form a group, $[G_a, G_b] = C_{ab}{}^c G_c, C_{c[d}{}^e C_{ab]}{}^c = 0$, even if the constraints do not fulfill the regularity conditions. This can be done with a single ghost pair $(\eta^a, \mathcal{P}_a)$ per constraint. What is the BRST cohomology if $G_a \equiv 0$ and $C_{ab}{}^c$ is taken to vanish (this model generalizes the example of the introduction)? What should the correct treatment yield in that case?

10.12. By dropping some constraints, one can (locally) replace any reducible set by an equivalent irreducible set. This exercise sketches a different method for achieving irreducibility. The idea behind this alternative method is to introduce new pure gauge variables upon which the original ineffective gauge transformations are made to act, hence becoming effective. For simplicity, we consider the explicit case of a reducible theory of the first stage (only $Z_{a_1}{}^{a_0}$, no higher order Z's).
(a) Introduce new pure gauge pairs (q^{a_1}, p_{a_1}) constrained by $p_{a_1} \approx 0$. Show that the reducible constraints $G_{a_0} \approx 0, p_{a_1} \approx 0$ are equivalent to the irreducible constraints $G_{a_0} + A_{a_0}{}^{a_1} p_{a_1} \approx 0$, where $A_{a_0}{}^{a_1}$ is such that the matrix $Z_{a_1}{}^{a_0} A_{a_0}{}^{b_1}$ is invertible. Verify that the new constraints are first class. Work out the explicit form of the gauge transformations with gauge parameter $\varepsilon^{a_0} = Z_{a_1}{}^{a_0} \mu^{a_1}$.
(b) Find an equivalent irreducible set of constraints for the abelian 2-form gauge field of Chapter 19 with constraints $\pi^{kl}{}_{,l} \approx 0$. [Answer: The extra pair is $\phi(\vec{x}), \pi(\vec{x})$, and the new constraints can be taken to be $\pi^{kl}{}_{,l} + \partial^k \pi = 0$.]
(c) Generalize the construction to higher order reducible theories.

CHAPTER ELEVEN

DYNAMICS OF THE GHOSTS—GAUGE-FIXED ACTION

The discussion of Chapters 8 through 10 has dealt with the algebraic and geometrical aspects of the BRST transformation. We now analyze in what sense this transformation defines a symmetry. For that purpose, one needs to determine the action in the extended phase space. This is the object of this chapter, which gives "physical life" to the ghosts.

11.1. BRST COHOMOLOGY AND THE POISSON BRACKET

11.1.1. BRST Observables

Theorem 8.3 establishes that the cohomology $H^*(s)$ of the BRST differential s is isomorphic to the cohomology of the longitudinal exterior derivative at ghost number $g \geq 0$ and vanishes for $g < 0$.

$$H^g(s) = \begin{cases} 0, & g < 0, \quad (11.1a) \\ H^g(d), & g \geq 0. \quad (11.1b) \end{cases}$$

In particular, one finds at ghost number zero,

$$H^0(s) = \{\text{classical observables}\}, \tag{11.2}$$

where the classical observables in (11.2) are the equivalence classes of weakly gauge-invariant functions that coincide on the constraint surface. Equation (11.2) exhibits the profound connection between gauge invariance and BRST invariance.

Because of (11.2), one calls "BRST observables" the ghost number zero functions $A(z, \eta, \mathcal{P})$ that are BRST invariant,

$$sA \equiv [A, \Omega] = 0, \tag{11.3a}$$

$$\text{gh}\, A = 0, \tag{11.3b}$$

with the understanding that two BRST observables that differ by a trivial BRST-exact one should be identified. The connection between a BRST observable $A(z, \eta, \mathcal{P})$ and the corresponding classical observable $A_0(z)$ is simply that the ghost number zero component $\overset{(0)}{A}$ of A is equal to A_0, *i.e.*,

$$A(z, \eta, \mathcal{P}) = A_0(z) + \sum_{p \geq 1} \overset{(p)}{A}(z), \qquad \text{antigh} \overset{(p)}{A} = p, \tag{11.4}$$

where the terms of antighost number ≥ 1 are determined recursively along the lines of §8.4.3 and §10.5.4. The details are worked out in Exercise 11.2. If A_0 is real, one can take also A to be real (Exercise 11.4),

$$A = A^*. \tag{11.5}$$

Because of (11.4), one says that A is a "BRST-invariant extension" of A_0.

A physical theory is not just defined by the set of physically relevant functions. One also needs a bracket structure among them and a Hamiltonian to generate the motion. The bracket structure determines the graded commutation rules in the quantum theory. It follows (i) that the set $H^0(s)$ of BRST observables should be endowed with a bracket structure; and (ii) that the isomorphism (11.2) should preserve the bracket structure.

That these properties hold is the content of the next two theorems. The equations of motion are analyzed in Sec. 11.2.

Theorem 11.1. *There is a natural Poisson algebra structure in* $H^0(s)$.

(A Poisson algebra $\mathcal{A}$ *is an associative algebra with a further bracket structure for which* $\mathcal{A}$ *is a graded Lie algebra and such that* $[AB, C] = A[B, C] + (-)^{\varepsilon_B \varepsilon_C}[A, C]B$.*)*

Proof. The BRST transformation is a canonical transformation. The Jacobi identity for the Poisson bracket implies then that s is not only a derivation for the associative product but also for the Poisson bracket,

$$s(AB) = A(sB) + (-)^{\varepsilon_B}(sA)B, \tag{11.6a}$$

$$s[A,B] = [A, sB] + (-)^{\varepsilon_B}[sA, B]. \tag{11.6b}$$

Hence, if A and B are BRST-observables, AB and $[A,B]$ are also BRST observables. Furthermore, if A' and B' are, respectively, in the same cohomological classes as A and B, then $A'B'$ and $[A',B']$ are, respectively, in the same cohomological classes as AB and $[A,B]$. This proves the theorem.

What is the relationship between the respective Poisson structures of $H^0(s)$ and of the set of gauge-invariant functions? Not surprisingly, they are the same.

Theorem 11.2. *If A and B are, respectively, BRST-invariant extensions of A_0 and B_0, then AB and $[A,B]$ are, respectively, BRST-invariant extensions of A_0B_0 and $[A_0, B_0]$.*

Proof. The theorem follows from the fact that the terms of antighost number ≥ 1 in A and B are at least bilinear in $\eta, \mathcal{P}$, since $\text{gh}\, A = 0 = \text{gh}\, B$. Hence, these terms cannot contribute to the zeroth-order piece of AB or $[A,B]$.

Remark. Because the constraints $G_{a_0} \approx 0$ are first class, they are gauge invariant. They thus possess BRST-invariant extensions. One possible choice for the extension is

$$G_{a_0} \to \bar{G}_{a_0} = [-\mathcal{P}_{a_0}, \Omega]. \tag{11.7a}$$

This choice explicitly shows that the BRST-invariant extensions of the constraints are BRST-exact, *i.e.*, belong to the same equivalence class as the trivial observable zero, as they should.

In the irreducible group case, with Ω given by (9.27), the extensions $\bar{G}_a$ fulfill

$$[\bar{G}_a, \bar{G}_b] = C_{ab}{}^c \bar{G}_c. \tag{11.7b}$$

In general, however, the extensions $\bar{G}_{a_0}$ do not obey any particular algebra.

11.1.2. What Is the Meaning of the Higher Cohomological Groups $H^k(s)$, $k > 0$?

Because $H^k(s)$ is in general different from zero for $k > 0$, one cannot say that the BRST complex provides a resolution of the algebra of classical observables. To recover the gauge-invariant functions, one must not only pass to the BRST cohomology; one must also impose, in general, the ghost number zero condition.

What is the physical meaning of the higher cohomological groups $H^k(s)$, $k > 0$? It seems that these groups, even though mathematically interesting, do not have a direct physical interpretation. [$H^1(s)$ and $H^2(s)$ appear to be related to the possible breaking of the gauge symmetry by quantum effects ("anomalies"), but this question will not be addressed in the book.]

11.1.3. Ghost Transformation Law under Global Symmetries

Let $Q_0(z)$ be the generator of a global symmetry. We have mentioned in §3.2.7 that Q_0 is first class,

$$[Q_0, G_{a_0}] \approx 0. \tag{11.8}$$

The transformation generated by Q_0 is defined in the original phase space. How should it be extended to the ghosts? For example, in a relativistic gauge theory, how should the ghosts transform under Poincaré transformations?

The answer to this question is obvious from our previous analysis. Since the symmetry generator Q_0 is gauge invariant, it should be extended in a BRST-invariant way,

$$Q_0 \rightarrow Q = Q_0 + \text{ghost terms}, \tag{11.9a}$$

$$[Q, \Omega] = 0, \tag{11.9b}$$

$$\operatorname{gh} Q = 0. \tag{11.9c}$$

Once Q has been determined, the ghost transformation properties simply follow from taking their brackets with Q. Furthermore, the transformation rules of the original variables get generically modified by ghost contributions induced by the extra terms that are added to Q to form a BRST-invariant generator.

To the redefinition

$$Q_0 \rightarrow Q_0 + t^{a_0} G_{a_0} \tag{11.10a}$$

of the global symmetry, which amounts to adding to $[z^A, Q_0]$ a gauge transformation, corresponds the redefinition

$$Q \rightarrow Q + [T, \Omega], \quad \operatorname{gh} T = -1, \quad T = -t^{a_0} \mathcal{P}_{a_0} + \text{"more,"} \tag{11.10b}$$

of the BRST-invariant generator Q.

Further aspects of the realization of global symmetries in the extended phase space are discussed in Exercises 11.6 through 11.8.

11.2. GHOST DYNAMICS; GAUGE FIXING. THE BRST FUNCTION AS THE GENERATOR OF A SYMMETRY

11.2.1. BRST-Invariant Hamiltonians

In the original phase space, the time evolution is generated by the extended Hamiltonian $H_0 + \lambda^{a_0} G_{a_0}$. Because $H_0(z)$ is gauge invariant,

$$[H_0, G_{a_0}] = V_{a_0}{}^{b_0} G_{b_0}, \tag{11.11}$$

it possesses a BRST-invariant extension H,

$$H_0 \to H, \qquad [H, \Omega] = 0. \tag{11.12}$$

Consider the time evolution generated by H in the extended phase space,

$$\dot{F} = [F, H]. \tag{11.13}$$

If F is the BRST-invariant extension of the gauge-invariant function $F_0(z)$, these equations are equivalent to the original equations of motion,

$$\frac{dF_0}{dt} \approx [F_0, H_0] \qquad (\text{since } [F_0, G_{a_0}] \approx 0.) \tag{11.14}$$

This is because the bracket is preserved in the correspondence that associates with a gauge-invariant function a BRST-invariant one. Furthermore, the higher order ghost terms in (11.13) contain no cohomological information and are just such that $\dot{F}$ is BRST invariant. Hence, one can adopt (11.13) as dynamical equations of motion in the extended phase space.

The BRST-invariant Hamiltonian may be used to define the time evolution of all variables of the extended phase space, even those that are not BRST invariant. In particular, one finds, for the ghosts associated with the constraints,

$$\dot{\eta}^{a_0} = [\eta^{a_0}, H] = \eta^{b_0} V_{b_0}{}^{a_0} + \text{“more,”} \tag{11.15a}$$

$$\dot{\mathcal{P}}_{a_0} = -V_{a_0}{}^{b_0} \mathcal{P}_{b_0} + \text{“more,”} \tag{11.15b}$$

where “more” in Eq. (11.15) stands for terms of higher antighost number.

The equations of motion (11.13) yield a nonambiguous time evolution in the extended phase space with no arbitrariness. For this reason, one says that the equations (11.13) are “gauge fixed.”

One finds from (11.15) and from the equations of motion for the higher order ghosts that when the ghosts and their momenta vanish initially, they vanish at all times. The equations of motion for the original canonical variables z^A reduce in that case to the dynamical equations of motion generated by $H_0 = H_E(\lambda^{a_0} = 0)$. One may thus say that Eqs. (11.13) are the equations of motion in the multiplier gauge $\lambda_{a_0} = 0$ (§3.4.2). The constraint equations $G_{a_0} = 0$ are, however, not fulfilled unless one also imposes $G_{a_0} = 0$ initially.

If one changes the BRST-invariant extension of the Hamiltonian by adding an Ω-exact term,

$$H \to H + [K, \Omega], \qquad \varepsilon(K) = 1, \qquad K^* = -K, \tag{11.16}$$

one does not modify the dynamics of BRST-invariant functions at the cohomological level. However, the time evolution of the ghosts and of the other non-BRST-invariant functions is generically changed. This is the reason that the function K is called the "gauge-fixing fermion." The transformation (11.16) replaces the gauge-invariant Hamiltonian H_0 by $H_0 + k^{a_0} G_{a_0}$, where $k^{a_0} \mathcal{P}_{a_0}$ is the term of antighost number one in K. This amounts to replacing the multiplier gauge $\lambda^{a_0} = 0$ by the multiplier gauge $\lambda^{a_0} = k^{a_0}$.

It should also be noted that for an open algebra of rank ≥ 2, the "gauge-fixed" Hamiltonian $H + [K, \Omega]$, with a definite choice for both H and K, generically involves multi-ghost interaction terms. These multighost vertices can appear both in H and in $[K, \Omega]$ and are necessary for securing BRST invariance. The functions $\overset{(2)}{H}{}^{ae}_{bc}$ determining $\overset{(2)}{H}$ already appeared in §3.2.5 when we analyzed the algebra of the gauge transformations.

Remark. We will impose the condition $\mathrm{gh}\, K = -1$. This implies conservation of the ghost number for all trajectories, since then $\mathrm{gh}\,(H + [K, \Omega]) = 0$. The condition of $K = -1$ may be lifted without changing the dynamics of the BRST-invariant functions at the cohomological level, but such a possibility does not appear to be useful in practice.

11.2.2. BRST Symmetry—Gauge-Fixed Action

It can now be seen in what sense the BRST transformation is a symmetry transformation of the theory. Its Poisson bracket with the "gauge-fixed" Hamiltonian $H + [K, \Omega]$ always vanishes, no matter what the gauge-fixing function K is. The "gauge-fixed dynamics" is accordingly always BRST invariant. The canonical action from which the dynamics in the extended phase space follows is called the gauge-fixed action and is given by

$$S_K[z^A, \eta^{a_k}, \mathcal{P}_{a_k}] = \int_{t_1}^{t_2} \left(\dot{z}^A a_A(z) + \sum_k \dot{\eta}^{a_k} \mathcal{P}_{a_k} - H - [K, \Omega]\right) dt. \quad (11.17)$$

Because the Hamiltonian is BRST invariant and because the BRST transformation is a canonical transformation, which leaves therefore the kinetic term invariant up to a boundary term at t_1 and t_2, the action (11.17) is also BRST invariant, up to that same boundary term.

It cannot be stressed enough that the choice of the gauge-fixed Hamiltonian H and of the gauge-fixed action (11.17) comes *after* the determination of the BRST symmetry. In the extended phase space, the BRST symmetry is intrinsic and takes the same form no matter what K is in (11.17).

11.2.3. Comments

With the introduction of the ghosts, the original constrained dynamics can be equivalently reformulated as an unconstrained dynamics in the extended phase space. Gauge invariance is recovered through the passage to the BRST cohomology.

The extended phase space is of dimension $2(n+m_0+m_1+\cdots+m_L)$, where n is the number of original conjugate pairs and m_i the number of ghost pairs of the i-th generation. When one passes to the BRST cohomology, one effectively goes to the reduced phase space of dimension $2(n-m_0')$ where m_0' is the number of independent constraints. This is the content of (11.2). In the counting of the degrees of freedom, the ghosts count thus with the sign $(-)^{i+1}$, since $m_0' = \Sigma(-)^i\, m_i$. The gauge degrees of freedom and the ghosts cancel in pairs of opposite statistics, and only the unpaired physical degrees of freedom remain. This reduction is consistent with the time evolution because the Hamiltonian is BRST invariant.

Finally, the interpretation of the ghosts η^{a_k} and of their momenta $\mathcal{P}_{a_k}$ as generators of appropriate differential complexes, which was helpful in understanding the BRST construction, disappears when discussing the ghost dynamics. Indeed, there does not appear to be a useful notion of time derivative of the geometric one-forms ω^a. As a matter of fact, the antighost number and the pure ghost number are not separately conserved by the dynamics unless "more" in Eq. (11.15) vanishes, so that the variables of the Koszul–Tate and the longitudinal complexes become mixed.

11.3. NONMINIMAL SOLUTIONS

11.3.1. Nonminimal Sector

It is often useful to enlarge further the extended phase space by adding new variables that do not modify $H^0(s)$. This is necessary for imposing derivative gauge conditions on the Lagrange multipliers, a step that may be required for implementing manifestly covariant gauges in relativistic theories. The enlargement of the extended phase space without modification of $H^0(s)$ is physically permissible, since the concept of classical observable is left unchanged.

The requirement that $H^0(s)$ shoud not be modified is not very restrictive and leaves many possibilities open for enlarging the extended phase space. We shall not explore this question in detail here, but, rather, we shall consider only the simpler possibility, which consists of adding new variables that cancel in pairs at the cohomological level. Namely, one of the variables in each pair is taken to be the BRST variation of the other. If one denotes the variables by α and β, one thus has

$$s\alpha = \beta, \qquad s\beta = 0. \tag{11.18a}$$

The definition (11.18a) leaves not only $H^0(s)$ unchanged but also the other cohomological groups $H^k(s)$, since (11.18a) takes the simple form of Theorem 8.2. (See also Exercise 11.9.) To get a canonical action for s, the variables α and β should have conjugate momenta π_α and π_β such that

$$s\pi_\beta = \pi_\alpha, \qquad s\pi_\alpha = 0. \tag{11.18b}$$

The BRST generator in the (α–β)-sector is then given by

$$\pi_\alpha \beta \tag{11.18c}$$

and is quadratic in the new variables.

The original variables define what is known as the *minimal sector.* The cohomologically trivial variables α, β, π_α, and π_β that have been added belong to the *nonminimal sector.* The total BRST charge is the sum of the original minimal solution, plus a cohomologically trivial nonminimal term of the form (11.18c),

$$\Omega = \Omega^{\text{Min}} + \Omega^{\text{Nonmin}}, \tag{11.19a}$$

$$\Omega^{\text{Nonmin}} = \sum \pi_\alpha \beta. \tag{11.19b}$$

The sum in (11.19b) ranges over all the nonminimal pairs.

Such an extension of the phase space has been encountered already in Sec. 10.1, where it was found that the BRST generator for an unconstrained system could be taken to be either $\Omega \equiv 0$ in the original phase space or $\Omega = \mathcal{P}\phi$ in the extended phase space containing also $\mathcal{P}$, η, π, and ϕ. More generally, *the introduction of trivial reducibility identities at any order amounts to adding cohomologically trivial pairs and, hence, to enlarging the nonminimal sector.* The example of Sec. 10.1 also shows that the distinction between minimal and nonminimal sectors possesses no fundamental character, since the solution $\Omega = \mathcal{P}\phi$ appears to be nonminimal in the formulation where the (empty) constraint $G \equiv 0 = 0$ is not included, while it becomes minimal when $G \equiv 0 = 0$ is introduced.

11.3.2. The Lagrange Multipliers as Canonical Variables

The nonminimal sector can be chosen so as to include the Lagrange multipliers λ^a appearing in the extended action (5.4),

$$S_E[z^A, \lambda^a] = \int [\dot{z}^A a_A(z) - H_0 - \lambda^a G_a]\, dt. \qquad (11.20)$$

We assume from now on that the constraints are irreducible. Similar considerations apply to the reducible case but need a further enlargement of the nonminimal sector to reach the propagating gauges defined below. The momentum conjugate to λ^a is denoted by b_a and is required to vanish,

$$b_a \approx 0. \qquad (11.21a)$$

The constraints (11.21a) are first class and generate arbitrary shifts in λ^a. Accordingly, the inclusion of λ^a and its conjugate b_a among the dynamical variables does not modify the physical content of the theory.

The fully extended phase space contains the original variables z^A, the multipliers λ^a, and their momenta b_a, the ghost pairs $(\eta^a, \mathcal{P}_a)$ associated with $G_a = 0$, and the ghost pairs$(\rho^a, \bar{C}_a)$ associated with $b_a = 0$. The variables $\bar{C}_a$ are known as the "antighosts" because they have negative ghost number. In the gauge considered below, the ghost momenta $\mathcal{P}_a$ are equal to the time derivatives of the antighosts [Eq. (11.28a)]. The ρ^a are the "antighost momenta." The nonvanishing Poisson brackets between the new variables are

$$[\rho^a, \bar{C}_b] = -\delta^a{}_b = [b_b, \lambda^a], \qquad (11.21b)$$

and are generated by the kinetic term

$$\int (\dot{\bar{C}}_a \rho^a + \dot{\lambda}^a b_a)\, dt. \qquad (11.21c)$$

The BRST charge is given by (11.19a),

$$\Omega = \Omega^{\text{Min}} + \Omega^{\text{Nonmin}}, \tag{11.22a}$$

$$\Omega^{\text{Min}} = \eta^a G_a + \text{“more,”} \tag{11.22b}$$

$$\Omega^{\text{Nonmin}} = (-i)^{\varepsilon_a + 1} \rho^a b_a, \tag{11.22c}$$

where the "more" in (11.22b) depends only on the variables $(z^A, \eta^a, \mathcal{P}_a)$ of the minimal sector. The phase $(-i)^{\varepsilon_a+1}$ appears in (11.22c) because we take λ^a to be imaginary if the constraints are fermionic so that $\lambda^a G_a$ is real. The b_a and the $\bar{C}_a$ are real, while ρ^a is imaginary or real according to whether G_a is real or imaginary.

The following class of gauge-fixing fermions K are found to be of great use in field theory,

$$K = i\,\bar{C}_a\,\chi^a - \mathcal{P}_a\,\lambda^a \tag{11.23a}$$

where the χ^a's are real functions of the z's. The corresponding gauge-fixed Hamiltonian reads

$$\begin{aligned} H_K &= H + [K, \Omega] \\ &= H + i\,\bar{C}_a[\chi^a, \Omega^{\text{Min}}] - \chi^a b_a (i)^{\varepsilon_a} \\ &\quad + \lambda^a G_a - \lambda^a \Big[\mathcal{P}_a, \sum_{p\geq 1} \overset{(p)}{\Omega}{}^{\text{Min}}\Big] \\ &\quad + i\,\mathcal{P}_a \rho^a (i)^{\varepsilon_a}, \end{aligned} \tag{11.23b}$$

where H is a particular BRST-invariant extension of H_0 that does not depend on the variables of the nonminimal sector.

The gauge-fixed Hamiltonian contains linearly the momenta b_a conjugate to the Lagrange multipliers. Accordingly, the momenta cannot all be eliminated from the action by means of their own equations of motion, and one cannot recover a pure Lagrangian formulation. For this reason, one sometimes replaces (11.23a) by

$$K = i\,\bar{C}_a \chi^a - \mathcal{P}_a \lambda^a + \tfrac{i}{2} \bar{C}_a\, g^{ab} b_b, \tag{11.24a}$$

which yields a gauge-fixed Hamiltonian quadratic in b_a,

$$\begin{aligned} H_K = H &+ i\,\bar{C}_a[\chi^a, \Omega^{\text{Min}}] - \chi^a b_a (i)^{\varepsilon_a} - \tfrac{1}{2} b_a\, g^{ab} b_b\, (-i)^{\varepsilon_a} \\ &+ \lambda^a G_a - \lambda^a \Big[\mathcal{P}_a, \sum_{p\geq 1} \overset{(p)}{\Omega}{}^{\text{Min}}\Big] + i\,\mathcal{P}_a\, \rho^a (i)^{\varepsilon_a}. \end{aligned} \tag{11.24b}$$

In (11.24), the "metric" g^{ab} can be any convenient c-number metric. (That "metric" can, in fact, even be taken to depend on the dynamical variables, but then there are extra terms generated by $[K, \Omega]$.) If g^{ab} is invertible—which can always be achieved if the b's are bosonic, since g^{ab}

is then symmetric—one can also eliminate the momenta b_a by means of their own equations of motion. Gauge-fixing fermions such that one can eliminate the momenta are said to define "propagating gauges."

As a final comment, we note that the Lagrange multipliers λ^a and their momenta may actually be already included among the canonical variables z^A in alternative formulations of the theory (see Exercises 4.1 and 4.2). In that case, the above discussion is unnecessary. This illustrates again that the distinction between minimal and nonminimal sectors is not fundamental.

11.3.3. The Faddeev-Popov Action

When the constraint algebra is closed, the expressions (11.23b) and (11.24b) simplify considerably. This is because Ω^{Min} reads

$$\Omega^{\text{Min}} = \eta^a G_a - \tfrac{1}{2}(-)^{\varepsilon_a}\eta^a\eta^b C_{ba}{}^c\mathcal{P}_c. \tag{11.25}$$

Furthermore, H_0 can be assumed to be strongly gauge invariant (Exercise 1.24),

$$[H_0, G_a] = 0, \tag{11.26}$$

so that $H = H_0$ is an acceptable BRST-invariant extension of H. One then gets

$$\begin{aligned}[\chi^a, \Omega^{\text{Min}}] &= [\chi^a, \eta^b G_b] \\ &= \delta_\eta \chi^a (-)^{\varepsilon_a},\end{aligned} \tag{11.27a}$$

and

$$\left[\mathcal{P}_a, \sum_{p\geq 1} \overset{(p)}{\Omega}{}^{\text{Min}}\right] = (-)^{\varepsilon_a}\eta^b C_{ba}{}^c\mathcal{P}_c. \tag{11.27b}$$

In (11.27a), $\delta_\eta\chi^a$ is the gauge variation of χ^a under a gauge transformation in which the infinitesimal gauge parameter ε^a is replaced by the ghost η^a. To be precise, $\delta_\eta\chi^a$ is obtained by (*i*) first writing $\delta_\varepsilon\chi^a$ in a form where ε stands to the extreme left; and (*ii*) replacing then ε by η. The sign factor $(-)^{\varepsilon_a}$ appears in (11.27a) because ε^a and η^a possess opposite parity properties.

The equations of motion $\delta S_K/\delta\mathcal{P}_a = 0$ and $\delta S_K/\delta\rho^a = 0$ can be solved for $\mathcal{P}_a$ and ρ^a,

$$\dot{\bar{C}}_a - i\,\mathcal{P}_a(i)^{\varepsilon_a} = 0, \tag{11.28a}$$

$$\dot{\eta}^a + (-)^{\varepsilon_c}\lambda^c\eta^b C_{bc}{}^a + i\,\rho^a(-i)^{\varepsilon_a} = 0. \tag{11.28b}$$

If one substitutes (11.28) back in S_K [= (11.17) + (11.21c)], one gets the gauge-fixed action,

$$S'_K[z^A, \lambda^a, b_a, \eta^a, \bar{C}_a] = S_E + S^{\text{gauge breaking}} + S^{\text{ghost}}, \qquad (11.29a)$$

$$S^{\text{gauge breaking}} = \int dt(\dot{\lambda}^a + (i)^{\varepsilon_a}\chi^a)b_a, \qquad (11.29b)$$

$$S^{\text{ghost}} = -i(i)^{\varepsilon_a}\int dt\, \bar{C}_a \delta_\eta[\dot{\lambda}^a + (i)^{\varepsilon_a}\chi^a], \qquad (11.29c)$$

in the case of (11.23), and

$$\tilde{S}_K[z^A, \lambda^a, b_a, \eta^a, \bar{C}_a] = S_E + \tilde{S}^{\text{gauge breaking}} + S^{\text{ghost}}, \qquad (11.30a)$$

$$\tilde{S}^{\text{gauge breaking}} = \int dt(\dot{\lambda}^a + (i)^{\varepsilon_a}\chi^a + \frac{(-i)^{\varepsilon_a}}{2}b^a)b_a, \qquad (11.30b)$$

in the case of (11.24). To reach (11.29c), we have made an integration by parts to get rid of the time derivative of $\bar{C}_a$. The boundary term at t_1 and t_2 is zero with the boundary conditions $\bar{C}_a(t_1) = \bar{C}_a(t_2) = 0$.

The action (11.30) can be further transformed by eliminating the momenta b_a by means of $\delta\tilde{S}_K/\delta b_a = 0$,

$$\frac{\delta\tilde{S}_K}{\delta b_a} = 0 \Leftrightarrow \dot{\lambda}^a + (i)^{\varepsilon_a}\chi^a + (-i)^{\varepsilon_a}b^a = 0. \qquad (11.31a)$$

This has the effect of replacing (11.30b) by

$$\tilde{S}'^{\text{gauge breaking}} = -\tfrac{1}{2}(i)^{\varepsilon_a}\int dt\,(\dot{\lambda}^a + (i)^{\varepsilon_a}\chi^a)(\dot{\lambda}_a + (i)^{\varepsilon_a}\chi_a). \qquad (11.31b)$$

The gauge-fixed actions (11.29) and (11.30)–(11.31) were derived first along different lines by Faddeev and Popov. These actions consist of three pieces:

(*i*) first, the original gauge-invariant action;
(*ii*) second, a gauge-breaking term, which determines the Lagrange multipliers and, hence, freezes the gauge invariance of S_E;
(*iii*) third, a Faddeev-Popov ghost term that is quadratic in the ghosts.

The gauge conditions enforced by the gauge breaking term are of the derivative type, since it follows from $\delta S_K/\delta b_a = 0$ that

$$\dot{\lambda}^a + (i)^{\varepsilon_a}\chi_a = 0 \qquad (11.32a)$$

in the case of (11.29b), and

$$\dot{\lambda}^a + (i)^{\varepsilon_a}\chi^a + (-i)^{\varepsilon_a}b^a = 0 \qquad (11.32b)$$

in the case of (11.30b). In that latter case, one finds $b_a = 0$ from the equations of motion and the boundary conditions for physical trajectories (see §11.3.5 below). Hence, (11.32b) reduces on-shell to the derivative gauge (11.32a). Derivative gauges can thus be implemented in the formalism once the Lagrange multipliers and their momenta are included among the canonical variables. These derivative gauges arise as equations of motion for the b-field. The gauge-breaking term contains the gauge conditions (11.32a) linearly in the case of (11.29b) and quadratically in the case of (11.30b)–(11.31b). Derivative gauges can be implemented in the same manner for gauge transformations with an open algebra, since the b-dependence of the gauge-fixed action is not affected by the higher order terms in the BRST generator.

If one compares (11.29c) with (11.32a), one sees that the Faddeev-Popov ghost term has the simple form $\bar{C}_a \delta_\eta F^a$, where F^a are the functions defining the gauge conditions through $F^a = 0$. This familiar form of the ghost action is justified under the above assumption that the gauge algebra is closed but is, in general, incorrect for an open gauge algebra, since nonquadratic ghost vertices are then generically generated through H and $[K, \Omega]$. Nonquadratic ghost vertices may even be generated in the group case, if one takes a gauge-fixing fermion K that is nonlinear in the $\bar{C}$'s and the $\mathcal{P}$'s ("nonlinear gauges").

11.3.4. Lagrangian Form of the BRST Symmetry—The BRST Generator as a Noether Charge

We have seen that in propagating gauges, one can eliminate the conjugate momenta by means of their own equations of motion. The resulting Lagrangian gauge-fixed action is invariant under the BRST transformation in which the momenta are replaced by their on-shell value. This form of the BRST transformation is known as the "Lagrangian BRST transformation."

Because the relation between the momenta and the velocities involves the gauge-fixing fermion K, the BRST transformation in Lagrangian form, where all momenta have been eliminated, is not universal and does depend on the gauge-fixing procedure.

From the invariance of the action under the BRST symmetry, one finds a conserved current. The corresponding charge, written in Hamiltonian form, is just Ω. This follows from standard methods, which are particularly straightforward in propagating gauges, since the Legendre transformation is invertible. Although the form of Ω as a function of the coordinates and the velocities is not universal, its phase space expression is.

Because the equations of motion for the momenta have been used to arrive at the Lagrangian version of the theory, there is no guarantee that the BRST transformation is nilpotent off-shell. Actually, it generically fails to be so (see Exercise 11.12). To achieve off-shell nilpotency, it is necessary to reinstate some, if not all, of the momenta. The momenta that are reintroduced are sometimes known under a different name. Instead of calling them "canonically conjugate momenta," the names "auxiliary field," "b-field," "third" or "extra" ghost, "Takanishi–Lautrup auxiliary field," or even "Stuckelberg field," are favoured in the literature on the Lagrangian form of the BRST symmetry.

11.3.5. Hamilton Principal Function and Ghosts

The gauge-fixed action and the original gauge-invariant action are defined in different functional spaces and so cannot be compared for arbitrary histories. It turns out, however, that their values coincide on-shell when appropriate boundary conditions on the ghosts and the momenta b_a are imposed. This fact implies that the on-shell value of the gauge-fixed action yields the Hamilton principal function.

To prove the on-shell equality of the gauge-fixed action with the Hamilton function, let us assume that the original canonical variables z^A split in conjugate pairs. Let us also assume, as in Chapter 5, §5.4.5, that the original gauge-invariant variational principle admits a unique solution (up to a gauge transformation) when the coordinates q^i are given at the time boundaries t_1 and t_2. Finally, assume that the derivative gauge (11.32a) is accessible.

Theorem 11.3. *Provided it is unique, the solution of the equations of motion generated by the gauge-fixed Hamiltonians (11.23b) or (11.24b) and obeying the boundary conditions*

$$q^i(t_1) = q^i_1, \quad q^i(t_2) = q^i_2, \tag{11.33a}$$

$$b_a(t_1) = b_a(t_2) = 0, \tag{11.33b}$$

$$\eta^a(t_1) = \eta^a(t_2) = 0, \tag{11.33c}$$

$$\bar{C}_a(t_1) = \bar{C}_a(t_2) = 0 \tag{11.33d}$$

(conjugate variables unspecified), is also a solution of the equations of motion of the original constrained system. This solution obeys further the derivative gauge condition

$$\dot{\lambda}^a + (i)^{\varepsilon_a}\chi^a = 0. \tag{11.34}$$

Proof. With $\eta^a(t) = 0$, $\bar{C}_a(t) = 0$, $\mathcal{P}_a(t) = 0$, $\rho^a(t) = 0$, and $b_a(t) = 0$, the boundary conditions (11.33b)–(11.33d) are fulfilled and the equations of motion reduce to

$$\dot{z}^A = [z^A, H_0] + [z^A, \lambda^a G_a], \tag{11.35a}$$

$$G_a = 0, \tag{11.35b}$$

$$\dot{\lambda}^a + (i)^{\varepsilon_a} \chi^a = 0. \tag{11.35c}$$

Equations (11.35a) and (11.35b) are just those of the original system. Let $z^A(t), \lambda^a(t)$ be a solution of (11.35a) and (11.35b), obeying the boundary conditions (11.33a) but not necessarily the gauge condition (11.35c). This solution is defined up to a gauge transformation vanishing at t_1 and t_2 to preserve the boundary conditions (11.33a) (see §5.4.5). By using such a gauge transformation, one can make λ^a fulfill the derivative gauge conditions (11.35c) (see §3.4.1). If one takes the solution of the original constrained system that obeys (11.33a) and (11.34) and takes, in addition, $\eta^a(t) = \bar{C}_a(t) = \mathcal{P}_a(t) = \rho^a(t) = b_a(t) = 0$, one gets a solution of the full system fulfilling the required boundary conditions. Because the solution is assumed to be unique, it is *the* solution. This proves the theorem.

Remarks. (*i*) The solution of the boundary value problem exists and is unique except for unfortunate choices of $t_2 - t_1$. This would not be true if K did not couple the minimal and nonminimal sectors. The choice $K = 0$ is, for instance, a bad choice for all $t_2 - t_1$, since the solution fails to be unique when it exists (more on this in Exercise 11.14).

(*ii*) Both the derivative gauge (11.34) and the constraints $G_a = 0$ arise on-shell and only on-shell. For an arbitrary competing history in the gauge-fixed variational principle, neither $G_a = 0$ nor (11.34) is fulfilled.

(*iii*) It is not necessary to assume that $G_a = 0$ initially. It is necessary, however, to impose $b_a = 0$ at both endpoints.

Theorem 11.4. *The Hamilton principal function* $W(q_2^i, t_2; q_1^i, t_1)$ *is equal to*

$$W(q_2^i, t_2; q_1^i, t_1) = \int (\dot{q}^i p_i + \dot{\lambda}^a b_a + \dot{\eta}^a \mathcal{P}_a + \dot{\bar{C}}_a \rho^a - H - [K, \Omega])\, dt \tag{11.36}$$

where the integral is evaluated on the classical history fulfilling the boundary condition (11.33).

Proof. This theorem is an immediate consequence of the previous one. At the extremum, $b_a = 0 = \bar{C}_a = \eta^a = \mathcal{P}_a = \rho^a$, and q^i, p_i obey the original equations of motion. Hence, (11.36) indeed reduces to the Hamilton principal function (5.53).

Remarks. (*i*) There is a major difference between (11.36) and (5.53). In (5.53), the extremum is not unique because the gauge is not fixed, so one must pick out one particular classical trajectory by hand. By contrast, in (11.36), the gauge is completely fixed and the extremum is unique. The integral does not depend on the choice of K—*i.e.*, on the choice of χ^a—because the action S_E is gauge invariant.

(*ii*) The boundary conditions (11.33) are BRST invariant: $sq^i = 0$ and $s\eta^a = 0$ with $\eta^a = 0$. Furthermore, $sb_a = 0$ and $s\bar{C}_a = b_a = 0$ with $b_a = 0$.

EXERCISES. CHAPTER ELEVEN

11.1. Check that the modification $A_0 \to A_0 + k^a G_a$ of the gauge-invariant function $A_0(z)$ can be accounted for by a change $A \to A + [K, \Omega]$ of the corresponding BRST-invariant extension.

11.2. (a) Let $\overset{(0)}{A}(z, \eta)$ be a D-closed q-form along the gauge orbits. Show that the BRST-invariant extensions of $\overset{(0)}{A}$ are explicitly determined by the equations

$$A = \overset{(0)}{A} + \sum_{p \geq 1} \overset{(p)}{A}, \quad \text{antigh}\, \overset{(p)}{A} = p, \quad \text{gh}\, \overset{(p)}{A} = q, \tag{i}$$

$$\delta \overset{(p+1)}{A} + \overset{(p)}{M} = 0 \tag{ii}$$

$$\overset{(p)}{M} = \sum_{k=0}^{p} [\overset{(p-k)}{A}, \overset{(k)}{\Omega}]_{\text{orig}} + \sum_{k=0}^{p} \sum_{s=0}^{k+q-1} [\overset{(k)}{A}, \overset{(p+s+1-k)}{\Omega}]_{\eta^{a_s} \mathcal{P}_{a_s}}. \tag{iii}$$

(b) Check explicitly that $\overset{(0)}{M} \approx 0$ and $\delta \overset{(p)}{M} = 0$ $(p \geq 1)$, so that Eq. (ii) can be solved recursively for $\overset{(p+1)}{A}$. (Hint: Proceed as in §10.5.4.)

(c) Prove that the BRST-invariant extensions of $\overset{(0)}{A}$ are determined up to a BRST-exact term. (Hint: Proceed as in §10.5.5.)

11.3. (a) Construct the BRST-invariant extension of a strongly gauge-invariant function when the constraints form a group. (Answer: $A = A_0$.)

(b) Show that the condition $[A_0, G_a] = 0$ (strongly), which can be imposed (see Exercise 1.24), does not completely fix the freedom $A_0 \to A_0 + k^a G_a$.

11.4. Check explicitly that the BRST-invariant extensions A of $\overset{(0)}{A}$ can be taken to be real when $\overset{(0)}{A}$ is real.

11.5. The first-class constrained system with Hamiltonian H_0 and constraints $G_a = 0$ can be replaced by the equivalent parametrized constrained system $G_\alpha = (p_t + H_0, G_a)$ with vanishing Hamiltonian $\mathcal{H}$.

(a) Relate the BRST charge $\bar{\Omega}$ associated with the first-class set G_α to the BRST-invariant extension H of H_0 and the BRST charge Ω associated with G_a.

(b) Using (a) and the remarks made in §4.2.3, develop the BRST theory for time-dependent first-class constraints.

11.6. Let $\overset{(0)}{Q}_\alpha$ be the generators of a symmetry group,

$$\big[\overset{(0)}{Q}_\alpha, G_a\big] = V_{\alpha a}{}^b G_b, \qquad \big[\overset{(0)}{Q}_\alpha, H\big] = k_\alpha{}^a G_a,$$
$$\big[\overset{(0)}{Q}_\alpha, \overset{(0)}{Q}_\beta\big] = D_{\alpha\beta}{}^\gamma \overset{(0)}{Q}_\gamma + W_{\alpha\beta}{}^a G_a.$$

Here, $D_{\alpha\beta}{}^\gamma$ are the structure constants of the symmetry algebra, while $V_{\alpha a}{}^b$, $k_\alpha{}^a$, and $W_{\alpha\beta}{}^a$ are phase space functions.

(a) Show that the BRST extensions Q_α fulfill

$$[Q_\alpha, Q_\beta] = D_{\alpha\beta}{}^\gamma Q_\gamma + [W_{\alpha\beta}, \Omega], \qquad [Q_\alpha, H] = [K_\alpha, \Omega],$$

for some $W_{\alpha\beta}$ and K_α.

(b) Derive the identities on $W_{\alpha\beta}$ and K_α that follow from the Jacobi identity.

(c) Discuss the conditions under which one can remove the "BRST cocycles" $W_{\alpha\beta}$ and K_α.

(d) Verify that if $[K_\alpha, \Omega] \neq 0$, then the gauge-fixed action is not invariant under the transformation generated by Q_α.

(e) Analyze how one can regain invariance of the gauge-fixed action by bringing in the Lagrange multipliers as canonical variables along the lines of §11.3.2.

11.7. (Same notations as for the above exercise). Assume that the global charges $\overset{(0)}{Q}_\alpha$ and the gauge generators G_a form a true algebra, *i.e.*, that the gauge structure functions $C_{ab}{}^c$, $V_{\alpha a}{}^b$, and $W_{\alpha\beta}{}^a$ are all constant.

(a) Observe that the gauge group G is an invariant subgroup of the group F generated by $\overset{(0)}{Q}_\alpha$ and G_a, and that the symmetry group Q is just F/G.

(b) Show that the "BRST cocycle" $W_{\alpha\beta}$ can be removed by transformations that preserve the group structure of F if and only if F is the semidirect product of G by Q ($W_{\alpha\beta}{}^a = 0$).

11.8. Consider a system with constraints $G_a = 0$ and global symmetry generators $\overset{(0)}{Q}_\alpha$. Introduce additional variables x^α, π_α, where x^α are coordinates on the global symmetry group manifold and π_α their conjugate momenta. Let $Y_\alpha(x)$ be a basis of left invariant vector fields on the group manifold, $[Y_\alpha, Y_\beta] = D_{\alpha\beta}{}^\gamma Y_\gamma$.

(a) Show that the original system is equivalent to the enlarged system with constraints

$$G_a \approx 0, \qquad G_\alpha = \overset{(0)}{Q}_\alpha + Y_\alpha{}^\beta \pi_\beta \approx 0,$$

where $Y_\alpha{}^\beta(x)$ are the components of the vectors Y_α. Check that the original Hamiltonian is still first class. The variables X^α, π_α "gauge" the global symmetry (in a trivial way, *i.e.*, without introducing gauge degrees of freedom).

(b) Construct the BRST charge for the enlarged system and compare it with the BRST charge of the original system.

11.9. Let $s = s_1 + s_2$ be a differential acting on the tensor product algebra $\mathcal{A}_1 \otimes \mathcal{A}_2$. The Kunneth formula relates the cohomology of s to the cohomologies of s_1 and s_2,

$$H^k(s) = \bigoplus_{p=0}^{k} [H^p(s_1) \otimes H^{k-p}(s_2)].$$

Use the Kunneth formula to show that cohomologically trivial pairs do not change the BRST cohomology.

11.10. In the formalism where the Lagrange multipliers λ^a are canonical variables, the extended action reads

$$S'_E[z^A, \lambda^a, b_a, \mu^a, \nu^a] = \int dt(\dot{z}^A a_A + \dot{\lambda}^a b_a - H_0 - \mu^a G_a - \nu^a b_a).$$

Show that the gauge condition $\lambda^a = \mu^a$ is admissible (*i.e.*, correctly freezes the freedom generated by $b_a = 0$) and that S'_E reduces to S_E in that gauge.

11.11. Consider a constrained system in which the Lagrange multipliers are incorporated as dynamical variables.

(a) Prove that locally in phase space there is a ghost number minus one "anti-BRST generator" $\Omega' = \sum\{\bar{C}_a G_a + i\mathcal{P}_a b_a (i)^{\varepsilon_a}\}$ + "more" such that $[\Omega, \Omega'] = 0 = [\Omega', \Omega']$. [Hints: Work out Ω' in the abelian case (one finds that the "more" is absent) and use the abelianization theorem.] Show that $\Omega' = [M, \Omega]$ with $\text{gh}\, M = -2$.

(b) Find the conditions on the gauge-fixing fermion K such that the gauge-fixed Hamiltonion is both BRST and anti-BRST invariant. (Hints: Use again the abelianization theorem and observe that $[[L, \Omega'], \Omega]$ is BRST and anti-BRST invariant.)

11.12. (a) Write explicitly the BRST transformaton s_L in Lagrangian form for the gauge-fixed actions (11.29) and (11.30).

(b) Show that these actions can be written as $S_E + \int dt\, s_L F$, where F is an appropriate function of the Lagrangian variables and their derivatives.

(c) Verify that if one eliminates b_a from (11.30) by means of its equation of motion (11.31a), the resulting BRST transformation fails to be off-shell nilpotent on $\bar{C}_a$. Furthermore, $s_L \bar{C}_a$ explicitly involve the gauge-fixing function χ^a. The b-variables are thus necessary to maintain off-shell nilpotency in the Lagrangian formalism.

11.13. Consider the Lagrangian

$$L = \sum_{i=1}^{n-1} \frac{1}{2}\left(q_i - \frac{dq_{i+1}}{dt}\right)^2$$

with n coordinates q_i ($i = 1, \ldots, n$), which was analyzed in §1.6.1. The gauge transformations are

$$\delta q_i = d^{n-i}\varepsilon / dt^{n-i},$$

and the system possesses no true degrees of freedom.

(a) Construct the BRST charge.

(b) Show that there exist gauge-fixing functions K for which (i) the Lagrangian BRST transformation just reads

$$\delta q_i = \frac{d^{n-i}C}{dt^{n-i}}$$

where C is the ghost associated with the last constraint—this is just a gauge transformation with $\varepsilon \to C$; and (ii) the ghost C obeys differential equations of motion of order n and the ghost Lagrangian takes the Faddeev–Popov form (11.29) and (11.30).

11.14. This exercise deals with the Hamilton–Jacobi theory of §11.3.5. We again assume throughout that the BRST-invariant Hamiltonian H is taken not to depend on the variables λ^a, b_a, $\bar{C}_a$, and ρ^a of the nonminimal sector, as is permissible.

(a) Show that the boundary conditions (11.33) define an ill-posed problem if K is taken not to involve the variables of the nonminimal sector (for instance, $K = 0$). Analyze, in particular, the case with $H_0 = 0$ and a single constraint $p = 0$.

(b) Show that the inclusion in K of the term $\lambda^a \mathcal{P}_a$ coupling the minimal and nonminimal sectors generically leads to a well-posed problem. (Hint: Analyze the "thin-sandwich problem" with $t_2 = t_1 + \varepsilon$.) Repeat the analysis in the particular case considered in (a).

CHAPTER TWELVE

THE BRST TRANSFORMATION IN FIELD THEORY

The BRST formalism developed so far may be formally applied to field theories as it stands by (i) regarding the indices $A, a_0, a_1, \ldots$ characterizing the fields, the constraints, and the reducibility functions as ranging over both a discrete set and the continuous set of all points *in space,* $A \rightarrow (A, x^k)$, $a_0 \rightarrow (a_0, x^k)$, etc...; (ii) replacing the summations over $A, a_0, a_1, \ldots$, by ordinary summations over the discrete indices *and* integrations over x^k; and (iii) replacing the ordinary derivatives with respect to the dynamical variables by functional derivatives,

$$\frac{\partial A}{\partial z^A} \rightarrow \frac{\delta A}{\delta z^A(x)}, \tag{12.1a}$$

with

$$\delta A = \int dx^k \, \delta z^A(x) \, \frac{\delta A}{\delta z^A(x)}. \tag{12.1b}$$

Although useful, this approach to the field theoretical case remains, however, rather formal as long as one does not specify precisely the functional space to which the relevant functionals should belong. This turns out, in general, to be a complicated task. Furthermore, the approach does not give any hint concerning the spacetime locality of the BRST formalism.

The purpose of this chapter is to reanalyze the BRST construction in the case of local field theories. We show that by using locality and making the assumption of finite rank, one can prove the existence of a BRST charge Ω that is a local functional. Namely, Ω is given by the spatial integral of a function of the variables of the extended phase space and of their spatial derivatives up to a finite order. As a result, the gauge-fixed action is not only local in time, as the Hamiltonian formalism guarantees, but it can also be taken to be local in space. Hence, it is local in spacetime. This fact makes the BRST formalism particularly attractive, since spacetime locality is a cornerstone of quantum field theory.

To undertake the analysis, it is necessary to establish some theorems on local functionals and to reformulate precisely the regularity conditions on the constraint functions in the case of field theory. These technical considerations are developed in Sec. 12.1 and Sec. 12.2, respectively. For simplicity, we assume that the topology of space is that of R^N, although the formalism can be extended to more complicated topologies.

12.1. LOCAL FUNCTIONALS AND NONINTEGRATED DENSITIES

Local functionals are given by the integral of functions of the fields and their derivatives up to some finite order ("nonintegrated densities"); *e.g.*,

$$A = \int a(z^A, \partial_k z^A, \ldots, \partial_{k_1} \ldots \partial_{k_s} z^A)\, dx \tag{12.2}$$

is a local functional. We shall in this chapter reserve capital letters (Latin or Greek) for local functionals and the corresponding small letters for the corresponding nonintegrated densities.

It was mentioned in Chapter 3 that the relationship between the functional derivative $\delta A/\delta z^A(x)$ of the local functional A and the variational derivative $\delta a/\delta z^A$ of the corresponding nonintegrated density a is

$$\frac{\delta A}{\delta z^A(x)} = \frac{\delta a}{\delta z^A}, \tag{12.3a}$$

where in $\delta a/\delta z^A$, the fields and their derivatives are evaluated at x. One has

$$\frac{\delta a}{\delta z^A} = \frac{\partial a}{\partial z^A} - \partial_k \frac{\partial a}{\partial(\partial_k z^A)} + \partial_k \partial_l \frac{\partial a}{\partial(\partial_k \partial_l z^A)} - \cdots,$$

$$\partial_k = \partial_k z^A \frac{\partial}{\partial z^A} + \partial_k \partial_s z^A \frac{\partial}{\partial(\partial_s z^A)} + \cdots. \tag{12.3b}$$

Theorem 12.1. *The variational derivative $\delta a/\delta z^A$ identically vanishes if and only if a is the sum of a "total divergence" and a constant,*

$$\frac{\delta a}{\delta z^A} = 0 \Longleftrightarrow a = \partial_k \alpha^k + C, \tag{12.4}$$

for some $\alpha^k(z^A, \partial_s z^A, \dots, \partial_{s_1} \dots \partial_{s_m} z^A)$ and C.

Proof. (a) That $a = \partial_k \alpha^k + C$ implies $\delta a/\delta z^A = 0$ is straightforward.

(b) To prove the converse, we introduce a derivation N in the algebra of nonintegrated densities that counts the powers of the fields,

$$N = \sum_{n \geq 0} \partial_{k_1} \dots \partial_{k_n} z^A \frac{\partial^L}{\partial(\partial_{k_1} \dots \partial_{k_n} z^A)}. \tag{12.5}$$

Homogeneous polynomials of order k in the fields and their derivatives are eigenstates of N with eigenvalue k. Making repeated integrations by parts, one easily shows

$$Na = z^A \frac{\delta^L a}{\delta z^A} + \partial_k \mathcal{V}^k \tag{12.6a}$$

with

$$\begin{aligned} \mathcal{V}^k = {} & z^A \Big(\frac{\partial^L a}{\partial(\partial_k z^A)} - \partial_s \frac{\partial^L a}{\partial(\partial_k \partial_s z^A)} + \cdots \Big) \\ & + \partial_s z^A \Big(\frac{\partial^L a}{\partial(\partial_k \partial_s z^A)} - \partial_m \frac{\partial^L a}{\partial(\partial_m \partial_k \partial_s z^A)} + \cdots \Big) \\ & + \cdots . \end{aligned} \tag{12.6b}$$

Hence, if $\delta a/\delta z^A = 0$, one finds

$$Na = \partial_k \mathcal{V}^k. \tag{12.7}$$

If a is homogeneous of degree m in the z^A and their derivatives, the proof is finished, since (12.7) implies that $a = \partial_k[(1/m)\nu^k]$ ($m \neq 0$) (if $m = 0$, then $a = \text{const}$). The result can be extended to arbitrary smooth functions by using the identity

$$t \frac{da}{dt}(tz) = (Na)(tz) \tag{12.8}$$

with $a(tz) \equiv a(tz^A, t\partial_k z^A, \dots, t\partial_{k_1} \dots \partial_{k_s} z^A)$. Therefore, one gets from (12.7) and (12.8),

$$t \frac{da}{dt}(tz) = \partial_k \mathcal{V}^k(tz), \tag{12.9}$$

and then

$$a(z) = \partial_k \alpha^k + C, \tag{12.10a}$$

$$\alpha^k = \int_0^1 \frac{1}{t} \mathcal{V}^k(tz), \quad C = a(0). \tag{12.10b}$$

The function $\mathcal{V}^k$ is at least linear in z^A and its derivatives [see (12.6b)] and, thus, of order $0(t)$ as $t \to 0$. Accordingly, the integral in (12.10b) is well defined. This completes the proof of the theorem. Note that the constant C can be absorbed in a total divergence, $C = \partial_1(Cx^1)$. This, however, requires an explicit dependence on the coordinates.

Remark. The previous proof can be applied separately to each field z^A. Thus, if the variational derivative of a with respect to one variable identically vanishes, then one can get rid of that variable by adding a total divergence to a.

Theorem 12.2. *If $A = \int a\, dx$ and $B = \int b\, dx$ are two local functionals that are equal for all field configurations,*

$$\int a\, dx = \int b\, dx, \tag{12.11a}$$

then

$$a = b + \partial_k j^k, \tag{12.11b}$$

where the surface integral $\oint j^k\, d\Sigma_k$ extended over the sphere at infinity vanishes for all the allowed field configurations.

Proof. The condition that the surface integral should vanish is obvious, so let us prove (12.11b) only.

If A equals B for all field configurations, the functional derivatives $\delta A/\delta z^A(x)$ and $\delta B/\delta z^A(x)$ are equal, *i.e.*,

$$\frac{\delta a}{\delta z^A} = \frac{\delta b}{\delta z^A}.$$

It follows from Theorem 12.1 that

$$a = b + \partial_k j^k + C.$$

Inserting this expression into $A = B$, one gets

$$\int (\partial_k j^k + C)\, dx = 0.$$

The surface term is found to vanish by considering the zero field—or any constant field—configuration, for which j^k is a constant vector density having, therefore, zero flux. Hence, the constant C must be zero. This proves the theorem.

Theorem 12.2 shows that a local functional determines the corresponding local densities only up to a total divergence. *One can thus identify the local functionals with the equivalence classes of local densities that differ by a local total divergence.*

Theorem 12.3. *If $\partial_k j^k = 0$ (identically), then there exists $\sigma^{km} = -\sigma^{mk}$ such that $j^k = \partial_m \sigma^{km} + C^k$ where the C^k's are constants,*

$$\partial_k j^k = 0 \Longleftrightarrow j^k = \partial_m \sigma^{km} + C^k, \quad \sigma^{km} = -\sigma^{mk}. \tag{12.12}$$

That the right-hand side of the implication implies the left-hand side is immediate. The proof of the converse needs the following.

Lemma 12.1. *Let $\phi^\alpha = (\phi^a, \phi^{N+1})$, $a = 1, \ldots, N$, $\alpha = 1, \ldots, N+1$, be $N+1$ independent fields and let m be a local function of the first N fields ϕ^a and their derivatives. If*

$$\sum_{a=1}^{N} \frac{\delta m}{\delta \phi^a} \phi^{a+1} = \partial_k \beta^k, \tag{12.13a}$$

then

$$\frac{\delta m}{\delta \phi^a} = 0, \tag{12.13b}$$

i.e.,

$$m = \partial_k n^k + C. \tag{12.13c}$$

Proof of Lemma 12.1. We first show $\delta m / \delta \phi^N = 0$. The dependence of the left-hand side of (12.13a) on the last field ϕ^{N+1} is linear, since m does not depend on ϕ^{N+1}. Accordingly, β^k can be assumed to be at most linear in ϕ^{N+1} and its derivatives,

$$\begin{aligned}\beta^k = \overset{(0)}{\beta}{}^k &+ b^k \phi^{N+1} + b^{ks} \partial_s \phi^{N+1} + \cdots \\ &+ b^{ks_1 \ldots s_u} \partial_{s_1} \ldots \partial_{s_u} \phi^{N+1}.\end{aligned} \tag{12.14a}$$

If (12.13a) is to hold identically, one must have

$$\frac{\delta m}{\delta \phi^N} \phi^{N+1} = \partial_k (b^k \phi^{N+1} + \cdots + b^{ks_1 \ldots s_u} \partial_{s_1} \ldots \partial_{s_u} \phi^{N+1}).$$

This implies, upon equating the coefficients of $\phi^{N+1}, \partial_s \phi^{N+1}, \ldots,$

$$\frac{\delta m}{\delta \phi^N} = \partial_k b^k \tag{12.14b}$$

$$0 = b^{(k_1 \ldots k_j)} + \partial_m b^{m k_1 \ldots k_j}, \qquad j = 1, 2, \ldots, u, \tag{12.14c}$$

$$0 = b^{(k_1 \ldots k_{u+1})}. \tag{12.14d}$$

Using (12.14c)–(12.14d), one finds successively $\partial_k b^k = -\partial_k \partial_s b^{sk} = + \partial_m \partial_k \partial_s b^{mks} = \cdots = (-)^u \partial_{k_1} \ldots \partial_{k_{u+1}} b^{k_1 \ldots k_{u+1}} = 0$ and, from (12.14b), $\delta m / \delta \phi^N = 0$. It thus follows that ϕ^N can be removed from m by adding a total divergence. Repeating the argument, one then gets $\delta m / \delta \phi^{N-1} = 0, \delta m / \delta \phi^{N-2} = 0, \ldots, \delta m / \delta \phi^1 = 0$.

Proof of Theorem 12.3. One proves the theorem by induction on the dimension.

(*i*) The theorem is true in one spatial dimension for any number of fields: the identity

$$\frac{dm}{dx} = \frac{\partial m}{\partial z^A} \frac{dz^A}{dx} + \cdots + \frac{\partial m}{\partial (d^u z^A / dx^u)} \frac{d^{u+1} z^A}{dx^{u+1}} = 0,$$

where $(d^u z^A / dx^u)$ is the last derivative of z^A occurring in m, implies successively $\partial m / \partial (d^u z^A / dx^u) = 0$, $\partial m / \partial (d^{u-1} z^A / dx^{u-1}) = 0$, ..., $\partial m / \partial z^A = 0$, *i.e.*, $m = C$.

(*ii*) If the theorem is true in n spatial dimensions, then it is also true in $n+1$ spatial dimensions. Straightforward integrations by parts in

$$\frac{\partial m^{n+1}}{\partial x^{n+1}} + \frac{\partial m^k}{\partial x^k} = 0 \qquad (k = 1, \ldots, n) \tag{12.15}$$

yield

$$\frac{\delta m^{n+1}}{\delta z^A} \frac{\partial z^A}{\partial x^{n+1}} + \frac{\delta m^{n+1}}{\delta (\partial z^A / \partial x^{n+1})} \frac{\partial^2 z^A}{(\partial x^{n+1})^2} + \cdots$$
$$+ \cdots + \frac{\delta m^{n+1}}{\delta \left(\partial^u z^A / (\partial x^{n+1})^u \right)} \frac{\partial^{u+1} z^A}{(\partial x^{n+1})^{u+1}} = \frac{\partial \bar{m}^k}{\partial x^k},$$

for some $\bar{m}^k$. Here, u is the order of the highest derivative of z^A with respect to the last coordinate x^{n+1} occurring in m^{n+1}, and $\delta / \delta z^A$ is the variational derivative in n dimensions $(\delta / \delta z^A = \partial / \partial z^A - \partial_k (\partial / \partial (\partial_k z^A) + \cdots)$, for which the successive derivatives z^A, $\partial z^A / \partial x^{n+1}$, $\partial^2 z^A / (\partial x^{n+1})^2, \ldots$ with respect to x^{n+1} are independent. Lemma 12.1 then show that there exists $\sigma^{n+1,k}$ such that $m^{n+1} = \partial_k \sigma^{n+1,k} + C^{n+1}$. Substitution of this relation in (12.15) and use of the theorem in n dimensions yield finally the desired result.

Remark. (a) One may use dual notations and replace the local densities by local n-forms in space, $a \to \alpha = a\, dx^1 \wedge dx^2 \wedge \cdots dx^n$. The addition of $\partial_k j^k$ to a then amounts to adding a d-exact n-form to α, where d is the spatial exterior d defined on local p-forms by the relations

$$dF = \partial_k F\, dx^k, \quad d(dx^k) = 0,$$
$$\partial_k F \equiv \frac{\partial F}{\partial z^A} \frac{\partial z^A}{\partial x^k} + \cdots. \tag{12.16}$$

In these dual notations, the equivalence classes $[a]$ of local densitites that differ by a total divergence become the cohomological classes $[\alpha]$ of (closed) n-forms modulo exact ones.

Similarly, the equivalence classes of conserved currents j^k that differ by the divergence of an antisymmetric tensor become the equivalence classes of closed $(n-1)$-forms modulo exact ones. Theorem 12.3 states that the cohomology of d in degree $n-1$ is spanned by the constant $(n-1)$-forms. [In the other lower degrees $k \leq n-2$, the cohomology of d is given similarly by the constant k-forms (Exercise 12.1).]

(b) Let $\bar{\partial}_k$ be the operator that differentiates only with respect to a subset of fields, *i.e.*, $\bar{\partial}_k F = (\partial F/\partial z^a)(\partial z^a/\partial x^k) + \cdots$, with $z^A = (z^a, z^\alpha)$. An argument similar to the proof of Theorem 12.3 shows that

$$\bar{\partial}_k j^k = 0 \Leftrightarrow j^k = \bar{\partial}_m \sigma^{km} + C^k(z^\alpha). \tag{12.17}$$

12.2. LOCAL COMPLETENESS AND REGULARITY CONDITIONS

12.2.1. Hamiltonian Definition of a Local Gauge Theory

A gauge theory in Hamiltonian form is local in spacetime if the following conditions hold:

(*i*) The Hamiltonian action is local, *i.e.*, takes the form

$$S_E[z^A, \lambda^{a_0}] = \int dt \int dx \, (\dot{z}^A \, a_A - h_0 - \lambda^{a_0} g_{a_0}), \tag{12.18}$$

where a_A, h_0, and the constraint functions g_{a_0} are ordinary functions of the fields z^A and their derivatives up to some finite order. The number of fields and of constraints at each space point is, furthermore, finite. The Hamiltonian H_0 is given by $\int dx \, h_0$. The constraints are $g_{a_0} = 0$.

(*ii*) The brackets $[z^A(x), z^B(y)]$ are local, *i.e.*, involve the delta function $\delta(x-y)$ and its derivatives up to some finite order and no primitive of $\delta(x-y)$ ("local commutativity"). This implies that the bracket of two local functionals is also a local functional, so that nonlocal terms cannot be generated through the operation of taking the bracket.

(*iii*) (Reducible theory). The reducibility identities take a local form, *i.e.*, can be expressed as local combinations of the g_{a_0} and their derivatives with coefficients that depend on the canonical variables and their derivatives up to a finite order.

It is only for local gauge theories that one should expect the BRST formalism to be local.

Some remarks on the conditions (*i*), (*ii*), and (*iii*) are gathered in Exercises 12.2 through 12.5.

12.2.2. Regularity Conditions

Appropriate regularity conditions played an important role in the development of the BRST theory of systems with a finite number of degrees of freedom. It is necessary to reformulate precisely these conditions in the case of field theory.

12.2.2a. Local Completeness of the Constraint Functions

We will first require that the constraint functions are not only local but are also locally complete. By this, it is meant that any function f of the fields and their derivatives that vanishes on the constraint surface is zero by virtue of $g_{a_0} = 0$, $\partial_k\, g_{a_0} = 0$, $\partial_k \partial_s\, g_{a_0} = 0$, ... only, without having to invoke the boundary conditions.

Consider, for instance, a theory with a single conjugate pair $(\phi(\sigma), \pi(\sigma))$ subject to the condition $\pi(\sigma) \to 0$ for $\sigma \to \pm\infty$. Let the constraint be

$$\pi' = 0. \tag{12.19}$$

The function $\pi(\sigma)$ vanishes on the constraint surface. However, one cannot infer $\pi = 0$ from (12.19) by local means, *i.e.*, by mere differentiation: one must use the boundary conditions. With different boundary conditions, π would not vanish. The constraint (12.19) is thus not locally complete. To achieve local completeness, one must replace (12.19) by

$$\pi = 0, \tag{12.20}$$

which is permissible due to the boundary conditions.

By replacing, if necessary, the original constraints g_{a_0} by their appropriate primitives, as above, or by adding extra constraints at the price of increasing the reducibility, one can always assume that the local completeness condition on g_{a_0} is fulfilled.

12.2.2b. Zero Is a Regular Value of the Map Defined by the Constraint Functions

A nonintegrated density depends on a finite number of variables, which are the fields z^A and their derivatives up to a finite order. At an arbitrary point, these variables are independent. It is thus natural to introduce the finite-dimensional space V^k of the fields and their derivatives up to order k, regarded as independent variables. For any (smooth) nonintegrated density f, there is a k such that f belongs to the functional space $C^\infty(V^k)$; *e.g.*, a in (12.2) is a smooth function in V^s. Furthermore, if f belongs to $C^\infty(V^k)$, it also belongs to $C^\infty(V^{k'})$ for $k' \geq k$. Hence, given a finite number of local functionals, one can find a finite k such that the corresponding nonintegrated densities all belong to the same $C^\infty(V^k)$.

The constraint equations

$$g_{a_0}(z^A, \partial_k z^A, \dots, \partial_{k_1} \dots \partial_{k_s} z^A) = 0 \tag{12.21}$$

define a submanifold in the finite-dimensional space V^s of the fields and their derivatives up to order s. Similarly, the equations

$$g_{a_0} = 0, \qquad \partial_k g_{a_0} = 0, \dots, \partial_{k_1} \dots \partial_{k_n} g_{a_0} = 0, \tag{12.22}$$

define a submanifold in V^{s+n}. We will assume that for each integer $n \geq 0$, the Eqs. (12.22) in V^{s+n} can be locally split into independent equations $F_{A_n} = 0$ and dependent equations $E_{\Delta_n} = 0$ in such a way that (i) the F_{A_n} can locally (in V^{s+n}) be taken as first coordinates of a new, regular coordinate system in V^{s+n}; and (ii) the equations $E_{\Delta_n} = 0$ are consequences of $F_{A_n} = 0$. This requirement is just the translation in the field theoretical case of the regularity conditions imposed in Chapter 1.

With the conditions of §12.3.2a and §12.3.2b, the following theorem holds.

Theorem 12.4. *If a function $f(z^A, \partial_k z^A, \dots, \partial_{k_1} \dots \partial_{k_p} z^A)$ vanishes when the constraints hold, then f can be written as a combination of the constraint functions and their derivatives,*

$$f = \lambda^{a_0} g_{a_0} + \lambda^{a_0 k} \partial_k g_{a_0} + \cdots + \lambda^{a_0 k_1 \dots k_q} \partial_{k_1} \dots \partial_{k_q} g_{a_0} = 0, \tag{12.23}$$

where the coefficients $\lambda^{a_0}, \lambda^{a_0 k}, \dots, \lambda^{a_0 k_1 \dots k_q}$ involve the fields z^A and their derivatives up to some finite order.

Proof. The proof is immediate and proceeds as in the finite-dimensional case. The function f belongs to $C^\infty(V^p)$ and, hence, to any $C^\infty(V^{p'})$ for $p' \geq p$. Only a finite number of equations

$$g_{a_0} = 0, \ \dots, \ \partial_{k_1} \dots \partial_{k_q} g_{a_0} = 0 \tag{12.24}$$

are relevant for determining the constraint surface induced in V^p (if one differentiates too much the functions g_{a_0}, one gets after a while equations that imply no new restriction in V^p but, rather, that constrain the derivatives of order $> p$). One has $q + s \geq p$. The function f belongs to $C^\infty(V^{q+s})$ and vanishes on the surface defined by (12.24) in V^{q+s}. Hence, it can be written as in (12.23) (Theorem 1.1 with ϕ_m replaced by $g_{a_0}, \dots, \partial_{k_1}, \dots, \partial_{k_q} g_{a_0}$). This proves Theorem 12.4.

12.2.3. Local Completeness of the Reducibility Functions

The constraints may not be independent, in which case there are relations among the function g_{a_0} and their derivatives. By the assumptions (*iii*) of §12.2.1, these take a local form,

$$\zeta_{a_1}{}^{a_0} g_{a_0} + \zeta_{a_1}{}^{a_0 k}\, \partial_k\, g_{a_0} + \cdots + \zeta_{a_1}{}^{a_0 k_1 \ldots k_t} \partial_{k_1} \ldots \partial_{k_t} g_{a_0} = 0\,, \qquad a_1 = 1, \ldots, m_1\,. \tag{12.25}$$

We will choose the reducibility identities (12.25) to be not only complete but also locally complete. That is, there is no local identity among $g_{a_0}, \partial_k\, g_{a_0}, \ldots$ that cannot be derived from (12.25) by algebraic manipulations and a finite number of differentiations.

To illustrate this requirement, consider again a single conjugate pair $(\phi(\sigma), \pi(\sigma))$ with constraints

$$g_1 \equiv \pi = 0, \quad g_2 \equiv \pi' = 0. \tag{12.26a}$$

The constraints are not independent. One has, for instance,

$$g_1' - g_2 = 0, \tag{12.26b}$$

or also

$$g_1'' - g_2' = 0. \tag{12.26c}$$

Both (12.26b) and (12.26c) are complete under the condition that $\pi(\sigma)$ and its derivatives go to zero as $\sigma \to \pm\infty$. However, only (12.26b) is locally complete. The relation (12.26c) is not locally complete because (12.26b), viewed as an equation for g_1 and g_2, cannot be inferred from (12.26c) without using the boundary conditions.

For convenience, we organize the reducibility identities as follows: for $1 \leq a_1 \leq m_1^{(0)}$, they involve only g_{a_0},

$$\zeta_{a_1}{}^{a_0} g_{a_0} = 0, \quad 1 \leq a_1 \leq m_1^{(0)}; \tag{12.27a}$$

for $m_1^{(0)} < a_1 \leq m_1^{(1)}$, they involve only g_{a_0} and $\partial_k\, g_{a_0}$,

$$\zeta_{a_1}{}^{a_0} g_{a_0} + \zeta_{a_1}{}^{a_0 k}\, \partial_k\, g_{a_0} = 0, \quad m_1^{(0)} < a_1 \leq m_1^{(1)}; \tag{12.27b}$$

for $m_1^{(u)} < a_1 \leq m_1^{(u+1)}$, they involve only g_{a_0} and its successive derivatives up to $\partial_{k_1} \ldots \partial_{k_{u+1}} g_{a_0}$,

$$\zeta_{a_1}{}^{a_0} g_{a_0} + \zeta_{a_1}{}^{a_0 k} \partial_k\, g_{a_0} + \cdots + \zeta_{a_1}{}^{a_0 k_1 \ldots k_{u+1}} \partial_{k_1} \ldots \partial_{k_{u+1}} g_{a_0} = 0, \qquad m_1^{(u)} < a_1 \leq m_1^{(u+1)}, \tag{12.27c}$$

etc..... At each stage, the $\zeta_{a_1}{}^{a_0}, \zeta_{a_1}{}^{a_0 k}, \zeta_{a_1}{}^{a_0 kl}, \ldots$ are taken so as to exhaust, together with the derivatives of the previous identities, all the identities on the g_{a_0}'s and their derivatives up to that order. The coefficients $\zeta_{a_1}{}^{a_0 k_1 \ldots k_m}$ are functions in V^{s+u+1} for $a_1 \leq m_1^{(u+1)}$.

Similar local completeness requirements should be imposed on the higher order reducibility functions $\zeta_{a_2}{}^{a_1}$, if any. For simplicity, we will, however, assume that the relations (12.27) are independent so that there are no nontrivial $\zeta_{a_2}{}^{a_1}$'s: the theory is a reducible theory of the first stage.

12.3. LOCALITY OF THE BRST CHARGE

12.3.1. Homology of δ modulo $\partial_k j^k$ as the Central Issue in the Problem of the Spacetime Locality of the BRST Formalism

We now have all the technical tools for investigating the locality of the BRST formalism. Our aim is to establish that the BRST charge is given by a local functional,

$$\Omega = \int dx\, \omega, \tag{12.28a}$$

where the nonintegrated density ω depends on the fields, the ghosts, their momenta, and their spatial derivatives up to a finite order.

The proof of the existence of the BRST charge given in Chapters 9 and 10 heavily relied on the acyclicity of the Koszul–Tate differential at strictly positive antighost numbers. This property is still true in the case of field theory, but there is no *a priori* reason that if A is a δ-closed local functional, it is equal to δB where B is also a local functional: B is known to exist but may be nonlocal. It is thus necessary to reanalyze carefully the equations for Ω in terms of the nonintegrated density ω.

The expansion of Ω according to antighost number

$$\Omega = \sum_{k \geq 0} \overset{(k)}{\Omega} \tag{12.28b}$$

yields for the nonintegrated density ω of (12.28a),

$$\omega = \sum_{k \geq 0} \overset{(k)}{\omega}. \tag{12.28c}$$

Furthermore, the nilpotency condition $[\Omega, \Omega] = 0$ is equivalent by virtue of Theorem 12.2 to the condition

$$\alpha(\omega, \partial_k \omega, \ldots) = \partial_k \mu^k \tag{12.29}$$

for (any of) the local integrand(s) α appearing in $[\Omega, \Omega] = \int \alpha$. If one can find a solution ω of (12.29) obeying the appropriate "initial" conditions

$$\overset{(0)}{\omega} = \eta^{a_0} g_{a_0}, \tag{12.30a}$$

$$\overset{(1)}{\omega} = \eta^{a_1}(\zeta_{a_1}{}^{a_0}\mathcal{P}_{a_0} + \cdots + \zeta_{a_1}{}^{a_0 k_1 \dots k_t}\partial_{k_1}\dots\partial_{k_t}\mathcal{P}_{a_0}) + \overset{(1)}{\omega}{}', \tag{12.30b}$$

where $\overset{(1)}{\omega}{}'$ does not depend on η^{a_1}, then it is guaranteed that the equations for the BRST charge Ω admit a solution given by a local functional.

Since the constraints are first class, the bracket

$$\left[g_{a_0}(x), \int dx' \, \eta^{b_0}(x') \, g_{b_0}(x')\right]$$

vanishes weakly. Theorem 12.4 guarantees then the existence of a local functional $\overset{(1)}{\Omega}{}' = \int \overset{(1)}{\omega}{}' \, dx$ linear in $\mathcal{P}_{a_0}$ such that

$$\int dx \, dx' \, \eta^{a_0}(x) \, \eta^{b_0}(x') \, [g_{a_0}(x), \, g_{b_0}(x')] = 2 \int \frac{\delta \overset{(1)}{\Omega}{}'}{\delta \mathcal{P}_{a_0}(x)} \, g_{a_0}(x) \, dx.$$

Accordingly, $\overset{(1)}{\Omega}$ not only exists but can also be chosen to be a local functional.

The remaining equations for the higher order terms $\overset{(n)}{\Omega}$ read [see (10.65d)]

$$\delta \overset{(p+1)}{\Omega} + \overset{(p)}{D} = 0 \tag{12.31a}$$

or, in terms of the corresponding nonintegrated densities,

$$\delta \overset{(p+1)}{\omega} + \overset{(p)}{d} = \partial_k \overset{(p)}{\mu}{}^k, \tag{12.31b}$$

for some $\overset{(p)}{\mu}{}^k$. The Koszul–Tate differential δ, defined by

$$\delta z^A = 0, \quad \delta \eta^{a_0} = 0, \quad \delta \eta^{a_1} = 0, \tag{12.32a}$$

$$\begin{aligned} \delta \mathcal{P}_{a_0} &= -g_{a_0}, \\ \delta \mathcal{P}_{a_1} &= -(\zeta_{a_1}{}^{a_0}\mathcal{P}_{a_0} + \cdots + \zeta_{a_1}{}^{a_0 k_1 \dots k_t} \, \partial_{k_1} \dots \partial_{k_t} \, \mathcal{P}_{a_0}), \end{aligned} \tag{12.32b}$$

is extended to the derivatives of the fields by using the rule

$$\delta \partial_k = \partial_k \delta. \tag{12.32c}$$

Because $\delta \overset{(p)}{D} = 0$, the nonintegrated density $\overset{(p)}{d}$ obeys the identity

$$\delta \overset{(p)}{d} = \partial_k \overset{(p-1)}{\nu}{}^k \tag{12.33a}$$

for some $\overset{(p-1)}{\nu}{}^{k}$. Furthermore, $p \geq 1$, *i.e.*,

$$\text{antigh} \overset{(p)}{d} \geq 1, \quad \text{pure gh} \overset{(p)}{d} \geq 2. \tag{12.33b}$$

The question of existence of $\overset{(p+1)}{\omega}$—and hence of locality of $\overset{(p+1)}{\Omega}$—is: do the identities (12.33) imply that $\overset{(p)}{d}$ is δ-exact up to a spatial divergence [Eq. (12.31b)]? The answer to this question is affirmative, as indicated by the following theorem.

Theorem 12.5 (on the homology of δ modulo $\partial_k j^k$). *Let f be a nonintegrated density with the following properties:*

$$\text{(i)} \quad \text{antigh}\, f \geq 1, \quad \textit{pure}\ \text{gh}\, f \geq 1; \tag{12.34a}$$

(ii) *f is δ-closed modulo a divergence, i.e.,*

$$\delta f = \partial_k \alpha^k. \tag{12.34b}$$

Then, f is δ-exact modulo a divergence,

$$f = \delta e + \partial_k \beta^k. \tag{12.34c}$$

Here, f, α^k, e and β^k are functions of the fields and their derivatives up to a finite order.

The proof of this theorem, which asserts that the cohomology of δ is still trivial in the space of local functionals, is given in the next two subsections.

12.3.2. Proof of Theorem 12.5: (i) Local Homology of δ

The first step in the proof of Theorem 12.5 lies in the following lemma.

Lemma 12.2. *Let f be a nonintegrated density of antighost number ≥ 1, which is δ-closed,*

$$\delta f = 0, \qquad \text{antigh}\, f \geq 1. \tag{12.35a}$$

Then, f is δ-exact,

$$f = \delta e, \tag{12.35b}$$

where e is also a local function of the fields and their derivatives. In other words, the homology of δ is trivial at antighost ≥ 1 in the space of local functions,

$$H_k^{\text{loc}}(\delta) = 0, \qquad k \geq 1. \tag{12.35c}$$

The proof of this lemma goes as in Chapter 10 because once the problem is formulated in terms of local functions, it becomes a finite-dimensional problem.

More precisely, let W^{s+n} be for each n the space containing: (i) the variables z^A and their derivatives up to order $s+n$; (ii) the ghost momenta $\mathcal{P}_{a_0}$ and their derivatives up to order n; and (iii) the ghost-of-ghost momenta $\mathcal{P}_{a_1}$ and their derivatives up to order n for $a_1 \leq m_1^{(0)}$; up to order $n-1$ for $m_1^{(0)} < a_1 \leq m_1^{(1)}, \ldots$; and up to order $n-u$ for $m_1^{(u-1)} < a_1 \leq m_1^{(u)}$.

The differential δ is well defined and nilpotent on the smooth functions on W^{s+n}. Furthermore, it enforces through its homology the following relations on z^A and its derivatives up to order $s+n$,

$$g_{a_0} = 0, \quad \partial_k \, g_{a_0} = 0, \quad \ldots, \quad \partial_{k_1} \ldots \partial_{k_n} \, g_{a_0} = 0. \tag{12.36a}$$

This is because W^{s+n} contains $\mathcal{P}_{a_0}$ and its derivatives up to order n.

The relations (12.36a) are not independent. As we have seen, a complete set of reducibility identities is given by

$$\zeta_{a_1}{}^{a_0} \, g_{a_0} = 0 \, , \; \partial_k(\zeta_{a_1}{}^{a_0} \, g_{a_0}) = 0, \, \ldots \, , \, \partial_{k_1} \ldots \partial_{k_n} \, (\zeta_{a_1}{}^{a_0} \, g_{a_0}) = 0 \qquad (a_1 \leq m_1^{(0)}) \tag{12.36b}$$

$$\zeta_{a_1}{}^{a_0} \, g_{a_0} + \zeta_{a_1}{}^{a_0 k} \partial_k \, g_{a_0} = 0 \, , \, \ldots \, , \, \partial_{k_1} \ldots \partial_{k_{n-1}} (\zeta_{a_1}{}^{a_0} \, g_{a_0} + \zeta_{a_1}{}^{a_0 k} \partial_k \, g_{a_0}) = 0 \qquad (m_1^{(0)} < a_1 \leq m_1^{(1)}) \tag{12.36c}$$

etc.... For each of these relations, there is an independent ghost-of-ghost momentum, which is either $\mathcal{P}_{a_1}$ or one of its appropriate derivatives. Hence, it follows from the analysis of Chapter 10 that δ is acyclic at antighost number $\neq 0$ in W^{s+n} for every n.

As an arbitrary function of z^A, $\mathcal{P}_{a_0}$, $\mathcal{P}_{a_1}$, and their derivatives up to a finite order belongs to a W^{s+n} for n great enough, we can conclude that

$$H_k^{\text{loc}}(\delta) = 0, \qquad k \neq 0, \tag{12.37}$$

as stated in the lemma.

Note that the result is unchanged if one includes the ghosts η^{a_0}, η^{a_1} and their derivatives, since these play a completely decoupled role in the analysis of the homology of δ.

12.3.3. Proof of Theorem 12.5: (*ii*) Homology of δ modulo $\partial_k j^k$

To analyze the homology of δ modulo $\partial_k j^k$, it is convenient to introduce an even derivation M that counts the number of derivatives of the ghosts,

$$M\partial_{k_1} \dots \partial_{k_n} z^A = 0, \quad M\partial_{k_1} \dots \partial_{k_n} \mathcal{P}_{a_i} = 0 \quad (n \geq 0), \tag{12.38a}$$

$$M\partial_{k_1} \dots \partial_{k_n} \eta^{a_i} = n\partial_{k_1} \dots \partial_{k_n} \eta^{a_i}. \tag{12.38b}$$

The derivative ∂_k can be split into a part of M-degree zero and a part of M-degree one,

$$\partial_k = \tilde{\partial}_k + \bar{\partial}_k, \tag{12.39a}$$

$$[M, \tilde{\partial}_k] = 0, \qquad [M, \bar{\partial}_k] = \bar{\partial}_k. \tag{12.39b}$$

The derivative $\bar{\partial}_k$ differentiates only with respect to the ghosts and their derivatives; the derivative $\tilde{\partial}_k$ sees only the other variables. One has separately

$$[\delta, \tilde{\partial}_k] = 0, \qquad [\delta, \bar{\partial}_k] = 0. \tag{12.39c}$$

Lemma 12.3. *The homology of δ modulo $\bar{\partial}_k j^k$ is trivial at strictly positive antighost number,*

$$\delta f = \bar{\partial}_k \alpha^k, \quad \text{antigh} f \geq 1 \Rightarrow f = \delta e + \bar{\partial}_k \beta^k. \tag{12.40}$$

Proof. The variables of W^{s+n} can be redefined locally in W^{s+n} in such a way that δ takes the standard form (8.30c) of Theorem 8.2, namely,

$$\delta x_i = 0, \qquad \delta y_\alpha = 0, \qquad \delta \mathcal{P}_\alpha = y_\alpha. \tag{12.41a}$$

Defining a contracting homotopy σ as in §8.3.2,

$$\sigma x_i = 0, \qquad \sigma y_\alpha = \mathcal{P}_\alpha, \qquad \sigma \mathcal{P}_\alpha = 0, \tag{12.41b}$$

and extending σ to the ghosts and their derivatives as

$$\sigma\, \partial_{k_1} \dots \partial_{k_n} \eta^{a_i} = 0 \qquad (n \geq 0), \tag{12.41c}$$

one gets

$$\sigma \bar{\partial}_k = \bar{\partial}_k \sigma, \qquad \sigma\delta + \delta\sigma = \bar{N}, \tag{12.42}$$

where $\bar{N}$ counts the number of y's and $\mathcal{P}$'s.

One can assume $\bar{N} f = \bar{n} f$, with $\bar{n} \neq 0$ because antigh $f > 0$. It follows from (12.42) and $\delta f = \bar{\partial}_k \alpha^k$ that

$$f = \delta e + \bar{\partial}_k \beta^k, \qquad e = \frac{1}{\bar{n}} \sigma f, \qquad \beta^k = \frac{1}{\bar{n}} \sigma \alpha^k.$$

The result (12.40) holds, therefore, in any open set where the variables of W^{s+n} can be redefined so as to achieve (12.41a). The result can then be extended to the whole of W^{s+n} by using an appropriate partition of unity (the functions of the partition of unity are annihilated by both $\bar{\partial}_k$ and δ). This proves the lemma.

We are now in a position to prove Theorem 12.5. If one could find a contracting homotopy such that $\sigma\partial_k = \partial_k\,\sigma$, the proof would be straightforward and would proceed as the proof of Lemma 12.3. However, there is, in general, no such contracting homotopy (see Exercise 12.8). Consequently, one must proceed differently. The idea of the proof is to use a perturbative argument based on Lemma 12.3 and on an expansion according to the M-degree.

Let f be such that $\delta f = \partial_k\,\alpha^k$. The local functions f and α^k can be expanded according to the number of derivatives of the ghosts,

$$f = \overset{(0)}{f} + \cdots + \overset{(s)}{f}, \tag{12.43a}$$

$$\alpha^k = \overset{(0)}{\alpha}{}^k + \cdots + \overset{(t)}{\alpha}{}^k, \tag{12.43b}$$

$$M\overset{(n)}{f} = n\overset{(n)}{f}, \qquad M\overset{(n)}{\alpha}{}^k = n\overset{(n)}{\alpha}{}^k. \tag{12.43c}$$

The expansions (12.43) both stop at some finite value of s and t because f and α^k are polynomials in the ghosts and are local, *i.e.*, contain a finite number of derivatives of the ghosts.

The equation $\delta f = \partial_k\alpha^k$ is equivalent to the set of equations

$$\delta\overset{(i)}{f} = \bar{\partial}_k\overset{(i-1)}{\alpha}{}^k + \tilde{\partial}_k\overset{(i)}{\alpha}{}^k. \tag{12.44}$$

Without loss of generality, one can assume $\overset{(i)}{\alpha} = 0$ for $i \geq s$, since otherwise the last equation (12.44) with $i = t+1$ reads $\bar{\partial}_k\overset{(t)}{\alpha}{}^k = 0$, which implies $\overset{(t)}{\alpha}{}^k = \bar{\partial}_m\overset{(t-1)}{\gamma}{}^{km}$ $(t \geq 1)$ or $\overset{(t)}{\alpha}{}^k = 0$ $(t = 0)$ by using Eq. (12.17). [The ghost-independent term is zero, since the pure ghost number of α does not vanish.] Replacing α^k by $\alpha^k - \partial_m\overset{(t-1)}{\gamma}{}^{km}$ leads then to a new α^k without component of degree t and still such that $\delta f = \partial_k\,\alpha^k$.

When $t = s-1$, Eq. (12.44) with $i = s$ is given explicitly by

$$\delta\overset{(s)}{f} = \bar{\partial}_k\overset{(s-1)}{\alpha}{}^k.$$

Because of Lemma 12.3, this implies

$$\overset{(s)}{f} = \delta\overset{(s)}{e} + \bar{\partial}_k\overset{(s-1)}{\beta}{}^k = \delta\overset{(s)}{e} + \partial_k\overset{(s-1)}{\beta}{}^k - \tilde{\partial}_k\overset{(s-1)}{\beta}{}^k.$$

Accordingly, the component $\overset{(s)}{f}$ of f can be removed by adding a δ-exact term and a total divergence. (The extra term $\tilde{\partial}_k\overset{(s-1)}{\beta}{}^k$ is of M-degree

$s-1$ and thus affects only $\overset{(s-1)}{f}$.) The problem is thus reduced by one unit of the M-degree. By repeating the same procedure at the lower M-degrees, one ultimately arrives at the desired result.

12.3.4. Locality of the Gauge-Fixed Action

Theorem 12.5 guarantees that the BRST generator Ω for a local Hamiltonian theory can be taken to be a local functional provided the rank of the theory is finite. If the rank of the theory is infinite, Ω may *a priori* contain derivatives of the fields of arbitrarily high order because the order of the derivatives in $\overset{(n)}{\Omega}$ may increase with n (the order of the derivatives in $\delta a/\delta z^A$ is greater than the order of the derivatives in a). However, there is no known reducible theory of the first stage that is of infinite rank, so we shall make the assumption that the theory has finite rank. With this assumption, Ω is a local functional.

Once the locality of Ω is established, the BRST symmetry is easily seen to be local. The locality of the gauge-fixed Hamiltonian is then demonstrated along the same lines. Using Theorem 12.5, one shows that the gauge-invariant Hamiltonian H_0 possesses at least one local BRST-invariant extension $H = \int h\,dx$ (see Exercise 12.9). The gauge-fixed Hamiltonians $H + [K, \Omega]$ are accordingly local functionals if K is taken to be a local functional. This shows that the gauge-fixed action in Hamiltonian form is also the integral of a local, gauge-fixed density.

The techniques of this chapter can be extended to cover the case when the constraints or the Hamiltonian have an explicit dependence on the coordinates, as would occur in an external background field (see Exercise 12.13). The same algebraic techniques will also be used in the analysis of the antifield formalism in Chapters 17 and 18, where it will be shown that the gauge-fixed action in Lagrangian form is a local functional.

EXERCISES. CHAPTER TWELVE

12.1. Prove by induction on the spatial dimension that $d\alpha = 0$ implies $\alpha = d\beta + C$ for any local k-form α, $k \neq n$. Here, d is the spatial exterior derivative defined by (12.16) in the space of local p-forms.

12.2. This exercise shows that conditions (i) and (ii) of §12.2.1 are independent.
(a) Show that the action $S[\phi, p, \lambda] = \int (\dot{\phi}\phi' + p\lambda)\, dt\, d\sigma$ obeys (i) but not (ii).
(b) Conversely, show that the local brackets $[\chi(\sigma), \chi(\sigma')] = d\delta(\sigma - \sigma')/d\sigma$ obey (ii) but cannot be derived from an action fulfilling (i).

12.3. This exercise and the next show that the Legendre transformation may not preserve spatial locality. Hence, if both the Hamiltonian and Lagrangian actions are required to be local, further conditions should be imposed besides (i), (ii), and (iii)

of §12.2.1, as these only guarantee the spatial locality of the Hamiltonian gauge-fixed action. Consider the local action $S = \frac{1}{2}\int dt\, d\sigma (\dot{\phi}')^2$. Calculate the momentum $\pi \equiv (\partial L/\partial\dot{\phi}) - (\partial L/\partial\dot{\phi}')'$. Show that the first-order action $S_H[\phi,\pi] = \int dt\, d\sigma\, (\pi\dot{\phi} - L)$ is not local in space.

12.4. Consider the local first-order action $S = \int dt\, d\sigma\, (\pi\dot{\phi} - \frac{1}{2}\pi'^2)$. Show that the action obtained by elimination of the momentum π by means of $\delta S/\delta\pi = 0$ is not local in space.

12.5. Consider the constraints $g(\sigma) \equiv \pi'(\sigma) = 0$ on the circle. Show that the constraints are reducible but that the reducibility identity is not local. [It reads $\int g(\sigma)\, d\sigma = 0$.]

Note: Nonlocal reducibility identities depend on the topology of the manifold or on the boundary conditions. They cannot be found by a local analysis of the constraints and can be accounted for in the BRST formalism either by introducing global ghosts of ghosts (one single such ghost of ghost in the above case) or by choosing appropriately the functional space for the ghosts (ghosts without zero mode for the above example).

12.6. Let $\phi(x)$ and $\pi(x)$ be such that $[\phi(x), \pi(y)] = \delta(x-y) = -[\pi(y), \phi(x)]$. One has, for arbitrary functionals,

$$[A, B] = \int dx \left(\frac{\delta A}{\delta\phi(x)} \frac{\delta B}{\delta\pi(x)} - \frac{\delta A}{\delta\pi(x)} \frac{\delta B}{\delta\phi(x)} \right).$$

Verify that if A, B are local functionals and f is a nonintegrated density, then

$$[f, B] = \sum_k \left[\frac{\partial f}{\partial\phi^{(k)}} \left(\frac{\delta b}{\delta\pi}\right)^{(k)} - \frac{\partial f}{\partial\pi^{(k)}} \left(\frac{\delta b}{\delta\phi}\right)^{(k)} \right],$$

$$[A, B] = \int dx\, \{a, b\}, \qquad \{a, b\} = \frac{\delta a}{\delta\phi}\frac{\delta b}{\delta\pi} - \frac{\delta a}{\delta\pi}\frac{\delta b}{\delta\phi}.$$

12.7. Show that the bracket $\{a, b\}$ defined in Exercise 12.6 among nonintegrated densities fulfills the Jacobi identity up to a total divergence. Find explicitly three nonintegrated densities a, b, c for which $\{\{a,b\},c\} + \{\{b,c\},a\} + \{\{c,a\},b\}$ does not vanish identically.

12.8. Consider the system $(\phi(\sigma), \pi(\sigma))$ with constraint $\pi' = 0$ (but $\pi \neq 0$).

(a) Verify that $\mathcal{P}$ is δ-closed modulo $d/d\sigma$.

(b) Show that $\mathcal{P}$ cannot be written as $\delta e + (d\beta/d\sigma)$. [Hint: Show that $\delta e + (d\beta/d\sigma)$ vanishes when one sets the derivatives of the fields equal to zero, while $\mathcal{P}$ does not.] There is no contradiction with Theorem 12.5 because the pure ghost number of $\mathcal{P}$ vanishes.

(c) Prove that if there were a constracting homotopy commuting with ∂_k, then $\mathcal{P}$ should be equal to $\delta e + d\beta/d\sigma$ for some e and β. Hence, even though there exist contracting homotopies commuting with $\bar{\partial}_k$, there are none commuting with ∂_k.

12.9. Let $\overset{(0)}{A} = \int \overset{(0)}{a}\, dx$ be a local functional that is weakly gauge invariant.

(a) From $[\overset{(0)}{A}, \int dx\, g_{a_0}\, \xi^{a_0}] \approx 0$ for arbitrary $\xi^{a_0}(x)$, infer that $\overset{(0)}{a}$ is weakly gauge invariant up to a divergence. (Hint: The invariance of $\overset{(0)}{A}$ can be written as $\int dx\, t_{a_0} \xi^{a_0} \approx 0$, where t_{a_0} differs from the gauge variation of $\overset{(0)}{a}$ by a divergence. Since ξ^{a_0} is arbitrary, this implies $t_{a_0} \approx 0$.)

(b) Use this information to show the existence of a local functional $\overset{(1)}{A}$ such that $\left[\overset{(0)}{A} + \overset{(1)}{A}, \Omega\right] =$ terms of antighost number ≥ 1.

(c) Prove the existence of BRST-invariant extensions $A = \int a\, dx$ of $\overset{(0)}{A}$ that are local functionals. (Hint: Complete $\overset{(0)}{A} + \overset{(1)}{A}$ by adding $\overset{(2)}{A} + \overset{(3)}{A} + \cdots$, and use Theorem 12.5 on the local homology of δ modulo $\partial_k\, j^k$ to show that $\overset{(2)}{A}, \overset{(3)}{A}, \ldots$ can be taken to be local functionals.)

12.10. Theorem 12.5 is equivalent to $H_k(\delta) = 0$ for $k > 0$. Here, δ acts in the space of local functionals $A = \int a\, dx$ with positive pure ghost number. The purpose of this exercise is to investigate $H_0(\delta)$.

(a) Let A be a local functional with $\text{antigh}\, A = 0$ and $\text{pure gh}\, A > 0$. Show that the weak equality $A \approx 0$ on A implies the weak equality $a \approx \partial_k\, j^k$ on a. [Hint: Proceed as for (a) of the previous exercise and use the arbitrariness of the ghosts.] At strictly positive ghost number, $H_0(\delta)$ is, thus, just the set of the equivalence classes of local functionals that coincide when the constraints hold.

(b) The requirement $\text{pure gh}\, A > 0$ is, in general, essential for deriving this result. Indeed, let the constraint be $\pi'(\sigma) = 0$, where σ belongs to the union of the disjoint intervals $[0, 1]$ and $[2, 3]$. One takes as boundary conditions $\pi(1) = -\pi(2)$. The integral $\int \pi(\sigma)\, d\sigma$ extended to the whole domain of σ vanishes on the constraint surface, and yet $\pi \not\approx (dj/d\sigma)$ with a local j. [Of course, $\int \pi(\sigma)\, d\sigma$ can be written as a combination of the constraints, albeit a nonlocal one.] Check these assertions.

(c) Point (b) shows the possible existence of local functionals that are δ-exact in the space of all (local and nonlocal) functionals but not δ-exact in the space of local functionals. The existence of such local functionals depends on the form of the constraints and on the global properties of the spatial manifold.

12.11. Repeat explicitly the analysis of this chapter for higher-order reducible theories.

12.12. A different proof of the acyclicity of δ modulo $\partial_k j^k$ can be given when the antighost number is strictly greater than the space dimension D (without restriction on the pure ghost number).

(a) Show that the equation $\delta a = \partial_k \alpha^k$ implies the "zig-zag equations"

$$\delta \alpha^k = \partial_n s^{kn}, \;\; \delta s^{kn} = \partial_t s^{knt}, \;\; \ldots .$$

(b) Prove $a = \delta b + \partial_k \beta^k$ by starting from the last of these equations and "going up the ladder." [Hint: The last of the equations is of the form $\delta(\text{something}) = 0$, where "something" is of strictly positive ghost number.]

12.13. Show that the analysis of this chapter goes through unchanged when the constraints or the Hamiltonian have an explicit coordinate dependence. [Hint: The explicit coordinate dependence does not modify $\bar{\partial}_k$ in (12.39) and the perturbative argument of §12.3.3 can thus be repeated.]

CHAPTER THIRTEEN

QUANTUM MECHANICS OF CONSTRAINED SYSTEMS: STANDARD OPERATOR METHODS

We have seen that the constraints appearing in the Hamiltonian description of a constrained system can be of two different types: they can be either first class or second class.

This distinction is quite important and carries through in the quantum theory. While second-class constraints can be enforced as operator equations once the Poisson bracket is replaced by the Dirac bracket, the incorporation of the first-class constraints is more subtle.

The purpose of this chapter is to explicitly describe and compare the usual (non-BRST) methods for taking into account both first-class and second-class constraints quantum-mechanically.

Because of the practical nature of the present-day quantization methods, we will develop here only the general ideas and illustrate them by means of simple examples.

13.1. QUANTIZATION OF SECOND-CLASS CONSTRAINTS

13.1.1. An Example

How should second-class constraints be incorporated into the quantum theory? In order to get some insight into this question, let us analyze the model of §1.3.2, with N pairs of canonical coordinates where the first pair (q^1, p_1) is constrained to be zero. The constraints are

$$\chi_1 = q^1 \approx 0 \tag{13.1a}$$

$$\chi_2 = p_1 \approx 0 \tag{13.1b}$$

and clearly tell us that the pair (q^1, p_1) does not describe a true physical degree of freedom. Only the $N-1$ pairs $(q^2, p_2), \ldots, (q^N, p_N)$ are to be thought of as physical. Therefore, only $(q^2, p_2), \ldots, (q^N, p_N)$ should be realized as nontrivial operators in the quantum theory. The variables q^1 and p_1, with no physical content, should be replaced by the zero operator.

This can consistently be done provided the commutators and the anticommutators of the quantum theory are defined to be $i\hbar$ times the corresponding (graded) Dirac bracket. The use of the Poisson bracket would indeed lead to contradictions, since one would then have

$$[\hat{p}_1, \hat{q}_1] = -i\hbar, \tag{13.2a}$$

which is inconsistent with

$$\hat{q}_1 = 0, \qquad \hat{p}_1 = 0. \tag{13.2b}$$

However, no such inconsistency arises if one uses instead the Dirac bracket, since then

$$[\hat{p}_1, \hat{q}_1] = i\hbar\, [p_1, q_1]_{DB} = 0. \tag{13.2c}$$

Furthermore, the Dirac brackets of the other canonical pairs are equal to their original Poisson bracket. Therefore the correspondence rules

$$(i) \quad [\hat{A}, \hat{B}]_\pm = i\hbar\, \widehat{[A, B]^*} \quad \text{(modulo factor ordering)},$$

$$(ii) \quad q_1 = 0, \quad p_1 = 0 \rightarrow \hat{q}_1 = 0, \quad \hat{p}_1 = 0,$$

yield a satisfactory theory in which only the $N-1$ canonical pairs $(q^2, p_2), \ldots, (q^N, p_N)$ are realized as nontrivial quantum operators.

13.1.2. Correspondence Rules in the General Case

The method we have followed in the simple example (13.1) turns out to be applicable to an arbitrary system of second-class constraints $\chi_\alpha \approx 0$.

The key correspondence rule is again that the graded commutator $[\hat{A}, \hat{B}]_{\pm}$ of two quantum operators should be equal to $i\hbar$ times the *Dirac* bracket $[A, B]^*$,

$$[\hat{A}, \hat{B}]_{\pm} = i\hbar\, \widehat{[A, B]^*} \tag{13.3}$$

(modulo factor ordering). The graded commutator $[\hat{A}, \hat{B}]_{\pm}$ is equal to the standard commutator $AB - BA$, unless the operators A and B are both odd, in which case it is equal to the anticommutator $AB + BA$. We shall systematically denote in the sequel the graded commutator by $[A, B]$, even when it stands for the anticommutator.

With the prescription (13.3), one can consistently enforce the second-class constraints as operator equations,

$$\hat{\chi}_{\alpha} = 0, \tag{13.4}$$

since one has $[A, \chi_{\alpha}]^* = 0$ for any A.

One way to realize in principle (13.3)–(13.4) is to solve classically the constraints $\chi_{\alpha} = 0$ for some of the variables, x^{α}, say, in terms of the other "independent" variables y^i.

$$\chi_{\alpha} = 0 \Leftrightarrow x^{\alpha} = f^{\alpha}(y^i). \tag{13.5}$$

Using (13.5), the Dirac brackets $[y^i, y^j]^*$ are functions of y^k only,

$$[y^i, y^j]^* = \sigma^{ij}(y^k), \tag{13.6}$$

and furthermore, the matrix σ^{ij} is invertible, since the χ_{α} are second class (see §2.3.1).

The quantum theory is defined by finding a representation of (13.6) for the independent operators $\hat{y}^i$

$$[\hat{y}^i, \hat{y}^j] = i\hbar\, \sigma^{ij}(\hat{y}^k). \tag{13.7}$$

If one wants the quantum theory to contain as many degrees of freedom as the classical one, this representation must be irreducible. Once a solution of (13.7) has been found, one can define the operators $\hat{x}^{\alpha}$ as $f^{\alpha}(\hat{y}^i)$, with an ordering that preserves the reality properties of x^{α}. This implies

$$\hat{\chi}_{\alpha} = 0,$$

with a definite ordering for $\hat{\chi}_{\alpha}$ once an ordering for $f^{\alpha}(\hat{y}^i)$ has been chosen.

13.1.3. Difficulties

The main difficulty one faces when quantizing second-class constraints by the above method is that one must find an explicit representation of the Dirac brackets (13.7) in the quantum theory. This is a complicated problem that possesses no general solution, except in very particular instances, *e.g.*, when the Dirac brackets are c-numbers, or when the independent variables y^i form an algebra in the Dirac bracket. [The realization of c-number anticommutators is covered by the theory of Clifford algebras, some aspects of which are discussed in Chapter 20 ("Complementary Material").]

In general, however, it is impossible to find an explicit solution of (13.7), even if one allows for departures ("Dirac bracket anomaly") of order $\hbar^2$,

$$[\hat{A}, \hat{B}] = i\hbar\, \widehat{[A, B]}^* + O(\hbar^2). \tag{13.8}$$

The problem of the representation of the Dirac bracket appears thus to be a serious one. This is the main reason we feel that the other treatment of the second-class constraints, which turns them into first class by adding extra variables, is of interest. Indeed, in that case one can consistently work with the Poisson bracket, which is simpler.

13.2. REDUCED PHASE SPACE QUANTIZATION OF FIRST-CLASS CONSTRAINTS

13.2.1. Description of the Method

We now turn to the question of quantizing first-class constraints and assume again, for simplicity of notations, that there is no second-class constraint left. We also do not write the * on the Dirac bracket if this one is used and refer to it as the Poisson bracket.

One may attempt to drop the first-class constraints at an early stage through reduced phase space quantization. This amounts to quantizing only the gauge-invariant functions. These functions are functions defined on the space of equivalence classes of gauge orbits, *i.e.*, on the reduced phase space, which has a regular Poisson bracket.

So, to carry out the reduced phase space quantization, one must find a complete set of gauge-invariant functions and define the quantum space as the irreducible representation space of the commutation/anticommutation relations of this complete set. Every state in that space is physical.

For instance, if one has a single first-class constraint

$$p_1 = 0 \tag{13.9}$$

with N canonical pairs $(q^1, p_1), \ldots, (q^N, p_N)$, a complete set of observables is given by $(q^2, p_2), \ldots (q^N, p_N)$ since (i) these are constants along the gauge orbits generated by (13.9), $q^1 \to q^1 + \varepsilon$, and (ii) every function F obeying $[F, p_1] \approx 0$ is weakly equal to a function of $q^2, p_2, \ldots,$

q^N, p_N only. The Poisson brackets of the functions $q^2, p_2, \ldots, q^N, p_N$ are canonical

$$[p_i, q^j] = -\delta_i{}^j \qquad (i \geq 2).$$

In the Schrödinger representation, one realizes the operators $\hat{q}^j$ and $\hat{p}_i$ as

$$\hat{q}^j\,\psi = q^j\psi \tag{13.10a}$$

$$\hat{p}_i\,\psi = \frac{\hbar}{i}\partial_i\,\psi, \tag{13.10b}$$

where the functions ψ are square-integrable functions of $q^2, \ldots, q^N$. There is no operator $\hat{q}^1$ or $\hat{p}_1$ in the reduced phase space quantization approach, since the pair (q^1, p_1) is not gauge invariant. Gauge invariance has been completely eliminated because one deals only with gauge-invariant objects. In that sense, the invariance is manifest.

13.2.2. Gauge Conditions

In practice, it is extremely difficult to find a complete set of observables. Indeed, this amounts to solving the differential equations

$$[F, G_a] \approx 0, \tag{13.11}$$

which may not be tractable.

It is then less difficult to reach the reduced phase space by a different method. That other method works when Gribov obstructions do not prevent the existence of a global canonical gauge fixing (see Appendix 2.A) and simply consists of imposing canonical gauge conditions

$$\chi_a = 0. \tag{13.12}$$

This is legitimate because any function of the canonical variables can be viewed, after complete gauge fixing, as the restriction in that gauge of a gauge-invariant function. Hence, once the gauge is fixed, one is effectively working with gauge-invariant functions. Furthermore, one finds that the Dirac bracket associated with the constraints $G_a = 0$ and the gauge conditions $\chi_a = 0$ is just the Poisson bracket of the corresponding gauge-invariant functions, so that the Dirac bracket yields the correct bracket in the reduced phase space (recall Exercise 1.18).

Hence, a complete set of independent gauge-fixed functions provides one with a complete set of gauge-invariant observables. With canonical gauge conditions, the reduced phase space quantization becomes identical to the quantization of second-class constraints.

If one considers again the simple model described by the single constraint (13.9), one finds that an appropriate gauge condition is

$$q^1 = 0.$$

It is easy to check that the reduced phase space quantization worked out in the previous subsection is equivalent to the quantization of the second-class constraint $p_1 = 0$, $q^1 = 0$ studied in §13.1.1.

13.2.3. Difficulties

The reduced phase space approach appears to be the most natural one at first sight, since only gauge-independent observables are realized as quantum-mechanical operators and every state in the Hilbert space is physical.

However, this approach may suffer from a number of drawbacks. First, as already mentioned in Chapter 2, the elimination of the gauge degrees of freedom—*i.e.*, the finding of a complete set of gauge-invariant observables—may spoil manifest invariance under an important symmetry. For instance, in the case of the free relativistic particle described by the constraint $p^2 + m^2 = 0$ (Exercise 4.1), a complete set of independent observables is given by p^i and $x^i - (p^i x^0 / \sqrt{m^2 + p^j p_j})$. These do not belong to a linear representation of the Lorentz group, and the unconstrained theory expressed in terms of them is not manifestly Lorentz invariant (manifest = linear). Second, in the case of field theory, the elimination of the gauge degrees of freedom generally destroys locality in space. Third, the brackets of the complete set of observables that one has found may be complicated functions of these observables, and their quantum-mechanical realization may again not be straightforward. Similarly, the Hamiltonian in terms of the independent degrees of freedom may be also quite complicated, and it may turn out to be impossible to give a quantum-mechanical definition of it.

For these reasons, it is important to develop alternative methods of quantization that do not meet these difficulties. The first alternative method is due to Dirac.

13.3. DIRAC QUANTIZATION OF FIRST-CLASS CONSTRAINTS

13.3.1. Formal Aspects

In the Dirac method, one does not try to eliminate the gauge degrees of freedom. Rather, one keeps all variables, including the pure gauge ones, as operators in a bigger linear space. No gauge condition is imposed, and every dynamical variable is realized quantum-mechanically. In particular, the constraints become operators acting nontrivially on the Dirac representation space.

Because the gauge degrees of freedom have been included, the representation space of the Dirac method carries unphysical information. One removes this unphysical information by imposing a condition that selects the "physical states."

The required condition must be such as to enforce gauge invariance in the quantum theory. Namely, every physical state should remain unchanged if one performs a transformation generated by the constraints. This leads to

$$\hat{G}_a|\psi\rangle = 0, \tag{13.13}$$

as this condition implies that the physical states are invariant under the finite gauge transformations in the connected component of the identity,

$$e^{i\varepsilon^a \hat{G}_a}|\psi\rangle = |\psi\rangle. \tag{13.14}$$

[The transformation properties of the physical states under gauge transformations that are not in the component of the identity ("large gauge transformations") are not dictated by the action principle, since no constraints are associated with them. Demanding invariance of the physical states under such transformations is thus an extra assumption.] The conditions (13.13) are linear, so that any linear combination of physical states is also physical.

In the model with $p_1 = 0$, the condition (13.13) reads, in the Schrödinger representation,

$$\frac{\hbar}{i}\frac{\partial \psi}{\partial q^1} = 0.$$

Physical states do not depend on the gauge degree of freedom q^1 and are, thus, obviously gauge invariant. They coincide with the states of the reduced phase space method. Similarly, physical operators may be taken not to depend on $\hat{q}^1$ and $\hat{p}_1$.To the extent that any set of first-class constraints can be (locally) abelianized, this indicates that the reduced phase space method and the Dirac approach are formally equivalent. In the general case of arbitrary constraints, however, the equivalence may break down because of factor-ordering problems (a simple ordering of the factors in one approach may not appear to be natural in the other) or nontrivial global issues. The existence of different quantum systems having the same classical limit is not peculiar to constrained dynamics but occurs already in the unconstrained case.

Further insight on the meaning of (13.13) is given in Exercises 13.5 through 13.9. In particular, the case of internal gauge symmetries is analyzed in Exercises 13.7 and 13.8.

13.3.2. Anomalies

The classical constraints G_a are first class,

$$[G_a, G_b] = C_{ab}{}^c\, G_c. \tag{13.15}$$

If this relation is preserved quantum mechanically,

$$[\hat{G}_a, \hat{G}_b] = i\hbar\, \hat{C}_{ab}{}^c\, \hat{G}_c, \tag{13.16}$$

with structure functions $\hat{C}_{ab}{}^c$ that may acquire corrections of order $\hbar$, then the integrability conditions of (13.13) are satisfied. Indeed,

$$[\hat{G}_a, \hat{G}_b]\, |\psi\rangle = 0 \tag{13.17}$$

automatically holds as a consequence of (13.13) and (13.16). However, the first-class condition (13.16) may be spoiled by extra terms of quantum mechanical origin,

$$[\hat{G}_a, \hat{G}_b] = i\hbar\, \hat{C}_{ab}{}^c\, \hat{G}_c + \hbar^2 \hat{D}_{ab}. \tag{13.18}$$

If this is the case, then one finds from (13.13) and (13.18) that the physical states $|\psi\rangle$ should also obey the extra conditions

$$\hat{D}_{ab}|\psi\rangle = 0, \tag{13.19}$$

as well as all further conditions that may follow from (13.13) and (13.19), by taking repeated (anti)commutators.

Now, the conditions (13.19) have no classical analog and restrict the physical subspace too much. In extreme cases, where $\hat{D}_{ab}$ is invertible, they imply $|\psi\rangle = 0$, and the theory, if quantized along the above lines, becomes empty. Thus, it does not appear to be legitimate to impose (13.19).

What happens is that whenever $\hat{D}_{ab} \neq 0$ in (13.18), the quantum operators $\hat{G}_a$ cease to be first-class and, hence, can no longer be interpreted as gauge generators. One says that gauge invariance is broken at the quantum level, and the operator $\hat{D}_{ab}$ is called a gauge anomaly. But if gauge invariance is broken by quantum effects, it is meaningless to search for gauge-invariant states, *i.e.*, to impose the conditions (13.13). Thus, one sees that the Dirac method is not directly applicable when there is a gauge anomaly. For that reason, we will assume that $\hat{D}_{ab}$ in (13.18) vanishes.

Another source of inconsistency in the condition $\hat{G}_a|\psi\rangle = 0$ may arise from the dynamical evolution. Indeed, one has classically

$$[H_0, G_a] = V_a{}^b\, G_b, \tag{13.20a}$$

and this implies quantum mechanically

$$[\hat{H}_0, \hat{G}_a] = i\hbar\, \hat{V}_a{}^b\, \hat{G}_b + \hbar^2 \hat{C}_a. \tag{13.20b}$$

If $\hat{C}_a \neq 0$, the Hamiltonian and the time evolution are not gauge invariant. A physical state would be mapped, by the Schrödinger equation, on a state that fails to be gauge invariant. Gauge invariance is again (dynamically) broken by quantum effects, and $\hat{C}_a$ is also called a gauge anomaly. We will assume throughout that $\hat{C}_a = 0$.

Finally, we mention that even when $\hat{D}_{ab}$ or $\hat{C}_a$ is nonzero, the BRST quantization to be discussed in the next chapter may turn out to be consistent (see, for instance, Exercises 13.8 and 14.19). This shows that it may be misleading to carry along the pure gauge degrees of freedom without including at the same time their BRST partners, *i.e.*, the ghosts.

13.3.3. Generally Covariant Systems

It is of interest to write down explicitly the physical state condition in the case of the parametrized systems of §4.2.1. The constraint reads

$$p_t + H_0(q^i, p_i) \approx 0. \tag{13.21}$$

In the Schrödinger representation, the states are functions of q^i, t and τ,

$$\psi = \psi(q^i, t, \tau). \tag{13.22}$$

The τ-dependence is determined by the Schrödinger equation,

$$i\hbar\,\frac{\partial\psi}{\partial\tau} = H_E\,\psi = u^0\Big(\frac{\hbar}{i}\,\frac{\partial\psi}{\partial t} + H_0\psi\Big), \tag{13.23}$$

and the constraint equation becomes

$$\frac{\hbar}{i}\,\frac{\partial\psi}{\partial t} + H_0\psi = 0. \tag{13.24}$$

Equations (13.23) and (13.24) show: (*i*) that the dependence of the physical wave functions on the laboratory time t is determined by the Schrödinger equation of the unparametrized formalism; and (*ii*) that the extended Hamiltonian H_E annihilates the physical states, so that these do not depend on the parameter time τ

$$\frac{\partial\psi}{\partial\tau} = 0. \tag{13.25}$$

This last result ($\partial\psi/\partial\tau = 0$) holds for any generally covariant theory with dynamical variables that transform as scalars under $\tau \to \bar{\tau}(\tau)$, since then H_E weakly vanishes.

Once (13.25) is taken into account, the physical wave functions of the parametrized theory reduce to the solutions of the Schrödinger equation of the unparametrized theory.

13.3.4. Scalar Product

We have not discussed as yet how the states should be normalized in the Dirac method. This is an important—and subtle—question, which needs to be resolved if one wants to be able to give a physical interpretation to the theory.

Consider first the case of the constraint $p_1 = 0$ of §13.2.1. The scalar product in the Dirac Hilbert space is given by

$$\langle f \mid g \rangle = \int dq^1\, dq^2 \dots dq^N\, f^*(q^1, \dots, q^N)\, g(q^1, \dots, q^N). \tag{13.26}$$

If f and g are physical states, the integrand of (13.26) does not depend on q^1. Hence, unless q^1 has a compact range, the integral (13.26) is infinite for physical states. This is not acceptable. To get a finite integral, one needs to eliminate the integral over q^1, which is not a physical variable. This can be achieved by inserting a gauge condition in (13.26). One thus defines the physical scalar product by

$$(f \mid g) = \int dq^1\, dq^2 \dots dq^N\, \delta(\chi)\, [\chi, p_1]\, f^* g, \tag{13.27a}$$

where χ is a function of $q^1, q^2, \dots, q^N$ such that

$$\chi = 0 \Leftrightarrow q^1 = q^1(q^2, \dots, q^N) \tag{13.27b}$$

(gauge condition). The factor $[\chi, p_1]$ is included so that the integral (13.27a) does not depend on the choice of gauge condition for physical states. One can formally rewrite (13.27a)

$$(f \mid g) = \langle f \mid \hat{\mu} \mid g \rangle, \tag{13.27c}$$

where $\hat{\mu}$ is the (singular) operator obtained by "putting hats" on $\mu(q^1, q^2, \dots, q^N) = \delta(\chi)\, [\chi, p_1]$.

The physical scalar product (13.27a) is explicitly equal to

$$(f \mid g) = \int dq^2 \dots dq^N\, f^*(q^2, \dots, q^N)\, g(q^2, \dots, q^N) \tag{13.27d}$$

for physical states and coincides with the scalar product of the reduced phase space method. With (13.27a), the inner product of physical states is finite, and the real, gauge-invariant functions are hermitian operators. In the momentum representation, the regularization (13.27) amounts to removing one δ-function of the constraints (Exercise 13.10).

The same scalar product difficulty appears if p_1 is fermionic. In that case, one finds that the unregularized scalar product (13.26) contains the fermionic $\delta(0)$ $(= 0)$ rather than the bosonic $\delta(0)$ $(= \infty)$ found previously. Accordingly, the scalar product also needs to be regularized by the insertion of an appropriate gauge condition (Exercise 13.11).

In the general case of arbitrary first-class constraints $G_a = 0$, the regularization of the scalar product requires more complicated redefinitions. The idea is to write the physical scalar product as

$$(f \mid g) = \langle f \mid \hat{\mu} \mid g \rangle \tag{13.28a}$$

where the (singular) operator $\hat{\mu}$ is hermitian

$$\hat{\mu} = \hat{\mu}^{\dagger} \tag{13.28b}$$

[so that $(f \mid g) = (g \mid f)^*$] and has the effect of restricting the integral to the physical degrees of freedom. When the constraints take the simple linear form

$$G_a \equiv p_a + \frac{\partial V}{\partial q^a} = 0, \tag{13.28c}$$

one can easily describe μ in terms of canonical gauge conditions $\chi_a(q^i) = 0$,

$$(f \mid g) = \int dq^i \prod_a \delta(\chi_a) \operatorname{sdet}[\chi_a, G_b]\, f^* g. \tag{13.28d}$$

For physical states, the expression (13.28d) does not depend on the choice of χ_a. Indeed, the product f^*g does not depend in that case on the gauge coordinates q^a. Furthermore, for $\chi_a(q^i) = 0$ to be good gauge conditions, they should be solvable for the q^a, $\chi_a = a_{ab}(q^i)(q^b - f^b)$ where f^b does not depend on q^a and $\operatorname{sdet} a_{ab} \neq 0$. Inserting this expression into (13.28d) shows that the physical scalar product is invariant under changes of the gauge conditions.

When the constraints are not of the simple formal expression (13.28c), the straightforward generalization of (13.28d) involves factor-ordering problems. Useful conclusions can, nevertheless, be obtained from the formal expression (13.28a)–(13.28b) of the regularized scalar product (see below and §14.5.2). An analysis of $\hat{\mu}$ from the viewpoint of BRST quantization will be discussed in §14.5.5.

One important property of $\hat{\mu}$ is that it is not invariant under redefinitions of the constraints $G_a \rightarrow \bar{G}_a = M_a{}^b G_b$. In the simple case

(13.28c)–(13.28d), one finds that for an infinitesimal alteration $\delta G_a = \varepsilon_a{}^b(q) G_b$ of the constraints, the insertion μ changes by

$$\delta\mu = \mu \operatorname{str} \varepsilon. \tag{13.29a}$$

We shall assume that in the general case, the operator orderings in the definitions of $\hat{\mu}$ and of $\hat{M}_a{}^b(\hat{q}, \hat{p})$ are handled in such a way that under the redefinition

$$G_a \to \bar{G}_a = \big(\delta_a{}^b + \varepsilon_a{}^b(q,p)\big) G_b, \tag{13.29b}$$

$\hat{\mu}$ transforms in a manner that generalizes (13.29a)

$$\delta\hat{\mu} = \tfrac{1}{2}(\operatorname{str}\hat{\varepsilon} \cdot \hat{\mu} + \hat{\mu} \operatorname{str}\hat{\varepsilon}) \tag{13.29c}$$

(up to terms that annihilate the physical states). This is consistent with the hermiticity of $\hat{\mu}$.

As will be seen in Sec. 13.4 below, there exists an alternative way of dealing with first-class constraints in which the scalar products of physical states are finite. In that case, there is no need for regularization and the physical product $(f \mid g)$ coincides with $\langle f \mid g \rangle$, $(f \mid g) = \langle f \mid g \rangle$.

13.3.5. A Different Derivation of the Physical Condition

The physical state condition was arrived at by insisting on gauge invariance of the physical states. There exists an alternative method of deriving the same conditions by using the other aspect played by the constraints, namely, that of restricting the dynamics to the constraint surface. This alternative derivation goes as follows.

The functions on the constraint surface can be described as the equivalence classes of phase space functions that differ by a combination $k^a G_a$ of the constraints. This identification must be implemented quantum-mechanically. The simplest possibility for doing so is to demand that the constraints annihilate the physical states. When this condition holds, it is manifest that the operators f and $f + k^a G_a$ can be identified because one then finds

$$f\,|\psi\rangle = f\,|\psi\rangle + k^a G_a |\psi\rangle$$

for any physical state $|\psi\rangle$ and for any choice of k^a. Hence, we are again led to (13.13).

13.3.6. Projected Kernel of Gauge-Invariant Operators

We shall assume for simplicity throughout this subsection that the canonical variables are all bosonic.

The kernel $A_0(q', q)$ in the coordinate representation of the operator $\hat{A}_0$ is defined by its action on the wave functions $\psi(q)$ through

$$(A_0\psi)(q') = \int dq\, A_0(q', q)\, \psi(q). \tag{13.30a}$$

One has

$$A_0(q', q) = \langle q' \,|\, \hat{A}_0 \,|\, q \rangle, \tag{13.30b}$$

since $\psi(q) = \langle q \,|\, \psi \rangle$ and $(A_0\psi)(q') = \langle q' \,|\, \hat{A}_0 \,|\, \psi \rangle$. The kernel of the product of the operators A_0 and B_0 is given by

$$(A_0 B_0)(q'', q) = \int dq'\, A_0(q'', q')\, B_0(q', q). \tag{13.30c}$$

One can view $A_0(q', q)$ for fixed q as the wave function of the vector $\hat{A}_0|q\rangle$. This wave function is, in general, not annihilated by the constraints, even if the operator $\hat{A}_0$ is gauge-invariant.

One can, however, introduce a different kernel for gauge-invariant operators. This kernel possesses the property of being annihilated by the constraints and is called the "projected kernel."

Let $|\psi_\alpha\rangle$ be a basis of physical states, $\hat{G}_a|\psi_\alpha\rangle = 0$. The basis is assumed to be orthonormal in the scalar product defined in the physical subspace

$$(\psi_\alpha \,|\, \psi_\beta) = \int dq\, \psi_\alpha^*(q)\, \mu\Big(q, \frac{\hbar}{i}\frac{\partial}{\partial q}\Big)\, \psi_\beta = \delta_{\alpha\beta}. \tag{13.31}$$

Here, $\mu[q, (\hbar/i)(\partial/\partial q)]$ is the coordinate representation expression of the insertion (13.28) needed to regularize the scalar product. As we have seen, this insertion has the effect of restricting the integration in (13.31) to an integration over the physical degrees of freedom.

The projected kernel $A_0^P(q', q)$ of the gauge-invariant operator A_0 is defined by

$$A_0^P(q', q) = \sum_{\alpha,\beta} \psi_\alpha(q')\, (\psi_\alpha \,|\, A_0 \,|\, \psi_\beta)\, \psi_\beta^*(q) \tag{13.32}$$

and carries information only about the action of A_0 on gauge-invariant states. If the state $|\psi\rangle$ is physical, the wave function $(A_0\psi)(q')$ is given by

$$(A_0\psi)(q') = \int dq\, A_0^P(q', q)\, \mu\Big(q, \frac{\hbar}{i}\frac{\partial}{\partial q}\Big)\psi(q). \tag{13.33a}$$

It follows from (13.33a) that the projected kernels are folded together as

$$(A_0 B_0)(q'', q) = \int dq'\, A_0^P(q'', q)\, \mu\Big(q', \frac{\hbar}{i}\frac{\partial}{\partial q'}\Big)\, B_0^P(q', q). \tag{13.33b}$$

The projected kernel may be regarded as a regularization of the kernel of the operator $\hat{P}\hat{A}_0\hat{P}$, where $\hat{P}$ is the orthogonal projection operator on the physical subspace, hence the terminology.

The interest of the projected kernel is twofold. (i) First, it is annihilated by the constraints acting on the first argument and their complex conjugates acting on the second argument,

$$G_a\left(q', \frac{\hbar}{i}\frac{\partial}{\partial q'}\right) A_0^P(q', q) = 0, \tag{13.34a}$$

$$G_a\left(q, \frac{\hbar}{i}\frac{\partial}{\partial q}\right) \left(A_0^P(q', q)\right)^* = 0. \tag{13.34b}$$

This is because $A_0^P(q, q')$ involves the wave function factors $\psi_\alpha(q')$ and $\psi_\beta^*(q)$.

(ii) Second, it makes direct contact with the reduced phase space quantization when the latter is equivalent to the Dirac method. Indeed, one can then respectively identify (a) the states $|\psi_\alpha\rangle$ with a basis in the Hilbert space $\mathcal{H}^R$ of the reduced phase space quantization, (b) $(\psi_\alpha \mid \psi_\beta)$ with the corresponding scalar product, and (c) $\hat{A}_0$ with an operator $\hat{A}_0^R$ in $\mathcal{H}^R$. Let q^*, p^* be canonical coordinates on the reduced phase space. The reduced kernel $A_0^R(q^{*\prime}, q^*)$ of A_0 is defined by

$$A_0^R(q^{*\prime}, q^*) = (q^{*\prime} \mid \hat{A}_0^R \mid q^*) \tag{13.35a}$$

Upon inserting the decomposition of the identity in $\mathcal{H}^R$, *i.e.*,$1^R = \sum_\alpha |\psi_\alpha)(\psi_\alpha|$, one gets

$$A_0^R(q^{*\prime}, q^*) = \sum_{\alpha,\beta} \psi_\alpha(q^{*\prime})\,(\psi_\alpha \mid A_0 \mid \psi_\beta)\,\psi_\beta^*(q^*). \tag{13.35b}$$

The projected and reduced kernels are thus the kernels of the same operator but in different representations. One representation is the q^*-coordinate representation in the reduced space $\mathcal{H}^R$. The other is the q-coordinate representation in the Hilbert space of the Dirac method.

Finally, assume that the constraints are abelian. In that case, the $\hat{G}_a$ can be diagonalized simultaneously. We denote by y_α the extra quantum number necessary to completely specify a basis,

$$\hat{G}_a|G_a, y_\alpha\rangle = G_a|G_a, y_\alpha\rangle, \;\; \hat{y}_\alpha|G_a, y_\alpha\rangle = y_\alpha|G_a, y_\alpha\rangle. \tag{13.36a}$$

A complete set of solutions of $\hat{G}_a|\psi\rangle = 0$ is given by the states $|G_a = 0\; y_\alpha\rangle$. The y_α can thus be thought of as quantum numbers in the reduced space $\mathcal{H}^R$. The regularized scalar product is given by

$$(G_a = 0\; y_\alpha \mid G_a' = 0\; y_\alpha') = \delta(y_\alpha - y_\alpha') \tag{13.36b}$$

and differs from $\langle G_a = 0\ y_\alpha \mid G'_a = 0\ y'_\alpha\rangle$ by the removal of $\delta(G_a - G'_a) = \delta(0-0) = \delta(0)$. It follows that if $\hat{A}_0$ commutes with the $\hat{G}_a$, then

$$\langle G'_a y'_\alpha \mid \hat{A}_0 \mid G_a = 0\ y_\alpha\rangle = \delta(G'_a)\,(G'_a = 0\ y'_\alpha \mid \hat{A}_0 \mid G_a = 0\ y_\alpha), \quad (13.36c)$$

since $\hat{A}_0$ leaves the quantum number G_a unchanged. This implies that the projected kernel of $\hat{A}_0$ is equal to

$$A_0^P(q', q) = \langle q' \mid \hat{A}_0\, \delta(\hat{G}_a) \mid q\rangle \qquad ([\hat{G}_a, \hat{G}_b] = 0) \qquad (13.37)$$

because in

$$\langle q' \mid \hat{A}_0\, \delta(\hat{G}_a) \mid q\rangle = \int dG_a\, dG'_a\, dy_\alpha\, dy'_\alpha\, \langle q' \mid G'_a y'_\alpha\rangle \langle G'_a y'_\alpha \mid \hat{A}_0\, \delta(\hat{G}_a) \mid G_a y_\alpha\rangle\, \langle G_a y_\alpha \mid q\rangle,$$

the integrals reduce to integrals over the complete set of gauge-invariant states $|G_a = 0\ y_\alpha\rangle$,

$$\langle q' \mid \hat{A}_0\, \delta(\hat{G}_a) \mid q\rangle = \int dy_\alpha\, dy'_\alpha \langle q' \mid G_a = 0\ y'_\alpha)(G_a = 0\ y'_\alpha \mid \hat{A}_0 \mid G_a = 0\ y_\alpha)\langle G_a = 0\ y_\alpha \mid q\rangle.$$

13.4. DIRAC–FOCK QUANTIZATION OF FIRST-CLASS CONSTRAINTS

13.4.1. Definition

The Dirac–Fock method implements the constraints differently and possesses the virtue of leading to finite scalar products for physical states.

Consider a system with N degrees of freedom, taken to be bosonic for simplicity, and with two first-class constraints given by

$$p_1 = 0, \qquad p_2 = 0. \qquad (13.38)$$

Combine them as follows:

$$a = p_1 + ip_2, \qquad a^* = p_1 - ip_2, \qquad (13.39a)$$

$$b = -\frac{i}{2}(q^1 + iq^2), \qquad b^* = \frac{i}{2}(q^1 - iq^2). \qquad (13.39b)$$

(It is crucial to have an even number of constraints.) The variables a, a^*, b, and b^* obey the Poisson bracket relations

$$[a, a] = [a, a^*] = [b, b] = [b, b^*] = [a, b] = [a^*, b^*] = 0, \quad (13.39c)$$

$$[a, b^*] = [b, a^*] = -i. \qquad (13.39d)$$

Furthermore, the constraints $p_1 = 0$, $p_2 = 0$ are equivalent to

$$a = 0, \qquad a^* = 0. \tag{13.39e}$$

Turn now to the quantum theory. The relations (13.39) become

$$[a, a] = [a, a^*] = [b, b] = [b, b^*] = [a, b] = [a^*, b^*] = 0, \tag{13.40a}$$

$$[a, b^*] = [b, a^*] = 1, \tag{13.40b}$$

for the commutators. These can be rewritten as

$$[a_\mu, a_\nu^*] = \eta_{\mu\nu}, \qquad \mu = 1, 2, \tag{13.40c}$$

$$[a_\mu, a_\nu] = [a_\mu^*, a_\nu^*] = 0, \tag{13.40d}$$

with

$$a_\mu = (a, b). \tag{13.40e}$$

Here, the metric $\eta_{\mu\nu}$ is indefinite and reads explicity

$$\eta_{\mu\nu} = \begin{pmatrix} 0 & 1 \\ 1 & 0 \end{pmatrix}. \tag{13.40f}$$

The oscillators a, b define a null basis. The commutation relations (13.40) are the standard creation-annihilation operator commutation relations, except that the metric is indefinite. This suggests using a Fock space representation, with a vacuum $|0\rangle$ of unit norm, annihilated by all destruction operators,

$$a_\mu \, |0\rangle = 0 \Leftrightarrow a|0\rangle = b\,|0\rangle = 0, \tag{13.41a}$$

$$\langle 0 \,|\, 0\rangle = 1. \tag{13.41b}$$

The states are obtained by acting on the vacuum with the creation operators a^* and b^*. Because the metric is indefinite, one finds that there are negative norm states in the Fock space. More precisely, $a^* - b^*$ creates negative norm states when acting an odd number of times and $a^* + b^*$ creates positive norm states, while a^* and b^* create states of zero norm.

The full representation space for all the dynamical variables is obtained by taking the direct product of the above Fock space by the standard representation space of the physical degrees of freedom $q^3, p_3, \ldots, q^N, p_N$,

$$f(q^3, \ldots, q^N), \tag{13.42a}$$

$$\hat{q}^i f = q^i f, \;\; \hat{p}_i \, f = \frac{\hbar}{i} \partial_i \, f, \qquad (i \geq 3), \tag{13.42b}$$

$$(f, g) = \int dq^3 \ldots dq^N f^* g. \tag{13.42c}$$

A general state is thus a combination of Fock space states with coefficients that are functions of $q^3, \dots, q^N$,

$$|\psi\rangle = \sum_\alpha f_\alpha(q^3, \dots, q^N)\,|\alpha\rangle. \tag{13.42d}$$

Here $\{\,|\alpha\rangle\}$ is a basis for the above Fock-representation space of the gauge degrees of freedom q^1, p_1, q^2, p_2. The inner product in the full space is given by

$$\langle\psi_1\,|\,\psi_2\rangle = \sum_{\alpha,\beta} \int dq^3 \dots dq^N f^{1*}_\alpha\, f^2_\beta\,\langle\alpha\,|\,\beta\,\rangle \tag{13.42e}$$

and clearly differs from the scalar product used in the Schrödinger representation discussed in the previous section,

$$\int dq^1 dq^2 dq^3 \dots dq^N f^* g, \tag{13.42f}$$

which is positive definite.

Accordingly, the Fock representation and the Schrödinger representation of the gauge degrees of freedom are unitarily inequivalent for bosonic variables, although, as we shall see in the next subsection, they lead to isomorphic physical subspaces. (For fermionic variables, the Fock and Schrödinger representations are equivalent; see §20.3.2d.)

13.4.2. Physical Subspace

The constraints $p_1 = 0$ and $p_2 = 0$ are classically equivalent to $a = 0$, $a^* = 0$. So, the statement that the states should be gauge invariant would seem to read

$$a|\psi\rangle = 0, \qquad a^*|\psi\rangle = 0 \qquad \text{(wrong)}. \tag{13.43}$$

Equations (13.43) are inconsistent, however, not because there is an anomaly (there is none, since $[a, a^*] = 0$), but because there is no Fock state that is annihilated by the creation operator a^*: If one adds an a^*-particle to a state, one cannot get zero; one simply increases the occupation number of a^* by one unit.

Therefore, one must weaken (13.43). The natural way to do so is to keep only the destruction part of (13.43),

$$a|\psi\rangle = 0. \tag{13.44}$$

This yields a consistent equation.

The replacement of (13.43) by (13.44) is quite a drastic step, and one should wonder whether (13.44) still guarantees full gauge invariance.

Since only half of the gauge generators annihilate the physical states, it is not clear that the theory is invariant under the full set of gauge transformations. It turns out, however, that the other half of the gauge invariance is recovered in a subtle way through the presence of null states.

Indeed, the equation $a|\psi\rangle = 0$ implies, since b^* is conjugate to a in the commutator, that a physical state cannot contain b^* modes. Accordingly, a general physical state can be written as

$$a|\psi\rangle = 0 \Rightarrow |\psi\rangle = f(q^3, \dots, q^N)\,|0\rangle + |ns\rangle \tag{13.45}$$

where $|ns\rangle$ is a state that contains at least one a^* oscillator and no b^* oscillator. The states $|ns\rangle$ possess the crucial property of being "null spurious states that decouple," *i.e.*, they are physical states that possess vanishing scalar product with every other physical state, including themselves.

$$a|\psi\rangle = 0 \Rightarrow \langle\psi\,|\,ns\rangle = 0. \tag{13.46}$$

This is because the states $|ns\rangle$ contain at least one a^* oscillator,

$$|ns\rangle = a^*|\chi\rangle, \tag{13.47}$$

so that

$$\langle\psi\,|\,ns\rangle = \langle\psi\,|\,a^*\,|\,\chi\rangle = 0. \tag{13.48}$$

Therefore, the null spurious states completely disappear from any physical matrix element, and *one can consistently factor them out.* In other words, one can identify two physical states $|P\rangle$ and $|P'\rangle$ that differ by a null spurious state,

$$|P\rangle \sim |P'\rangle + |ns\rangle. \tag{13.49}$$

After the null states are factored out, one remains with states that depend only on the physical degrees of freedom. The inner product in the quotient space is given by

$$\begin{aligned}\langle\psi_1\,|\,\psi_2\rangle &= \int dq^3 \dots dq^N\, f_1^*(q^3, \dots, q^N)\, f_2(q^3, \dots, q^N)\, \langle 0\,|\,0\rangle \\ &= \int dq^3 \dots dq^N\, f_1^*\,(q^3, \dots, q^N)\, f_2(q^3, \dots, q^N),\end{aligned} \tag{13.50}$$

as it should. It is manifestly positive definite. Thus, the physical state condition (13.44) effectively removes two degrees of freedom from the theory and leaves us with a physical space with no negative norm state. In the space of equivalence classes of (13.49), the operator a^* acts trivially, since $a^*|P\rangle$ is a null spurious state whenever $|P\rangle$ is a physical state. Hence, a^* is equivalent to the zero operator.

A "quantum gauge condition" is a condition that selects a representative in each equivalence class of physical states, *i.e.*, that freezes the "gauge freedom" (13.49). One such condition is

$$b|\psi\rangle = 0. \tag{13.51a}$$

It implies

$$|\psi\rangle = f(q^3, \ldots, q^N)\,|0\rangle. \tag{13.51b}$$

In the gauge (13.51), the bijective correspondence between the states of the Dirac–Fock approach and the states of the reduced phase space method given by functions of $q^3, \ldots, q^N$ is manifest.

13.4.3. Conclusions

The main feature of the Fock representation is that one can consistently define in the full linear space of states an inner product that reduces to the appropriate scalar product on the physical subspace. No factorization of an infinite volume is ever necessary when computing probabilities. This result is achieved by allowing negative norm states in the Fock space. The negative norm states, however, do not obey the physical state condition and so do not appear in the physical spectrum. These features, explicitly verified here in the bosonic case, also hold for fermionic constraints (Exercise 13.13).

The Fock representation is available when the variables that are constrained by $G_a = 0$ can be naturally grouped in pairs as in (13.39) (see Exercise 13.15). It has been applied successfully to a wide range of important physical systems, including electromagnetism and string theories. One imposes the analog of the condition (13.44),

$$\hat{G}_a^{(-)}\,|\psi\rangle = 0 \tag{13.52}$$

on the physical states, where $\hat{G}_a^{(-)}$ is the destruction part of the constraints.

Two requirements should be fulfilled in order to have an anomaly-free quantum theory with complete gauge invariance. First of all, the constraints $\hat{G}_a^{(-)}$ should remain first class quantum-mechanically. Second, the physical subspace should contain null spurious states that decouple in order to recover the second half of the gauge invariance.

If this latter condition is not fulfilled, gauge invariance is lost in the quantum theory, even though the conditions $\hat{G}_a^{(-)}\,|\psi\rangle = 0$ may be consistent and guarantee that the constraints hold in the mean, $\langle\psi_1\,|\,\hat{G}_a\,|\,\psi_2\rangle = 0$. Because of the absence of null spurious states, not all gauge degrees of freedom would be eliminated from the quantum theory. It would then be clearly illegitimate to impose a gauge condition like (13.51a), and equivalence with the reduced phase space method would be lost. Such a situation occurs in string theory below the critical dimension.

Finally, we mention that in spite of its successes, the Dirac–Fock method possesses one unsatisfactory feature. This is that there exist classically first-class functions that cease to be first class quantum-mechanically when half of the constraints are dropped. Such operators fail to map the physical subspace on itself even though they are classically observable. For example, in the case of the simple model (13.38)–(13.39), the operator

$$A_0 = a^* b^* + ab \tag{13.53}$$

vanishes weakly at the classical level but is not proportional to a only [for any ordering of the factors: there is no ordering ambiguity in (13.53)]. Furthermore, A_0 acting on the vacuum yields the state $a^* b^* |0\rangle$, which is not annihilated by a.

An extra prescription is thus necessary to handle operators like A_0 in the quantum theory. This issue is avoided in the BRST formalism, where a physical operator is associated with each classical observable (up to factor ordering), even in the Fock representation (see §14.4.3). From that point of view, the inclusion of the ghosts through the BRST symmetry appears again to be more satisfactory.

EXERCISES. CHAPTER THIRTEEN

13.1. Consider the example of §13.1.1. Show that the quantum states can be taken to be functions of $q_2, \ldots, q_N$. Find an explicit representations of all the operators $\hat{q}_1, \hat{p}_1, \ldots, \hat{q}_N, \hat{p}_N$.

13.2. Show that for a system with second-class constraints, it is inconsistent to realize the canonical variables as operators obeying commutation/anticommutation relations following from the Poisson bracket even if one weakens the conditions $\hat{\chi}_\alpha = 0$ as $\hat{\chi}_\alpha |\psi\rangle = 0$.

13.3. Let z^A $(A = 1, \ldots, 2n)$ be $2n$ bosonic variables. Show that the Dirac brackets $[z^A, z^B]^* = C^{AB}$, where C^{AB} is a constant antisymmetric matrix of rank $2m \leq 2n$, can be represented in the space $L^2(R^m)$ of square integrable functions of m variables. [Hint: Make a linear transformation $z^A \to z'^A = L^A{}_B z^B$ that turns C^{AB} into the canonical form $C = \left(\begin{smallmatrix} D & 0 \\ 0 & 0 \end{smallmatrix}\right)$ where D is the $2m$ by $2m$ matrix $D = \left(\begin{smallmatrix} 0 & I \\ -I & 0 \end{smallmatrix}\right)$.]

13.4. Quantize the parametrized system with constraint $p_0 + H_0 \approx 0$ in the gauge $t = 0$ and in the gauge $t = \tau$. Verify explicitly that the gauge-invariant extensions of q^i and p_i in one gauge are related to the gauge-invariant extensions of q^i and p_i in the other gauge by a τ-dependent unitary transformation (the same unitary transformation as the one that relates the Heisenberg and Schrödinger pictures).

13.5. Consider the constraints $p_i - \partial V/\partial q^i = 0$.

(a) Show that the gauge transformations acting on q^i are $\delta_\varepsilon q^i = \varepsilon^i$.

(b) Write the Dirac physical state condition in the Schrödinger representation and prove that the states are *not* invariant under the transformations of (a). Rather, they are invariant only up to a phase.

(c) Show that the Dirac quantization is physically equivalent to the reduced phase space quantization.

(d) Conclusion: the action of the gauge transformations on states in the quantum theory may be subtle. In the present case, the action of a gauge transformation is to shift the argument of ψ *and* to multiply the wave function by a definite phase.

13.6. When the constraints contain powers of the momenta higher than the first, the action of the gauge transformations on quantum states is even more complicated than the one discussed in the previous problem. The complication can be argued, however, to be due to the choice of basis in Hilbert space. Let q_α, p_α be operators obeying $[q_\alpha, p_\beta] = i\hbar\, \delta_{\alpha\beta}$ and let $\tilde{q}_\alpha = U q_\alpha U^{-1}, \tilde{p}_\alpha = U p_\alpha U^{-1}$ where U is a unitary operator, $U = \exp -iM, M^\dagger = M$. Let $|q_\alpha\rangle$ be the basis of the eigenstates of q_α and $|\tilde{q}_\alpha\rangle = U|q_\alpha\rangle$. Assume that the constraints are $G_i \equiv p_i = 0$ for some of the momenta p_i. The gauge transformations read $\delta|\psi\rangle = \varepsilon^i G_i|\psi\rangle$. Show that the wave functions $\psi(\tilde{q}) = \langle \tilde{q}\,|\,\psi\rangle$ in the $\tilde{q}$-representation transform as

$$\delta\psi(\tilde{q}) = \varepsilon^i \int \langle \tilde{q}\,|\,q\rangle \frac{\hbar}{i}\frac{\partial}{\partial q^i} \langle q\,|\,\tilde{q}'\rangle\, \psi(\tilde{q}')\, d\tilde{q}'\, dq,$$

while they simply transform as

$$\delta\psi(q) = \varepsilon^i \frac{\hbar}{i}\frac{\partial}{\partial q^i}\, \psi(q)$$

in the q-representation.

13.7. Let the constraints be linear and homogeneous in the momenta,

$$G_a = \xi_a{}^i\,(q)\, p_i$$

with

$$[\xi_a, \xi_b]^i \equiv \xi_a{}^j\, \xi_b{}^i{}_{,j} - \xi_b{}^j\, \xi_a{}^i{}_{,j} = C_{ab}{}^c\,(q)\, \xi_c{}^i.$$

The q's and the p's are bosonic. We assume the vector fields $\xi_a{}^i$ to be linearly independent at each point and adopt conventions appropriate to a left action. The gauge transformations for q^i are $\delta_\varepsilon\, q^i = \xi_a{}^i\,(q)\, \varepsilon^a$ and define orbits in the configuration space Q. The reduced configuration space is the quotient space Q/G obtained by identifying all the points lying on the same gauge orbits. The purpose of this exercise and the next is to show that in this case, "quantization and reduction commute" for the subalgebra under commutation of observables $X(q,p) = X^i(q)\, p_i$ that are linear and homogeneous in the momenta. (Other observables need further examination.)

We first regard, in this exercise, the wave functions $\psi(q)$ as scalars in configuration space. The ordering ambiguity in $X(q,p)$ is then resolved by taking for X the operator $-iX^i\, \partial/\partial q^i$ with p_i to the right of $X^i(q)$, so that $X\psi$ is the scalar $-i\mathcal{L}_X\, \psi \equiv -iX^i\, \psi_{,i}$ ($\mathcal{L}_X\psi$ is the Lie derivative along X^i of the scalar ψ). We also take $\hbar = 1$.

(a) Show that the wave functions $\psi(q)$ obeying the constraints define wave functions ψ' in the quotient space Q/G and, conversely, that any wave function ψ' in Q/G define a physical wave function ψ in Q. (Verify that the constraints $G_a\psi = 0$ imply strict invariance under the gauge transformations $\delta_\varepsilon q^i = \xi_a{}^i \varepsilon^a$.)

(b) Prove that the "observables" $X \equiv X^i(q)\, p_i$, $[X^i(q)\, p_i, G_a] \approx 0$, induce well-defined vector fields X' in Q/G and conversely.

(c) Verify that $(X\psi)' = X'\psi'$.

(d) Check that the formalism is invariant under redefinitions of the constraints of the form $G_a \to \bar{G}_a = M_a{}^b(q)\, G_b$.

(e) Conclusion: in this case, there is equivalence between the Dirac approach ("quantizing and then imposing the constraints") and the reduced phase space method.

13.8. (Continuation of previous exercise.) When the wave functions are regarded as scalars, a measure in Q/G is necessary in order to compute probabilities. One then finds that the vector field X' is formally hermitian if and only if it leaves the measure invariant.

One may alternatively view the wave functions as densities of weight $\frac{1}{2}$. Then, an additional measure is not needed to compute probabilities, since the product of two wave functions itself defines a measure. The observable $Z^\alpha p_\alpha$ linear in the momenta becomes the Lie derivative operator along Z acting on densities of weight $\frac{1}{2}$ (up to $-i$), $Z\psi = -iZ^\alpha \partial_\alpha \psi - (i/2) Z^\alpha{}_{,\alpha}\psi = -i\mathcal{L}_Z\psi$. With this prescription, the operator $\hat{Z}$ is always formally hermitian.

So, a physical wave function should be a density of weight $\frac{1}{2}$ in the reduced configuration space Q/G. The physical operators linear in the momenta should be the vector fields on Q/G acting on the wave functions through the Lie derivative.

How can one reach densities of weight $\frac{1}{2}$ on the reduced configuration space Q/G in the Dirac formalism where one imposes the constraints after quantization? One may think that the appropriate starting point is given by the densities of weight $\frac{1}{2}$ on Q. This is, however, incorrect.

(a) Show that the equations $\mathcal{L}_{\xi_a}\psi \equiv \xi_a{}^i\, \psi_{,i} + \frac{1}{2}\xi_a{}^i{}_{,i}\, \psi = 0$ where $\psi(q^i)$ is a density of weight $\frac{1}{2}$ on Q are consistent if and only if $C_{ab}{}^c{}_{,i}\, \xi_c{}^i = 0$. [Hint: Use $[\mathcal{L}_{\xi_a}, \mathcal{L}_{\xi_b}] = \mathcal{L}_{[\xi_a, \xi_b]}$ and $\mathcal{L}_{\lambda\xi_a}\, \psi = \lambda\, \mathcal{L}_{\xi_a}\psi + \frac{1}{2}(\mathcal{L}_{\xi_a}\lambda)\, \psi$]. So, in general, one finds an "anomaly" ($C_{ab}{}^c{}_{,i}\, \xi_c{}^i \neq 0$).

(b) Prove that the equations $\mathcal{L}_{\xi_a}\psi = 0$ are *not* invariant under redefinitions $\xi_a{}^i \to M_a{}^b\, \xi_b{}^i$.

(c) Let $\psi(q)$ be an object on Q that transforms as a density of weight $\frac{1}{2}$ for changes of coordinates of Q and as a density of weight $-\frac{1}{2}$ for changes of basis $\xi_a \to M_a{}^b(q)\, \xi_b$,

$$q \to q'(q), \tag{i}$$

$$\xi_a \to M_a{}^b\, \xi_b, \tag{ii}$$

$$\psi(q) \to \left|\frac{\partial q}{\partial q'}\right|^{1/2} |\det M_a{}^b|^{-1/2}\, \psi(q(q')). \tag{iii}$$

For such an object, one can define a Lie derivative $\mathcal{L}_X\psi$ along a vector X obeying $\mathcal{L}_X\xi_a \equiv [X, \xi_a] = \alpha_a{}^b\, \xi_b$ by the formula

$$\mathcal{L}_X\psi = X^i\, \psi_{,i} + \tfrac{1}{2}X_{,i}{}^i\, \psi + \tfrac{1}{2}(\mathrm{tr}\,\alpha)\, \psi.$$

Discuss the meaning of the various terms. Show that $\mathcal{L}_X\psi$ transforms as ψ. Verify $[\mathcal{L}_X, \mathcal{L}_Y]\psi = \mathcal{L}_{[X,Y]}\psi$. Check that $\mathcal{L}_{\lambda X}\, \psi = \lambda\, \mathcal{L}_X\, \psi$ when X is tangent to the gauge orbits, $X = X^a \xi_a$.

(d) Show that if one adds the correction term $(\frac{1}{2}\operatorname{tr}\alpha)\psi$ to the equations of (a), *i.e.*, if one takes as constraint equations

$$\mathcal{L}_{\xi_a}\psi \equiv \xi_a{}^i\,\psi_{,i} + \tfrac{1}{2}\xi_a{}^i{}_{,i}\,\psi + \tfrac{1}{2}C_{ab}{}^b\,\psi = 0 \qquad \text{(iv)}$$

where ψ has the variance (iii), then one gets constraint equations that (*i*) are consistent (no anomaly); (*ii*) are covariant under changes of vectors $\xi_a, \xi_a \to M_a{}^b\,\xi_b$; and (*iii*) define a density of weight $\frac{1}{2}$ on the reduced configuration space Q/G. Furthermore, these equations are equivalent to the equations of the previous exercise. [Hint: For proving this last statement, redefine ψ as $\psi = \mu f$ where μ is a particular solution of (iv) and where f is a scalar.]

(e) Equations (iv) do not express gauge invariance in the naive sense (*i.e.*, mere invariance under shift of the argument). The extra correction term in (iv), which kills the anomaly found in (a), may appear *ad hoc*. However, this term is, in fact, induced by the ghosts of the BRST formalism. Furthermore, the transformation law (iii) is really the transformation law of an ordinary density of weight $\frac{1}{2}$ but in the space $\psi(q^i, \eta^a)$, *which includes the ghosts*. (See Exercise 14.19).

13.9. Verify that Eq. (13.13) is the Klein–Gordon equation in the case of the relativistic particle of Exercise 4.1.

13.10. Consider the constraint $p = 0$ where p is a bosonic variable. Analyze the quantum mechanics in the momentum representation where the states are functions of p. Show that in order to get finite physical amplitudes, one needs to replace one delta function $\delta(p)$ by 1 in $\int dp\,\psi_1^*(p)\,\psi_2(p)$ and that the regularized scalar product can be written as $\langle p = 0\,|\,2\pi\,\delta(\hat{q})\,|\,p = 0\rangle$. [Note: This coincides with (13.27a) if one multiplies (13.27a) by the normalization factor 2π, which is just a matter of conventions.]

13.11. Consider the fermionic, hermitian constraint $G = 0$. Show that if any solution of $G|\psi\rangle = 0$ can be written as $|\psi\rangle = G|\chi\rangle$ for some $|\chi\rangle$, then the scalar product of two physical states vanishes. [Note: This property holds for $G = ip_1$, where p_1 is a fermionic canonical momentum. In that case, the insertion of $\delta(q^1) = q^1$ (q^1 fermionic) regularizes the scalar product.]

13.12. Let $\phi_{\alpha_1} = 0$ be the primary first-class constraints of a constrained Hamiltonian system, $\phi_{\alpha_2} = 0$ be the secondary first-class constraints of the second generation, etc.... Let H_0 be the (first-class) Hamiltonian.

(a) Show that the conditions $\phi_{\alpha_1}|\psi\rangle = 0$ and $i\hbar\,\partial|\psi\rangle/\partial t = H\,|\psi\rangle$ imply $\phi_{\alpha_i}|\psi\rangle = 0$ for all i.

(b) Observe that the physical states are invariant under the gauge transformations generated by $G = \sum \mu^{\alpha_i}\,\phi_{\alpha_i}$, where the coefficients μ^{α_i} are completely arbitrary.

(c) Conclusion: primary and secondary constraints play equivalent roles in the quantum theory, as already indicated in §3.3.7.

13.13. Repeat the Dirac–Fock analysis of Sec. 13.4 in the case of fermionic constraints $\pi_1 = 0, \pi_2 = 0$.

(a) Prove that, again, the single condition $a|\psi\rangle = 0$ is satisfactory.

(b) Show that the zero-norm difficulty described in Exercise 13.11 is solved in the Dirac–Fock quantization method.

13.14. Let the constraints be $p_1 + A(q^\alpha, p_\alpha) = 0$ and $p_2 = 0$ with $\alpha = 3, 4, \ldots, N$. Define oscillator variables as in (13.39) and verify that the constraints become $a + A = 0, a^* + A = 0$. Show that the general solution of the Dirac–Fock condition $(a + A)|\psi\rangle = 0$ is given, up to null states, by $|\psi\rangle = e^{-Ab^*}|0\rangle|f\rangle$. Here, $|0\rangle$ is the Fock vacuum annihilated by a and b, while $|f\rangle$ is a state in the representation space for the last $N - 2$ degrees of freedom q^α, p_α ($\alpha = 3, \ldots, N$). [Hint: The unitary transformation $\exp(Ab - Ab^*)$ brings the constraints to the form (13.39e). Note that the null states are of the form $(a^* + A)^k e^{-Ab^*}|0\rangle|g\rangle$, $k \neq 0$.]

13.15. Consider the free relativistic particle of Exercise 4.1, with constraint $p^2 + m^2 = 0$. To get an even number of constraints, one adds the Lagrange multiplier e ("einbein") as a canonical variable (Exercise 4.2). For states with positive energy ($p^0 > 0$), the constraint $p^2 + m^2 = 0$ can be linearized for one of the momenta, either by replacing $p^2 + m^2 = 0$ by $p^0 - \sqrt{\vec{p}^{\,2} + m^2} = 0$, or by making the invertible canonical transformation $Q = x^0/2p_0, P = (p_0)^2$. Use the results of the previous exercise to show that the physical spectrum of the Dirac–Fock quantization is (as expected) isomorphic to the space of functions of the 3-momentum $\vec{p}$. [Note: To apply the Dirac–Fock method, one needs to pair some of the variables to form oscillator variables. In the present case, this procedure possesses the unpleasant feature of spoiling manifest covariance, since p^0 and $\vec{p}$ are treated differently (p^0 is paired with p_e).]

13.16. Consider the system with second-class constraints $q^1 \approx 0, p_1 \approx 0$. Replace it by a classically equivalent first-class system with new variables, as in §1.4.3. Verify that the systems remain equivalent after quantization.

13.17. (Holomorphic Quantization of Second-class Constraints.)

Consider the second-class system $q^1 \approx 0, p_1 \approx 0$ of the previous exercise. Define oscillator variables in the standard manner, $\sqrt{2}\,a = q^1 + ip_1, \sqrt{2}\,a^* = q^1 - ip_1$. Show that the single condition $a|\psi\rangle = 0$ leads to a quantum theory equivalent to that of §13.1.1. Observe that there is no null spurious state because a is not a null oscillator (a and a^* are conjugate to one another, $[a, a^*] = -i$ because the original constraints are second class). Extend the analysis to more general second-class systems whose constraints can be holomorphically split into conjugate oscillator variables as above.

13.18. Observe that the reduced phase space quantization method—and, hence, any other equivalent quantization method—is invariant under the redefinitions (1.44) of the first- and second-class constraints.

CHAPTER FOURTEEN

BRST OPERATOR METHOD—QUANTUM BRST COHOMOLOGY

14.1. GENERAL FEATURES

14.1.1. States and Operators

We have argued at length in the previous chapters that the most natural and powerful description of the dynamics of a gauge system is that which treats the ghosts on the same footing as the "original" dynamical variables z^A. To implement this same viewpoint in quantum mechanics, one must realize not only the z's, but also the ghosts η and their conjugate momenta $\mathcal{P}$ as linear operators in a Hilbert space. Thus, the Hilbert space should not only yield a representation of the brackets $[z^A, z^B]$, but it should also yield a representation of the ghost commutation relations

$$\mathcal{P}_{a_k}\,\eta^{a_s} - (-)^{(\varepsilon_{a_k}+k+1)(\varepsilon_{a_s}+s+1)}\,\eta^{a_s}\,\mathcal{P}_{a_k} = -i\,\delta_{sk}\,\delta_{a_k}{}^{a_s}. \tag{14.1}$$

In the Hilbert space where the ghosts are realized as operators, the BRST charge becomes also a linear operator. Since the Poisson bracket of two anticommuting functions becomes upon quantization an anticommutator, the nilpotency property of Ω reads,

$$[\Omega, \Omega] = 2\,\Omega^2 = 0. \tag{14.2}$$

Furthermore, since Ω was real in the classical theory, one now demands that it should be a self-adjoint operator

$$\Omega^* = \Omega, \tag{14.3}$$

where * denotes the hermitian conjugate in the nondegenerate inner product that makes the real classical variables hermitian operators. That inner product is itself denoted by (ψ, χ), the bracket notation being reserved for the states of the Dirac method. As a consequence of (14.2) and (14.3), the Hilbert space inner product must contain negative norm states. Again, the hermiticity properties must be taken with a grain of salt because, as we shall see, the norm of the physical states may need regularization.

By the same arguments used in the classical theory, one defines a BRST observable as a linear operator A, which commutes with the BRST charge

$$[A, \Omega] = 0. \tag{14.4}$$

Here, the word "commutes" is again used in a generalized sense; the bracket in (14.4) is to be understood as an anticommutator when A is anticommuting.

It will be assumed that one can find a charge Ω satisfying the nilpotency and hermiticity conditions (14.2) and (14.3). Unlike the situation in the classical case, there is no *a priori* guarantee that this can always be done starting from a classical theory for which $[\Omega, \Omega] = 0$, since the question of ordering of the factors comes in crucially.

14.1.2. Ghost Number

The ghost number of an operator is defined as in the classical theory. The ghost number operator is thus given by

$$\mathcal{G} = \sum_k i(k+1)\,\eta^{a_k}\,\mathcal{P}_{a_k} + \text{constant} \tag{14.5}$$

and is such that

$$[\mathcal{G}, z^A] = 0, \tag{14.6a}$$

$$[\mathcal{G}, \eta^{a_k}] = (k+1)\,\eta^{a_k}, \qquad [\mathcal{G}, \mathcal{P}_{a_k}] = -(k+1)\,\mathcal{P}_{a_k}. \tag{14.6b}$$

The constant in (14.5) is real and reflects the ordering ambiguity. Any operator A can be decomposed in components of definite ghost number

$$A = \sum_g A_g, \qquad [\mathcal{G}, A_g] = g\, A_g, \qquad g \in Z, \tag{14.7}$$

where g is an integer.

One can adjust the constant in (14.5) so that $\mathcal{G}$ is formally antihermitian

$$\mathcal{G} = \frac{i}{2} \sum_k (k+1)\, (\eta^{a_k}\, \mathcal{P}_{a_k} - (-)^{\varepsilon_{a_k}+k} \mathcal{P}_{a_k} \eta^{a_k}), \tag{14.8a}$$

$$\mathcal{G}^* = -\mathcal{G}. \tag{14.8b}$$

Another choice, which is also useful, is to take the constant term in (14.5) equal to zero,

$$\mathcal{G}' = i \sum_k (k+1)\, \eta^{a_k}\, \mathcal{P}_{a_k}. \tag{14.9}$$

The constant by which the operators $\mathcal{G}$ and $\mathcal{G}'$ differ is usually infinite in the case of field theory where there is an infinite number of constraints. In that instance, $\mathcal{G}$ and $\mathcal{G}'$ cannot both be well-defined operators simultaneously.

It will be assumed in the sequel that the linear space of states W splits as a sum of eigenspaces of $\mathcal{G}$ with definite real ghost number

$$W = \sum_p W_p, \tag{14.10a}$$

$$\mathcal{G}\, \psi_p = p\, \psi_p, \qquad \psi_p \in W_p. \tag{14.10b}$$

This property always holds for bosonic constraints, as well as for fermionic constraints when the corresponding bosonic ghosts are quantized in the Fock representation (Exercises 14.3 and 14.4). For bosonic ghosts quantized in the standard Hilbert space $L^2(R^m)$, the decomposition (14.10a) cannot be correct, since the existence of real eigenvalues for the antihermitian operator $\mathcal{G}$ acting only on the ghost variables conflicts with the positiveness of the scalar product of $L^2(R^m)$. The discrete spectrum of $\mathcal{G}$ in $L^2(R^m)$ is empty. We shall exclude the case of bosonic ghosts quantized in $L^2(R^m)$ from the general discussion and come back to it in §14.5.3 (see pp. 435 and 436).

Contrary to g in (14.7), the ghost number p of a state need not be an integer. However, the difference in the ghost numbers of two different states must be an integer. Otherwise, one could construct operators with noninteger ghost numbers. Accordingly, one has

$$p = p_0 + k, \qquad p_0 \in R\,, \;\; k \in Z. \tag{14.10c}$$

The constant p_0 is defined only up to an integer, and so one can take $0 \leq p_0 < 1$.

Theorem 14.1.

(a) The scalar product of two states ψ_p and $\psi_{p'}$ with respective ghost numbers p and p' vanishes if $p + p' \neq 0$,

$$(\psi_p, \psi_{p'}) = 0, \qquad p + p' \neq 0. \tag{14.11}$$

(b) The constant p_0 in (14.10c) is equal to 0 or $\frac{1}{2}$. That is, the ghost number of the states is either an integer or a half-integer.

Proof. (a) From $p'(\psi_p, \psi_{p'}) = (\psi_p, \mathcal{G}\,\psi_{p'}) = -(\mathcal{G}\,\psi_p, \psi_{p'}) = -p\,(\psi_p, \psi_{p'})$, it follows that $(\psi_p, \psi_{p'})$ vanishes unless $p + p' = 0$. Note that the argument assumes $(\psi_p, \psi_{p'})$ to be finite.

(b) Because the scalar product is nondegenerate, there must be states with ghost number $-p$ if there are states with ghost number p. Hence, by (14.10c), $-p_0$ must differ from p_0 by an integer, *i.e.*,

$$2\,p_0 = \text{integer}.$$

This proves (b).

When $p_0 = \frac{1}{2}$, one says that there is "fractionalization of the ghost number." This occurs when the number of independent constraints is odd. [See Exercises 14.3 and 14.4.]

It follows from the theorem that the states with nonvanishing ghost number have zero norm when that norm is well defined.

14.1.3. Physical State Condition

The description of the system in terms of the extended phase space has redundant variables. This redundancy was already present in the formulation of the gauge theory in terms of the z^A's, and it becomes even larger when one has the ghosts. The whole point of the BRST formalism is that by enlarging the redundancy, the description becomes ultimately more transparent, and, in a sense, the two redundancies cancel each other.

To make the redundancies cancel each other, one must bring in a condition that will select the physical subspace. This condition must be the analog of the demand that the gauge generators G_a annihilate the physical states in the formalism without ghosts.

Now, we have seen that the statements of BRST invariance and gauge invariance are equivalent in the classical theory. Accordingly, one should require that the physical states be BRST invariant,

$$\Omega\psi = 0. \tag{14.12}$$

This condition possesses the following good properties:

(*i*) it is linear and hence it selects a subspace;
(*ii*) BRST observables, obeying $[A, \Omega] = 0$, map the physical subspace onto itself;
(*iii*) trivial observables of the form $[K, \Omega]$ have vanishing matrix elements between physical states,

$$(\psi_1, [K, \Omega] \psi_2) = 0 \tag{14.13}$$

if ψ_1, ψ_2 obey (14.12). This is a much desired property, since we have seen that $[K, \Omega]$ is classically unobservable. Note that (14.13) needs the hermiticity of Ω to hold.

14.1.4. Quantum BRST Cohomology

Equation (14.12) by itself does not completely describe the physical states. The true physical states are obtained by a further identification of solutions of (14.12) and are also subject to a ghost number condition. We discuss in this subsection how the solutions of (14.12) should be identified. The analysis of the ghost number condition is postponed to § 14.2.5.

Because the operator Ω is nilpotent quantum-mechanically, one can define the quantum BRST cohomology for the operators along exactly the same lines as in the classical theory. So, one defines again

$$[A, \Omega] = 0 \Leftrightarrow A \text{ is BRST-closed,} \tag{14.14a}$$
$$A = [B, \Omega] \Leftrightarrow A \text{ is BRST-exact.} \tag{14.14b}$$

where $[\ \ ,\ \ ]$ is the graded commutator. Since the graded commutator obeys the same Jacobi identity as the graded Poisson bracket, BRST-exact operators are BRST-closed, and one defines the "quantum operator cohomology" $H^*_{\text{op}}(\Omega)$ as the set of equivalence classes of BRST-closed operators modulo BRST-exact ones.

Similarly, the nilpotency of Ω implies that any state of the form $\Omega\chi$ obeys (14.12)

$$\Omega(\Omega\chi) = \Omega^2\chi = 0. \tag{14.15}$$

Therefore, one defines

$$\Omega\psi = 0 \Leftrightarrow \psi \text{ is BRST-closed,} \tag{14.16a}$$
$$\psi = \Omega\chi \Leftrightarrow \psi \text{ is BRST-exact.} \tag{14.14b}$$

and one introduces the "quantum state cohomology" $H^*_{\text{st}}(\Omega)$ as the set of equivalence classes of BRST-closed states moduly BRST-exact ones,

$$H^*_{\text{st}}(\Omega) = \frac{\operatorname{Ker}\Omega}{\operatorname{Im}\Omega}. \tag{14.17}$$

We have seen that in classical mechanics, the BRST observables A and $A + [K, \Omega]$ should be identified. To make possible the operator identification $A \sim A + [K, \Omega]$ in quantum mechanics, the states $A\psi$ and $(A + [K, \Omega,])\psi$ should be identified whenever ψ obeys the physical state condition (14.12). In that case, $[\Omega, K]\psi$ reduces to $\Omega K\psi$. But $K\psi$ can be an arbitrary state, since K itself is arbitrary. Hence, $\Omega\chi$ should be identified with zero for any χ. We thus come to the important conclusion that *physical states are given by the equivalence classes of the BRST state cohomology*. The identification of $\Omega\chi$ with zero is permissible because the state $\Omega\chi$ has vanishing scalar product with any physical state

$$(\psi, \Omega\chi) = (\Omega\psi, \chi) = 0 \quad \text{iff} \quad \Omega\psi = 0 \tag{14.18}$$

and, hence, does not contribute to physical amplitudes. This property could actually have been used to infer that ψ and $\psi + \Omega\chi$ should be identified.

Because the BRST-invariant extensions of the constraints are given by

$$\bar{G}_{a_0} = [-\mathcal{P}_{a_0}, \Omega], \tag{14.19a}$$

they are null operators. As a result, they produce a null state when acting on a BRST-closed state

$$\bar{G}_{a_0}|\psi\rangle = \Omega\big((-)^{\varepsilon_{a_0}+1}\mathcal{P}_{a_0}|\psi\rangle\big) \quad \text{if} \quad \Omega|\chi\rangle = 0. \tag{14.19b}$$

Therefore, the gauge generators $\bar{G}_{a_0}$ possess a trivial action on the space of equivalence classes of physical states.

14.1.5. Anomalies

A crucial feature of the BRST formalism is the nilpotency of the BRST charge. This feature enabled one to infer that the physical subspace is the quotient space $H^*(\Omega)$ on which the gauge group acts trivially.

Since the passage to the cohomological space $H^*(\Omega)$ is only meaningful because Ω is nilpotent, one can say that the nilpotency of Ω is the quantum expression of the gauge invariance. If the condition $\Omega^2 = 0$ fails to hold due to quantum effects, not all gauge degrees of freedom disappear from the physical spectrum and the quantum theory is anomalous. For this reason, we will assume $\Omega^2 = 0$ throughout. This condition can be fulfilled even if the quantum constraints $G_a(z^A)$ have an anomaly (Exercise 14.19). Similarly, we will assume that the quantum Hamiltonian is BRST invariant to avoid anomalies that may arise from a non-BRST-invariant time evolution.

14.2. ANALYSIS OF QUANTUM BRST COHOMOLOGY: GENERAL THEOREMS

The central question in the BRST operator quantization of gauge systems is to determine the BRST cohomology. This section is devoted to proving general theorems on $H^*_{\rm op}(\Omega)$ and $H^*_{\rm st}(\Omega)$.

In the proof of these theorems, we will proceed as if the space of states were finite dimensional. In other words, we will not pay any attention to the functional subtleties arising from the infinite dimensionality of the space of states. Our main goal is to stress the characteristic features of the BRST formalism itself, not of Hilbert space theory.

We first compare the operator cohomology with the state cohomology. This is done by constructing an appropriate basis in which Ω takes a simple form ("Jordan basis").

14.2.1. Jordan Canonical Form of the BRST Charge: Operator Cohomology versus State Cohomology

14.2.1a. State Cohomology

Let g_α be a basis of the vector space $\operatorname{Im}\Omega$, *i.e.*, of the subspace onto which the entire space of states is mapped. For each g_α, there exists (at least) one f_α such that

$$g_\alpha = \Omega f_\alpha. \tag{14.20}$$

The vectors $\{g_\alpha, f_\alpha\}$ are linearly independent. This easily follows from the linear independence of the g_α's and from the nilpotency of Ω, which implies $\Omega\, g_\alpha = 0$.

Consider now the subspace $\operatorname{Ker}\Omega$, *i.e.*, the subspace of vectors annihilated by Ω. Complete the set $\{g_\alpha\}$ by adding vectors e_i such that $\{e_i, g_\alpha\}$ form a basis of $\operatorname{Ker}\Omega$. The vectors e_i do not belong to $\operatorname{Im}\Omega$; hence, they cannot be written as the Ω of something.

The set $\{e_i, g_\alpha, f_\alpha\}$ is a basis of the entire space of states. Indeed, for an arbitrary vector χ, $\Omega\chi$ belongs to $\operatorname{Im}\Omega$ and can be written as $\Omega\chi = \lambda^\alpha\, g_\alpha$. The vector $\chi - \lambda^\alpha\, f_\alpha$ belongs then to $\operatorname{Ker}\Omega$ and can thus be expanded in terms of e_i and g_α.

In the basis $\{e_i, g_\alpha, f_\alpha\}$, the operator Ω takes a simple form, containing one-dimensional and two-dimensional "Jordan blocks."

The two-dimensional Jordan blocks are given by $\{g_\alpha, f_\alpha\}$,

$$\Omega\, f_\alpha = g_\alpha, \qquad \Omega\, g_\alpha = 0. \tag{14.21a}$$

The one-dimensional blocks are given by $\{e_i\}$,

$$\Omega\, e_i = 0. \tag{14.21b}$$

The ambiguity in the construction of the basis $\{e_i, g_\alpha, f_\alpha\}$ is exhausted by

$$g_\alpha \to \bar{g}_\alpha = A_\alpha{}^\beta\, g_\beta, \tag{14.22a}$$
$$f_\alpha \to \bar{f}_\alpha = A_\alpha{}^\beta\, f_\beta + \nu_\alpha{}^\beta\, g_\beta + \nu_\alpha{}^i\, e_i, \tag{14.22b}$$
$$e_i \to \bar{e}_i = A_i{}^j\, e_j + \mu_i{}^\alpha\, g_\alpha, \tag{14.22c}$$

where $A_\alpha{}^\beta$ and $A_i{}^j$ are invertible (Exercise 14.8).

The description of the BRST state cohomology is very simple in the Jordan canonical basis. The condition $\Omega\psi = 0$ removes the vectors f_α and leaves one with only e_i and g_α. The further passage to the quotient $\operatorname{Ker}\Omega/\operatorname{Im}\Omega$ eliminates, then, the vectors g_α, which are BRST trivial. Hence, the two-dimensional blocks are completely removed, and the BRST cohomology is isomorphic with the subspace spanned by the e_i.

14.2.1b. Operator Cohomology

Just as the BRST-state cohomology is simply described in the basis $\{e_i, g_\alpha, f_\alpha\}$ where the BRST charge takes the Jordan canonical form, the BRST-operator cohomology is also easily analyzed in that same basis. Let A be a BRST-invariant operator. The condition $[A, \Omega] = 0$ implies

$$A\, e_i = A_i{}^j\, e_j + A_i{}^\alpha\, g_\alpha, \tag{14.23a}$$
$$A\, g_\alpha = A_\alpha{}^\beta\, g_\beta, \tag{14.23b}$$
$$A\, f_\alpha = A_\alpha{}^\beta\, f_\beta + M_\alpha{}^i\, e_i + M_\alpha{}^\beta\, g_\beta. \tag{14.23c}$$

On the other hand, the operator B is BRST null if and only if

$$B\, e_i = B_i{}^\alpha\, g_\alpha, \tag{14.24a}$$
$$B\, g_a = B_\alpha{}^\beta\, g_\beta, \tag{14.24b}$$
$$B\, f_\alpha = B_\alpha{}^\beta\, f_\beta + N_\alpha{}^i\, e_i + N_\alpha{}^\beta\, g_\beta. \tag{14.24c}$$

One has indeed $B = [K, \Omega]$ with K given, for instance, by

$$K\, e_i = B_i{}^\alpha\, f_\alpha, \tag{14.25a}$$
$$K\, g_\alpha = B_\alpha{}^\beta\, f_\beta + N_\alpha{}^i\, e_i + N_\alpha{}^\beta\, g_\beta, \tag{14.25b}$$
$$K\, f_\alpha = 0. \tag{14.25c}$$

Hence, the only cohomologically nontrivial part of a BRST-invariant operator is given by $A_i{}^j$, *i.e.*, by its action on the one-dimensional Jordan blocks that determine the state cohomology. This establishes the following simple theorem.

Theorem 14.2. *The BRST-operator cohomology is isomorphic with the algebra of linear operators acting on the BRST-state cohomology.*

These cohomologies, therefore, determine each other. In particular, if one can prove a decoupling theorem stating that the physical states are isomorphic to some well-defined "transverse" states, one automatically knows that the set of physical operators is isomorphic to the set of "transverse" operators, and vice-versa.

14.2.1c. Lefschetz Trace Formula

The vector space W containing all the states decomposes according to ghost number as in (14.10). Let A be an arbitrary ghost number zero operator. The operator A maps W_p on W_p and so induces an operator A_p in W_p. One defines the W-trace of A as

$$\mathrm{tr}_W A = \sum_p (-)^p \,\mathrm{tr}\, A_p, \tag{14.26}$$

where $\mathrm{tr}\, A_p$ is the trace of A_p in W_p. If p is a half-integer, one replaces $(-)^p$ by $(-)^{p+1/2}$ in (14.26). The subspaces differing by one unit of ghost number contribute to (14.26) with the opposite sign.

If A is BRST-invariant, it also induces a well-defined operator $\tilde{A}$ in the space $H^*_{\mathrm{st}}(\Omega)$ of the state cohomology. This space is isomorphic to the space E generated by the one-dimensional Jordan blocks $\{e_i\}$ and decomposes according to ghost number as

$$E = \sum_p E_p. \tag{14.27}$$

One defines the E-trace of the BRST-invariant operator A as

$$\mathrm{tr}_E A = \sum_p (-)^p \,\mathrm{tr}\, \tilde{A}_p, \tag{14.28}$$

where $\mathrm{tr}\, \tilde{A}_P$ is the trace of $\tilde{A}_p$ in E_p.

Theorem 14.3. *(Lefschetz trace formula)*

$$\mathrm{tr}_W A = \mathrm{tr}_E A. \tag{14.29}$$

Proof. The proof is straightforward and relies on the formula (14.23), which shows that the coefficient of g_β in Ag_α is equal to the coefficient of f_β in Af_α. Because the ghost numbers of g_α and f_α differ by one unit, one finds that the sum over the corresponding diagonal elements cancels out. The W-trace (14.26) reduces then to the expression

$$\begin{aligned} \text{tr}_W A &= \sum_p (-)^p (A_p)_i{}^i \\ &= \text{tr}_E A, \end{aligned}$$

which is the desired formula.

14.2.2. Duality Formula for the Operator Cohomology

In classical mechanics, there is, in general, no relationship between the BRST cohomological classes at positive and negative ghost numbers. This is because the BRST cohomology at negative ghost number is determined by the Koszul-Tate differential and is, accordingly, trivial, while the BRST cohomology at positive ghost number is determined by the longitudinal exterior derivative along the gauge orbits and may be nontrivial.

The linear structure of quantum mechanics, with the observables being realized as linear operators acting on states, implies, however, new features. The most striking one is that the BRST cohomology at negative ghost number becomes dual to the BRST cohomology at positive ghost number.

Indeed, suppose that there exists a nontrivial BRST-invariant operator A with $\text{gh}\, A = g$, $g > 0$. This implies that there exist nontrivial BRST-invariant vectors e, e' with

$$Ae = e', \qquad \mathcal{G}e = ke, \qquad \mathcal{G}e' = (k+g)\,e'. \tag{14.30}$$

The ghost numbers k and $k+g$ of e and e' differ by the ghost number g of A.

Because of (14.30) and our analysis of §14.2.1, one can define a nontrivial operator A' of ghost number $-g$. This operator A' can be taken, for instance, to map e' on e and all other basis vectors on zero. Hence, whenever there exists a nontrivial BRST-closed operator at ghost number $+g$, there exists also a nontrivial BRST-closed operator at ghost number $-g$.

More precisely, if we decompose the space E of the one-dimensional Jordan blocks according to ghost number, one finds that a BRST-invariant operator of ghost number g defines a linear map from E_k to E_{k+g} for each k. Since the vector space of linear maps from E_k to E_{k+g} is dual to the vector space of linear maps from E_{k+g} to E_k, one gets the duality formula

$$H^g_{\text{op}}(\Omega) = \left(H^{-g}_{\text{op}}(\Omega)\right) \qquad \text{(operator cohomology)}\,, \tag{14.31}$$

and thus the operator cohomologies at $\pm g$ are isomorphic.

An immediate consequence of (14.31) is that the relation between the quantum and classical theories is bound to be subtle when the classical BRST cohomology $H^g(s)$ is nontrivial for $g > 0$. Indeed, (14.31) is compatible with the classical results only if $(\operatorname{Ker}\Omega/\operatorname{Im}\Omega)^g = 0$ for $g \neq 0$. If the classical cohomology at positive ghost number is nontrivial, then the quantization procedure must either introduce new cohomology at negative ghost number or remove the cohomology at positive ghost number or do a mixture of both.

14.2.3. (Pseudo-)Unitary Representations of the BRST-Ghost Number Algebra

The above proof of the duality formula for the operator cohomology uses very little (nilpotency of BRST operator and linear structure of quantum mechanics) and, in particular, does not rely on the existence of a scalar product. This is to be noticed because, as we have already experienced and emphasized in the previous chapter, considerations based on the existence of a scalar product are heuristic when there is no normalizable solution of the physical state condition. There are instances, however, when the physical states possess a finite norm. One can then extend the duality formula (14.31) to the state cohomology. [Without an extra structure, there is, in general, no duality formula for the state cohomology, since the subspaces E_p of E in (14.27) may be completely unrelated for different values of p.]

To prove the duality theorem for the state cohomology, it is necessary to first establish what are the unitary representations of the BRST-ghost number algebra with nondegenerate scalar product.

Let W_k be the eigenspace of $\mathcal{G}$ at ghost number k. One has

$$W_k = E_k \oplus G_k \oplus F_k \tag{14.32}$$

where E_k contains the one-dimensional Jordan blocks $\{e_i\}$ of ghost number k, while G_k contains the vectors g_α and F_k contains the vectors f_α, both also at ghost number k.

One has

$$G_k = \Omega\, F_{k-1}, \tag{14.33}$$

since Ω increases the ghost number by one unit. The mapping from F_{k-1} to G_k defined by (14.33) is invertible by construction.

By Theorem 14.1, we know that the scalar product defines a nondegenerate pairing between W_k and W_{-k}. We want to show that this induces a nondegenerate pairing betwen E_k and E_{-k}. This will be done by first proving that the scalar product pairs G_k with F_{-k}, which are therefore isomorphic.

The states in G_k are BRST-exact and can only have nonvanishing scalar products with states in F_{-k}. Furthermore, since the scalar product is nondegenerate, the dimension of F_{-k} is at least equal to the dimension of G_k. Suppose that it is strictly greater. Then, the dimension of $G_{-k+1} = \Omega\, F_{-k}$ would be strictly greater than the dimension of F_{k-1} ($G_k = \Omega\, F_{k-1}$), and one could find states in G_{-k+1} that would be orthogonal to all states in W_{k-1}, leading to a contradiction. Hence,

$$\dim F_{-k} = \dim G_k. \tag{14.34}$$

Because $\dim W_k = \dim W_{-k}$, we can already infer from (14.34) that $\dim E_k = \dim E_{-k}$, *i.e.*, that the spaces of the one-dimensional Jordan blocks at ghost numbers $\pm k$ are isomorphic.

Let ${g_{\alpha_k}}^k$ be a basis of G_k, $k \geq 0$ and $k \neq \frac{1}{2}$. By linear redefinition of ${f_{\alpha_k}}^{-k}$, one can achieve

$$({g_{\alpha_k}}^k, {f_{\beta_k}}^{-k}) = \delta_{\alpha_k \beta_k}, \qquad k \geq 1. \tag{14.35}$$

One also finds,

$$\begin{aligned}({g_{\alpha_{-k}}}^{-k},\, {f_{\beta_{-k}}}^{k}), &= (\Omega\, {f_{\alpha_{-k}}}^{-k-1},\, {f_{\beta_{-k}}}^{k}),\\ &= ({f_{\alpha_{-k}}}^{-k-1},\, {g_{\beta_{-k}}}^{k+1}),\\ &= \delta_{\alpha_{-k}\beta_{-k}} \qquad (k \geq 0),\end{aligned}$$

so that (14.35) holds for all k's $\neq \frac{1}{2}$

$$({g_{\alpha_k}}^k,\, {f_{\beta_k}}^{-k}) = \delta_{\alpha_k \beta_k} \qquad (k \neq \tfrac{1}{2}). \tag{14.36}$$

For $k = \frac{1}{2}$, one can arrange that $({g_{\alpha_{1/2}}}^{1/2}, {f_{\beta_{1/2}}}^{-1/2}) = \pm\delta_{\alpha_{1/2}\beta_{1/2}}$. In the same way, one gets

$$({g_{\alpha_k}}^k,\, {g_{\alpha_{-k}}}^{-k}) = 0, \tag{14.37a}$$

$$({g_{\alpha_k}}^k,\, {e_{i_k}}^{-k}) = 0, \tag{14.37b}$$

if ${e_{i_k}}^k$ is a basis of E_k, since the g_α's are BRST null vectors.

By adding to ${e_{i_k}}^k$ an appropriate linear combination of ${g_{\alpha_k}}^k$, one can reach

$$({e_{i_k}}^k,\, {f_{\alpha_k}}^{-k}) = 0. \tag{14.37c}$$

This implies that the matrix $({e_{i_k}}^k, {e_{j_k}}^{-k})$ is nondegenerate, since, otherwise, one could find an e-vector at ghost number k orthogonal to W_{-k}, in contradiction with the fact that the scalar product is nondegenerate. By linear redefinition, one can assume

$$(e_{i_k}{}^k, e_{j_k}{}^{-k}) = \delta_{i_k j_k} \qquad (k \neq 0), \tag{14.38a}$$
$$(e_{i_0}{}^0, e_{j_0}{}^0) = \pm\delta_{i_0 j_0} \qquad (k = 0). \tag{14.38b}$$

The last step in reaching the canonical form of the matrix of the scalar product is to use the remaining freedom

$$f_{\alpha_{-k}}{}^k \rightarrow f_{\alpha_{-k}}{}^k + \nu_{\alpha_{-k}}{}^{\beta_k} g_{\beta_k}{}^k \tag{14.39a}$$

to achieve

$$(f_{\alpha_{-k}}{}^k, f_{\beta_k}{}^{-k}) = 0 \tag{14.39b}$$

[for $k = 0$, use $(\alpha, \beta) = (\beta, \alpha)^*$].

We have thus proved:

Theorem 14.4. *The most general (pseudo-)unitary representation of the BRST-ghost number algebra*

$$\Omega^2 = 0, \qquad [\Omega, \mathcal{G}] = \Omega, \tag{14.40a}$$
$$\Omega^* = \Omega, \qquad \mathcal{G}^* = -\mathcal{G}, \tag{14.40b}$$

is the direct orthogonal sum of the following building blocks,

(i) Singlet:

$$\Omega e = 0, \qquad \mathcal{G} e = 0, \tag{14.41a}$$
$$(e, e) = \pm 1, \tag{14.41b}$$

(ii) Non-null doublet:

$$\Omega e = 0, \qquad \mathcal{G} e = ke, \qquad k \neq 0, \tag{14.42a}$$
$$\Omega e' = 0, \qquad \mathcal{G} e' = -ke', \tag{14.42b}$$
$$(e, e') = 1, \tag{14.42c}$$

(iii) Null doublet at ghost number $\pm\frac{1}{2}$*:*

$$\Omega f = g, \qquad \Omega g = 0, \tag{14.43a}$$
$$\mathcal{G} f = -\tfrac{1}{2} f, \qquad \mathcal{G} g = \tfrac{1}{2} g, \tag{14.43b}$$
$$(f, g) = \pm 1, \tag{14.43c}$$

(iv) Quartet:

$$\Omega f = g, \qquad \Omega g = 0, \tag{14.44a}$$
$$\Omega f' = g', \qquad \Omega g' = 0, \tag{14.44b}$$
$$\mathcal{G} f = (k-1) f, \qquad \mathcal{G} g = kg, \tag{14.44c}$$
$$\mathcal{G} f' = -kf', \qquad \mathcal{G} g' = (-k+1) g', \tag{14.44d}$$
$$(f, g') = (f', g) = 1. \tag{14.44e}$$

(The scalar products that are not written are zero.)

14.2.4. Duality Formula for the State Cohomology

It results from Theorem 14.4 that E_k and E_{-k} are dual to each other, since the BRST scalar product defines a nondegenerate pairing between these subspaces [see (14.38a)],

$$E_{-k} = E_k^*. \tag{14.45}$$

Thus, the BRST-state cohomologies at ghost numbers k and $-k$ are isomorphic

$$H^k_{\text{st}}(\Omega) \simeq H^{-k}_{\text{st}}(\Omega) \qquad \text{(state cohomology)}. \tag{14.46}$$

It should be stressed that although E_k and E_{-k} are isomorphic, there is no canonical isomorphism between them. Hence, given a state in E_k, there is no unique state in E_{-k} $(= E_k^*)$ dual to it. As is well known, adopting a definite isomorphism between E_k and E_{-k} is equivalent to defining a nondegenerate scalar product on E_k itself. For this effective scalar product, the ghost number operator $\mathcal{G}$ cannot be antihermitian (unless $k = 0$).

14.2.5. Physical States and Ghost Number

The classical observables are characterized not only by the condition of BRST invariance but also by the ghost number zero condition. To incorporate this restriction into the quantum theory, it is necessary to impose a further condition on the physical states. Otherwise, BRST-closed operators connecting states with different values of the ghost number—and having thus nonvanishing ghost number—would be allowed.

One may wish to consider an extension of the formalism in which the condition that the physical operators should be of ghost number zero is abandoned. Even though this may incorporate further interesting cohomology, we shall not explore that possibility in this book. We shall restrict, instead, the physical operators to ghost number zero throughout, as the classical observables. This approach leaves enough room to develop a consistent quantum theory.

The simplest possibility for taking into account the restriction $[A, \mathcal{G}] = 0$ for the physical operators is to demand that the states possess all the same definite ghost number g. In view of Theorem 14.1, that value should be zero, since the scalar product induced on the subspaces $W_{p\neq 0}$ is identically zero. We thus take as a further postulate

$$\mathcal{G}\psi = 0 \tag{14.47}$$

for the physical states, besides $\Omega\psi = 0$. That is, the physical states are the elements of the BRST-state cohomology at ghost number zero.

The condition (14.47) is quite satisfactory for systems quantized in the Fock representation (Sec. 14.4). However, if applied blindly to systems quantized in the Schrödinger representation (or in any equivalent representation), it leads to physically incorrect results because the interesting cohomology is then formed by nonnull doublets at the extreme values of the ghost number (Sec. 14.5). These values are $\pm m/2$, where m is the number of constraints (Exercises 14.3 and 14.4). There may even be no cohomology at all at ghost number zero. This would occur, for instance, if the ghost number is a half-integer. However, one recovers a uniform treatment by including the Lagrange multipliers for all the constraints as dynamical variables. This does not modify the BRST–Fock quantization (§14.4.3) but has the virtue of bringing the cohomology at $\pm m/2$ to ghost number zero. Indeed, the Lagrange multipliers contribute their own BRST cohomology at ghost number $\pm m/2$ and form, with the states at $\pm m/2$ of the original system, ghost number zero singlets according to the formula

$$H^0(\Omega) = \left(H^{-m/2}(\Omega_1) \otimes H^{m/2}(\Omega_2)\right) \oplus \left(H^{m/2}(\Omega_1) \otimes H^{-m/2}(\Omega_2)\right). \tag{14.48}$$

Here, $\Omega = \Omega_1 + \Omega_2$, where Ω_1 is the original BRST charge and Ω_2 the one for the Lagrange multiplier sector. [We have assumed for simplicity that the cohomology of either Ω_1 or Ω_2 is concentrated at $\pm m/2$. If not, there may be extra contributions to $H^0(\Omega)$ arising by taking the tensor product of $H^p(\Omega_1)$ with $H^{-p}(\Omega_2)$, $p \neq \pm m/2$.]

14.2.6. No Negative Norm State Criterion

It follows from (14.41) that the inner product induced on $H^0_{\text{st}}(\Omega)$ by the original scalar product is positive definite if and only if there is no singlet of negative norm. This condition is equivalent to the demand that the signature of the scalar product induced on E_0—which is also equal to the signature of the scalar product induced on $W_0 = E_0 \oplus G_0 \oplus F_0$, since $G_0 \oplus F_0$ contributes an equal number of + and − signs by (14.44)—should be equal to the dimension of E_0 ($\equiv H^0_{\text{st}}(\Omega)$). As the spaces W_k with $k \neq 0$ also do not contribute to the signature, the absence of negative norm singlets can be further rephrased as

$$\dim H^0_{\text{st}}(\Omega) = \text{signature}_W(\ ,\ ) \tag{14.49}$$

where the signature is evaluated in the full space W.

The condition (14.49) is fulfilled when the quantization procedure preserves the following key property of the classical BRST cohomology:

$$H^0_{\text{op}}(\Omega) = \{\text{gauge-invariant operators}\}. \tag{14.50}$$

In that case, the space E_0 of the physical states is isomorphic with the space of states of the reduced phase space method (*i.e.*, with the space of states in a physical gauge), whose metric is positive definite. The isomorphism (14.50), which holds classically, is expected to hold also quantum-mechanically in the absence of quantum subtleties (factor-ordering difficulties, nontrivial topology, ...).

It turns out, indeed, that the condition (14.50) is verified by simple models quantized in the Fock representation (Sec. 14.4). It is violated, however, when the states at ghost number zero are obtained by combining nonnull doublets as in (14.48). In that case, each state is at least duplicated ("doubling") because $H^{-p}(\Omega_1) \otimes H^p(\Omega_2)$ is isomorphic to $H^p(\Omega_1) \otimes H^{-p}(\Omega_2)$ by the duality formulas. Furthermore, the metric induced in (14.48) is of mixed signature; that is, each state comes once with a positive norm and once with a negative norm. The negative norm states must then be eliminated by an extra condition over and above $\Omega\psi = 0$, $\mathcal{G}\psi = 0$. This is necessary if a probabilistic interpretation of the theory is to be given (Exercise 14.16).

14.3. TIME EVOLUTION

14.3.1. Schrödinger Equation

The time evolution in the big linear space W containing all the states is determined by the Schrödinger equation,

$$i\hbar \frac{\partial \psi}{\partial t} = H\,\psi. \tag{14.51a}$$

In (14.51a), H is one of the BRST-invariant extensions of the gauge-invariant Hamiltonian H_0,

$$[H, \Omega] = 0. \tag{14.51b}$$

The solution of (14.51a) can be written as

$$\psi(t) = U_H(t)\,\psi_0, \tag{14.51c}$$

where $U_H(t)$ is the evolution operator

$$U_H(t) = \exp -\frac{i}{\hbar}\,Ht. \tag{14.51d}$$

Theorem 14.5.

(a) The operator $U_H(t)$ is BRST invariant and of ghost number zero,

$$[U_H(t), \Omega] = 0, \qquad [U_H(t), \mathcal{G}] = 0, \tag{14.52}$$

and so maps the physical subspace on itself.

(b) The evolution operators $U_H(t)$ and $U_{H+[K,\Omega]}(t)$ differ by a BRST-exact operator,

$$U_{H+[K,\Omega]} = U_H + [M, \Omega]. \tag{14.53}$$

Accordingly, they are identical in cohomology and have the same matrix elements between physical states.

Proof. Part (a) is obvious. Part (b) is proved by Taylor expansion of the exponential in (14.51d). This is explicitly worked out in Exercise 14.13.

Note that for generically covariant systems, the evolution operator is BRST exact. This corresponds to the fact that the time evolution moves then the physical states within their BRST cohomological classes.

14.3.2. Unitarity in the Physical Subspace

Theorem 14.6.

(a) The operator U_H is unitary,

$$U_H^* U_H = U_H U_H^* = I. \tag{14.54}$$

(b) The induced evolution operator in the physical space is unitary for the induced scalar product.

Proof. Again, the proof is immediate. Unitarity in W follows from hermicity of H, $H^* = H$. Unitarity in cohomology follows from (14.52) and unitarity in W. (For this last point to make sense, the scalar product must be finite in cohomology.)

If the scalar product in $H^*_{\text{st}}(\Omega)$ is not positive definite, then, as we have pointed out, extra conditions are required to get a satisfactory physical theory. This may be done by selecting a subspace through

$$P_m \psi = 0, \tag{14.55a}$$

a demand consistent with the time evolution if

$$[P_m, H] = 0, \tag{14.55b}$$

since then

$$[P_m, U] = 0. \tag{14.55c}$$

Finally, we note that the trace of the evolution operator in the full space of states is equal to its trace in cohomology on account of the Lefschetz formula.

14.4. BRST QUANTIZATION IN THE FOCK REPRESENTATION

14.4.1. BRST Charge and Ghost Number Operator

We first illustrate the BRST formalism in the case of the Fock representation. It was shown in Chapter 13 that in this representation, the scalar products of physical states are finite. This is made possible because only half of the constraints are directly enforced on the physical states, namely, $G_a^{(-)}\psi = 0$. The other half is recovered through the presence of null states.

We shall again analyze the Fock representation in the case of abelian bosonic constraints. This simple example illustrates all the main points. To define the Fock representation, the constraints must be grouped in pairs. In the case of a single pair of constraints $p_1 = 0, p_2 = 0$, the BRST charge is given by

$$\Omega = p_1\, \eta^1 + p_2\, \eta^2, \tag{14.56a}$$

and the antihermitian ghost number operator $\mathcal{G}$ reads

$$\mathcal{G} = \tfrac{i}{2}(\eta^1 \mathcal{P}_1 - \mathcal{P}_1 \eta^1 + \eta^2 \mathcal{P}_2 - \mathcal{P}_2 \eta^2). \tag{14.56b}$$

One defines oscillators a, a^*, b, b^* for the original variables through

$$a = p_1 + ip_2, \qquad a^* = p_1 - ip_2, \tag{14.57a}$$

$$b = -\frac{i}{2}(q^1 + iq^2), \qquad b^* = \frac{i}{2}(q^1 - iq^2), \tag{14.57b}$$

[see (13.39)] and for the ghosts

$$c = \tfrac{1}{2}(\eta^1 + i\eta^2), \qquad c^* = \tfrac{1}{2}(\eta^1 - i\eta^2), \tag{14.57c}$$

$$\bar{c} = i(\mathcal{P}_1 + i\mathcal{P}_2), \qquad \bar{c}^* = i(\mathcal{P}_1 - i\mathcal{P}_2). \tag{14.57d}$$

(Recall that the anticommuting momenta are antihermitian.) The c's obey the anticommutation rules

$$[c, c] = [\bar{c}, \bar{c}] = [c, c^*] = [\bar{c}, \bar{c}^*] = [c, \bar{c}] = [c^*, \bar{c}^*] = 0, \tag{14.58a}$$

$$[c, \bar{c}^*] = [\bar{c}, c^*] = 1. \tag{14.58b}$$

and are therefore null fermionic oscillators. In terms of the oscillator variables, the BRST charge and ghost number are, respectively,

$$\Omega = c^* a + a^* c, \tag{14.59a}$$
$$\mathcal{G} = c^* \bar{c} - \bar{c}^* c. \tag{14.59b}$$

In the case of m pairs of constraints $p_{1i} = 0$, $p_{2i} = 0$ $(i = 1, \ldots, m)$, each pair brings in its own set of oscillators. The same transformations yield thus

$$\Omega = \sum_i (c_i^* a_i + a_i^* c_i), \tag{14.60a}$$

$$\mathcal{G} = \sum_i (c_i^* \bar{c}_i - \bar{c}_i^* c_i), \tag{14.60b}$$

with

$$[a_i, a_j] = [a_i, a_j^*] = [b_i, b_j] = [b_i, b_j^*] = [a_i, b_j] = [a_i^*, b_j^*] = 0, \tag{14.61a}$$
$$[a_i, b_j^*] = \delta_{ij} = [b_i, a_j^*], \tag{14.61b}$$
$$[c_i, c_j] = [\bar{c}_i, \bar{c}_j] = [c_i, c_j^*] = [\bar{c}_i, \bar{c}_j^*] = [c_i, \bar{c}_j] = [c_i^*, \bar{c}_j^*] = 0, \tag{14.61c}$$
$$[c_i, \bar{c}_j^*] = \delta_{ij} = [\bar{c}_i, c_j^*]. \tag{14.61d}$$

The vacuum ψ_0 is annihilated not only by a_i and b_i but also by c_i and $\bar{c}_i$. The Fock space is generated by acting on the vacuum with the creation operators a_i^*, b_i^*, $\bar{c}_i^*$, and c_i^*. It contains negative norm states. A general state is of the form $f(q^\alpha_{\text{phys}}, a_i^*, b_i^*, \bar{c}_i^*, c_i^*)\psi_0$ where the q^α_{phys} are the unconstrained physical degrees of freedom appearing in addition to the oscillator variables.

14.4.2. Quartet Mechanism

The cohomology of the BRST charge (14.60a) is easily evaluated by introducing the occupation number operator N that counts the gauge and ghost modes,

$$N = \sum_i (a_i^* b_i + b_i^* a_i + \bar{c}_i^* \, c_i + c_i^* \, \bar{c}_i). \tag{14.62}$$

The spectrum of N is given by the nonnegative integers, and the states $f(q^\alpha_{\text{phys}})\psi_0$ containing no oscillators are the only eigenvectors with eigenvalue zero,

$$N\psi = 0 \Rightarrow \psi = f(q^\alpha_{\text{phys}})\psi_0. \tag{14.63}$$

Now, the operator N is BRST invariant,

$$[N, \Omega] = 0, \tag{14.64}$$

and thus one can investigate the BRST cohomology in a definite eigenspace of N. Actually, N is even BRST exact,

$$N = [K, \Omega] \tag{14.65a}$$

with

$$K = \sum_i (b_i^* \, \bar{c}_i + \bar{c}_i^* \, b_i). \tag{14.65b}$$

This implies that if a BRST-invariant state contains $n \neq 0$ gauge or ghost modes,

$$N\psi = n\psi, \tag{14.66a}$$

then it is BRST exact,

$$\psi = \frac{N}{n} \psi = \frac{1}{n}(K\Omega + \Omega K)\,\psi = \Omega\Big(\frac{K}{n}\,\psi\Big). \tag{14.66b}$$

The only nontrivial cohomological class is thus given by the states $f(q^\alpha_{\text{phys}})\psi_0$ $(n = 0)$ at ghost number zero. Only the physical degrees of freedom survive in cohomology. The decoupling of the ghost and gauge modes (which form "quartets") is known as the "quartet mechanism" and constitutes one of the most powerful tools for investigating the BRST cohomology. A different way to compute the BRST cohomology is worked out in Exercise 14.17. Note that the duality between cohomologies at positive and negative ghost numbers is manifest, since one has

$$\left(\frac{\operatorname{Ker}\Omega}{\operatorname{Im}\Omega}\right)^g = 0, \qquad g \neq 0, \tag{14.67}$$

for both the state and the operator cohomologies.

14.4.3. Comments

The nontrivial state cohomology of the above model lies at ghost number zero. This makes the condition $\mathcal{G}\psi = 0$ not only permissible but actually superfluous, since it is a consequence of $\Omega\psi = 0, \psi \sim \psi + \Omega\chi$. The inclusion of the Lagrange multipliers and of their ghosts would not change the conclusions provided these are also quantized in the Fock representation. In that case, one simply increases the number of quartets, which disappear in cohomology. Similar considerations apply in the reducible case, or whenever one adds nonminimal variables quantized in the Fock representation.

The scalar product of two physical states is given by

$$(\chi, \psi) = \int g^*(q^\alpha_{\text{phys}})\, f(q^\alpha_{\text{phys}})\, dq^\alpha_{\text{phys}}, \tag{14.68}$$

since $(\psi_0, \psi_0) = 1$. There is thus complete equivalence with the Dirac–Fock quantization without ghosts and with the reduced phase space quantization (in those simple cases).

There is, however, a major conceptual advantage over the Dirac-Fock quantization. It was pointed out in §13.4.3 that without the introduction of the ghosts, some operators associated with gauge-invariant functions did not map the physical subspace defined by $G_a^{(-)}\psi = 0$ on itself. An explicit example was constructed [Eq. (13.53)]. The difficulty disappears in the BRST formalism because the physical states are annihilated by the full Ω and not just half of it. Indeed, even though the states $f(q^\alpha_{\text{phys}})\,\psi_0$ are annihilated only by the negative frequency part of the constraints, they are solutions of the full equation $\Omega\psi = 0$ because the positive frequency part of the constraints is multiplied in Ω by the destruction ghost operators, which annihilate the vacuum and, hence, also $f(q^\alpha_{\text{phys}})\,\psi_0$. Accordingly, any BRST-invariant operator maps the physical space on itself, and any BRST-null operator maps a physical state on a null state.

If one comes back to the example of §13.4.3, one finds that the BRST-invariant extension of (13.53) is

$$\begin{aligned} A &= a^*\, b^* + a\, b - \bar{c}^*\, c^* + \bar{c}\, c \\ &= [B, \Omega] \end{aligned}$$

with

$$B = b^*\, \bar{c}^* + b\, \bar{c}.$$

The action of A on the vacuum is manifestly a null state

$$\begin{aligned} A\,\psi_0 &= (a^*\, b^* - \bar{c}^*\, c^*)\,\psi_0 \\ &= \Omega\,(b^*\, \bar{c}^*\, \psi_0), \end{aligned}$$

even though $A_0\,\psi_0$ was *not* even a physical state in the formalism without ghosts. So, the "ghost leg" in A is essential and compensates the "pure gauge leg" A_0. The inconsistencies found in §13.4.3 are removed. Any BRST-invariant operator can consistently be applied to the physical states.

Finally, we note that the BRST–Fock method can be easily extended to fermionic constraints. This is because the bosonic a_i and the fermionic c_i play a symmetric role in the BRST charge (14.60a). The BRST charge for fermionic constraints takes the same form but with a_i playing now the role of the ghosts and c_i the role of the constrained variables. The BRST–Fock method is particularly useful in field theory, where it has been applied with success to electromagnetism and string theory.

14.5. BRST QUANTIZATION AND SOLUTIONS OF THE CONSTRAINT EQUATIONS $G_a|\psi\rangle = 0$

14.5.1. Quantum Constraints and Ordering of Ω

We investigate in this section how the direct enforcement of all the constraints can be implemented in the BRST formalism. That is, we develop here the BRST analog of the treatment of Sec. 13.3, where the conditions $G_a|\psi\rangle = 0$ were imposed on the physical states. This approach is permissible in the Schrödinger—or any equivalent (*e.g.*, momentum)—representation of the commutation relations and involves a regularization of the scalar product. Our main result is to rewrite the projected kernel of a gauge-invariant operator as the matrix element of its BRST-invariant extensions between appropriate BRST-invariant states. This is done by expressing the regularized scalar product in terms of the BRST scalar product. The analysis is somewhat heuristic, unlike the case of the Fock representation, since it involves the manipulation of states with infinite or ill-defined norm. We shall assume for simplicity that the constraints are irreducible so that there is only one type of ghosts. The reducible case can be treated along similar lines but requires a further extension of the nonminimal sector introduced in §14.5.4 below.

To compare the Dirac quantization without ghosts with the BRST quantization, it is necessary to isolate the terms involving no ghost momentum in Ω. This we do by rewriting the given BRST operator in η–$\mathcal{P}$ order.

$$\Omega = \eta^a G_a + \text{“more”} \tag{14.69}$$

where “more” contains at least two η's and one $\mathcal{P}$, with the $\mathcal{P}$'s standing to the right of the η's. We define the quantum constraint operators G_a of the Dirac method to be the coefficients of the ghosts η^a in (14.69).

This definition of the operators G_a has the following properties.

(*i*) The G_a's are first class quantum-mechanically,

$$[G_a, G_b] = iC_{ab}{}^c G_c \tag{14.70}$$

with G_c to the right of $C_{ab}{}^c$ in (14.70). This follows from $\Omega^2 = 0$. So, the necessary consistency conditions for the equations $G_a|\psi\rangle = 0$ are automatically fulfilled when Ω is nilpotent (as we assume). The operators $C_{ab}{}^c$ in (14.70) may differ from the corresponding classical structure functions by quantum corrections of order $\hbar$ $(= 1)$.

(*ii*) The G_a's themselves also pick up, in general, nontrivial quantum corrections arising from the η–$\mathcal{P}$ ordering of the higher order ghost terms. These corrections would not have been anticipated in the

Dirac treatment without ghosts and are necessary for (14.70) to hold (see Exercises 14.19 and 13.8).

(*iii*) The G_a's may be nonhermitian, even though Ω is. This is because the monomials $\eta^{a_1} \dots \eta^{a_p} \mathcal{P}_{b_1} \dots \mathcal{P}_{b_{p-1}}$ are not hermitian. The G_a's are hermitian, though, in the case of abelian constraints, or in the case of unimodular groups ($C^a{}_{ab} = 0$). For non-unimodular groups ($C^a{}_{ab} \neq 0$), $G_a^\dagger \neq G_a$ (see again Exercises 14.19 and 13.8).

In the same fashion, one associates with any BRST-invariant operator A of the BRST Hilbert space (with gh $A = 0$) the operator A_0 of the Dirac Hilbert space that is equal to the ghost independent term in the η–$\mathcal{P}$ ordering of A. With this definition, one has

$$[A_0, G_a] = A_a{}^b G_b \tag{14.71}$$

with again G_b standing to the right.

14.5.2. Redefinitions of the Constraints

When one performs a unitary transformation in the Hilbert space of the BRST quantization, the operators G_a get modified. Consider, for instance, the infinitesimal transformation generated by the hermitian, even, generator

$$C = \tfrac{1}{2}[\eta^a \, \varepsilon_a{}^b \, \mathcal{P}_b - (-)^{\varepsilon_a \varepsilon_b} \mathcal{P}_b \, \varepsilon_a{}^b \, \eta^a] \tag{14.72a}$$

$$= \eta^a \, \varepsilon_a{}^b \, \mathcal{P}_b + \frac{i}{2} \varepsilon_a{}^a (-)^{\varepsilon_a}, \tag{14.72b}$$

where $\varepsilon_a{}^b(q,p)$ is such that $(\varepsilon_a{}^b)^\dagger = (-)^{\varepsilon_a \varepsilon_b + \varepsilon_b} \varepsilon_a{}^b$. Classically, (14.72) modifies the constraints as $G_a \rightarrow \bar{G}_a = (\delta_a{}^b + \varepsilon_a{}^b) G_b$. (Recall the proof in §10.5.5 of uniqueness of Ω up to a canonical transformation). Quantum-mechanically, the modification of the operator G_a is more subtle. One finds that the coefficient of η^a in the quantum BRST charge transforms as

$$G_a \rightarrow \bar{G}_a = (\delta_a{}^b + \varepsilon_a{}^b) G_b + \tfrac{1}{2}[G_a, \varepsilon_b{}^b (-)^{\varepsilon_b}] \tag{14.73a}$$

for

$$\Omega \rightarrow \bar{\Omega} = \Omega - [\Omega, iC]. \tag{14.73b}$$

This formula shows, in particular, that the $\bar{G}_a$'s may be nonhermitian even if the G_b's are. Similarly, the operator A_0 transforms as

$$A_0 \rightarrow \bar{A}_0 = A_0 + \tfrac{1}{2}[A_0, \varepsilon_b{}^b (-)^{\varepsilon_b}]. \tag{14.74}$$

Let $|\psi_\alpha\rangle$ be a complete set of physical states of the Dirac method,

$$G_a |\psi_\alpha\rangle = 0. \tag{14.75}$$

These states are, in general, not annihilated by the transformed constraints $\bar{G}_a$. However, the transformed states $|\bar{\psi}_\alpha\rangle$, with

$$|\bar{\psi}_\alpha\rangle = |\psi_\alpha\rangle - \tfrac{1}{2}\varepsilon_b{}^b(-)^{\varepsilon_b}|\psi_\alpha\rangle, \tag{14.76a}$$

are annihilated by the new constraint operators,

$$\bar{G}_a\,|\bar{\psi}_\alpha\rangle = 0. \tag{14.76b}$$

The transformation $|\psi\rangle \rightarrow |\psi\rangle - (\frac{1}{2})\varepsilon_b{}^b(-)^{\varepsilon_b}|\psi\rangle$ maps the space spanned by the $|\psi_\alpha\rangle$ on the space spanned by the $|\bar{\psi}_\alpha\rangle$. It is not a unitary transformation in the Hilbert space without ghosts, unless the supertrace of $\varepsilon_a{}^b$ vanishes.

So, a unitary transformation in the BRST space induces a nonunitary transformation in the space of the Dirac method. This is quite allright because the scalar product in the space of the Dirac method is not the physical scalar product. The physical scalar product involves an insertion of gauge conditions, which can formally be written as

$$(\psi_\alpha\,|\,\psi_\beta) = \langle\psi_\alpha\,|\,\mu\,|\,\psi_\beta\rangle. \tag{14.77}$$

The insertion μ restricts the integration in the scalar product to an integration over the physical degrees of freedom and depends on the form of the constraints (§13.4.4). If one combines the transformation formula (13.29c) for μ with (14.74) and (14.76a), one finds that the physical expectation values of gauge-invariant operators are invariant,

$$\langle\bar{\psi}_\alpha\,|\,\bar{\mu}\,\bar{A}_0\,|\,\bar{\psi}_\beta\rangle = \langle\psi_\alpha\,|\,\mu\,A_0\,|\,\psi_\beta\rangle. \tag{14.78}$$

14.5.3. BRST Cohomology at Ghost Number $\pm$m/2

To relate the physical states of the Dirac method with the physical states of the BRST method, it is necessary to expand the states belonging to the BRST Hilbert space in powers of the ghosts,

$$\psi = |\psi_0\rangle + |\psi_a\rangle\eta^a + |\psi_{ab}\rangle\eta^a\eta^b + \cdots + |\psi_{1\ldots m}\rangle\eta^1\ldots\eta^m. \tag{14.79}$$

The states $|\psi_0\rangle, |\psi_a\rangle, \ldots |\psi_{1\ldots m}\rangle$ belong to the Dirac Hilbert space. The ghosts and their momenta act on ψ as

$$\hat{\eta}^a\psi = \eta^a\psi, \qquad \hat{\mathcal{P}}_a\psi = \frac{1}{i}\frac{\partial\psi}{\partial\eta^a}, \qquad (\hbar = 1). \tag{14.80}$$

We take for definiteness the constraints to be bosonic and, thus, the ghosts η^a to be anticommuting, so that (14.80) defines a Clifford algebra with $2m$ generators. [A more detailed analysis of the irreducible representation of (14.80) is given as complementary material in Chapter 20 (Sec. 20.3).] Remarks concerning the case of commuting ghosts are gathered at the end of the subsection.

The state $|\psi_0\rangle$ is at ghost number $-m/2$, while $|\psi_{1\dots m}\rangle\eta^1\dots\eta^m$ is at ghost number $m/2$,

$$\mathcal{G}|\psi_0\rangle = -\frac{m}{2}|\psi_0\rangle \Leftrightarrow \mathcal{G}'|\psi_0\rangle = 0, \tag{14.81a}$$

$$\mathcal{G}|\psi_{1\dots m}\rangle\eta^1\dots\eta^m = \frac{m}{2}|\psi_{1\dots m}\rangle\eta^1\dots\eta^m \Leftrightarrow \mathcal{G}'|\psi_{1\dots m}\rangle\eta^1\dots\eta^m = m|\psi_{1\dots m}\rangle\eta^1\dots\eta^m. \tag{14.81b}$$

Furthermore, one has

$$\hat{\mathcal{P}}_a|\psi_0\rangle = 0, \qquad \hat{\eta}^a|\psi_{1\dots m}\rangle\eta^1\dots\eta^m = 0. \tag{14.82}$$

The existence of nontrivial BRST cohomology at ghost number $\neq \pm m/2$ depends on the detailed structure of the theory. It will not concern us here because it does not appear to be related to the states of the Dirac method without ghosts. By contrast, the cohomology at ghost number $\pm m/2$ can be completely characterized without knowledge of the explicit form of the constraints and is in direct correspondence with the solutions of $G_a|\psi\rangle = 0$.

Theorem 14.7. *The state $|\psi_0\rangle$ is BRST closed if and only if $G_a|\psi_0\rangle=0$; i.e.,*

$$\Omega|\psi_0\rangle = 0 \Leftrightarrow G_a|\psi_0\rangle = 0. \tag{14.83}$$

Proof. The proof follows from $\Omega|\psi_0\rangle = G_a|\psi_0\rangle\eta^a$ and the independence of the ghosts.

Because the states $|\psi_0\rangle$ are at the most negative value of the ghost number, they cannot be BRST-exact. Hence, there is a bijective correspondence between the Dirac physical states and the equivalence classes of the BRST state cohomology at ghost number $-m/2$.

Theorem 14.8. *The states $|\psi_{1\dots m}\rangle\eta^1\dots\eta^m$ are BRST closed for any choice of $|\psi_{1\dots m}\rangle$,*

$$\Omega|\psi_{1\dots m}\rangle\eta^1\dots\eta^m = 0. \tag{14.84}$$

Proof. Each term in Ω contains one more η than there are $\mathcal{P}$'s and therefore annihilates the product $\eta^1\dots\eta^m$ $[(\eta^a)^2 = 0]$.

Are the states $|\psi_{1\dots m}\rangle\eta^1\dots\eta^m$ BRST-exact? To shed some light on this subtle question, it is useful to analyze first the case of a single constraint $p = 0$. The BRST charge reads then $\Omega = p\eta$. The state

$\psi(q, \eta) = a(q) + b(q)\eta$ is BRST closed if $a =$ const. The addition of a BRST-exact state to ψ amounts to modifying b as

$$b(q) \rightarrow b(q) + \frac{df}{dq}. \tag{14.85}$$

Now, any function $b(q)$ can be written as $(-df/dq)$ with f equal to $-\int b(u)\, du$. Therefore, if one does not restrict the functional class of f, one finds that any state at ghost number 1/2 is exact. So, it would appear that $H^{1/2}_{\mathrm{st}}(\Omega) = 0$.

This point of view, which allows any f in (14.85), can be developed consistently but is not the most useful point of view for the subsequent analysis. Rather, it is preferable to retain the notion that BRST-exact states are null. In the present case, this means that one should not allow in (14.85) functions f that change the scalar product of b with the BRST-closed states at ghost number $-1/2$. Since the scalar product of $b(q)\,\eta$ with $a =$ const. is given by $\int a^* b(q)\, \eta\, d\eta\, dq = a^* \int b(q)\, dq$, one sees that $b(q)\eta$ is BRST-exact if and only if $b(q)$ "has no zero mode," $\int b(q)\, dq = 0$. More generally, the BRST classes at ghost number 1/2 are characterized by the integral $\int b(q)\, dq$. Consequently, $H^{1/2}_{\mathrm{st}}(\Omega)$ is isomorphic to $\mathbb{C}$ and hence also to $H^{-1/2}_{\mathrm{st}}(\Omega)$, in agreement with the duality theorem for the state cohomology.

Similar considerations apply to the case of arbitrary bosonic first-class constraints. If one insists that the states $\Omega|\chi\rangle$ added to the BRST-closed states $|\psi_{1\ldots m}\rangle\eta^1 \ldots \eta^m$ do not change their scalar products with the states at ghost number $-m/2$, as we shall do here, then the duality theorem for the state cohomology, which uses explicitly the scalar product, holds. The cohomology at ghost number $+m/2$ is thus isomorphic to the cohomology at ghost number $-m/2$, that is, to the space of physical states of the Dirac method. Hence, these states appear in the BRST formalism both at ghost number $-m/2$ (directly) and at ghost number $m/2$ (by duality).

The analysis of the cohomology at ghost number $\pm m/2$ proceeds in the same fashion for fermionic constraints and bosonic ghosts quantized in $L^2(R^m)$. In that case, the discrete spectrum of the ghost number operator $\mathcal{G}$ is empty. However, the eigenstates of the ghosts with eigenvalue zero, and of their momenta with same eigenvalue, are, respectively, at ghost number $-m/2$ and $+m/2$ (Exercise 14.4). Even though these states are not square integrable, they are quite useful (as are the solutions of $G_a|\psi\rangle = 0$, which are not square integrable for bosonic constraints). They have overlap equal to one, $\int \delta(\eta)\, e^{i\eta\mathcal{P}}\, d\eta = 1$, and so may be thought of as belonging to dual spaces. They can be used to form solutions of $\Omega\psi = 0$ in the same manner as in Theorems 14.7

and 14.8. Thus, the states annihilated by the ghost momenta and by the constraints are BRST-closed states at ghost number $m/2$, while those annihilated by the ghosts are also BRST-closed but now at ghost number $-m/2$.

14.5.4. Forming Ghost Number Zero States

For definiteness, we pursue the analysis in the explicit case of bosonic constraints. The states constructed in the previous subsection do not fulfill the ghost number zero condition $\mathcal{G}\psi = 0$ and as a result (Theorem 14.1) have zero or ill-defined norm. To get states whose norm can be regularized in a manner compatible with the BRST scalar product, one needs to form ghost number zero states by including the Lagrange multipliers λ^a, their conjugate momenta b^a, and the associated system of additional antighosts $\rho^a, \bar{C}_a$, as in §11.3.2. The introduction of this nonminimal sector adds the simple term $\rho^a b_a$ to the BRST charge corresponding to the new abelian constraints $b_a = 0$.

Since there is only one state $\psi(\lambda^a)$ annihilated by all the b_a's, the BRST cohomology at ghost number $\pm m/2$ in the Lagrange multiplier sector is isomorphic to $\mathbb{C}$. The BRST class at ghost number $-m/2$ contains the state annihilated by b_a and $\bar{C}_a$. The BRST class at ghost number $m/2$ contains the states annihilated by ρ^a.

By taking the tensor product of the Lagrange multiplier cohomology with the cohomology for the original system, one gets states at ghost number zero. Of particular interest are the states $\psi_{q,\eta=0,b=0,\bar{C}=0}$ ($=\psi_{q,\eta=0} \otimes \psi_{b=0,\bar{C}=0}$) obtained by combining the states $\psi_{q,\eta=0}$ at ghost number $m/2$ with the states $\psi_{b=0,\bar{C}=0}$ at ghost number $-m/2$

$$\hat{q}^i \psi_{q,\eta=0,b=0,\bar{C}=0} = q^i \psi_{q,\eta=0,b=0,\bar{C}=0}, \qquad \hat{\eta}^a \psi_{q,\eta=0,b=0,\bar{C}=0} = 0, \tag{14.86a}$$
$$b_a \psi_{q,\eta=0,b=0,\bar{C}=0} = 0, \qquad \bar{C}_a \psi_{q,\eta=0,b=0,\bar{C}=0} = 0, \tag{14.86b}$$
$$\Omega \psi_{q,\eta=0,b=0,\bar{C}=0} = 0. \tag{14.86c}$$

These states are BRST-closed. Furthermore, since the coordinate eigenstates form a complete set, the states $\psi_{q,\eta=0}$ completely exhaust the BRST cohomology at ghost number $m/2$; thus, the BRST cohomological classes defined by $\psi_{q,\eta=0,b=0,\bar{C}=0}$ are in bijective correspondence with the solutions of $G_a|\psi\rangle = 0$. It should be stressed that the states (14.86) are constructed straightforwardly, without having to solve explicitly the constraints.

The scalar products of the states (14.86) are still ill defined

$$(\psi_{q,\eta=0,b=0,\bar{C}=0}, \psi_{q',\eta=0,b=0,\bar{C}=0}) = 0 \cdot \delta(0) \cdot \delta(q-q').$$

The singular term $\delta(0)$ arises from the norm of the eigenstate of $\hat{b}$ with eigenvalue $b = 0$, while 0 arises from the norm of the eigenstate of η and $\bar{C}$ with eigenvalue 0. However, they can be regularized by using the "quantum gauge freedom" of the BRST formalism. As shown in the next subsection, the scalar products

$$\left(\psi_{q,\eta=0,b=0,\bar{C}=0}, \exp i\,[K,\Omega]\,\psi_{q',\eta=0,b=0,\bar{C}=0}\right)$$

yield the projected kernel $I_0^P(q,q')$ of the identity when K is suitably chosen.

14.5.5. BRST Formalism and Projected Kernels

The interest of the states (14.86) lies not only in the fact that they are BRST closed irrespective of the precise form of the constraints but also in their direct connection with the projected kernel of §13.3.6. Let A_0 be a gauge-invariant operator with BRST-invariant extension A. Its projected kernel is defined by

$$A_0^P(q',q) = \sum_{\alpha,\beta} \psi_\alpha(q')\,(\psi_\alpha \,|\, A_0 \,|\, \psi_\beta)\,\psi_\beta^*(q). \tag{14.87}$$

[The states $|\psi_\alpha\rangle$ are assumed to be orthonormal in the regularized scalar product. If $(\psi_\alpha|\psi_\beta) = g_{\alpha\beta}$, one must replace (14.87) by $\sum_{\alpha,\beta,\gamma,\delta} \times \psi_\alpha(q')\,g^{\alpha\beta}(\psi_\beta \,|\, A_0 \,|\, \psi_\gamma)\,g^{\gamma\delta}\,\psi_\delta^*(q)$.]

Theorem 14.9. *The matrix element* $(\psi_{q',\eta=0,b=0,\bar{C}=0}, A\psi_{q,\eta=0,b=0,\bar{C}=0})$ *is proportional to the projected kernel of* A_0

$$A_0^P(q',q) = \sigma(\psi_{q',\eta=0,b=0,\bar{C}=0}, A\psi_{q,\eta=0,b=0,\bar{C}=0}), \tag{14.88}$$

where $\sigma = 1, -1, i,$ *or* $-i$ *[see Eq. (14.89d) below].*

Since Theorem 14.9 is quite important for the path integral, we give two different demonstrations.

First Proof of Theorem 14.9.

(a) Both the right- and left-hand sides of (14.88) transform identically under changes of constraints generated by (14.72). Hence, it is enough to check (14.88) in one representation of the constraint surface. Indeed, the matrix elements of the operator $\bar{A} = A - [A, iC]$ are equal to

$$\begin{aligned}
&(\psi_{q',\eta=0,b=0,\bar{C}=0}, \bar{A}\psi_{q,\eta=0,b=0,\bar{C}=0}) - (\psi_{q',\eta=0,b=0,\bar{C}=0}, A\psi_{q,\eta=0,b=0,\bar{C}=0}) \\
&\quad = \int d\bar{q}\, d\tilde{q}\,\big[-\tfrac{1}{2}\delta(q'-\bar{q})\langle\tilde{q}\,|\,\varepsilon_a{}^a(-)^{\varepsilon_a}\,|\,q\rangle - \tfrac{1}{2}\delta(q-\tilde{q})\langle q'\,|\,\varepsilon_a{}^a(-)^{\varepsilon_a}\,|\,\bar{q}\rangle\big] \\
&\qquad \times (\psi_{\bar{q},\eta=0,b=0,\bar{C}=0}, A\psi_{\tilde{q},\eta=0,b=0,\bar{C}=0}).
\end{aligned}$$

This is because the action of $-iC$ on the state $\psi_{q,\eta=0,b=0,\bar{C}=0}$ is given by

$$-iC\,\psi_{q,\eta=0,b=0,\bar{C}=0} = -\frac{1}{2}\int d\tilde{q}\,\psi_{\tilde{q},\eta=0,b=0,\bar{C}=0}\,\langle\tilde{q}\,|\,\varepsilon_a{}^a(-)^{\varepsilon_a}\,|\,q\rangle$$

because of (14.72b) and $\eta^a\,\psi_{\tilde{q},\eta=0,b=0,\bar{C}=0} = 0$. But this is precisely the transformation law of the projected kernel, since the new states $|\bar{\psi}_\alpha\rangle$ are related to the old states $|\psi_\alpha\rangle$ by the formula (14.76a); and the physical expectation values are unchanged [formula (14.78)].

(b) By the abelianization theorem, one can assume the G_a to be abelian, $[G_a, G_b] = 0$. In that case, the theorem is immediate. Indeed, one can then take A_0 to strongly commute with G_a, $[A_0, G_a] = 0$, so that A_0 is BRST closed (Exercises 1.24 and 11.3). Furthermore, A_0 may be assumed not to depend on the variables of the nonminimal sector. Choosing for A the following BRST extension of A_0,

$$A = A_0 \exp i\,[K,\Omega], \qquad K = -\lambda^a\mathcal{P}_a \tag{14.89a}$$

[the operator $\exp i[K,\Omega]$ differs from unity by a BRST-exact operator (Exercise 14.13)], *i.e.*, explicitly,

$$A = A_0\,\exp i\,(\lambda^a G_a - i\mathcal{P}_a\rho^a), \tag{14.89b}$$

one gets

$$\sigma(\psi_{q',\eta=0,b=0,\bar{C}=0}, A\psi_{q,\eta=0,b=0,\bar{C}=0}) = \langle q'\,|\,\delta(G_a)A_0\,|\,q\rangle. \tag{14.89c}$$

To arrive at (14.89c), we have used the relations

$$(\psi_{\eta=0,\bar{C}=0}, [\exp\mathcal{P}_a\rho^a]\psi_{\eta=0,\bar{C}=0}) = \\ (\psi_{\eta=0,\bar{C}=0}, \psi_{\mathcal{P}=0,\rho=0}) = \sigma^{-1}\ (=\pm 1, \mp i) \tag{14.89d}$$

and

$$\langle b=0\,|\exp i\hat{\lambda}^a\hat{G}_a\,|\,b=0\rangle = \delta(\hat{G}_a).$$

Formula (14.89c) ends the proof of the theorem, since its right-hand side is the projected kernel of A_0 [Eq. (13.37)].

Second Proof of Theorem 14.9. (Not based on abelianization.)

The space $H_{\mathrm{st}}^{m/2}(\Omega)$ of Theorem 14.8 is dual to the space $H_{\mathrm{st}}^{-m/2}(\Omega)$ of Theorem 14.7. The states $|\psi_\alpha\rangle$ of the Dirac method are a basis of $H_{\mathrm{st}}^{-m/2}(\Omega)$. The representatives of the dual basis of $H_{\mathrm{st}}^{m/2}(\Omega)$ are of the form $|\psi^\alpha\rangle\eta^1\ldots\eta^m$. The pairing is defined by means of the BRST scalar product, so

$$\int d\eta\langle\psi_\alpha\,|\,\psi^\beta\rangle\eta^1\ldots\eta^m = \langle\psi_\alpha\,|\,\psi^\beta\rangle = \delta_\alpha{}^\beta. \tag{14.90a}$$

The states $\psi_{q\,\eta=0} \equiv |q\rangle\eta^1 \dots \eta^m$ are in $H^{m/2}_{\rm st}(\Omega)$. Thus, one has, up to exact states,

$$|q\rangle\eta^1 \dots \eta^m = \sum_\alpha \psi^*_\alpha(q)\,|\psi^\alpha\rangle\eta^1 \dots \eta^m. \tag{14.90b}$$

Let ψ^α be the states $|\psi^\alpha\rangle\eta^1 \dots \eta^m \otimes \psi_{b=0,\bar{C}=0}$. One finds from (14.90b) and the fact that the states $\psi_{b=0,\bar{C}=0}$ are BRST closed that, again up to exact states

$$\psi_{q',\eta=0,b=0,\bar{C}=0} = \sum_\alpha \psi^*_\alpha(q')\,\psi^\alpha. \tag{14.90c}$$

The amplitude $\left(\psi^\alpha, \exp\, i[K,\Omega]\,\psi^\beta\right)$ is finite for K of the form (14.89a). We define the regularized scalar product $(\psi_\alpha\,|\,\psi_\beta)$ to be proportional to the inverse of $\left(\psi^\alpha, \exp\, i[K,\Omega]\psi^\beta\right)$,

$$(\psi_\alpha\,|\,\psi_\beta)\,g^{\beta\rho} = \delta_\alpha{}^\rho, \qquad \sigma\left(\psi^\beta, \exp\, i[K,\Omega]\,\psi^\rho\right) = g^{\beta\rho}. \tag{14.91}$$

The phase σ is given by (14.89d). The scalar product (14.91) coincides with the definition given for simple constraints in §13.3.4 (Exercise 14.23) and is manifestly invariant under redefinitions of the constraints, since such redefinitions are unitary transformations in the BRST Hilbert space. That is, $\langle\bar{\psi}_\alpha\,|\,\bar{\mu}\,|\,\bar{\psi}_\beta\rangle = \langle\psi_\alpha\,|\,\mu\,|\,\psi_\beta\rangle$. It follows from (14.91) that $(\psi^\alpha, A\psi^\beta) = g^{\alpha\rho}(\psi_\rho\,|\,A_0\,|\,\psi_\sigma)\,g^{\sigma\beta}$ because one has $A|\psi_\alpha\rangle = A_0|\psi_\alpha\rangle$ for the states $|\psi_\alpha\rangle$ annihilated by $\mathcal{P}_a$. Using this relation and (14.90c) yields then

$$\begin{aligned}\sigma&(\psi_{q',\eta=0,b=0,\bar{C}=0}, A\psi_{q,\eta=0,b=0,\bar{C}=0})\\ &= \sum_{\alpha,\beta,\rho,\gamma} \psi_\alpha(q')\,g^{\alpha\beta}(\psi_\beta\,|\,A_0\,|\,\psi_\gamma)\,g^{\gamma\delta}\,\psi^*_\delta(q),\end{aligned}$$

as claimed.

Remarks. (i) The matrix element $(\psi_{q',\eta=0,b=0,\bar{C}=0}, A\psi_{q,\eta=0,b=0,\bar{C}=0})$ does not formally depend on which BRST-invariant extension of A_0 one takes for A, since A is BRST-invariant and sandwiched between BRST-closed states. It should be stressed that the argument is quite formal, however, in that the state $\psi_{q,\eta=0,b=0,\bar{C}=0}$ is not normalizable (see Exercise 14.22).

(ii) It is crucial that one takes K in (14.89a) to mix the minimal and nonminimal sectors. If this were not the case, the expectation value of A would be ill defined and should be regularized by taking a different K and using the formal independence of the amplitude on K. Such a difficulty does not arise in the Fock representation, where the scalar products are well defined and directly equal to physical amplitudes. In that case, any BRST-invariant extension of A_0 can be used to compute physical amplitudes.

(*iii*) It is easy to check directly that the expectation value

$$(\psi_{q',\eta=0,b=0,\bar{C}=0}, A\psi_{q,\eta=0,b=0,\bar{C}=0})$$

is a solution of the constraint equation

$$G_a\left(q', \frac{\hbar}{i}\frac{\partial}{\partial q'}\right)(\psi_{q',\eta=0,b=0,\bar{C}=0}, A\psi_{q,\eta=0,b=0,\bar{C}=0}) = 0 \tag{14.92}$$

and of the complex conjugate equation at the other endpoint. Indeed, the closedness of the states in (14.84) and the BRST invariance of A imply

$$\left(\psi_{q',\eta=0,b=0,\bar{C}=0}, [-\mathcal{P}_a, \Omega]\, A\psi_{q,\eta=0,b=0,\bar{C}=0}\right) = 0,$$

i.e.,

$$\left(\psi_{q',\eta=0,b=0,\bar{C}=0}, (G_a + \text{“}\eta\text{–}\mathcal{P}\text{-terms”})\, A\psi_{q,\eta=0,b=0,\bar{C}=0}\right) = 0.$$

The hermiticity of η^a and the relation $\eta^a\psi_{q',\eta=0,b=0,\bar{C}=0}$ lead then to (14.92). If Ω were not hermitian, Eq. (14.92) would still hold, but the amplitude would obey a different equation at the other endpoint, whose form is explicitly worked out in Exercise 14.24.

(*iv*) Again, one can add further nonminimal variables without changing the amplitudes (14.88). Indeed, the new amplitude with the new variables added factorizes, with appropriate choices of K, into the product of the amplitude (14.88) times unity.

(*v*) A BRST representation of the ordinary (nonprojected) kernel of gauge-invariant operators is worked out in Exercise 14.25. It does not need the introduction of the nonminimal sector of §14.5.4.

EXERCISES. CHAPTER FOURTEEN

14.1. Let (,) be an indefinite scalar product, *i.e.*,

(*i*) $(\psi_1, \psi_2) \in \mathbb{C}$;

(*ii*) $(\psi_1, \psi_2)^* = (\psi_2, \psi_1)$;

(*iii*) $(\psi_1, \alpha\,\psi_2 + \beta\,\psi_3) = \alpha(\psi_1, \psi_2) + \beta(\psi_1, \psi_3)$.

[The requirement of positivity $(\psi_1, \psi_1) \geq 0$ with $(\psi_1, \psi_1) = 0$ iff $\psi_1 = 0$ is not imposed.]

(a) Show that in a basis h_k, with $\psi_1 = {\psi_1}^k h_k, \psi_2 = {\psi_2}^k h_k$, one has

$$(\psi_1, \psi_2) = \sum_{k,l} \psi_1^{*k}\, \psi_2^l\, g_{kl}$$

where the matrix g is hermitian in the sense of matrices, *i.e.*, equal to the complex conjugate of its transpose,

$$g^*_{lk} = (g^\dagger)_{kl} = g_{kl} \qquad [g_{kl} = (h_k, h_l)].$$

(b) An operator A is hermitian for the scalar product $(\ ,\ )$ $(A = A^*)$ iff

$$(\chi, A\,\psi) = (A\,\chi, \psi).$$

This implies that $(\psi, A\psi)$ is real. Let M_A be the matrix representation of A in the basis h_k, *i.e.*, $Ah_i = h_k(M_A)^k{}_i$. Show that the equation $A = A^*$ is equivalent to

$$g\,M_A = (M_A)^\dagger\,g$$

(*i.e.*, gM_A is a hermitian matrix).

(c) Investigate how the matrices g and M_A transform under a change of basis $h_k \to h'_k = h_l\,S^l{}_k$ and verify that the matrix equations of (a) and (b) are invariant.

(d) Show that there is a basis in which all the hermitian operators are represented by hermitian matrices if and only if the scalar product is positive definite.

14.2. Let A be a hermitian operator in a Hilbert space with a nonpositive definite and nondegenerate scalar product.

(a) Show that A may possess eigenstates with complex eigenvalues and that the complex eigenvalues of A, if any, come in complex conjugate pairs.

(b) Prove that the eigenstates of A with eigenvalue not purely real have zero norm. [Hint: Show that if $A\psi = \alpha\psi$ and $A\chi = \beta\chi$, then $(\psi, \chi) = 0$ for $\alpha^* \neq \beta$.]

(c) Is A always diagonalizable? (Answer: No; see, for instance, the Jordan form of Ω.)

14.3. Let $G_a = 0$ be irreducible bosonic constraints $(a = 1, \ldots, m)$. The ghosts are fermionic, and the Fock representation of the ghost commutation relations, available when m is even, is then equivalent to the Schrödinger representation. Show that the spectrum of the anti-hermitian ghost number operator $\mathcal{G}$ given by (14.8) is $-m/2, -m/2+1, \ldots, m/2$. Accordingly, there is fractionalization of the ghost number for an odd number of constraints. Show that the spectrum of $\mathcal{G}'$ given by (14.9) is $0, 1, \ldots, m$.

14.4. Let $G_a = 0$ be irreducible fermionic constraints $(a = 1, \ldots, m)$. The ghosts are then even.

(a) Show that the monomials $(\eta^1)^{n_1}(\eta^2)^{n_2}\ldots(\eta^m)^{n_m}$ of the coordinate representation have ghost number (14.8) equal to $m/2 + n_1 + n_2 + \cdots + n_m$.

(b) Show that the monomials $(\mathcal{P}_1)^{n_1}(\mathcal{P}_2)^{n_2}\ldots(\mathcal{P}_m)^{n_m}$ of the momentum representation have ghost number $-m/2 - n_1 - n_2 - \cdots - n_m$. [Note: The monomials of (a) and (b) are not square integrable and, hence, do not belong to $L^2(R^m)$. However, they do belong to the representation space constructed in Exercise 14.15 below (with λ replaced by η and π by $\mathcal{P}$). In that space, the spectrum of the ghost number (14.8) is thus $\ldots, -m/2-1, -m/2, m/2, m/2+1, m/2+2, \ldots$. By contrast, the discrete spectrum of $\mathcal{G}$ in $L^2(R^m)$ is empty, although the states with $\mathcal{G} = -m/2$ (eigenstate of η with eigenvalue zero) and with $\mathcal{G} = m/2$ (eigenstate of $\mathcal{P}$ with eigenvalue zero) can be approximated by wave packets in $L^2(R^m)$.]

(c) Show that in the Fock representation with indefinite metric (available when m is even), the eigenvalues of $\mathcal{G}$ are all the integers.

14.5. Let $G_a = 0$ be constraints that form a Lie algebra, $[G_a, G_b] = C_{ab}{}^c G_c$. Show that if $\Omega^2 = 0$ (no BRST anomaly), then the BRST extension $\bar{G}_a$ of the constraints also close according to the same Lie algebra quantum-mechanically.

14.6. Show that one can consistently identify ψ with $(K\Omega + \Omega K)\psi$ for all K's if and only if: (*i*) $\Omega\psi = 0$; (*ii*) $\psi \sim \psi + \Omega\chi$. [Hints: Once (*i*) is established, the proof of (*ii*) is given in the text. To prove (*i*), observe that if $\Omega\psi \neq 0$, then ψ and $\Omega\psi$ are linearly independent. Accordingly, one can define the action of linear operators independently on ψ and $\Omega\psi$. It then follows that $(K\Omega + \Omega K)\psi$ can be any state, so the identification $\psi \sim (K\Omega + \Omega K)\psi$ would make the space of states collapse to a single state.]

14.7. Let V be a vector space with a semipositive scalar product. Let $\chi \in V$ be a zero-norm vector, $(\chi, \chi) = 0$. Show that $(\psi, \chi) = 0$ for any $\psi \in V$, so that χ can be factored out. Prove that, conversely, if any zero-norm vector in V is orthogonal to all other vectors in V, then the scalar product in V is positive (or negative) semidefinite.

14.8. Prove formula (14.22).

14.9. Check explicitly formulas (14.23) through (14.25).

14.10. Show that there is no projection operator P on the subspace $\Omega\psi = 0$ that is hermitian for the scalar product $(\ ,\ )$. Write explicitly a (nonhermitian) projection operator in terms of the basis e_i, g_α and f_α of §14.2.1.

14.11. When analyzing the quantum BRST cohomology, it is sometimes useful to introduce an auxiliary positive definite scalar product $\langle\psi, \chi\rangle$. We denote in this exercise the hermitian conjugate in $\langle\ ,\ \rangle$ with a dagger, $\langle\psi_1, A\psi_2\rangle = \langle A^\dagger\psi_1, \psi_2\rangle$ [while $(\psi_1, A\psi_2) = (A^*\psi_1, \psi_2)$]. Let $\Omega^\dagger$ be the hermitian conjugate of Ω. Define the Laplace operator Δ by $\Delta = (\Omega + \Omega^\dagger)^2 = \Omega\Omega^\dagger + \Omega^\dagger\Omega$. A state of the form $\Omega^\dagger\chi$ is said to be co-exact, whereas a state annihilated by Δ is harmonic.

(a) Show that the harmonic states are both closed and co-closed, *i.e.*,

$$\text{Ker } \Delta = \text{Ker } \Omega \cap \text{Ker } \Omega^\dagger.$$

(b) Prove the Hodge decomposition theorem

$$W = \text{Im } \Omega \oplus \text{Im } \Omega^\dagger \oplus \text{Ker } \Delta,$$

where the sum is a direct sum of orthogonal subspaces (in $\langle\ ,\ \rangle$). (Hint: Verify that Im Ω is orthogonal to Im $\Omega^\dagger$ and that Ker Δ is the orthogonal complement to Im $\Omega \oplus$ Im $\Omega^\dagger$.)

(c) Using the Hodge decomposition, prove the isomorphism of $H^*(\Omega)$ with Ker Δ.

(d) Let A be a hermitian, positive, semidefinite operator, $A^\dagger = A$, $\langle\psi, A\psi\rangle \geq 0$. Show that $\langle\psi, A\psi\rangle$ vanishes if and only if $A\psi = 0$. (Hint: Expand ψ along the eigenstates of A.)

(e) Let $\Delta = \sum_i \Delta_i$, where Δ_i are hermitian, positive, semidefinite operators. Show that the equation $\Delta\,\psi = 0$ is equivalent to the set of equations $\Delta_i\,\psi = 0$.

(f) Show that the vectors e_i, g_α, and f_α of §14.2.1.a can be chosen so that $\Delta e_i = 0$ and $f_\alpha \in \text{Im}\,\Omega^\dagger$.

14.12. (Alternative Proof of the Duality Formula.) Same notations as for Exercise 14.11.

(a) Prove the existence of a basis $\{h_k\}$ of W such that (i) the ghost number is diagonal; and (ii) the matrix g of the scalar products (see Exercise 14.1) squares to one, $g^2 = 1$.

(b) Define an auxiliary inner product so that $\langle h_k, h_l \rangle = \delta_{kl}$ (see Exercise 14.11). Show that with this choice one has not only $g\,\Omega = \Omega^\dagger g$ (Exercise 14.1) but also $\mathcal{G}^\dagger = \mathcal{G}$, $g\,\mathcal{G} = -\mathcal{G}\,g$, $\Omega\,g = g\,\Omega^\dagger$. Furthermore, $\Omega^\dagger$ is of ghost number minus one.

(c) Verify that $\psi' = g\,\psi$ defines an isomorphism between the subspaces W_p and W_{-p} of respective ghost numbers $+p$ and $-p$.

(d) Using the results of the previous exercise, show that this isomorphism induces an isomorphism in cohomology. [Hint: Prove first that $[\Delta, g] = 0$ and that the harmonic states can be assumed to have definite ghost number ($[\Delta, \mathcal{G}] = 0$).]

14.13. In this exercise, $A, B, C, \ldots$ are BRST-closed operators.

(a) Check that $A + B$, AB, λA are BRST-invariant and that AB is BRST exact if either A or B is BRST exact. So, the algebraic operations go over to cohomology without problem.

(b) Show that $\exp A$ is BRST invariant.

(c) Prove that $\exp\,[K, \Omega]$ differs from the identity by a BRST-exact operator, which is formally given by

$$\exp\,[K, \Omega] = I + [L, \Omega],$$

$$L = \left(\sum_{n \geq 0} \frac{1}{(n+1)!}\,[K, \Omega]^n \right) K$$

$$= \left(\frac{\exp\,[K, \Omega] - 1}{[K, \Omega]} \right) K.$$

(d) Show that $\exp\,(A + [K, \Omega]) = (\exp A) + [M, \Omega]$. [Hint: Use the Campbell–Hausdorff formula and (c).]

14.14. Let $\mathcal{L}$ be the group of invertible linear transformations that commute with Ω, and let $\mathcal{E}$ be the subset of invertible transformations that differ from the identity by a BRST-null operator . ($B \in \mathcal{E}$ iff B^{-1} exists, and $B = I + [K, \Omega]$ for some K)

(a) Show that $\mathcal{E}$ is an invariant subgroup of $\mathcal{L}$.

(b) Prove that the factor group $\mathcal{L}/\mathcal{E}$ is isomorphic with the set of invertible linear transformations acting on the quotient space $H^*_{\text{st}}(\Omega)$.

14.15. Consider the single constraint $\pi = 0$ for the Lagrange multiplier λ. This multiplier belongs to the real line. Quantize the system by using the following realization of the commutation relations. The operators λ and π are taken to act on $\mathbb{C}[\lambda] \oplus \mathbb{C}[\pi]$ (direct sum of polynomials in λ and polynomials in π), a basis of which is given by $\{f_n, \bar{f}_n\}$, where

$$\begin{aligned} f_n &= \lambda^n, & \bar{f}_n &= \pi^n \qquad (n \geq 0), \\ \lambda\, f_n &= f_{n+1}, & \lambda\, \bar{f}_n &= in\, \bar{f}_{n-1}, \\ \pi\, f_n &= -in\, f_{n-1}, & \pi\, \bar{f}_n &= \bar{f}_{n+1}. \end{aligned}$$

The scalar product is taken to be

$$\begin{aligned} (f_n, f_m) &= (\bar{f}_n, \bar{f}_m) = 0, \\ (\bar{f}_n, f_m) &= (-i)^n\, n!\, \delta_{n,m} = (f_m, \bar{f}_n)^*. \end{aligned}$$

This provides a realization of the Heisenberg algebra $[\lambda, \lambda] = 0 = [\pi, \pi]$, $[\lambda, \pi] = i$ by means of hermitian operators, $(f, \lambda g) = (\lambda f, g)$, $(f, \pi g) = (\pi f, g)$. The representation of the algebra itself is reducible, but the scalar product pairs the irreducible components. Prove that the BRST-state cohomology is given by $H_{\text{st}}^{\pm 1/2}(\Omega) = \mathbb{C}$. Show that the representation can be extended to many multipliers acting on $\mathbb{C}[\lambda^i] \oplus \mathbb{C}[\pi_i]$ and that the cohomology of the BRST charge $\Omega = \pi_i \eta^i$ is then given by $H_{\text{st}}^{\pm m/2}(\Omega) = \mathbb{C}$ (other cohomology groups = 0).

14.16. Consider the Lagrangian $L(\phi, \dot{\phi}) \equiv 0$ for an angular variable ϕ belonging to the circle S^1. Quantize the system in the Schrödinger representation where states are functions of ϕ and of the single ghost η. Show that the BRST cohomology is given by $H_{\text{st}}^{\pm 1/2}(\Omega) = \mathbb{C}$. Check that by combining the model with the Lagrange multiplier system of Exercise 14.15, one gets $H_{\text{st}}^0(\Omega^{\text{total}}) = \mathbb{C} \oplus \mathbb{C}$. Verify that the induced scalar product on H_{st}^0 is of mixed signature $(+-)$ and that a subspace with positive definite metric can be selected in a way compatible with the time evolution. (Note: There is no unique subspace with positive definite metric, but they are all one-dimensional.)

14.17. This exercise provides a different way to compute the cohomology of the BRST operator (14.59a),

$$\Omega = c^* \, a + a^* \, c.$$

This alternative method exploits the duality theorems and Lefschetz trace formula. One can ignore in the analysis the physical degrees of freedom, which are decoupled.

(a) Show that $H_{\text{st}}^g(\Omega)$ vanishes for $g < 0$. [Hint: Consider the "Koszul differential" defined by $\delta a^* = 0$, $\delta \bar{c}^* = a^*$, $\delta b^* = 0$, $\delta c^* = 0$ on functions $\psi(a^*, \bar{c}^*, b^*, c^*)$ in the holomorphic representation.]

(b) By duality, $H_{\text{st}}^{g>0}(\Omega) = 0$. Hence, only $H_{\text{st}}^0(\Omega)$ need be determined. By Lefschetz trace formula and (a), one has $\dim H_{\text{st}}^0(\Omega) = \text{tr}_W \, I$. The trace $\text{tr}_W \, I$ is ill defined because W is infinite dimensional, but since Ω commutes with the occupation number, one can proceed level by level. For a definite value of the occupation number, $\text{tr}_W \, I$ is finite. Show that $\text{tr}_W \, I = 1$ for zero occupation number (vacuum) and zero otherwise. (Count the states at ghost number -1, 0, and 1.) This implies that only the vacuum defines a nontrivial BRST-closed state.

(c) Repeat (a) and (b) for $\Omega = \sum_i (c_i^* a_i + a_i^* c_i)$.

14.18. In this exercise and the next, the constraints are bosonic, irreducible, and linear homogeneous in the momenta, $G_a = \xi_a{}^i(q) \, p_i$. One quantizes the system in the Schrödinger representation where the wave functions depend on q^i and η^a, $\psi = \psi(q^i, \eta^a)$. The theory is of rank one, and there is an ordering ambiguity in the quantum BRST charge.

(a) Show that Ω can be identified with $-id$ if one orders the p_i and the ghost momenta $\mathcal{P}_a$ to the right of q^i and η^a. Here, d is the exterior derivative operator along the gauge orbits in configuration space. This proves, in particular, that Ω is nilpotent quantum-mechanically.

(b) Show that the 0-forms annihilated by the constraints are BRST-closed. Verify that these 0-forms are just the Dirac states of Exercise 13.7.

(c) Show that if the constraints have a further inhomogeneous piece independent of p_i, then one can identify Ω with $-ie^F \, de^{-F}$.

14.19. Take now the hermitian symmetric ordering for Ω, $\xi_a{}^i(q)p_i \to \frac{1}{2}(\xi_a{}^i(q)p_i + p_i\xi_a{}^i(q))$, $\eta\eta\mathcal{P} \to \frac{1}{2}(\eta\eta\mathcal{P}+\mathcal{P}\eta\eta)$. This amounts to treating ψ as a density of weight $\frac{1}{2}$ in the space of the q's and the η's; that is, ψ transforms with the square root of the superdeterminant (Sec. 20.2) of the transformation $q, \eta \to q', \eta'$.

(a) Show that this leads to the transformation rule (*iii*) of Exercise 13.8.

(b) Prove that the hermitian ordering for Ω leads to a nilpotent operator. Check that the states independent of η^a are solutions of $\Omega\psi = 0$ if and only if they obey Eq. (iv) of Exercise 13.8, with the ghost correction term included. Hence, the approach of this exercise is equivalent to that of Exercise 13.8. (Recall that the conventions of Exercise 13.8 for the Lie bracket are adopted to a left action of vector fields.)

(c) Prove that the formalism is invariant under redefinitions of the constraints of the form $G_a \to a_a{}^b(q)\, G_b$. To that end: (*i*) Show that the unitary transformation implementing the change of constraint functions is given by $\hat{q}^{i\prime} = \hat{q}^i, \hat{p}_i' = \hat{p}_i + M_i{}^a{}_b(\hat{q})\hat{\eta}^b\hat{\mathcal{P}}_a + \frac{i}{2}\operatorname{tr} M_i$, $\hat{\eta}'^a = (a^{-1})^a{}_b\,\hat{\eta}^b$, $\mathcal{P}_b' = (a)^a{}_b\hat{\mathcal{P}}_a$ with $M_i{}^a{}_b = ((\partial_i a)a^{-1})^a{}_b$. We denote the unitary transformation by U. (*ii*) Prove that $\psi' = U\psi$ solves $\Omega'\psi' = 0$ and is annihilated by $\mathcal{P}_b$ if ψ solves $\Omega\psi = 0$ and is annihilated by $\mathcal{P}_b$.

14.20. Given the BRST operator Ω (with $\Omega^2 = 0, \Omega^* = \Omega$), one can rewrite it in η-$\mathcal{P}$ order by using the ghost (anti)commutation relations. One then gets Eq. (14.69). One can also rewrite Ω in $\mathcal{P}$-η order. Show that the coefficient of η^a is then equal to $G_a^\dagger$.

14.21. Consider the BRST charge for a reducible system. Show that η-$\mathcal{P}$ ordering of Ω yields again consistent quantum constraints G_{a_0} obeying $Z_{a_1}{}^{a_0}G_{a_0} = 0$ (in that order).

14.22. (a) Replace K by αK in (14.89a) with $\alpha > 0$. Verify explicitly that the amplitude $(\psi_{q,\eta=0,b=0,\bar{C}=0}, A\psi_{q',\eta=0,b=0,\bar{C}=0})$ does not depend on α.

(b) Assume that there is a single constraint. Show that the amplitude changes sign if one replaces α by $-\alpha$.

(c) Conclusion: One has $d(\psi_{q,\eta=0,b=0,\bar{C}=0}, A\psi_{q',\eta=0,b=0,\bar{C}=0})/d\alpha = 0$, except at $\alpha = 0$ where there is a discontinuity. Note that for $\alpha = 0$ ($K = 0$), the amplitude is ill defined [it contains $0\,\delta(0)$].

14.23. Consider the constraint $p = 0$. Check explicitly that the regularized scalar product (14.91) coincides with that defined in §13.3.4. Repeat the exercise for the constraint $q = 0$. [Hint: In the first case, the dual state $f(q)\eta$ must fulfill $\int f(q)\,dq = 1$. In the second case, it must fulfill $f(q = 0) = 1$.]

14.24. Although we by no means advocate a nonhermitian BRST charge, it is of interest to examine what it would lead to. Assume then that $\Omega^2 = 0$ but $\Omega^* \neq \Omega$.

(a) Show that the coefficients G_a of η^a in the η-$\mathcal{P}$ ordering of Ω and $\bar{G}_a$ of η^a in the η-$\mathcal{P}$ ordering of Ω^* both form a set of first-class constraint operators.

(b) Verify that the states $\psi_{q,b=0,\eta=0,\bar{C}=0}$ are annihilated by Ω and Ω^*.

(c) Show that the amplitude $A(q', q) \equiv (\psi_{q',b=0,\eta=0,\bar{C}=0}, A\psi_{q,b=0,\eta=0,\bar{C}=0})$, where A is BRST-invariant, fulfills $G_a(q', -i\,\partial/\partial q')\,A(q', q) = 0$ and $\bar{G}_a(q, -i\,\partial/\partial q)\,A^*(q', q) = 0$.

(d) Express $A(q', q)$ in terms of complete sets of solutions of $G_a|\psi\rangle = 0$ and of $\bar{G}_a|\psi\rangle = 0$.

Conclusion: With $\Omega^* \neq \Omega$, the amplitude is not gauge invariant (*i.e.*, annihilated by the same G_a or G_a^*) in both arguments.

14.25. (a) Show that the ordinary kernel $A_0(q', q) = \langle q' \,|\, A_0 \,|\, q\rangle$ of the gauge-invariant operator A_0 ($[A_0, G_a] = A_a{}^b G_b$) is equal to the matrix element $(\psi_{q',\eta=0}, A\psi_{q,\mathcal{P}=0})$ with $\psi_{q,\eta=0} \equiv |q\rangle|\eta = 0\rangle$, $\psi_{q,\mathcal{P}=0} \equiv |q\rangle|\mathcal{P} = 0\rangle$. [Note: The ordinary kernel is not annihilated by the constraints (in contrast to the projected kernel). This follows from the fact that the state $\psi_{q,\mathcal{P}=0}$ is not BRST invariant.]

(b) Show that the amplitude $(\psi_{q',\mathcal{P}=0}, A\psi_{q,\eta=0})$ is equal to the kernel of $\bar{A}_0$, where $\bar{A}_0^\dagger$ is the leading term in the η–$\mathcal{P}$ ordering of A^*. (If A is η–$\mathcal{P}$ ordered, A^* naturally comes $\mathcal{P}$-η ordered. The η–$\mathcal{P}$ reordering modifies the leading term.)

14.26. Compute the regularized scalar product (14.91) in the case of the free relativistic particle with constraint $p^2 + m^2 = 0$. [Hint: The dual states to $\psi_{\vec{P},\varepsilon}(p^\mu) = \delta(\vec{p} - \vec{P}\,)\,\delta(p^0 - \varepsilon\omega(\vec{P}\,))$ are given by $\delta(\vec{p} - \vec{P}\,')\,\theta(\varepsilon' p^0)\,\eta$.]

CHAPTER FIFTEEN

PATH INTEGRAL FOR UNCONSTRAINED SYSTEMS

Path integral methods are extremely useful in the analysis of quantum gauge theories. The remaining part of this book aims at developing these methods.

Our first task is to review the path integral approach in the case of unconstrained systems. We assume that the reader has some familiarity with the subject, so we will only survey here the salient points, with an emphasis on aspects that are less often discussed: the Schwinger–Dyson equation, the path integral for Bose systems with an indefinite Hilbert space, and the path integral for Fermi systems. The first elements of the antifield formalism will also be explained.

We will not try to give a rigorous presentation of the path integral. Many formal manipulations will thus be allowed without attempting to provide a mathematical justification. Experience shows that when dealing with the path integral, it is best not to try to be rigorous too early. This has, of course, its dangers.

15.1. PATH INTEGRAL METHOD OF BOSE SYSTEMS—BASIC FEATURES

15.1.1. Path Integral as a Kernel

The usual application of the path integral is to yield a representation of the amplitude to go from q_i at time t_i to q_f at time t_f,

$$K(q_f, t_f; q_i, t_i) \equiv \langle q_f | U(t_f, t_i) | q_i \rangle, \tag{15.1a}$$

$$U(t_f - t_i) = \exp -\frac{i}{\hbar} H(t_f - t_i), \tag{15.1b}$$

[kernel of the evolution operator $U(t_f - t_i)$ in the coordinate representation]. If $\psi(q_i, t_i)$ is the wave function at time t_i, then the wave function at time t_f is given by

$$\psi(q_f, t_f) = \int dq_i\, K(q_f, t_f; q_i, t_i)\, \psi(q_i, t_i). \tag{15.1c}$$

The amplitude (15.1a) can be rewritten as

$$K(q_f, t_f; q_i, t_i) = \langle q_f, t_f \,|\, q_i, t_i \rangle, \tag{15.1d}$$

where the states $|q, t\rangle$ form a basis of eigenstates of the Heisenberg operators $q(t)$ at time t. That is, one can view the amplitude (15.1a) as the matrix element of the unit operator but in a mixed representation.

The searched-for path integral representation of K is obtained by inserting the decomposition of the identity,

$$1 = \int dq\, |q\rangle\, \langle q| \tag{15.2}$$

in

$$U(t_f, t_i) = \exp -\frac{iH}{\hbar}(t_f - t_N) \exp -\frac{iH}{\hbar}(t_N - t_{N-1}) \dots \exp -\frac{iH}{\hbar}(t_1 - t_i) \tag{15.3a}$$

with

$$t_\alpha = t_i + \alpha\varepsilon, \qquad \varepsilon = \frac{(t_f - t_i)}{N+1}, \qquad \alpha = 1, \dots, N. \tag{15.3b}$$

One gets

$$K(q_f, t_f; q_i, t_i) = \int dq_N \dots dq_1\, K(q_f, t_f; q_N, t_N) \dots$$
$$K(q_{\beta+1}, t_{\beta+1}; q_\beta, t_\beta) \dots K(q_1, t_1; q_i, t_i). \tag{15.4a}$$

By using then the decomposition of unity in the momentum representation, the matrix element $K(q_{\beta+1}, t_{\beta+1}; q_\beta, t_\beta)$ can be further decomposed as

$$K(q_{\beta+1}, t_{\beta+1}; q_\beta, t_\beta) = \int \frac{dp_\beta}{(2\pi\hbar)^{n/2}}\, e^{(i/\hbar)p_\beta q_{\beta+1}} \langle p_\beta | U(\varepsilon) | q_\beta \rangle, \tag{15.4b}$$

since

$$\langle q_{\beta+1} \,|\, p_\beta \rangle = \frac{1}{(2\pi\hbar)^{n/2}} \exp \frac{i}{\hbar}\, p_\beta\, q_{\beta+1}. \tag{15.4c}$$

Here, n is the number of degrees of freedom. (We collectively denote all the coordinates and the momenta by q and p. The index β refers to the time.) For small ε, one finds

$$\begin{aligned} \langle p_\beta \,|\, U(\varepsilon) \,|\, q_\beta \rangle &= \left\langle p_\beta \middle| I - \frac{i}{\hbar}\, \varepsilon H \middle| q_\beta \right\rangle \\ &= \frac{1}{(2\pi\hbar)^{n/2}} \exp \left[-\frac{i}{\hbar}\, p_\beta\, q_\beta - \frac{i}{\hbar}\, \varepsilon\, H(p_\beta, q_\beta) \right], \end{aligned} \tag{15.4d}$$

where we have defined $H(p_\beta, q_\beta)$ as

$$H(p_\beta, q_\beta) = \frac{\langle p_\beta | H | q_\beta \rangle}{\langle p_\beta | q_\beta \rangle}. \tag{15.4e}$$

The function $H(p, q)$ given by (15.4e) is the p-q symbol of the Hamiltonian operator $\hat{H}$, that it, it is equal to the phase space function obtained by ordering all the $\hat{p}$'s to the left of the $\hat{q}$'s in the given $\hat{H}$ and then dropping the hats. Indeed, one has

$$\langle p \,|\, f(\hat{p})\, g(\hat{q}) \,|\, q \rangle = f(p)\, g(q)\, \langle p \,|\, q \rangle \tag{15.5}$$

for any functions f and g. The p-q symbol may be not real even if $\hat{H}$ is hermitian. For instance, if $\hat{H} = \frac{1}{2}(\hat{p}\hat{q} + \hat{q}\hat{p}) = \hat{p}\hat{q} + i\hbar/2$, the p-q symbol of $\hat{H}$ is $H(p, q) = pq + i\hbar/2$ and fails to be real by terms of order $\hbar$.

With (15.4d), the integral (15.4b) can be rewritten as

$$K(q_{\beta+1}, t_{\beta+1}; q_\beta, t_\beta) = \int \frac{dp_\beta}{(2\pi\hbar)^n} \exp \frac{i}{\hbar} \left[p_\beta (q_{\beta+1} - q_\beta) - \varepsilon\, H(p_\beta, q_\beta) \right], \tag{15.6}$$

leading, in the limit $\varepsilon \to 0$, to the expression

$$K(q_f, t_f; q_i, t_i) = \lim_{N \to \infty} \int \left(\prod_{\alpha=1}^{N} \frac{dq_\alpha\, dp_\alpha}{(2\pi\hbar)^n} \right) \frac{dp_i}{(2\pi\hbar)^n} \exp \frac{i}{\hbar}\, S_H^\varepsilon \tag{15.7a}$$

with

$$S_H^\varepsilon = \sum_{\alpha=0}^{N} [p_\alpha(q_{\alpha+1} - q_\alpha) - \varepsilon\, H(p_\alpha, q_\alpha)]\,, \quad p_0 \equiv p_i\,,\; q_0 \equiv q_i\,,\; q_{N+1} \equiv q_f. \tag{15.7b}$$

One can view the q_α's and the p_β's in (15.7) as defining a skeletonized path in phase space. More precisely, the q_β's can be thought of as the values of $q(t)$ at the times t_β, while the p_β's can be viewed as the values of $p(t)$ at the midpoint times $\frac{1}{2}(t_\beta + t_{\beta+1})$. With these identifications, $p_\beta(q_{\beta+1} - q_\beta)$ is an approximation of $\int_{t_\beta}^{t_{\beta+1}} dt\, p\dot{q}$. Moreover, $\varepsilon H(p_\beta, q_\beta)$ is an approximation of $\int_{t_\beta}^{t_{\beta+1}} H(p,q)\, dt$ in which q is evaluated at the bottom of the time interval and p is evaluated slightly later. One can rewrite the integral (15.7) as a symbolic sum over phase space paths

$$K(q_f, t_f; q_i, t_i) = \int [Dq][Dp] \exp \frac{i}{\hbar}\, S_H[q(t), p(t)] \tag{15.8a}$$

with

$$S_H = \int_{t_i}^{t_f} (p\dot{q} - H)\, dt, \qquad q(t_f) = q_f, q(t_i) = q_i. \tag{15.8b}$$

15.1.2. Comments

The path integral (15.8) involves three ingredients:

(*i*) The class of paths over which one sums; this class is determined by the arguments of $K(q_f, t_f; q_i, t_i)$ and contains all the paths that start from q_i at the initial time t_i and end at q_f at the final time t_f, without restriction on the momenta at the endpoints.

(*ii*) The integration measure $[Dq]\,[Dp]$, which is formally a product over time of the phase space Liouville measures at each time.

(*iii*) The integrand $\exp{(i/\hbar)}\, S_H[q(t), p(t)]$, which gives the amplitude of each individual path. The functional S_H is the action adapted to the given boundary conditions and the chosen skeletonization of the paths. It differs from the classical action at most by terms of order $\hbar$ in H, since it is the p–q symbol of $\hat{H}$, rather than the classical Hamiltonian itself, that appears in (15.8b). However, the precise form of the "quantum corrections" depends on the manner in which the symbolic expression (15.8a) is defined. If, for instance, we had skeletonized (15.8) not as in (15.7) but, rather, had chosen to evaluate the q's slightly later that the p's in $\int_{t_\beta}^{t_{\beta+1}} H(p,q)\, dt$, then it would have been necessary to replace the p–q symbol of $\hat{H}$ by its q–p symbol to get the same amplitude (see Exercise 15.1). Keeping the same H in S_H while changing the evaluation prescriptions of the symbolic expression (15.8) leads to a different evolution operator associated with a different Hamiltonian operator $\hat{H}$. That Hamiltonian operator differs from the original Hamiltonian operator only in the ordering of the $\hat{q}$'s and the $\hat{p}$'s and so has the same function H as classical limit. The operator ambiguities that arise when passing from the classical Hamiltonian H to the operator $\hat{H}$ do not disappear when the evolution operator is expressed as a formal path integral involving only classical concepts. These ambiguities reflect themselves through the precise way in which the symbolic formula (15.8) is to be evaluated.

Since the exact form of the $\hbar$-corrections to the classical Hamiltonian depends on the evaluation prescriptions for (15.8), we shall not write them explicitly in the subsequent formal derivations, and we shall use for S_H the original classical action.

Because of the boundary conditions, the coordinates and the momenta are not treated symmetrically in the path integral. That it should be so is clear on physical grounds, since the coordinates and the momenta do not commute and, hence, cannot be specified simultaneously at the endpoints.

In the momentum representation of the kernel $K(p_f, t_f; p_i, t_i) \equiv \langle p_f | U(t_f, t_i) | p_i \rangle$, one finds that the roles of q and p are reversed, with the action S_H replaced by the appropriate action adapted to giving the p's at the endpoints [Exercise 15.2].

A more symmetrical formulation of the path integral arises if one considers the Weyl symbol of the evolution operator rather than its matrix elements in a definite representation (see §15.4.2 and Exercise 15.21). The q's and the p's are also treated more symmetrically in the spectral function

$$\operatorname{tr} U(t_f, t_i) = \int dq\, K(q, t_f; q, t_i), \tag{15.9}$$

which is obtained by setting $q_f = q_i = q$ and integrating over q. The functions $q(t)$ over which one sums in (15.9) are thus periodic in time with period $t_f - t_i$. If one defines p_{N+1} [which does not appear in the integral (15.7a)] as $p_{N+1} = p_i$, one can rewrite the spectral function (15.9) as a sum over periodic paths,

$$\operatorname{tr} U(t_f, t_i) = \int_{\text{periodic paths } q(t), p(t)} [Dq\, Dp] \exp \frac{i}{\hbar} S_H[q(t), p(t)]. \tag{15.10}$$

The spectral function contains the information about the energy levels and their degeneracies (Exercise 15.3). Its analytic continuation to imaginary times yields the partition function.

15.1.3. Quantum Averages of Functionals

It is useful to consider operator insertions in the transition amplitude. Let $\hat{A}_1, \hat{A}_2, \ldots, \hat{A}_k$ be k operators and $\tau_1, \tau_2, \ldots, \tau_k$ be k intermediate times such that $t_f > \tau_{i_1} > \cdots > \tau_{i_k} > t_i$. The insertion of the operators $\hat{A}_j$ acting at τ_j is defined as the expectation value between the initial state $|i\rangle$ and the final state $|f\rangle$ of the product

$$\hat{U}(t_f-\tau_{i_1})\,\hat{A}_{i_1}\,\hat{U}(\tau_{i_1}-\tau_{i_2})\,\hat{A}_{i_2}\,\hat{U}(\tau_{i_2}-\tau_{i_3})\ldots\hat{U}(\tau_{i_{k-1}}-\tau_{i_k})\,\hat{A}_{i_k}\,\hat{U}(\tau_{i_k}-t_i) \tag{15.11}$$

in which the operators $\hat{A}_{i_j}$ are ordered from right to left according to increasing times, $t_f > \tau_{i_1} > \cdots > \tau_{i_k} > t_i$. By going to the Heisenberg picture, this expectation value can be rewritten as

$$\langle f'|\,T(\hat{A}_1(\tau_1)\,\hat{A}_2(\tau_2)\ldots\hat{A}_k(\tau_k)\,|i\rangle \tag{15.12}$$

where: (i) $\hat{A}_j(\tau_j)$ is the Heisenberg operator $\hat{A}_j$ at time τ_j

$$\hat{A}_j(\tau_j) = U^{-1}(\tau_j - t_i)\,\hat{A}_j\,U(\tau_j - t_i);$$

(ii) T is the time-ordered product; and (iii) $|f'\rangle = U^{-1}(t_f - t_i)\,|f\rangle$. If $|f\rangle$ is invariant under time translations, like the vacuum in field theory, then $|f'\rangle = |f\rangle$.

The expectation value (15.12) possesses a path integral representation. As in the previous subsection, we choose to work in the coordinate representation, *i.e.*, we take $|i\rangle$ and $|f\rangle$ to be coordinate eigenstates. We denote the expectation value (15.12) by

$$\langle A_1(\tau_1)\,A_2(\tau_2)\ldots A_k(\tau_k)\rangle \tag{15.13}$$

without explicit reference to either the given initial and final states, or the time-ordering symbol. Note that we *do not* normalize the amplitude (15.12) by dividing by the expectation value of 1, so that $\langle 1\rangle \neq 1$ but is rather given by

$$\langle 1\rangle = K(q_f, t_f; q_i, t_i). \tag{15.14}$$

To derive the path integral representation of the operator insertion (15.11), one first p-q orders the operators $\hat{A}_j$, and one slightly splits the products $f(\hat{p})\,g(\hat{q})$ at time τ_j as $f(\hat{p})$ at time $\tau_j + \eta$ times $g(\hat{q})$ at time τ_j, where η is an infinitesimally small number. By taking a subdivision of the time interval $[t_i, t_f]$ that includes the intermediate times τ_j and $\tau_j + \eta$, one then easily finds that the expectation value of the operator insertion (15.11) possesses the path integral representation

$$\langle A_1(\tau_1)\dots A_k(\tau_k)\rangle = \int A_1[q(t),p(t)](\tau_1)\dots A_k[q(t),p(t)](\tau_k)$$
$$\exp\frac{i}{\hbar}S_H[q(t),p(t)]\,[Dq]\,[Dp] \qquad (15.15)$$

where $A_j[q(t),p(t)](\tau_j)$ is by definition the functional of the phase space paths that is equal to $A_j(q,p)$ evaluated at time τ_j. The phase space function $A_j(q,p)$ is itself the p-q symbol of the Schrödinger picture operator $\hat{A}_j(\hat{q},\hat{p})$. If $\hat{A}_j = \hat{q}$, then $A_j[q(t),p(t)](\tau_j) = q(\tau_j)$. The derivation of (15.15) is direct and proceeds as the derivation of (15.8). It relies on the formulas

$$\langle p\,|\,f(\hat{p})\,\hat{U}(\varepsilon)\,|\,q\rangle = f(p)\langle p\,|\,\hat{U}(\varepsilon)\,|\,q\rangle,$$
$$\langle p\,|\,\hat{U}(\varepsilon)\,g(\hat{q})\,|\,q\rangle = g(q)\langle p\,|\,\hat{U}(\varepsilon)\,|\,q\rangle. \qquad (15.16)$$

The p-q symbol of $\hat{A}_j$ appears in (15.15) due to the particular skeletonization through which the symbolic expression (15.15) is arrived at. Other skeletonizations would lead to different symbols. Therefore, we shall again not emphasize the distinction between $A_j(q,p)$ and the classical limit of $\hat{A}_j(\hat{q},\hat{p})$.

The expectation value of the chronological product $T[\hat{A}_1(\tau_1)\dots \hat{A}_k(\tau_k)]$ is thus equal to the functional (or "quantum") average of the classical functional $A_1A_2\dots A_k$ over all phase space paths that join q_i at time t_i to q_f at time t_f. These paths are weighted by $\exp(i/\hbar)S_H$. The path integral naturally orders the operators from right to left according to increasing times. This explains, in particular, why we have to evaluate p slightly later than q when p-q ordering $\hat{H}$ (§15.1.1).

The concept of time ordering can be generalized to expressions involving the time derivatives of the Heisenberg operators $\hat{p}(\tau_1)$ and $\hat{q}(\tau_2)$, $t_i < \tau_{i_1} < \tau_{i_2} < t_f$. However, because these time derivatives refer to infinitesimally close but different times, they must be handled with care in a time ordering. For instance, $T[\dot{\hat{p}}(\tau_1)\,\hat{q}(\tau_2)]$ stands for the limit $\varepsilon \to 0$ of

$$T\Big[\frac{\hat{p}(\tau_1+\varepsilon/2)-\hat{p}(\tau_1-\varepsilon/2)}{\varepsilon}\,\hat{q}(\tau_2)\Big].$$

This limit is equal to $d/d\tau_1\{T[\hat{p}(\tau_1)\,\hat{q}(\tau_2)]\}$ (the derivative is outside T), *i.e.*, explicitly

$$[\dot{\hat{p}}(\tau_1)\,\hat{q}(\tau_2)] = \dot{\hat{p}}(\tau_1)\,\hat{q}(\tau_2)\,\theta(\tau_1-\tau_2)+\hat{q}(\tau_2)\,\dot{\hat{p}}(\tau_1)\,\theta(\tau_2-\tau_1)-i\hbar\,\delta(\tau_1-\tau_2). \qquad (15.17)$$

The extra term $-i\hbar\,\delta(\tau_1 - \tau_2)$ arises from the derivative of the step functions $\theta(\tau_1 - \tau_2)$ and $\theta(\tau_2 - \tau_1)$ occurring in

$$T[\hat{p}(\tau_1)\,\hat{q}(\tau_2)] = \hat{p}(\hat{\tau}_1)\,\hat{q}(\tau_2)\,\theta(\tau_1 - \tau_2) + \hat{q}(\tau_2)\,\hat{p}(\tau_1)\,\theta(\tau_2 - \tau_1)$$

as well as from the noncommutativity of $\hat{q}$ and $\hat{p}$ at equal times. The time ordering of expressions involving time derivatives of $\hat{q}(\tau)$ and $\hat{p}(\tau)$ is thus defined so as to commute with $d/d\tau$, $T(d/d\tau) = (d/d\tau)T$.

The above derivation of the path integral (15.15) can be extended to more complicated functionals. One finds that the expectation value $\langle T[\hat{F}]\rangle \equiv \langle F\rangle$ of an arbitrary operator functional of $\hat{q}(t)$ and $\hat{p}(t)$ is equal to the quantum average

$$\langle F\rangle = \int [Dq]\,[Dp]\,F[q(t),p(t)]\,e^{(i/\hbar)S_H[q(t),p(t)]}, \tag{15.18}$$

where the functional F is the classical limit of $\hat{F}$.

15.1.4. Equations of Motion—Schwinger–Dyson Equations

In classical mechanics, two functionals $F[q(t),p(t);t]$ and $G[q(t), p(t);t]$ that coincide when $q(t)$ and $p(t)$ obey the classical equations of motion,

$$\dot{q} = [q, H], \tag{15.19a}$$

$$\dot{p} = [p, H], \tag{15.19b}$$

must be identified, since there is no way to distinguish them. One can define an equivalence relation among functionals by

$$F \sim G \quad \text{iff} \quad F - G = 0 \quad \text{on-shell.} \tag{15.20a}$$

This is equivalent to

$$F \sim G \quad \text{iff} \quad F - G = \int \lambda^A(t)\,\frac{\delta S}{\delta z^A(t)}\,dt, \quad z^A \equiv (q,p). \tag{15.20b}$$

By means of the equations of motion, one can express any functional as a function of the initial data. Hence, the equivalence classes of functionals are exhausted by the "one-time functionals" $A_\tau[q(t),p(t),t] \equiv A_\tau\big(q(\tau),p(\tau),t\big)$.

Similarly, in quantum mechanics, the Heisenberg-picture operators obey the (appropriately ordered) equations of motion (15.19). These equations can be rewritten as

$$T\Big(\widehat{\frac{\delta S_H}{\delta z^A(t)}}\Big) = 0, \tag{15.21}$$

since $T\hat{\dot{z}}^A = \hat{\dot{z}}^A$ and since $\widehat{\partial H/\partial z^A}$ are one-time operators $[T\hat{A}(t) = \hat{A}(t)$ if $\hat{A}$ refers to one time only]. The existence of equations of motion for $\hat{q}(t)$ and $\hat{p}(t)$ implies that not all the operator functionals of $\hat{q}(t)$ and $\hat{p}(t)$ are independent, so that one can also define an equivalence relation among them.

It turns out to be useful to identify two operator functionals that are equal after being T-ordered.

$$\hat{F} \sim \hat{G} \quad \text{iff} \quad T[\hat{F}] = T[\hat{G}]. \tag{15.22}$$

This defines an equivalence relation among the classical functionals F and G of which the expectation values of $T[\hat{F}]$ and $T[\hat{G}]$ are the quantum averages. The form of this equivalence relation is more subtle than (15.20) because the time ordering of a product is not always equal to the product of the time orderings, so $T[\int \lambda^A(t)\, \delta S/\delta z^A(t)\, dt]$ may be different from zero even though (15.21) holds.

For instance, in the case of the free particle, the equations of motion are $\dot{\hat{p}}(t) = 0$. Therefore, the functional $q(\tau)\, \dot{p}(\tau')$ is classically equivalent to zero. Yet, one gets from (15.17) that the expectation value of $T[\hat{q}(\tau)\, \dot{\hat{p}}(\tau')]$ is given by

$$\langle q(\tau)\, \dot{p}(\tau')\rangle = -i\hbar\, \langle 1\rangle\, \delta(\tau' - \tau) \neq 0. \tag{15.23}$$

The explicit form of the quantum equivalence relation defined among the functionals by (15.22) is most easily obtained by making the change of integration variables

$$z^A(t) \to z^A(t) + \zeta^A(t) \tag{15.24a}$$

in the path integral $\langle F\rangle$. Here, the $\zeta^A(t)$ are arbitrary functions of time. One gets

$$\langle F\rangle = \int [Dz^A]\, F[z^A(t) + \zeta^A(t)]\, \exp \frac{i}{\hbar}\, S_H[z^A(t) + \zeta^A(t)] \tag{15.24b}$$

because the integration measure is invariant under (15.24a). Expanding (15.24b) to first order in $\zeta^A(t)$, one then finds

$$\left\langle F \frac{\delta S_H}{\delta z^A(t)} \right\rangle + \frac{\hbar}{i} \left\langle \frac{\delta F}{\delta z^A(t)} \right\rangle = 0. \tag{15.25}$$

Equations (15.25) are very important. They are known as the "Schwinger–Dyson equations" because they lead to the usual Schwinger–Dyson equations for the Green functions if one takes for F the product of the "fields" q, $F = q(t_1)\, q(t_2) \ldots q(t_m)$ (Exercise 15.7).

Equations (15.25) reproduce (15.23) for $z^A(t) = q(\tau')$ and $F = q(\tau)$. They contain thus the information about the commutation relation $[q,p] = i\hbar$. They also embody the Heisenberg equations of motion and can, accordingly, be used as a starting point for a complete formulation of quantum mechanics. The correctness of the Schwinger–Dyson equations justifies the postulate made above that the path integral "measure" $[Dz]$ is translation-invariant.

Equations (15.25) indicate that the expectation values of the functional $F\,\delta S_H/\delta z^A(t)$, which is proportional to the equations of motion, is not always zero. It is only so if $\delta F/\delta z^A(t)$ vanishes, *i.e.*, if F does not depend on z^A at that time t. In that case, there is no time-ordering problem in $F\,\delta S_H/\delta z^A(t)$. Otherwise, $\langle F\,\delta S_H/\delta z^A(t)\rangle$ differs from zero. There is an extra correction term of order $\hbar$ in (15.25) reflecting the noncommutativity of the operators in the time ordering.

It follows from the Schwinger–Dyson equations that the functionals fall into quantum equivalence classes defined by

$$F \sim G \quad \text{iff} \quad F - G = \int \left[\lambda^A(t)\,\frac{\delta S_H}{\delta z^A(t)} + \frac{\hbar}{i}\,\frac{\delta \lambda^A(t)}{\delta z^A(t)}\right] dt \qquad (15.26)$$

for some functional λ^A of the z^A. Two functionals in the same equivalence class possess the same quantum average. In the limit $\hbar \to 0$, the equivalence classes (15.26) reduce to the classical equivalence classes defined by (15.20). One can thus say that the relations (15.26) are the quantum transcription of the classical equivalence relation (15.20). The description of (15.26) by means of the cohomology of an appropriate nilpotent operator will be analyzed below (Sec. 15.5).

Remark. For one-time functions of q and p and of their time derivatives at time τ, the coefficients λ^A in (15.26) contain $\delta(t-\tau)$ and its derivatives. The functional derivatives $\delta\lambda^A(t)/\delta z^A(t)$ that appear when one uses the equations of motion to remove the time derivatives of q and p contain then the products $\delta(t-\tau)\,\delta(t-\tau)$, $\delta(t-\tau)\,\dot{\delta}(t-\tau)$, etc., so that the integral $\int dt\,[\delta\lambda^A(t)/\delta z^A(t)]$ in (15.26) is singular and proportional to $\delta(0)$, $\dot{\delta}(0)$, $\ddot{\delta}(0)$, etc.... If such singularities occur in the actual calculation of a physical process, they must either cancel each other or arise from improper manipulations. A convenient way to avoid such singularities is to slightly displace the time arguments of the operators according to the order in which they occur. One can then freely use the equations of motion without correction terms, since the times are now different, so that $\delta z^A(\tau+\varepsilon)/\delta z^A(\tau) = 0$ (see Exercise 15.9).

15.1.5. Stationary Phase Method—Lagrangian Path Integral

If the momenta appear quadratically in the Hamiltonian, the integral over p in (15.8) is Gaussian and can be performed exactly by repeated use of the formula

$$\int_{-\infty}^{+\infty} e^{-ax^2+bx}\, dx = \sqrt{\frac{\pi}{a}}\, e^{b^2/4a}. \tag{15.27}$$

("The value of a Gaussian integral is equal, up to numerical factors that can usually be omitted, to the value of the integrand evaluated at the extremum times the determinant of the quadratic part to the power $-\frac{1}{2}$.") One thus finds that (15.8) becomes the Lagrangian path integral

$$\int [Dq]\, M(q) \exp \frac{i}{\hbar} \int dt\, L(q, \dot{q}), \tag{15.28}$$

since the value of the momenta at the extremum just yields the inverse Legendre transformation.

In the skeletonized form of the path integral, the measure factor $M(q)$ is given by

$$M(q) = \left(\frac{m}{2\pi i\hbar\varepsilon}\right)^{n(N+1)/2} \prod_{\alpha} \sqrt{g(q_\alpha)} \tag{15.29a}$$

if H is equal to

$$H = \frac{1}{2m}\, g^{ij}(q)\, p_i\, p_j + \text{terms of order zero or one in } p. \tag{15.29b}$$

In (15.29a), g is the determinant of the metric g_{ij}.

It is customary to absorb the first factor of (15.29a), which does not depend upon the dynamical variables, into $[Dq]$. The second factor is, in the limit $N \to \infty$, the product over time of a function of q evaluated at time t and is called a "local measure." The path integral becomes

$$\int [Dq]\, \mu_1[q] \exp \frac{i}{\hbar} \int dt\, L(q, \dot{q}), \tag{15.30a}$$

where the local measure $\mu_1[q]$, which does not depend on $\hbar$, reads

$$\mu_1[q(t)] = \prod_t \sqrt{g(t)}. \tag{15.30b}$$

One sometimes rewrites (15.30a) formally as

$$\int [Dq] \exp \frac{i}{\hbar} \int dt\, w(q, \dot{q}) \tag{15.31a}$$

where

$$w(q, \dot{q}) = L(q, \dot{q}) + \frac{\hbar}{i}\, \delta(0)\, \ell n \sqrt{g}. \tag{15.31b}$$

This last expression follows from the exponentiation of $\prod_t \sqrt{g}(t) = \exp \sum_t \ell n \sqrt{g}(t)$ and the observation that*

$$\sum_t f(t) = \lim_{\varepsilon \to 0} \frac{1}{\varepsilon} \sum_i \varepsilon\, f(i) = \lim_{\varepsilon \to 0} \frac{1}{\varepsilon} \int f(t)\, dt = \delta(0) \int f(t)\, dt.$$

One can thus view the measure μ_1 as a quantum correction of order $\hbar$ to the classical Lagrangian. This correction is of the same $\hbar$-order as the corrections resulting from the ordering of H (§15.1.2) but is, however, singular.

The above manipulations are manifestly very formal, as they involve infinite expressions. However, it must be recalled that the original Hamiltonian path integral was nonsingular to begin with, so the path integral (15.31a) really gives a finite answer. This means that the $\delta(0)$-singularity in w is compensated by an equal and opposite singularity arising from the path integral of L. The two terms in the right side of expression (15.31b) for w cannot be considered separately, unless one has consistent regularization prescriptions that set equal to finite and opposite values both the $\delta(0)$-singularity in w and the compensating singularity arising from the path integration.

If the dependence of the Hamiltonian on the momenta is not quadratic, then the path integral over p cannot be performed exactly. One has to resort to approximate methods, the most familiar of which is the stationary phase method. The basic formula of the stationary phase method gives the asymptotic expansion of the integral of a rapidly oscillating function and reads

$$\int F(x) \exp \frac{iS(x)}{\hbar}\, d^n x$$
$$= (2\pi\hbar)^{n/2} (\det -iA)^{-1/2} \exp \frac{iS(x_0)}{\hbar} \left[F(x_0) + 0(\hbar)\right]. \tag{15.32a}$$

Here, x_0 is the extremum point of $S(x)$,

$$\left.\frac{\partial S}{\partial x}\right|_{x_0} = 0, \tag{15.32b}$$

---

* By comparing the formal discretization of $f(x) = \int f(y)\, \delta(x-y)\, dy$, which reads $f(i) = \sum \varepsilon\, f(j)\, \delta(x_i - x_j)$, with $f(i) = \sum_j f(j)\, \delta_{ij}$ where δ_{ij} is the Kronecker delta, one gets the formal relation $\delta(0) = \lim_{\varepsilon \to 0} \frac{1}{\varepsilon}$.

and A is the matrix of the second derivatives of S at x_0,

$$A_{\alpha\beta} = \left. \frac{\partial^2 S}{\partial x^\beta \, \partial x^\alpha} \right|_{x_0} . \tag{15.32c}$$

If there are many extrema, one has to sum over all of them. The formula (15.32a) is applicable only if $\det A \neq 0$.

If one evaluates the integral over the momenta in (15.8) by means of (15.32) (with $F = 1$), one gets the Lagrangian path integral

$$K(q_f, t_f; q_i, t_i) = \int [Dq] \, \mu[q] \, \exp \frac{i}{\hbar} \int dt \, L(q, \dot{q}), \tag{15.33a}$$

where the measure $\mu[q]$ is given by an expansion in powers of $\hbar$,

$$\mu[q] = \mu_1 + \hbar\mu_2 + \hbar^2\mu_3 + \cdots . \tag{15.33b}$$

In the Gaussian case, only μ_1 is present and explicitly given by (15.30b), but, in general, the higher order terms $\mu_2, \mu_3, \ldots$ come in.

The measure can again be formally exponentiated into the action to yield

$$K(q_f, t_f; q_i, t_i) = \int [Dq] \, \exp \frac{i}{\hbar} \int dt \, w \tag{15.34a}$$

with

$$w = L_0 + \hbar L_1 + \hbar^2 L_2 + \cdots . \tag{15.34b}$$

The first term L_0 in w is the classical Lagrangian

$$L = L_0, \tag{15.34c}$$

and the "quantum corrections" $L_1, L_2, \ldots$ arise from the measure.

The formulas (15.32) can also be applied to the quantum averages. One gets, in a fashion similar to the derivation of (15.34)

$$\langle F \left[q(t), p(t) \right] \rangle = \int [Dq] \, \mathcal{F}[q(t)] \, \exp \frac{i}{\hbar} \int dt \, w \tag{15.35a}$$

where

$$\mathcal{F} \left[q(t) \right] = F_0 + \hbar \, F_1 + \hbar^2 F_2 + \cdots . \tag{15.35b}$$

The first term F_0 in the expansion (15.35b) is the value of the functional $F \left[q(t), p(t) \right]$ at the extremum for $p(t)$,

$$F_0[q(t)] = F \left[q(t), p(t) = \frac{\partial L}{\partial \dot{q}}(t) \right]. \tag{15.35c}$$

The other terms arise from the $\hbar$-corrections in (15.32a).

The terms $L_1, L_2, \ldots$ in (15.34b), and also $F_1, F_2, \ldots$ in (15.35b) if F is a one-time functional, are singular [cf. (15.31b)]. Again, these manifest singularities cancel hidden singularities that explicitly appear when one actually performs the path integration. These two types of singularities cannot be thought of independently, and if one regularizes one type of singularity, the other must be regularized in the same way (Exercise 15.9).

To conclude, one sees that the path integral can be rewritten in Lagrangian form by means of the stationary phase method (when applicable). The Lagrangian path integral generically contains singular correction terms of order at least $\hbar$ to both the classical Lagrangian and the functionals of which one takes the quantum average. These correction terms are not universal; *i.e.*, they depend on the detailed form of the Hamiltonian and of the averaged functionals. Furthermore, they are of the same $\hbar$-order as the correction terms reflecting the ordering ambiguity in the Hamiltonian (§15.1.2).

15.2. PATH INTEGRAL IN THE HOLOMORPHIC REPRESENTATION (BOSE SYSTEMS)

15.2.1. Definition of Holomorphic Representation

The path integral considered so far stands for a matrix element of the evolution operator and yields directly the amplitude for a definite process. There exist other useful applications of the path integral in which the sum-over-paths gives the kernel (or the symbol) of the evolution operator in different representations. The most widely known of these applications deals with the holomorphic representation and is briefly reviewed here. Another application, namely, the path integral representation of the Weyl symbol of $U(t_f - t_i)$, will be analyzed in Sec. 15.4 when we turn to Fermi systems.

The Fock representation of the commutation relations is based on the use of oscillator variables obeying

$$[a, a^*] = 1 \tag{15.36}$$

(we set $\hbar = 1$ from now on). The destruction operator a annihilates the vacuum

$$a|0\rangle = 0, \tag{15.37}$$

and every state can be represented as a function $f(a^*)$ of the creation operator, namely, that precise function $f(a^*)$ with which one has to act on the vacuum to get the state in question.

In that representation, the scalar product reads

$$\langle f_1 \,|\, f_2 \rangle = \int \left(f_1(a^*)\right)^* f_2(a^*)\, e^{-a^* a}\, \frac{da^*\, da}{2\pi i} \tag{15.38a}$$

where the integral (15.38a) is the two-dimensional integral in the p-q plane obtained by replacing a, a^* and $da^*\, da$ by

$$a^* = \frac{1}{\sqrt{2}}\,(q - i\,p), \qquad a = \frac{1}{\sqrt{2}}\,(q + i\,p), \tag{15.38b}$$

$$\frac{da^*\, da}{2\pi i} = \frac{dq\, dp}{2\pi}. \tag{15.38c}$$

The normalized states $|n\rangle \equiv |\psi_n\rangle$ with occupation number n are given by the monomials

$$\psi_n(a^*) = \frac{(a^*)^n}{\sqrt{n!}}, \tag{15.39a}$$

$$\langle \psi_n \,|\, \psi_m \rangle = \delta_{n,m}, \tag{15.39b}$$

and form a basis of Fock space. The operators a^* and a are represented as

$$a^* = \text{multiplication by } a^*, \tag{15.40a}$$

$$a = \frac{d}{da^*}, \tag{15.40b}$$

and can easily be checked not only to fulfill (15.36) but also to be hermitian conjugate to each other.

A Fock-space operator $\hat{A}$ can be represented by its normal symbol. By definition, the normal symbol of $\hat{A}$ is the function $K(a^*, a)$ such that if one "puts hats" on a^* and a in K, keeping all the a^*'s to the left of the a's, one gets the given operator $\hat{A}$. One can also describe $\hat{A}$ by means of its matrix elements in the Fock basis (15.39),

$$A_{nm} = \langle n|\, \hat{A}\, |m\rangle, \tag{15.41a}$$

and collect all these matrix elements to form a function of a^* and a,

$$A(a^*, a) = \sum_{n,m} A_{nm}\, \frac{(a^*)^n}{\sqrt{n!}}\, \frac{a^m}{\sqrt{m!}}. \tag{15.41b}$$

In particular, one has

$$A(0,0) = \langle 0|\, \hat{A}\, |0\rangle. \tag{15.41c}$$

The function $A(a^*, a)$ of the two complex variables a^* and a, which are not necessarily conjugates of each other, is known as the kernel of A in the holomorphic representation.

The following properties are immediate (Exercise 15.13),

$$(\hat{A}f)(a^*) = \int A(a^*, \alpha)\, f(\alpha^*)\, e^{-\alpha^* \alpha}\, \frac{d\alpha^*\, d\alpha}{2\pi i}, \tag{15.42a}$$

$$(A_1\, A_2)(a^*, a) = \int A_1(a^*, \alpha)\, A_2(\alpha^*, a)\, e^{-\alpha^* \alpha}\, \frac{d\alpha^*\, d\alpha}{2\pi i} \tag{15.42b}$$

(the convolution of kernels corresponds to the product of the operators), and

$$A(a^*, a) = e^{a^* a} K(a^*, a). \tag{15.42c}$$

In (15.42c), K and A are, respectively, the normal symbol and the kernel of the same operator $\hat{A}$.

15.2.2. Path Integral

By repeated application of the convolution formula (15.42b), one easily gets a path integral representation for the kernel of the evolution operator. Indeed, for small ε, the normal symbol of $\hat{U}(\varepsilon)$ is given by

$$1 - ih(a^*, a)\varepsilon = \exp -ih(a^*, a)\varepsilon \tag{15.43a}$$

where h is the normal symbol of the Hamiltonian. Hence, the kernel of $\hat{U}(\varepsilon)$ is equal to

$$\exp\left[a^* a - ih(a^*, a)\,\varepsilon\right]. \tag{15.43b}$$

By dividing the time interval $t_f - t_i$ into small pieces and using the convolution formula (15.42b), one then finds, along the lines of §15.1.1, that the kernel of the evolution operator can be written as

$$\begin{aligned} &U(t_f - t_i)\,(a^*, a) \\ &\quad = \lim_{N\to\infty} \int \exp\left\{\left[a_N^* a_{N-1} - a_{N-1}^* a_{N-1} + \cdots - a_1^* a_1 + a_1^* a_0\right]\right. \\ &\qquad \left. - i\left[h(a_N^*, a_{N-1}) + \cdots + h(a_1^*, a_0)\right]\varepsilon\right\} \prod_{\rho=1}^{N-1} \frac{da_\rho^*\, da_\rho}{2\pi i} \qquad (15.44a) \\ &\quad \equiv \int \exp i\, S_H'[a^*(t), a(t)]\,[Da\, Da^*], \qquad (15.44b) \end{aligned}$$

where we have set $a_N^* = a^*$ and $a_0 = a$. The sum in (15.44b) is taken over all paths that obey

$$a^*(t_f) = a^*, \qquad a(t_i) = a \tag{15.44c}$$

without restriction on $a^*(t_i)$ or $a(t_f)$. Furthermore, the action $S_H'[a^*(t), a(t)]$ is the action appropriate to the boundary conditions (15.44c), namely,

$$\begin{aligned} S_H'[a^*(t), a(t)] = &\int_{t_i}^{t_f} \left[\frac{1}{2i}\left(\dot{a}^* a - a^* \dot{a}\right) - h(a^*, a)\right] dt \\ &+ \frac{1}{2i}\left(a^*(t_f)\, a(t_f) + a^*(t_i)\, a(t_i)\right). \end{aligned} \tag{15.44d}$$

The kinetic term of (15.44d) leads to the Poisson bracket $[a, a^*]_{PB} = -i$, in agreement with (15.36).

The study of time-ordered operator insertions can also be repeated. One finds that the kernel of $T[\hat{F}]$, still denoted by $\langle F \rangle$, is given by the functional average

$$\langle F \rangle(a^*, a) = \int F[a^*(t), a(t)] \exp iS'_H [a^*(t), a(t)] \, [Da \, Da^*]. \qquad (15.45)$$

The vacuum expectation value of $T[\hat{F}]$ is obtained by taking $a^*(t_f) = 0$ and $a(t_i) = 0$ as boundary conditions. The quantum average obeys a Schwinger–Dyson equation that is identical to the one discussed in §15.1.4. Because the path integral orders again the operators chronologically, one can implement the normal ordering in an operator acting at time τ by evaluating a^* slightly later than a in that operator [see, for instance, h in (15.44)]. The extension of the analysis to many bosonic oscillators is straightforward and left as an exercise.

15.3. PATH INTEGRAL FOR SYSTEMS WITH INDEFINITE METRIC

15.3.1. Introduction

The commutation relations defining the "Heisenberg algebra,"

$$[\hat{q}, \hat{p}] = i, \qquad \hat{q}^* = \hat{q}, \; \hat{p}^* = \hat{p}, \qquad (15.46)$$

admits the well-known representation $\hat{q}$ = multiplication by q, $\hat{p} = -i \, d/dq$, in the Hilbert space of square-integrable functions. In that representation, the spectra of both $\hat{q}$ and $\hat{p}$ are real.

There exists another representation of the Heisenberg algebra in which $\hat{q}$ and $\hat{p}$ have a purely imaginary spectrum. That alternative representation takes place in a Hilbert space with an indefinite metric and is useful for the quantization of the pure gauge degrees of freedom in the Fock representation considered in §13.4.1.

15.3.2. Coordinate Representation

To describe the alternative representation of (15.46), it is convenient to replace $\hat{q}$ and $\hat{p}$ by the operators $\hat{Q}, \hat{P}$,

$$\hat{q} = i\hat{Q}, \qquad \hat{p} = -i\hat{P}, \qquad (15.47a)$$

which have a real spectrum. These new operators obey the same commutation relation as $\hat{q}$ and $\hat{p}$ do, but they are anti-hermitian,

$$[\hat{Q}, \hat{P}] = i, \qquad Q^* = -Q, \; P^* = -P. \tag{15.47b}$$

As shown by Pauli (1943), the representation space for (15.47b) can be taken to be the space of functions $\psi(Q)$ with scalar product defined by

$$\langle \psi_1 \,|\, \psi_2 \rangle \equiv \int_{-\infty}^{+\infty} dQ \, \psi_1^*(-Q) \, \psi_2(Q). \tag{15.48a}$$

The integral is taken over the real line. The operators $\hat{Q}$ and $\hat{P}$ are, respectively,

$$\hat{Q} = \text{multiplication by } Q, \tag{15.48b}$$

$$\hat{P} = \frac{1}{i} \frac{\partial}{\partial Q}. \tag{15.48c}$$

It follows from (15.48a) that even functions of Q have a positive norm, while odd functions of Q have a negative norm.

The eigenstates of $\hat{Q}$ are given by $\delta(Q - Q_0)$, and one has from (15.48)

$$\begin{aligned} \langle Q' \,|\, Q'' \rangle &= \int dQ \, \delta(-Q - Q') \, \delta(Q - Q'') \\ &= \delta(Q' + Q''). \end{aligned} \tag{15.49}$$

This leads to the decomposition of unity

$$1 = \int dQ \, |Q\rangle \, \langle -Q|; \tag{15.50a}$$

i.e.,

$$|\psi\rangle = \int dQ \, |Q\rangle \, \psi(Q) \tag{15.50b}$$

with

$$\psi(Q) = \langle -Q \,|\, \psi \rangle. \tag{15.50c}$$

Similarly, the wave functions of the states $|P\rangle$ with momentum P are given by

$$\langle -Q \,|\, P \rangle = \frac{1}{\sqrt{2\pi}} \, e^{iPQ}, \qquad \frac{1}{i} \frac{\partial}{\partial Q} \left(\frac{1}{\sqrt{2\pi}} \, e^{iPQ} \right) = P \, \frac{1}{\sqrt{2\pi}} \, e^{iPQ}. \tag{15.51a}$$

This implies

$$\langle Q \,|\, P \rangle = \frac{1}{\sqrt{2\pi}} \, e^{-iPQ}, \qquad \langle P \,|\, Q \rangle = \frac{1}{\sqrt{2\pi}} \, e^{iPQ}. \tag{15.51b}$$

One also finds

$$\langle P' \mid P'' \rangle = \delta(P' + P''), \tag{15.51c}$$

$$1 = \int dP \, |P\rangle \, \langle -P|. \tag{15.51d}$$

The decomposition of the identity yields

$$\begin{aligned}(A\psi)(Q) &\equiv \langle -Q| \, \hat{A} \, |\psi\rangle \\ &= \int dQ' \, A(Q, Q') \, \psi(Q').\end{aligned} \tag{15.52a}$$

where the kernel $A(Q, Q')$ is equal to

$$\langle -Q| \, A \, |Q'\rangle = A(Q, Q'). \tag{15.52b}$$

Finally, if $h(P, Q)$ is the P-Q symbol of the operator $\hat{H}$, one has

$$h(P, Q) = \frac{\langle -P| \, \hat{H} \, |Q\rangle}{\langle -P \mid Q\rangle}. \tag{15.52c}$$

The P-Q symbol of an operator is defined as in the usual positive definite case, since it does not require the concept of a scalar product. It is related to the p-q symbol $H(p, q)$ of $\hat{H}$ as

$$h(P, Q) = H(-iP, iQ). \tag{15.52d}$$

15.3.3. Path Integral in the Coordinate Representation

The above formulas enable one to repeat directly the path integral derivations of §15.1.1 and §15.1.3. One finds, in particular, that the kernel of the evolution operator in the coordinate representation is given by

$$\begin{aligned}K(Q_f, t_f; Q_i, t_i) &\equiv \langle -Q_f| \, U(t_f - t_i) \, |Q_i\rangle \\ &= \int [DQ] \, [DP] \, \exp i \, \tilde{S}_H[Q(t), P(t)]\end{aligned} \tag{15.53a}$$

with

$$\tilde{S}_H[Q(t), P(t)] = \int_{t_i}^{t_f} [P\dot{Q} - h(P, Q)] \, dt. \tag{15.53b}$$

The sum is taken over all paths with $Q(t_i) = Q_i$, $Q(t_f) = Q_f$, without restriction on $P(t_i)$ or $P(t_f)$, and both P and Q are real and range from $-\infty$ to $+\infty$.

The path integral can be rewritten in terms of the original variables q, p appearing in the classical theory as

$$K(q_f, t_f; q_i, t_i) = \int [Dq]\,[Dp]\,\exp i\,S_H[q(t), p(t)], \qquad (15.54a)$$

$$S_H[q(t), p(t)] = \int_{t_i}^{t_f} [p\dot{q} - H(p, q)]\,dt, \qquad (15.54b)$$

where the p-q symbol $H(p, q)$ differs from the original classical Hamiltonian at most by terms of order $\hbar$, which reflect the ordering ambiguity (if any).

The expression (15.54a) looks like (15.8a). But in (15.54a), one integrates over imaginary paths, while in (15.8a) one integrates over real paths. The precise meaning of the integration over the imaginary paths in (15.54a) is that one must replace in the given classical action q by iQ and p by $-iP$ and then integrate over real $Q(t)$ and $P(t)$. Furthermore, the boundary conditions for the paths in (15.54a) are also imaginary and read

$$q(t_i) = iQ_i, \qquad q(t_f) = iQ_f. \qquad (15.54c)$$

If the integral over $p(t)$ is Gaussian, it can be performed exactly. One ends up in that case with the Lagrangian path integral

$$\int [Dq]\,\mu[q]\,\exp i \int L(q, \dot{q})\,dt, \qquad (15.54d)$$

in which the sum is taken over imaginary paths $q(t)$. We stress that the rotation of the integration contour has nothing to do with going to imaginary times but is rather related to a different choice of the representation of the commutation relations.

15.3.4. Holomorphic Representation

The alternative representation (15.48) of the commutation relations (15.46) can also be described in terms of oscillator variables. To that end, one defines

$$\hat{a} = \frac{i}{\sqrt{2}}(\hat{Q} + i\hat{P}) = \frac{1}{\sqrt{2}}(\hat{q} - i\hat{p}), \qquad (15.55a)$$

$$\hat{a}^* = \frac{i}{\sqrt{2}}(\hat{Q} - i\hat{P}) = \frac{1}{\sqrt{2}}(\hat{q} + i\hat{p}). \qquad (15.55b)$$

The operators $\hat{a}$ and their adjoint $\hat{a}^*$ fulfill the commutation relations for a bosonic oscillator with the "wrong" sign,

$$[\hat{a}, \hat{a}^*] = -1. \qquad (15.55c)$$

The occupation number N is

$$\hat{N} = -\hat{a}^*\hat{a}, \qquad [N, \hat{a}^*] = \hat{a}^*, \qquad [N, \hat{a}] = -\hat{a}. \tag{15.55d}$$

The wave function of the vacuum annihilated by $\hat{a}$ is given by $(\pi)^{1/4} \exp(-Q^2/2)$ [$= (\pi)^{1/4} \exp(q^2/2)$]. Those of the excited states read

$$\psi_n(Q) = H_n(Q)\, e^{-Q^2/2}, \tag{15.56a}$$

where $H_n(Q)$ are the (appropriately normalized) Hermite polynomials. Because these polynomials are even or odd functions of Q according to whether n is even or odd, one finds from (15.48a) that

$$\langle \psi_n \mid \psi_m \rangle = (-)^n\, \delta_{n,m}. \tag{15.56b}$$

This shows that the oscillator $\hat{a}^*$ creates negative norm states from the vacuum and that the representation space defined by (15.48) is the same as the pseudo-Fock space introduced in §13.4.1 (with a single negative metric oscillator: $a^*_{\text{here}} = a^* - b^*$ of Chapter 13).

Rather than using the Q-representation, one can describe the states as functions of a^* if one takes for the vacuum the function 1 and associates with any state $f(\hat{a}^*)|0\rangle$ the function $f(a^*)$. The operator $\hat{a}^*$ is then the operator of multiplication by a^*, while a becomes

$$a = -\frac{d}{da^*}, \tag{15.57}$$

with the minus sign, in order to fulfill (15.55c). One can also define the normal symbol of any operator without difficulty, again by "dropping the hats" from the normal-ordered expression. The scalar product is then given by

$$\langle f_1 \mid f_2 \rangle = \int \big(f_1(a^*)\big)^* f_2(a^*)\, e^{a^* a}\, \frac{da^*\, da}{2\pi i}, \tag{15.58a}$$

where the integral is taken over q and p imaginary, or, what is the same, Q and P real.

More precisely, the integral (15.58) is defined by (i) rewriting the integrand in terms of Q and P by means of $\sqrt{2}\, a = i(Q + iP)$, $\sqrt{2}\, a^* = i(Q - iP)$ [see formulas (15.55a) and (15.55b)]; (ii) replacing $da\, da^*/2\pi i$ by $dQ\, dP/2\pi$; and (iii) integrating over real Q and P. This yields

$$\langle \psi_n \mid \psi_m \rangle = (-)^n\, \delta_{n,m}, \tag{15.58b}$$

as it should. [It is important to take the complex conjugate of $\big(f_1(a^*)\big)^*$, which yields a function of a, before expressing a and a^* in terms of the real integration variables Q and P, since performing the operations in reversed order would produce extra incorrect minus signs incompatible with (15.58b). For instance, $(a^*)^* = a$ becomes $(i/\sqrt{2}\,)(Q + iP)$ and not

$[(i/\sqrt{2})(Q - iP)]^* = -(i\sqrt{2})(Q + iP)]$. From now on, the integration over oscillator variables will always be understood in the sense (*i*)–(*iii*).

The analysis of the holomorphic representation proceeds then as in the case with a positive definite scalar product. The kernel of the operator $\hat{A}$ is defined by

$$A(a^*, a) = \sum_{n,m} A_{nm} \frac{(a^*)^n}{\sqrt{n!}} \frac{a^m}{\sqrt{m!}} (-)^{n+m} \tag{15.59a}$$

with

$$A_{nm} = \langle n| \hat{A} |m\rangle \ . \tag{15.59b}$$

The minus sign $(-)^{n+m}$ is included so as to make the following equations hold (Exercise 15.16),

$$(\hat{A}f)(a^*) = \int A(a^*, \alpha) \, f(\alpha^*) \, e^{\alpha^* \alpha} \, \frac{d\alpha^* \, d\alpha}{2\pi i} \,, \tag{15.60a}$$

$$(A_1 A_2)(a^*, a) = \int A_1(a^*, \alpha) \, A_2(\alpha^*, a) \, e^{\alpha^* \alpha} \, \frac{d\alpha^* \, d\alpha}{2\pi i} \,, \tag{15.60b}$$

$$A(a^*, a) = e^{-a^* a} \, K(a^*, a) \ , \tag{15.60c}$$

where $K(a^*, a)$ is the normal symbol of $\hat{A}$. These formulas are identical with those of the standard holomorphic representation (§15.2.1) with the sole exceptions that the exponential $e^{-\alpha^* \alpha}$ is replaced by $e^{\alpha^* \alpha}$ and that the integration ranges are different.

15.3.5. Path Integral in the Holomorphic Representation

The formal similarity between (15.60) and (15.42) enables one to write down the functional averages in the case of an indefinite metric by repeating exactly the same steps as in the standard case. One gets that the kernel of $T[\hat{F}]$ is given by

$$\int F[a^*(t), a(t)] \, \exp i \, \bar{S}_H \, [a^*(t), a(t)] \, [Da] \, [Da^*] \tag{15.61a}$$

where the sum is taken over all trajectories fulfilling

$$a^*(t_f) = a^* \ , \qquad a(t_i) = a \ . \tag{15.61b}$$

In (15.61a), the action $\bar{S}_H$ is the action appropriate to the boundary conditions (15.61b) and to the change of sign of the commutation relations,

$$\bar{S}_H[a^*(t), a(t)] = \int_{t_i}^{t_f} \left[\frac{1}{2i}\,(a^*\dot{a} - \dot{a}^*a) - h(a^*, a)\right] dt$$
$$-\frac{1}{2i}\big(a^*(t_f)\,a(t_f) + a^*(t_i)\,a(t_i)\big). \quad (15.61c)$$

The domain of integration for $a^*(t)$ and $a(t)$ is such that $p(t)$ and $q(t)$ range over imaginary values. We stress again that this "rotation of the integration contours" is completely independent from the rotation to imaginary times necessary for computing the partition function in statistical mechanics. It is already present in the real time formalism studied here. It arises because the creation operator a^* creates negative norm states. The effect of the rotation of the integration contours is to make the exponential factor e^{a^*a} in (15.58) decaying rather than blowing and, thus, to replace $+\infty$ by ± 1 in the calculation of the norms $\langle \psi_n \mid \psi_m \rangle$.

15.4. PATH INTEGRAL FOR FERMIONS

15.4.1. Path Integral in the Holomorphic Representation

The derivation of the path integral for a fermionic system is straightforward in the holomorphic representation. It relies on the Berezin integral reviewed in Chapter 20 ("Complementary Material," Sec. 20.2). A mere comparison of Exercise 15.19 with Eqs. (15.41)–(15.42) shows that the formulas characterizing the holomorphic representation for fermions are identical with the formulas characterizing the holomorphic representation for bosons, provided one takes proper care of the orders of the factors and replaces the measure $da^*\,da/2\pi i$ by $da^*\,da$. The kernel of the operator $T[\hat{F}]$ is thus equal to the path integral

$$\int F[a^*(t), a(t)] \exp iS'_H[a^*(t), a(t)][Da^*\,Da]. \quad (15.62a)$$

The sum is taken over all paths that obey

$$a^*(t_f) = a^*, \qquad a(t_i) = a. \quad (15.62b)$$

Furthermore, $S'_H[a^*(t), a(t)]$ is given by

$$S'_H[a^*(t), a(t)] = \int_{t_i}^{t_f} \left[\frac{1}{2i}(\dot{a}^*a - a^*\dot{a}) - h(a^*, a)\right] dt$$
$$+\frac{1}{2i}\big(a^*(t_f)a(t_f) + a^*(t_i)a(t_i)\big). \quad (15.62c)$$

The measure $[Da^*\, Da]$ is the product over time of the invariant Liouville measure $da^*\, da$, without the factor $(2\pi i)^{-1}$. We recall that for fermionic operators, the time ordering $T[\hat{O}_1(t_1)\, \hat{O}_2(t_2)]$ is equal to

$$\hat{O}_1(t_1)\, \hat{O}_2(t_2)\, \theta(t_1 - t_2) - \hat{O}_2(t_2)\, \hat{O}_1(t_1)\, \theta(t_2 - t_1). \tag{15.63}$$

In (15.62), we have taken fermionic oscillators with positive norms, $[a, a^*] \equiv aa^* + a^*a = 1$. In the indefinite metric case, the only differences are: (i) the sign of the kinetic term in S'_H is changed in order to reproduce $[a, a^*] = -1$; and (ii) the sign of the boundary term is also changed. The derivation of these properties is direct and left as an exercise (Exercise 15.20). It is much easier than in the bosonic case. The reason is that there is no difficulty associated with the rotation of the integration contour, as the Berezin integral is always convergent. Furthermore, the integration variables a^* and a are independent and do not need to be reexpressed in terms of q and p.

The extension to a mixed system of many Bose and Fermi oscillators is also straightforward. Finally, the invariance of the Berezin integral under translations of the integration variables leads to the Schwinger–Dyson equation

$$\frac{i}{\hbar}\left\langle \frac{\delta^L S}{\delta \phi^i(t)}\, F \right\rangle + \left\langle \frac{\delta^L F}{\delta \phi^i(t)} \right\rangle = 0, \tag{15.64}$$

where ϕ^i is any dynamical variable.

15.4.2. Path Integral for the Weyl Symbol of the Evolution Operator

The holomorphic representation is not available if the phase space is odd-dimensional, since it is then impossible to split all the variables into conjugate pairs. One can nevertheless develop a useful path integral that can be applied to any number of Fermi variables and that yields the Weyl symbol of the evolution operator. That path integral is again the "sum" over all histories satisfying appropriate boundary conditions of $(i/\hbar)$ times the classical action. The classical action includes the correct boundary term that makes well defined the variational principle appropriate to the boundary conditions at hand.

15.4.2a. Action Principle

Consider then a quantum Fermi system with Fermi variables θ_A described in the classical limit by the even action (7.6) analyzed in §7.1.2,

$$S[\theta_A(t)] = \int_{t_i}^{t_f} \left[-\frac{i}{2}\,\eta^{AB}\dot{\theta}_A\theta_B - H(\theta)\right] dt + \frac{i}{2}\,\eta^{AB}\theta_A(t_f)\,\theta_B(t_i), \quad (15.65)$$

$A = 1, \ldots, n$.

The classical solutions to the equations of motion are those histories which extremize action (15.65) under the boundary conditions

$$\delta\xi_A = 0, \qquad \xi_A = \tfrac{1}{2}[\theta_A(t_f) + \theta_A(t_i)]. \quad (15.66)$$

This action (15.65) leads to the Poisson brackets

$$[\theta_A, \theta_B] = -i\,\eta_{AB} \quad (15.67a)$$

and to the first-order evolution equations

$$\dot{\theta}_A = [\theta_A, H] \quad (15.67b)$$

without additional restrictions on the boundary values $\theta_A(t_i)$ and $\theta_A(t_f)$ [besides (15.66)]. Such restrictions, which would be inconsistent with (15.67b), would arise if the term $(i/2)\,\eta^{AB}\theta_A(t_f)\,\theta_B(t_i)$ were not included in (15.65). This boundary term will emerge automatically from the path integral.

15.4.2b. Weyl Correspondence Rule

In the quantum theory, the Poisson bracket of Fermi variables becomes $(i\hbar)^{-1}$ times the anticommutator. Accordingly, the dynamical variables θ_A become, upon quantization, elements of a Clifford algebra obeying

$$\hat{\theta}_A\hat{\theta}_B + \hat{\theta}_B\hat{\theta}_A = i[\theta_A, \theta_B] = \eta_{AB} \qquad (\hbar = 1), \quad (15.68)$$

and the phase space functions $f(\theta)$ become operators in that algebra. Some elements of the representation theory of (15.68) are given in Chapter 20. The transition from the classical function $f(\theta)$ to the corresponding operator is, however, ambiguous because θ_A^2, which is zero classically, does not vanish quantum-mechanically. This problem is the analog of the usual ordering problem of the commuting case. To eliminate the ambiguity, we adopt the "Weyl correspondence rule," defined as follows: expand any classical function $f(\theta)$ in powers of θ_A:

$$f(\theta) = \overset{(0)}{f} + \overset{(1)}{f}_A\,\theta^A + \overset{(2)}{f}_{AB}\,\theta^A\theta^B + \cdots .$$

The coefficients of this expansion are not unique, unless one requires that they be totally antisymmetric in their lower indices, which we shall do. Weyl's rule consists in taking antisymmetric f's and "putting hats" to the θ's. The function $f(\theta)$ is called the "Weyl symbol" of $\hat{f}$ or just its "symbol."

The operator $\hat{f}$ determined according to Weyl rules by the function $f(\theta)$ can be written as

$$\hat{f} = \int \hat{\Omega}(\rho)\, \tilde{f}(\rho)\, d^n \rho, \tag{15.69a}$$

$$\hat{\Omega}(\rho) = \exp -i\hat{\theta} \cdot \rho, \tag{15.69b}$$

$$\hat{\theta} \cdot \rho = \hat{\theta}_A\, \rho^A = \eta^{AB} \hat{\theta}_A\, \rho_B. \tag{15.69c}$$

In (15.69) the ρ^A are Grassmann variables anticommuting with both θ_A and $\hat{\theta}_A$, and $\tilde{f}$ is the fourier transform of f, explicitly given by

$$\tilde{f}(\rho) = \frac{1}{\varepsilon_n} \int \exp\,(i\theta \cdot \rho) \cdot f(\theta)\, d^n \theta. \tag{15.70a}$$

One has

$$f(\theta) = \int \exp\,(-i\theta \cdot \rho)\, \tilde{f}(\rho)\, d^n \rho. \tag{15.70b}$$

Here, $\varepsilon_n = 1$ for n even and $\varepsilon_n = -i$ for n odd.

The Weyl ordering associates to a given classical function a unique quantum operator. However, even after that order is adopted, the inverse correspondence is not well defined when the total number of θ's is odd because $\hat{\theta}_1 \ldots \hat{\theta}_n$ commutes with all operators and is accordingly a multiple of the identity. In that case, by appropriate multiplications with $\hat{\theta}_1 \ldots \hat{\theta}_n$, any operator can be represented by an even symbol, and the correspondence between operators and phase space functions becomes bijective.

From the product law

$$\hat{\Omega}(\rho_1)\, \hat{\Omega}(\rho_2) = \exp\,[\tfrac{1}{2}\rho_1 \cdot \rho_2]\, \hat{\Omega}(\rho_1 + \rho_2) \tag{15.71}$$

for the operators $\hat{\Omega}$, one easily infers the following multiplication rule for the symbols

$$\hat{f}_1 \hat{f}_2 = \hat{f} \rightarrow f(\theta) = \int W(\theta_1, \theta_2, \theta)\, f_1(\theta_1)\, f_2(\theta_2)\, d^n\theta_1\, d^n\theta_2, \tag{15.72a}$$

where

$$W(\theta_1, \theta_2, \theta) = \det \eta \cdot \varepsilon_n \cdot \left(\frac{1}{2i}\right)^n \exp -2(\theta_1 \cdot \theta + \theta \cdot \theta_2 + \theta_2 \cdot \theta_1). \tag{15.72b}$$

This rule will be repeatedly applied below.

15.4.2c. Path Integral Representation of the Evolution Operator

If ε is sufficiently small, the symbol of the evolution operator $\hat{U}(\varepsilon) = \exp -i\hat{H}\varepsilon$ can be approximated by $\exp -iH\varepsilon$, where H is the symbol of $\hat{H}$. The identity $\exp \hat{A} = \{\exp(\hat{A}/N)[\exp(\hat{A}/N)(\exp(\hat{A}/N)\ldots)]\}$ (N times) can then be used to write the symbol of $\hat{U}(T)$, with $T = t_f - t_i$, as

$$
\begin{aligned}
U(\xi, T) = {} & (\varepsilon_n \det \eta_{AB})^N \left(\frac{i}{2}\right)^{nN} \\
& \times \int \left[\exp \sum_{\nu=1}^{N} -2(\xi_\nu \cdot \eta_\nu + \eta_\nu \cdot \xi_{\nu+1} + \xi_{\nu+1} \cdot \xi_\nu) - iH(\eta_\nu)\frac{T}{N}\right] \\
& \times \prod_{\nu=1}^{N} (d^n \eta_\nu \, d^n \xi_\nu).
\end{aligned} \tag{15.73}
$$

Here $\xi_{N+1} = \xi$ is held fixed, and N is such that T/N is sufficiently small. When the number N of time slices is even, the term quadratic in the ξ_ν's in the exponent of (15.73) is nondegenerate. The integration over the ξ_ν's is thus Gaussian. By performing this integration, one can transform (15.73) into a more suggestive form. The derivation proceeds as follows. Theξ-integral in (15.73) (with N even) is equal to the value of the integrand at the extreme, multiplied by the square root of the determinant of the quadratic part. This latter is $(-2i)^{nN/2}(\varepsilon_n \det \eta)^{N/2}$. The extremum equations are explicitly

$$
\begin{aligned}
\xi_{\nu+1} - \xi_{\nu-1} &= \eta_\nu - \eta_{\nu-1}, \qquad \nu = 2, \ldots, N-1, \\
\xi_2 &= \eta_1, \\
\xi_{N-1} &= \xi + \eta_{N-1} - \eta_N.
\end{aligned} \tag{15.74}
$$

To solve these equations for the ξ_ν's, we introduce $N+1$ new variables θ_ν ($\nu = 1, \ldots, N+1$) implicitly defined by

$$
\begin{aligned}
\eta_\nu &= \tfrac{1}{2}(\theta_{\nu+1} + \theta_\nu), \qquad \nu = 1, \ldots, N, \\
\xi &= \tfrac{1}{2}(\theta_1 + \theta_{N+1}).
\end{aligned} \tag{15.75a}
$$

The change of variables $(\eta_\nu, \xi) \rightarrow (\theta_\nu)$ is acceptable, as it can indeed be inverted. Its inverse transformation reads

$$
\theta_\nu = \sum_{\mu=0}^{N} (-)^\mu \, \eta_{\nu+\mu}, \tag{15.75b}
$$

where the identification $\eta_{N+1} = \xi$ and $\eta_{N+1+\lambda} = \eta_\lambda$ are made in the summation.

In terms of the new variables, the solution to (15.74) reads

$$\xi_\nu = \tfrac{1}{2}(\theta_\nu + \theta_1). \tag{15.76}$$

The solution (15.76) yields,

$$\begin{aligned} U(\xi,T) = {} & (i)^{nN/2}\Big(\frac{1}{2}\Big)^{nN/2}(\varepsilon_n \det\eta)^{N/2} \\ & \times \int \exp i\Big[-\frac{i}{2}\sum_{\nu=1}^{N}\Big(\frac{\theta_{\nu+1}-\theta_\nu}{\varepsilon}\Big)\cdot\Big(\frac{\theta_{\nu+1}+\theta_\nu}{2}\Big)\varepsilon \\ & + \frac{i}{2}\theta_{N+1}\cdot\theta_1 - \varepsilon H\Big(\frac{\theta_{\nu+1}+\theta_\nu}{2}\Big)\Big]\cdot\prod_\nu d^n\eta_\nu, \end{aligned} \tag{15.77}$$

where $\varepsilon = T/N$ and where it is understood that the θ's are expressed in terms of the variables η_ν (to be integrated over) and ξ (held fixed).

To make contact with the classical action, one divides the time interval $[t_i, t_f]$ in N intervals of length ε and interpret θ_ν as the value of θ at the time $t_\nu = t_1 + (\nu-1)\varepsilon$ ($t_i = t_1$, $t_f = t_{N+1}$). The η's are thus "intermediate averages", whereas ξ is half of the sum of the θ's at the endpoints, as in (15.66). In the limit of large N, the above formula for the evolution operator can symbolically be rewritten as a path integral

$$U(\xi,T) = \int \exp i\Big[\int_{t_i}^{t_f}\Big(-\frac{i\dot\theta\cdot\theta}{2} - H(\theta)\Big)\,dt + \frac{i}{2}\theta_f\cdot\theta_i\Big]\,\mathcal{D}\theta(t). \tag{15.78}$$

The integral must be carried over all trajectories $\theta(t)$ satisfying the end-point conditions $\theta_f + \theta_i = 2\xi$. The measure $\mathcal{D}\theta(t)$ contains the numerical factors in front of (15.77). The exponent of (15.78) is the action (15.65) with the correct boundary term adapted to the boundary conditions (15.66). We have accordingly succeeded in getting the usual connection between the Weyl symbol of $\hat{U}(T)$ and the classical action.

A similar path integral representation for the Weyl symbol of the evolution operator of a bosonic system can be derived. It is based on the action principle (7.8) (Exercise 15.21) and treats the coordinates and the momenta more symetrically.

15.4.3. Example: Spin-$\frac{1}{2}$ in a Magnetic Field

An interesting application of the path integral is given by a spin-$\frac{1}{2}$ in a constant magnetic field. The system is classically described by three anticommuting variables $\theta_k(t)$ with brackets

$$[\theta_k, \theta_l] = -i\delta_{kl}. \tag{15.80a}$$

Upon quantization, the θ's become proportional to the Pauli matrices

$$\hat{\theta}_k \hat{\theta}_l + \hat{\theta}_l \hat{\theta}_k = \delta_{kl} \rightarrow \hat{\theta}_k = \frac{1}{\sqrt{2}} \sigma_k. \tag{15.80b}$$

If we take the constant magnetic field to be along z, the classical Hamiltonian is

$$H = -2igB\theta_1\theta_2. \tag{15.81a}$$

The variables θ_1 and θ_2 oscillate with frequency

$$\omega = 2gB, \tag{15.81b}$$

while θ_3 stays constant.

The quantum Hamiltonian reads

$$\hat{H} = gB\sigma_z, \tag{15.82a}$$

whose exponentiation yields the evolution operator

$$\hat{U}(T) = \cos\frac{\omega T}{2} - i\sigma_z \sin\frac{\omega T}{2}. \tag{15.82b}$$

The operator $\hat{U}$ has a period in time equal to twice the period of θ_1 and θ_2. The reason is that $\hat{U}$ acts on spin one-half states, while the θ_k form a vector. The even Weyl symbol of $\hat{U}$ is given by

$$U(\xi_1, \xi_2, \xi_3, T) = \cos\frac{\omega T}{2} - 2\,\xi_1\xi_2 \sin\frac{\omega T}{2}. \tag{15.82c}$$

The action is equal to

$$S = \int_0^T \Big(-\frac{i}{2}\,\dot{\theta}_k\,\theta_l\,\delta^{kl} - H(\theta)\Big)\,dt + \frac{i}{2}\,\delta^{kl}\,\theta_k(t_f)\,\theta_l(t_i). \tag{15.83}$$

The path integral is Gaussian and can be evaluated exactly. The extremum fulfilling the boundary conditions (15.66) is given by

$$\begin{aligned}
\theta_1 &= \Big(\xi_1 + \xi_2 \tan\frac{\omega T}{2}\Big)\cos\omega T + \Big(\xi_1 \tan\frac{\omega T}{2} - \xi_2\Big)\sin\omega T,\\
\theta_2 &= \Big(\xi_2 - \xi_1 \tan\frac{\omega T}{2}\Big)\cos\omega T + \Big(\xi_1 + \xi_2 \tan\frac{\omega T}{2}\Big)\sin\omega T,\\
\theta_3 &= \xi_3.
\end{aligned}$$

There is a singularity at $T = \pi/\omega + 2\pi n/\omega$ corresponding to the fact that the classical problem is then ill-posed: for a half period, there is no solution for θ, unless $\xi_1 = 0 = \xi_2$, in which case there is an infinity of them. This classical singularity disappears in the quantum theory.

At the extremum the action reduces to the boundary term,

$$S(\text{extremum}) = 2\,i\,\xi_1\xi_2 \tan\frac{\omega T}{2}. \tag{15.84a}$$

The Weyl symbol of the evolution operator is thus equal to

$$U(\xi_1, \xi_2, \xi_3; T) = D^{1/2} \exp\Big(-2\,\xi_1\xi_2 \tan\frac{\omega T}{2}\Big) \tag{15.84b}$$

$$= D^{1/2}\Big[1 - 2\,\xi_1\xi_2 \tan\frac{\omega T}{2}\Big], \tag{15.84c}$$

where D is the determinant of the quadratic part and does not depend on ξ_i.

If one compares (15.84c) with (15.82c), one finds that (15.84c) gives the correct ξ_i-dependence. One furthermore learns that the determinant is given by

$$D^{1/2} = \cos\frac{\omega T}{2}. \tag{15.84d}$$

The importance of the boundary terms should be stressed again. Without them, one would have found the incorrect answer $S(\text{extremum}) = 0$.

15.4.4. Ghost Transition Amplitude

As a final illustration of the path integral for fermions, we consider a system with two fermionic conjugate pairs $(C, \bar{P})$ and $(\bar{C}, P)$,

$$C^2 = 0, \qquad \bar{C}^2 = 0, \qquad \bar{P}^2 = 0, \qquad P^2 = 0, \tag{15.85a}$$

$$C\bar{P} + \bar{P}C = -i, \qquad \bar{C}P + P\bar{C} = -i, \tag{15.85b}$$

$$C^* = C, \qquad P^* = -P, \qquad \bar{P}^* = -\bar{P}, \qquad \bar{C}^* = \bar{C}, \tag{15.85c}$$

which we quantize in the Schrödinger representation where states are functions of two anticommuting variables (see §20.3.2b),

$$f = f_0 + f_1 C + f_2 \bar{C} + f_3 C\bar{C}. \tag{15.86}$$

The scalar products $i \int f^* g\, d\bar{C}\, dC$ of the basis vectors $\{1, C, \bar{C}, C\bar{C}\}$ with each other are given by (in that order)

$$\begin{pmatrix} 0 & 0 & 0 & i \\ 0 & 0 & i & 0 \\ 0 & -i & 0 & 0 \\ -i & 0 & 0 & 0 \end{pmatrix}. \tag{15.87}$$

The state $C\bar{C}$ is eigenvector of the operators $\hat{C}$ and $\hat{\bar{C}}$ with the eigenvalue zero

$$\hat{C}(C\bar{C}) = \hat{\bar{C}}(C\bar{C}) = 0 \tag{15.88}$$

and will be denoted by $|\psi_{C=\bar{C}=0}\rangle$ hereafter. This state was already considered in §14.5.4 (with $C \to \eta$, $\bar{P} \to \mathcal{P}$, $\bar{C} \to \bar{\mathcal{C}}$, $P \to \rho$).

The kernel $A(C, \bar{C}; C', \bar{C}')$ of the operator A in the $C - \bar{C}$ representation is defined by the integral relations

$$(Af)(C, \bar{C}) = \int dC'\, d\bar{C}'\, A(C, \bar{C}; C', \bar{C}')\, f(C', \bar{C}'), \tag{15.89a}$$

and obeys the composition formula

$$(AB)(C, \bar{C}; C', \bar{C}') = \int dC''\, d\bar{C}''\, A(C, \bar{C}; C'', \bar{C}'')B(C'', \bar{C}''; C', \bar{C}'). \tag{15.89b}$$

One easily checks that $A(0,0;0,0)$ is equal to the matrix element occurring in Theorem 14.9 $\langle \psi_{C=\bar{C}=0} \,|\, A \,|\, \psi_{C=\bar{C}=0} \rangle$,

$$A(0,0;0,0) = \langle \psi_{C=\bar{C}=0} \,|\, A \,|\, \psi_{C=\bar{C}=0} \rangle, \tag{15.89c}$$

and that the kernel of an operator and its $(P, \bar{P}–C, \bar{C})$ symbol are related by a Fourier transformation,

$$A(C, \bar{C}; C', \bar{C}') = \int d\bar{P}\, dP \exp i[(C - C')\bar{P} + (\bar{C} - \bar{C}')P] \times A(P, \bar{P}; C, C'). \tag{15.89d}$$

Using (15.89b) and (15.89d), one proves, exactly as in the bosonic case (§15.1.1), that the kernel of the evolution operator is given by the path integral

$$U(C_f, \bar{C}_f; C_i, \bar{C}_i; t_f - t_i) = \int [DC\, D\bar{C}\, DP\, D\bar{P}]\, e^{iS}, \tag{15.90a}$$

$$S = \int_{t_i}^{t_f} (\dot{C}\bar{P} + \dot{\bar{C}}P - H)\, dt, \tag{15.90b}$$

where the trajectories are required to fulfill the boundary conditions

$$C(t_f) = C_f, \qquad \bar{C}(t_f) = \bar{C}_f, \tag{15.90c}$$

$$C(t_i) = C_i, \qquad \bar{C}(t_i) = \bar{C}_i. \tag{15.90d}$$

In particular, one finds

$$\langle \psi_{C=\bar{C}=0} \,|\, \hat{U}(t_f - t_i) \,|\, \psi_{C=\bar{C}=0} \rangle = \int [DC\, D\bar{C}\, DP\, D\bar{P}]\, e^{iS}, \tag{15.91a}$$

where the path integral is to be carried over all paths that obey the boundary conditions $C = 0 = \bar{C}$ at both endpoints,

$$C_f = \bar{C}_f = C_i = \bar{C}_i = 0. \tag{15.91b}$$

This result can be obviously extended to the case where several pairs of conjugate Fermi variables are present, by generalizing the definition (15.89a) of the kernel of an operator in the coordinate representation. Even though not equal to a (complex) matrix element for arbitrary choices of the fermionic coordinates, the fermionic kernel plays exactly the same role as the kernel in the coordinate representation for a bosonic system. The concept of projected kernel (§13.3.6) can also be easily generalized to the fermionic case.

For two "fermionic free particles," the Hamiltonian is given by

$$H = i\bar{P}P, \tag{15.92}$$

and the equations of motion read

$$\dot{C} = iP, \qquad \dot{P} = 0, \tag{15.93a}$$
$$\dot{\bar{C}} = -i\bar{P}, \qquad \dot{\bar{P}} = 0. \tag{15.93b}$$

This system corresponds to the ghost Lagrangian $\dot{\bar{C}}\dot{C}$ that arises in the dynamics of the simplest ghost systems when the Lagrange multipliers are included among the canonical variables (see §11.3.2).

By direct computation, one finds

$$\begin{aligned}
&\int DC\, D\bar{C}\, DP\, D\bar{P}\, e^{iS} \\
&\quad = \langle \psi_{C=\bar{C}=0} |\, \exp[\bar{P}P(t_f - t_i)]\, | \psi_{C=\bar{C}=0} \rangle \\
&\quad = \langle \psi_{C=\bar{C}=0} |\, I + \bar{P}P(t_f - t_i)\, | \psi_{C=\bar{C}=0} \rangle \\
&\quad = t_f - t_i,
\end{aligned} \tag{15.94}$$

where the sum is taken over all paths with $C(t_2) = \bar{C}(t_2) = C(t_1) = \bar{C}(t_1) = 0$. So the path integral for two fermionic free particles scales with time like $t_f - t_i$, while, as is known, it scales like $(t_f - t_i)^{-1}$ for bosonic ones.

15.5. A FIRST BITE AT THE ANTIFIELD FORMALISM

15.5.1. Koszul–Tate Differential Associated with the Stationary Surface

We now come back to the Schwinger–Dyson equation. Our purpose is to reformulate in cohomological terms the equivalence relation enforced among the functionals by the path integral. To that end, it is necessary to introduce new concepts (antifields, antibracket, ...) that turn out to be not only useful in the present context but also crucial in subsequent developments dealing with gauge theories.

The starting point is the path integral yielding the quantum average of the functional A,

$$\langle A \rangle = \int [D\phi]\, \alpha[\phi]\, e^{(1/\hbar)W[\phi]}, \tag{15.95a}$$

$$W[\phi] = S[\phi] + \hbar M_1[\phi] + O(\hbar^2), \tag{15.95b}$$

$$\alpha[\phi] = A[\phi] + \hbar \alpha_1[\phi] + O(\hbar^2). \tag{15.95c}$$

This path integral can be the Hamiltonian path integral, in which case ϕ^i stands for the canonical variables (q,p), (a,a^*), or θ, or it can be the Lagrangian path integral, in which case ϕ^i stands for the coordinates only. We have reinstated $\hbar$ to keep track of the different orders in $\hbar$. The first term $S[\phi]$ in $W[\phi]$ is the original classical action, while the higher order terms are the corrections to the naive measure $[D\phi] \equiv \prod_t d\phi$ (if any; see §15.1.5). Similarly, the higher order terms in (15.95c) stand also for possible "quantum corrections."

From the invariance of the measure $[D\phi]$ under translations of the fields, $[D\phi] = [D(\phi + \eta)]$, one gets the Schwinger–Dyson equation

$$\left\langle \frac{\delta^L W}{\delta \phi^i} \lambda + \frac{\hbar}{i} \frac{\delta^L \lambda}{\delta \phi^i} \right\rangle = 0, \tag{15.96}$$

which leads to the identification

$$\alpha \sim \alpha + \frac{\delta^L W}{\delta \phi^i} \lambda^i + \frac{\hbar}{i} \frac{\delta^L \lambda^i}{\delta \phi^i}. \tag{15.97a}$$

We use again the condensed notations introduced in §3.1.3, where the summation over i involves also an integration over t (x for fields). In these condensed notations, ϕ^i stand for an entire history. We further recall that if λ is a local function of the ϕ's and their derivatives at a given time, *e.g.*, $\lambda = \phi^1(\tau)$, then the functional derivatives of λ at the same time are singular and the $\hbar$-order terms in the formulas (15.96) and (15.97a) are thus infinite ($\sim \delta(0), \dot{\delta}(0), \ddot{\delta}(0)$, etc., ...).

To zeroth order in $\hbar$, Eq. (15.97a) amounts to identifying the functionals that coincide on-shell

$$A \sim A + \frac{\delta^L S_0}{\delta \phi^i} \lambda^i, \tag{15.97b}$$

where $S_0[\phi^i] \equiv S[\phi^i]$. We will call from now on "stationary surface" the surface defined in the infinite-dimensional space of all the histories $\phi^i(t)$ ($\equiv \phi^i$ in condensed notations) by the equations of motion.

The equivalence relation (15.97b) is analogous to the equivalence relation encountered in the discussion of constrained systems, where it was found that two functions coinciding on the constraint surface must be identified.

Accordingly, one can, by similar arguments, define a differential complex that provides a homological resolution of the algebra of functions on the stationary surface. If one ignores the fact that the relevant spaces of histories are infinite dimensional rather than finite dimensional, the construction is particularly simple because the equations of motion are independent. There is no redundancy, since there is no gauge symmetry.

Thus, for each equation of motion—*i.e.*, for each variable ϕ^i ("field")—one introduces one "antifield" ϕ_i^* of opposite Grassmann parity, and one defines

$$\delta\phi_i^* = -\frac{\delta^L S_0}{\delta\phi^i}, \qquad \delta\phi^i = 0, \tag{15.98a}$$

$$\varepsilon(\phi_i^*) = \varepsilon(\phi_i) + 1, \tag{15.98b}$$

$$\text{antigh}\,\phi_i^* = 1 \tag{15.98c}$$

[compare with (9.6)]. We use the terminology "antighost number" for the grading of δ to emphasize the similarity with the Hamiltonian case, although "antifield number" would be perhaps more appropriate. Similarly, we use the same notation for δ, even though $\delta(\text{here}) \neq \delta(\text{there})$.

The derivation δ is nilpotent. Its homology is trivial at antighost number $k > 0$,

$$H_k(\delta) = 0, \qquad k > 0, \tag{15.98d}$$

while $H_0(\delta)$ is equal to the quotient algebra of all the functionals $A[\phi]$ modulo the functionals that vanish on the stationary surface.

The proof of these statements is a mere repetition of the proofs given in the analysis of the Koszul-Tate differential associated with the constraint surface for the case when the constraints are irreducible (Sec. 9.1).

As will be seen in Chapters 17 and 18 on the antifield formalism, it turns out that δ is just the antifield-BRST symmetry s in the case of a theory without gauge invariance. For this reason, we will also denote δ by s in the rest of this chapter,

$$\delta F \equiv sF \qquad \text{(theories without gauge invariance)}. \tag{15.99}$$

15.5.2. Antibracket

Because the number of fields and antifields are equal, one can naturally define an odd bracket structure in the space of fields and antifields. This is done by declaring that the antifield ϕ_i^* associated with the equation $\delta S_0/\delta\phi^i = 0$ is conjugate to ϕ^i. So one sets,

$$(\phi^i, \phi_j^*) = \delta^i{}_j, \tag{15.100a}$$

an expression that reads explicitly

$$\big(\phi^i(t), \phi^*_j(t')\big) = \delta^i{}_j\, \delta(t - t') \tag{15.100b}$$

(classical mechanics) or

$$\big(\phi^i(x), \phi^*_j(x')\big) = \delta^i{}_j\, \delta(x - x') \tag{15.100c}$$

(field theory).

The "antibracket" (,) is extended to arbitrary functionals $A[\phi^i, \phi^*_i]$, $B[\phi^i, \phi^*_i]$ as follows,

$$(A, B) = \frac{\delta^R A}{\delta \phi^i}\frac{\delta^L B}{\delta \phi^*_i} - \frac{\delta^R A}{\delta \phi^*_i}\frac{\delta^L B}{\delta \phi^i}. \tag{15.100d}$$

Because the parity of ϕ^*_i is opposite to the parity of ϕ^i, the antibracket possesses unusual properties. The most striking ones are:

(*i*) the antibracket carries ghost number +1, *i.e.*,

$$\mathrm{gh}\,(A, B) = \mathrm{gh}\, A + \mathrm{gh}\, B + 1 \tag{15.101a}$$

where we have defined $\mathrm{gh}\, A = -\mathrm{antigh}\, A$ (for later comparison, the complete formula is gh A = pure gh A – antigh A, but there is no ghost and hence no pure ghost number here because we deal with a theory without gauge invariance);

(*ii*) it is odd, *i.e.*,

$$\varepsilon\big((A, B)\big) = \varepsilon_A + \varepsilon_B + 1; \tag{15.101b}$$

(*iii*) it obeys symmetry properties that are opposite to those of the Poisson bracket,

$$(A, B) = -(-)^{(\varepsilon_A+1)(\varepsilon_B+1)}(B, A). \tag{15.101c}$$

So, in particular,

$$(\mathrm{Boson}_1, \mathrm{Boson}_2) = (\mathrm{Boson}_2, \mathrm{Boson}_1), \tag{15.101d}$$

$$(\mathrm{Fermion}, \mathrm{Boson}) = -(\mathrm{Boson}, \mathrm{Fermion}), \tag{15.101e}$$

$$(\mathrm{Fermion}_1, \mathrm{Fermion}_2) = -(\mathrm{Fermion}_2, \mathrm{Fermion}_1). \tag{15.101f}$$

A further important property of the antibracket is the Jacobi identity

$$(-)^{(\varepsilon_A+1)(\varepsilon_C+1)}\big(A, (B, C)\big) + \text{“cyclic”} = 0. \tag{15.102}$$

One has also

$$(AB, C) = A(B, C) + (-)^{\varepsilon_B(\varepsilon_C+1)}(A, C)B, \tag{15.103a}$$

$$(A, BC) = (A, B)C + (-)^{\varepsilon_B(\varepsilon_A+1)}B(A, C). \tag{15.103b}$$

The relations (15.102) and (15.103) are direct consequences of the definitions.

Because of (15.101) and (15.102), the bracket (15.100) defines an odd symplectic structure.

It should be stressed that an arbitrary bosonic functional A has, in general, nonvanishing antibracket with itself. However, by the Jacobi identity, $\big((A, A), A\big) = 0$. Furthermore, if A and B depend only on the fields ϕ^i, one has $(A, B) = 0$. This means that the antibracket structure in the space of ghost number zero functionals is trivial.

By means of the antibracket, one can rewrite the Koszul differential as

$$sF = (F, S_0). \tag{15.104a}$$

The nilpotency of s is equivalent to

$$(S_0, S_0) = 0. \tag{15.104b}$$

The transformation defined by s $[F \to F + \zeta sF, \mathrm{gh}\, \zeta = -1, \varepsilon(\zeta) = 1]$ is thus a canonical transformation in the antibracket.

15.5.3. Schwinger–Dyson Operator

The differential s implements the identification (15.97b) through its cohomology. Can one define a nilpotent operator σ that would implement the quantum equivalence relation (15.97a)? The answer is yes, but the searched-for operator fails to be a derivation.

That σ cannot be a derivation follows from the fact that the product of two σ-exact functionals should not, in general, be σ-exact. Indeed, the statement that A is σ-exact should be equivalent to $T\hat{A} = 0$, since the functional average gives the expectation value of $T\hat{A}$. But if $\hat{A}$ and $\hat{B}$ are such that $T\hat{A} = T\hat{B} = 0$, it is in general not true that $T\,\widehat{AB} = 0$. Rather, $T\,\widehat{AB}$ is of order $\hbar$.

To describe σ, it is useful to introduce the operator Δ defined by

$$\Delta\alpha = (-)^{\varepsilon_i + 1} \frac{\delta^R}{\delta\phi^i} \frac{\delta^R \alpha}{\delta\phi_i^*}, \qquad \varepsilon(\Delta) = 1, \tag{15.105a}$$

for any $\alpha(\phi, \phi^*)$. One has

$$\Delta^2 = 0, \tag{15.105b}$$

$$\Delta(\alpha, \beta) = (\alpha, \Delta\beta) - (-)^{\varepsilon_\beta}(\Delta\alpha, \beta), \tag{15.105c}$$

$$\Delta(\alpha\beta) = \alpha\, \Delta\beta + (-)^{\varepsilon_\beta}\, (\Delta\alpha)\beta + (-)^{\varepsilon_\beta}\, (\alpha, \beta). \tag{15.105d}$$

Even though denoted by the same letter, the operator (15.105a) should not be confused with the auxiliary differential introduced in §10.4.3. Because of (15.105d), Δ is not a derivation for the product of functionals $\alpha[\phi, \phi^*]$. The antibracket measures how much Δ fails to be a derivation.

When α is a local functional, $\Delta\alpha$ contains the delta function and its derivatives evaluated at coincident arguments, since Δ involves second functional derivatives. Accordingly, $\Delta\alpha$ needs to be regularized in that case. How this should be done will not be addressed here.

The interest of Δ is that it reproduces the second term in the Schwinger–Dyson equation. Namely, if $\beta[\phi, \phi^*]$ is linear in the antifields,

$$\beta = \phi_i^* \, \lambda^i[\phi] \, (-)^{\varepsilon_i + \varepsilon_\beta}, \qquad \varepsilon(\lambda^i) = \varepsilon_\beta + \varepsilon_i + 1, \tag{15.106a}$$

then

$$\Delta\beta = +\frac{\delta^L \lambda^i}{\delta\phi^i}. \tag{15.106b}$$

So the required σ is given by

$$\sigma\alpha = (\alpha, W) - i\hbar\Delta\alpha, \tag{15.107}$$

since one finds:

$$\text{(i)} \qquad \sigma\alpha[\phi] = 0, \tag{15.108a}$$

$$\text{(ii)} \qquad \sigma(\phi_i^* \, \lambda^i \, (-)^{\varepsilon_i + \varepsilon_\beta}) = \frac{\delta^L W}{\delta\phi^i} \lambda^i - i\hbar \, \frac{\delta^L \lambda^i}{\delta\phi^i}. \tag{15.108b}$$

With this definition, one gets

$$\sigma^2 = 0. \tag{15.109}$$

The identification $\alpha \sim \alpha + \sigma\beta$ coincides with (15.97a), and the Schwinger–Dyson equation becomes simply

$$\langle \sigma\beta \rangle = 0. \tag{15.110}$$

To zeroth order in $\hbar$, σ coincides with s,

$$\sigma = s + O(\hbar). \tag{15.111}$$

The nilpotent operator σ can be called the "Schwinger–Dyson operator." Because of its close connection with the antifield-BRST operator of gauge theories (Chapter 18), it will also be referred to as the "quantum BRST operator." The passage from s to σ preserves nilpotency but not the derivation property. Note, however, that σ acts as a derivation in the space of functionals $\alpha[\phi^i]$ with zero ghost number, where it reduces to the zero operator.

For further comparison with the antifield-BRST formalism, one should also observe that (W, W) and ΔW both vanish, since W does not involve the antifields. So, one has

$$\tfrac{1}{2}(W, W) = i\hbar \, \Delta W. \tag{15.112}$$

In the presence of a gauge freedom, it will, in general, not be true that (W, W) and ΔW separately vanish. However, Eq. (15.112) still holds and is known as the "quantum master equation." This equation plays a central role in the formalism (see Chapter 18).

15.5.4. Geometric Interpretation of Δ and of the Antibracket

If one changes the field variables as $\phi^i \to \phi'^j = \phi'^j(\phi^i)$, the antifields ϕ^*_i transform as

$$\phi^{*\prime}_i = \phi^*_k \frac{\delta^R \phi^k}{\delta \phi'^i} \tag{15.113a}$$

in order to leave both (15.98a) and (15.100a) invariant. This suggests identifying the antifields ϕ^*_i with the vector fields $\delta^R/\delta\phi^i$ defined in the (infinite-dimensional) space of all field histories $\phi^i(t)$ [$\phi^i(x)$ in field theory], since these vector fields transform in the same manner,

$$\frac{\delta^R}{\delta \phi'^i} = \left(\frac{\delta^R}{\delta \phi'^k}\right) \cdot \left(\frac{\delta^R \phi^k}{\delta \phi'^i}\right). \tag{15.113b}$$

With this identification, the product $\phi^*_i \phi^*_j$ becomes the bivector $(\delta/\delta\phi^*_i)(\delta/\delta\phi^*_j)$, and, more generally, a function(al) $A(\phi, \phi^*)$ becomes a p-vector field tangent to the space of all field histories.

On any supermanifold M, there is a natural operation that can be defined among p-vector fields. This operation generalizes the Lie bracket and is known as the Schouten bracket. The Schouten bracket (A, B) of a p-vector A of parity ε_A with a q-vector B of parity ε_B is a $(p+q-1)$-vector of parity $\varepsilon_A + \varepsilon_B + 1$ defined by the following conditions,

$$(A, B) = -(-)^{(\varepsilon_A+1)(\varepsilon_B+1)}(B, A), \tag{15.114a}$$

$$(A, BC) = (A, B)C + (-)^{\varepsilon_B(\varepsilon_A+1)}B(A, C), \tag{15.114b}$$

$$(A, B + C) = (A, B) + (A, C), \tag{15.114c}$$

$$(A, B) = \mathcal{L}_B A, \tag{15.114d}$$

if A is a 0-vector (scalar) and B a 1-vector, and

$$(A, B) = [A, B] \qquad \text{(Lie bracket of } A \text{ and } B\text{)}, \tag{15.114e}$$

if A and B are 1-vectors (see Exercise 15.26).

The conditions (15.114a)–(15.114c) are manifestly fulfilled by the antibracket. A direct calculation shows that (15.114d) and (15.114e) hold as well. Hence, once the functions $A(\phi, \phi^*)$ are identified with p-vector fields, *the antibracket becomes the Schouten bracket.*

To define an invariant integration theory on a supermanifold, it is necessary to introduce a superdensity of weight one ("supermeasure") (see Sec. 20.2). Given a (super)density μ, a second natural operation can be defined among p-vectors, that of taking the divergence. In the coordinates ϕ^i, this operation reads explicitly

$$\operatorname{div} F = -\frac{1}{\mu}\frac{\delta^R}{\delta\phi^i}\Big(\mu(-)^{\varepsilon_i}\frac{\delta^R F}{\delta(\delta^R/\delta\phi^i)}\Big), \tag{15.115}$$

where the p-vector F is an antisymmetric polynomial of degree p in the basis vectors $\delta^R/\delta\phi^i$. [The minus sign is conventional and inserted to make (15.116) below hold.] If the measure μ is equal to one on the coordinates ϕ^i, one finds by comparison of (15.115) with (15.105a) that the identification $\phi^*_i = \delta^R/\delta\phi^i$ yields

$$\operatorname{div} F = \Delta F. \tag{15.116}$$

Hence, the operator Δ can be identified with that of taking the divergence.

Because the identification (15.116) requires $\mu = 1$, the operator Δ has an invariant meaning only under changes of coordinates with unit superdeterminant. Under a change of coordinates with $\operatorname{sdet}(\delta\phi/\delta\phi') \neq 1$, the measure μ, if originally equal to unity, becomes different from one. It is a different measure that is equal to one in the new coordinates. The new operator Δ' defined in the new coordinates is thus not equal to the old operator Δ. The transformation properties of Δ will be investigated in more detail in Chapter 18.

The geometric interpretation of Δ enables one to rewrite the Schwinger–Dyson equation more geometrically. Indeed, the Stokes theorem states that the integral of a divergence is zero (when the integrand vanishes on the boundary). By (15.116), this is equivalent to

$$\int \Delta F\, d\phi = 0, \tag{15.117}$$

but this equation is precisely identical to (15.110), since

$$(\sigma\beta)\exp\frac{i}{\hbar}W = \Delta\Big(\frac{\hbar\beta}{i}\exp\frac{i}{\hbar}W\Big). \tag{15.118}$$

One can thus say that the Schwinger–Dyson equation is just the Stokes formula in the space of all field histories.

Finally, we note that in the bosonic case, the integration theory on the manifold M can be reformulated in terms of p-forms. Upon replacing the p-vectors by their dual $n-p$ forms, one finds that Δ becomes the exterior derivative on M (in dual terminology, taking the divergence becomes taking the exterior derivative). This interpretation is, however, not available in the graded case.

15.5.5. The Antibracket Does Not Define a Measure

In general, there is no natural measure on the manifold of the fields ϕ^i, unless that manifold has some extra structure. One may then ask the question of whether there is a natural measure in the manifold of the fields ϕ^i and the antifields ϕ^*_i. The similarities between the antibracket and the Poisson bracket suggest that there may perhaps be one. This turns out, however, to be incorrect. Indeed, there is no measure that is left invariant by the canonical transformations in the antibracket because these transformations do not possess unit superdeterminant. (The precise definition of the superdeterminant is given in §20.2.2.]

Theorem 15.1. *The superdeterminant of the infinitesimal canonical transformation*

$$\phi'^i = \phi^i + (\phi^i, F), \tag{15.119a}$$

$$\phi^{*\prime}_i = \phi^*_i + (\phi^*_i, F), \tag{15.119b}$$

$$\mathrm{gh}\, F = -1, \qquad \varepsilon(F) = 1, \tag{15.119c}$$

is equal to

$$\frac{\delta^R(\phi', \phi'^*)}{\delta(\phi, \phi)} = 1 - 2\Delta F, \tag{15.119d}$$

where Δ is the operator (15.105a).

The proof follows from a direct evaluation.

The operator Δ has thus yet another interpretation. It measures by how much infinitesimal canonical transformations fail to leave the volume element $d\phi\, d\phi^*$ invariant.

The reason that $\delta(\phi', \phi'^*)/\delta(\phi, \phi)$ is not equal to unity is that there is no compensation between the field sector and the antifield sector. Rather, things add up, as the next theorem explicitly indicates.

Theorem 15.2. *If the canonical transformation $(\phi, \phi^*) \to (\phi', \phi'^*)$ is such that $\phi = \phi(\phi', \phi'^*)$ can be inverted for ϕ', then*

$$\mathrm{sdet}\, \frac{\delta^R \phi}{\delta \phi'} = \mathrm{sdet}\, \frac{\delta^R \phi^*}{\delta \phi'^*} = J^{1/2}, \tag{15.120a}$$

where

$$J = \mathrm{sdet}\, \frac{\delta^R(\phi, \phi^*)}{\delta(\phi', \phi'^*)}. \tag{15.120b}$$

Proof. It follows from Theorem 20.2 demonstrated in Sec. 20.2 that

$$\text{sdet}\,\frac{\delta^R(\phi,\phi^*)}{\delta(\phi',\phi'^*)} = \text{sdet}\,\frac{\delta^R\phi}{\delta\phi'}\left(\text{sdet}\,\frac{\delta^R\phi^{*\prime}}{\delta\phi^*}\right)^{-1}.$$

To prove Theorem 15.2, one thus needs to show

$$\text{sdet}\,\frac{\delta^R\phi}{\delta\phi'} = \left(\text{sdet}\,\frac{\delta^R\phi^{*\prime}}{\delta\phi^*}\right)^{-1}.$$

It then follows that $\text{sdet}\,(\delta\phi/\delta\phi') = J^{1/2}$ and, by a similar reasoning, that $\text{sdet}\,\delta\phi^*/\delta\phi'^* = J^{1/2}$.

Now a finite canonical transformation in the antibracket such that $\phi = \phi(\phi', \phi'^*)$ can be inverted for ϕ' takes the form (Exercise 15.30)

$$\phi^i = \frac{\delta K}{\delta\phi_i^*}, \qquad \phi_j'^* = \frac{\delta K}{\delta\phi'^j}, \tag{15.121}$$

where the generating functional $K[\phi'^j, \phi_i^*]$ is odd and of ghost number minus one. It follows that

$$\frac{\delta^R\phi^i}{\delta\phi'^j} = \frac{(\delta^R)^2 K}{\delta\phi'^j\,\delta\phi_i^*} = (-)^{\varepsilon_j(\varepsilon_i+1)}\frac{\delta^R\phi_j'^*}{\delta\phi_i^*}.$$

If one denotes by A, B, C, D and $\bar{A}, \bar{B}, \bar{C}, \bar{D}$ the respective blocks of the supermatrices $\delta^R\phi/\delta\phi'$ and $\delta^R\phi'^*/\delta\phi^*$, as in §6.3.2, one finds, since the parities of ϕ^i and ϕ_i^* are opposite,

$$A = \bar{D}^T, \qquad B = -B^T, \qquad C = C^T, \qquad D = \bar{A}^T.$$

Therefore, using (20.24),

$$\text{sdet}\,\frac{\delta^R\phi}{\delta\phi'} = \left(\text{sdet}\,\frac{\delta^R\phi^{*\prime}}{\delta\phi^*}\right)^{-1},$$

as required.

Note that in the case of an ordinary canonical transformation, there is instead compensation of $\partial q/\partial q'$ and $\partial p'/\partial p$,

$$\text{sdet}\,\frac{\partial q}{\partial q'} = \text{sdet}\,\frac{\partial p'}{\partial p}$$

because q and p have the same parity (see §20.2.5).

EXERCISES. CHAPTER FIFTEEN

15.1. Derive a path integral expression for the kernel of the evolution operator in which the q-p symbol of the Hamiltonian appears. (Hint: Insert the decomposition of unity $1 = \int |p\rangle\, dp\, |p\rangle$ to the right of $U(\varepsilon)$ in $\langle q_{\beta+1}|\, U(\varepsilon)\, |q_\beta\rangle$, rather than to the left.) Show that the p_β's are now evaluated earlier than the q_β's in the path integral.

15.2. Derive the path integral for the kernel of the evolution operator in the momentum representation

$$\left\langle p_f \middle| \exp -\frac{i}{\hbar} H(t_f - t_i) \middle| p_i \right\rangle$$

(a) either directly, by repeating steps (15.3)–(15.8) leading to the path integral representation of $\langle q_f | \exp -(i/\hbar)\, H(t_f - t_i)\, | q_i \rangle$ but adapted to the momentum representation, or

(b) by Fourier-transforming (15.8),

$$\left\langle p_f \middle| \exp -\frac{i}{\hbar} H(t_f - t_i) \middle| p_i \right\rangle = \int \frac{dq_f\, dq_i}{(2\pi\hbar)^n}\, e^{-(i/\hbar) p_f q_f + (i/\hbar) p_i q_i} \times \left\langle q_f \middle| \exp -\frac{i}{\hbar} H(t_f - t_i) \middle| q_i \right\rangle .$$

Show in particular that the action that appears in the integral is now

$$S_H - p_f q_f + p_i q_i = \int (-\dot{p} q - H)\, dt.$$

15.3. Show that the spectral function is equal to

$$\operatorname{tr} U(t_f - t_i) = \sum_n N_n \exp\left(-\frac{iE_n t}{\hbar} \right), \qquad t = t_f - t_i,$$

where E_n are the energy levels and N_n their degeneracies.

15.4. Verify directly from its path integral representation (15.8) that $K(q_f, t_f; q_i, t_i)$ is a solution of the Schrödinger equation. [Hint: Use the identity

$$(2\pi\hbar)^{-1} \int dq\, dp [\exp \tfrac{i}{\hbar}\, p(q_f - q)]\, H(p, q)\, G(q) = H\left(\frac{\hbar}{i} \frac{\partial}{\partial q_f}, q_f \right) G(q_f). \qquad]$$

15.5. Let $z^A(q, p)$ be noncanonical coordinates in phase space. Show that the path integral can be formally written as

$$\int \prod_t \frac{|\det \sigma_{AB}|^{1/2}}{(2\pi\hbar)^n}\, dz^A \exp \frac{i}{\hbar} S_H,$$

where σ_{AB} is the symplectic form in the coordinate system z^A.

15.6. Prove explicitly formula (15.18) in the case when $\hat{F} = \hat{A}(\tau)$ is a one-time operator involving only $\hat{q}$.

15.7. Compute $\langle \hat{q}(\tau)\, \hat{q}(\tau') \rangle / \langle 1 \rangle$ for the harmonic oscillator. Verify that it is a Green function for the differential operator $d^2/dt^2 + \omega^2$ (*i*) either directly or (*ii*) by means of the Schwinger–Dyson equation. (Note: The expression for $\langle 1 \rangle$ is explicitly worked out in Feynman and Hibbs 1965.)

15.8. Consider the path integral in the noncanonical coordinates of Exercise 15.5. Exponentiate $(\det \sigma_{AB})^{1/2}$ into the action to get the path integral of $\exp (1/\hbar) W$, with $W = \int (a_A \dot{z}^A - H - \hbar/2i\, \delta(0) \operatorname{tr} \ell n\, \sigma_{AB})\, dt$. Show that in $T[\delta \hat{W} / \delta z^A(t)]$, the explicit $\delta(0)$ is cancelled by the implicit $\delta(0)$-singularity arising from the T-ordering of $\sigma_{AB}\dot{z}^A$.

15.9. The purpose of this exercise is to compute $\langle p^2(t)\rangle$ in the case of the free particle in the coordinate representation.

(a) Perform the time slicing of the Hamiltonian path integral and show, by evaluating the p-integrals, that $\langle p^2(t)\rangle$ is equal to the Lagrangian functional average of $\langle m^2\dot{q}^2(t) + (m\hbar/i)\,\delta(0)\rangle$.

(b) Show that the Lagrangian functional average of $\langle \dot{q}^2(t)\rangle$ is singular and that the singularity exactly compensates the $\delta(0)$-term found in (a).

(c) Evaluate the q-integrals and show that $\langle p^2(t)\rangle = [\bar{p}^2 + (m\hbar/iT)]\langle 1\rangle$, where $\bar{p}$ is the classical value of the momentum. Here, $T = t_f - t_i$.

(d) Verify that one can consistently drop the $\delta(0)$-term of (a) if, at the same time, one regularizes $\langle \dot{q}(t)\rangle$ as $\lim_{\eta\to 0}\langle \dot{q}(t+\eta)\dot{q}(t)\rangle$.

15.10. (Fradkin representation of the Lagrangian integration measure)In the Hamiltonian path integral, make the change of integration variables $p \to a$ defined by

$$p = \frac{\partial L}{\partial \dot{q}}(q, \dot{q} + a) = \frac{\partial L}{\partial a}(q, \dot{q} + a).$$

(a) Show that

$$[Dq][Dp] = [Dq][Da]\prod_t \det\left|\frac{\partial^2 L(q, \dot{q} + a)}{\partial a\,\partial a}\right|.$$

(b) Prove that

$$p\dot{q} - H = L(q, \dot{q} + a) - \frac{\partial L}{\partial a}(q, \dot{q} + a)a.$$

[Hint: Use $H[q, p(q,u)] = p(q,u)u - L(q,u)$, with $p(q,u) \equiv (\partial L/\partial u)(q,u)$.]

(c) By using the results of (a) and (b), as well as $\det M = \exp \operatorname{tr} \ell n\, M$, show that the Hamiltonian path integral can be rewritten as

$$\int Dq\,\mu[q]\,\exp\frac{i}{\hbar}\int dt\, L(q, \dot{q}),$$

where the measure $\mu[q]$ reads

$$\mu[q] = \int Da\,\exp\frac{i}{\hbar}\int dt\Big[L(q, \dot{q} + a) - \frac{\partial L}{\partial a}(q, \dot{q} + a)a - L(q, \dot{q}) - i\hbar\,\delta(0)\operatorname{tr}\ell n\,\frac{\partial^2 L(q, \dot{q} + a)}{\partial a\,\partial a}\Big].$$

Show that the leading $\hbar$-order term in the exponential of the path integral for $\mu[q]$ vanishes at the extremum for a.

15.11. Consider the integral

$$\int dx\,dy\,dz\,\exp -\tfrac{1}{2}[x^2 + y^2 + (z - yx)^2].$$

(a) Verify that both the integrand and the measure are invariant under

$$\delta x = \varepsilon y, \qquad \delta y = -\varepsilon x, \qquad \delta z = \varepsilon(y^2 - x^2). \tag{i}$$

(b) Perform the integral over y, which is Gaussian. One gets

$$\int dx\, dz\, \mu(x,z) \exp S(x,z).$$

Compute explicitly the measure $\mu(x,z)$ and the "action" $S(x,z)$.

(c) Show that $S(x,z)$ is invariant under the transformation (i) in which one replaces y by its extremum value $y(x,z)$,

$$\bar{\delta}x = \varepsilon y(x,z), \qquad \bar{\delta}z = \varepsilon[y^2(x,z) - x^2],$$

but that the measure $\mu(x,z)$ *is not invariant.* Rather,

$$\bar{\delta}\mu \equiv (\mu\, \bar{\delta}x)_{,x} + (\mu\, \bar{\delta}z)_{,z}$$

is proportional to the derivative $\partial S/\delta z$ of S with respect to z and vanishes only "on shell." Hence, a symmetry may not survive as such after integration over some variables.

15.12. The considerations of Exercise 15.11 apply to the integration over the momenta in the Hamiltonian path integral (§15.1.5). Assume that the Hamiltonian path integral is invariant under the transformation

$$\delta q(t) = Q(q,p)(t), \tag{i}$$

$$\delta p(t) = P(q,p)(t). \tag{ii}$$

Define

$$\delta_0 q = Q_0(q,\dot{q}) \equiv Q\left(q, \frac{\partial L}{\partial \dot{q}}\right). \tag{iii}$$

(a) Show that the Lagrangian $L(q,\dot{q}) \equiv L_0(q,\dot{q})$ is invariant under (iii).

(b) By differentiating the identities expressing the invariance of S_H and $[Dq][Dp]$ under (i)–(ii), as well as $\delta S_H/\delta p(t) = 0$ for $p = \partial L/\partial \dot{q}$, show that the next term L_1 in w [Eq. (15.34b)] is strictly invariant under (iii) if S_H is quadratic in p and $Q(q,p)$ is at most linear. Verify, however, that in general L_1 is only invariant on-shell.

(c) Show that if $\delta_0 L_1$ does not identically vanish, one can modify the transformation law for q,

$$\delta' q = Q_0 + \hbar Q_1,$$

so that the variation of L_1 under Q_0 is compensated by the variation of L_0 under Q_1. Continue the analysis to the next orders.

(d) Conclusion: When integrating over the momenta, symmetries may get corrected by (singular) terms of order at least $\hbar$.

15.13. Verify formulas (15.39), (15.42), and (15.44).

15.14. Derive the formula giving the normal symbol of the product of two operators in terms of the normal symbols of the individual operators.

15.15. Extend the path integral (15.44) to many oscillators.

15.16. Verify (15.60). [Hint: For (15.60c), compute $A(a^*, a)$ in the case of operators that are monomials, $\hat{A} = (\hat{a}^{*k})(\hat{a})^l$].

15.17. Compute the holomorphic kernel of the evolution operator for the Hamiltonian $h(a^*, a) = \omega a^* a$ (i) with standard commutation relation $[a, a^*] = 1$. (answer:

$U(t_f - t_i) = \exp[a^* a\, e^{-i\omega(t_f - t_i)}]$); and (*ii*) with "wrong" commutation relation $[a, a^*] = -1$ (answer: $U(t_f - t_i) = \exp[-a^* a\, e^{i\omega(t_f - t_i)}]$). [Hint: The path integral is Gaussian and can thus be evaluated exactly. Note that the extremum does not fulfill $a^*(t) = [a(t)]^*$ or $a^*(t) = [-a(t)]^*$ because there is no relationship between $a^*(t_f) = a^*$ and $a(t_i) = a$.]

Comments: The change of sign in the commutation relation has the effect of changing the sign of the spectrum of H [this is why ω is replaced by $-\omega$ in $e^{-\omega(t_f - t_i)}$].

15.18. Consider the action

$$S = \int \left(-\frac{1}{2}\dot{q}^2 \right) dt.$$

Compute the propagator

(*i*) in the positive definite metric case (leading to a negative spectrum for H); and

(*ii*) in the indefinite metric case (leading to a positive spectrum for H).

15.19. In the holomorphic representation for a positive metric Fermi oscillator, where the vacuum $|0\rangle$ is represented by the function 1 and the filled state $|1\rangle \equiv a^*|0\rangle$ by a^* (see §20.3.2.a), one defines the kernel $A(a^*, a)$ of an operator $\hat{A}$ by means of

$$A(a^*, a) = \sum_{\substack{n=0,1\\ m=0,1}} \langle n \,|\, \hat{A} \,|\, m\rangle \,(a^*)^n\, a^m.$$

Show that

$$(Af)(a^*) = \int A(a^*, \alpha)\, f(\alpha^*)\, e^{-\alpha^* \alpha}\, d\alpha^*\, d\alpha,$$

$$(A_1 A_2)(a^*, a) = \int A_1(a^*, \alpha)\, A_2(\alpha^*, a)\, e^{-\alpha^* \alpha}\, d\alpha^*\, d\alpha,$$

$$A(a^*, a) = e^{a^* a}\, K(a^*, a),$$

where K is the normal symbol of $\hat{A}$.

15.20. Derive the explicit form of the path integral for a Fermi oscillator with indefinite metric. (Hint: See §20.3.2.c and previous exercise.)

15.21. Let q, p be a bosonic dynamical system with Hamiltonian H. The Weyl correspondence rule treats p and q completely symmetrically, *i.e.*, associates with $q^\alpha p^\beta$ the operator $[1/(\alpha + \beta)!][\sum x^{i_1} x^{i_2} \ldots x^{i_{\alpha+\beta}}]$ where α of the x's are $\hat{q}$'s, β of them are $\hat{p}$'s, and the sum extends over all permutations of $\alpha + \beta$ elements.

(a) Show that the Weyl symbol $U(q, p, T)$ of the evolution operator is equal to

$$\int \exp\left(i\bar{S}[q(t), p(t)]\right) \prod_t \frac{dq\, dp}{\pi}, \tag{i}$$

where $\bar{S}$ is the action (7.8c) and where the sum is extended over all phase space trajectories such that

$$q(t_i) + q(t_f) = 2q, \qquad p(t_i) + p(t_f) = 2p$$

[boundary conditions (7.8a)–(7.8b)]. [Hint: The derivation of (i) proceeds as the derivation of (15.78). The composition kernel adapted to the Weyl symbols is given by

$$\hat{g}_1\hat{g}_2 \to \int W(z^1, z^2, z)\, g_1(z^1)\, g_2(z^2)\, dz^1\, dz^2,$$

$$W(z^1, z^2, z) = \frac{1}{(\pi)^2} \exp\left[2i(z^1 \circ z + z \circ z^2 + z^2 \circ z^1)\right],$$

where $\circ$ is defined by (7.10a).]

(b) By using the equation

$$\langle q''|\hat{M}|q'\rangle = (2\pi)^{-1} \int M(\tfrac{1}{2}q' + \tfrac{1}{2}q'', p)\, \exp[-ip(q' - q'')]\, dp,$$

relating the Weyl symbol M to the kernel $\langle q''|\hat{M}|q'\rangle$ of $\hat{M}$ in the coordinate representation, derive the path integral for $\langle q''|\hat{M}|q'\rangle$ in terms of the Weyl symbol of H (rather than its p-q symbol). Show that the coordinates must be evaluated at the midpoint, *i.e.*, $H[q(t), p(t)] \to H[(q_k + q_{k+1})/2, p_k]$. Check that the extra factor $\prod_t (2^{-1})$ in the measure comes from the change of integration variables $\sigma_k \to q_k$, with $\sigma_k = \frac{1}{2}(q_{k+1} + q_k)$ [see (15.75a)].

15.22. Compute the Weyl symbol of the evolution operator for a free particle, $H = (1/2m)p^2$, by means of the path integral of Exercise 15.21.

15.23. Same question for a harmonic oscillator,

$$H = \tfrac{1}{2}(p^2 + q^2).$$

[Hint: The integral is Gaussian and can be evaluated exactly. The determinant of the quadratic part is the same as in the fermionic case but now comes "downstairs," so the singularity at half a period remains. This singularity corresponds to the fact that the Weyl reordering of the well-defined evolution operator at half a period leads to divergent series, *i.e.*, that the Weyl symbol of $U(t)$ does not exist for T equal to half a period.]

15.24. Derive the path integral (15.91) directly from (15.78) by making the change of variables $\sqrt{2}\,\theta_1 = C + i\bar{P}$, $\sqrt{2}\,\theta_2 = C - i\bar{P}$, $\sqrt{2}\,\theta_3 = \bar{C} + iP$, $\sqrt{2}\,\theta_4 = \bar{C} - iP$.

15.25. Let $(A, [\ \ ,\ \ ])$ be a graded Lie algebra. Define a new parity $\bar{\varepsilon}(x) = \varepsilon(x) + 1$ (shift ε by one unit). Show that the bracket fulfills, with respect to the new parity, the algebraic properties (15.101b), (15.101c), and (15.102) of the antibracket.

15.26. Let B be a p-vector field on a manifold M. Prove that the formulas

$$D_B F = dF \,\lrcorner\, B \qquad \text{(contraction of } dF \text{ with } B) \qquad \text{(i)}$$

$$D_B X = \mathcal{L}_X B \qquad \text{(ii)}$$

where F and X are, respectively, 0-vector (scalar) and 1-vector and define a unique derivation in the algebra of q-vector fields on M. (Hint: Proceed as in Exercise 8.10.) Show that $\varepsilon(D_B) = \varepsilon_B + 1$. Verify that the (graded) commutator of two derivations of the type (i),(ii) is a derivation of the same type, with

$$[D_A, D_B] = D_{(A,B)}.$$

Show that the Jacobi identity (15.102) for the antibracket is equivalent to the Jacobi identity for the commutator of the corresponding derivations.

15.27. Write down the components of the Schouten bracket (A, B) in terms of those of A and B. Check explicitly that (A, B) transforms as a $(p+q-1)$-vector.

15.28. Prove formula (15.118).

15.29. Show that the matrix of the Poisson brackets of the coordinates defines a 2-vector σ that has a vanishing Schouten bracket with itself, $(\sigma, \sigma) = 0$.

15.30. Show that any canonical transformation in the antibracket possesses a fermionic generating function K, which can be taken to be a function of ϕ^* and ϕ' if the transformation $\phi = \phi(\phi', \phi'^*)$ can be inverted for ϕ' [formula (15.121)]. [Hint: Introduce the closed, odd symplectic 2-form associated with the antibracket and proceed as in the analysis of standard canonical transformations ($p\, dq = p'\, dq' + dS$).]

CHAPTER SIXTEEN

PATH INTEGRAL FOR CONSTRAINED SYSTEMS

When one writes down the transition amplitude for a system with gauge freedom as the naive sum over all histories of the exponential of the gauge-invariant action, one faces the problem that the path integral diverges because of the integration over the pure gauge degrees of freedom.

There exist two methods for solving this difficulty. The philosophies behind each of these methods are radically different.

The first method replaces the path integral over all the variables by a path integral over the gauge-invariant degrees of freedom only. The elimination of the pure gauge modes can be done by reformulating the theory in the reduced phase space, which, in practice, amounts to imposing canonical gauge conditions.

The idea of the second method is not to eliminate the gauge degrees of freedom, which are usually necessary for preserving manifest covariance and locality. Rather, one sums over all of them, as well as over the ghosts. To get a nondivergent path integral, one substitutes for the gauge-invariant action one without gauge invariance. The principle of gauge invariance under the gauge transformations in the connected component of the identity is replaced by the equivalent principle of BRST invariance. The action that appears in the path integral is one of the BRST-invariant extensions of the gauge-invariant action analyzed in Chapter 11.

The action of the BRST path integral is "gauge fixed" in the sense that it does not possess any gauge invariance. Yet, this does not mean that the integration range in the sum-over-paths is restricted to a definite gauge slice, except for particular choices of the gauge-fixing fermions yielding a delta function of the gauge conditions. For generic gauge-fixing fermions, however, all the trajectories consistent with the boundary conditions are included in the path integral so that the sum involves gauge-related histories. The gauge degrees of freedom are cancelled by the ghosts through the principle of BRST invariance, rather than by hand through the imposition of a canonical gauge.

We develop in this chapter both approaches to the path integral for gauge systems after a brief discussion of the path integral for systems with second-class constraints.

16.1. PATH INTEGRAL FOR SECOND-CLASS CONSTRAINTS

16.1.1. Derivation of the Path Integral

The derivation of the path integral for second-class constrained systems is straightforward. Indeed, the relevant phase space is then the constraint surface

$$\chi_\alpha(x^\mu) = 0 \tag{16.1}$$

equipped with the Dirac bracket. The x^μ stand here for all the phase space variables and have c-number Poisson brackets. The path integral is thus a sum over all paths obeying (16.1), and the integration measure is the formal product over time of the Liouville measure associated with the Dirac bracket (§20.2.5 and Exercise 15.5).

If one denotes by y^i a complete set of independent coordinates on the surface of the second-class constraints,

$$\chi_\alpha = 0 \Leftrightarrow x^\mu = x^\mu(y^i), \tag{16.2}$$

one obtains

$$\text{Path integral} = \int [Dy^i] \prod_t (\text{sdet}\, \sigma_{ij})^{1/2} \exp iS[y^i(t)], \tag{16.3a}$$

where $S[y^i(t)]$ is the action in terms of the y's, with the surface term appropriate to the boundary conditions at hand, and where $\sigma_{ij}(y)$ is the symplectic structure on $\chi_\alpha = 0$, *i.e.*, the matrix inverse to $[y^i, y^j]^*$,

$$\sigma_{ij}[y^i, y^j]^* = \delta_i{}^j. \tag{16.3b}$$

The path integral (16.3a) can be rewritten in terms of the original constrained variables x^μ by means of the identity

$$\int \delta(\chi_\alpha)\, f(x^\mu)\, (\text{sdet}\, C_{\alpha\beta})^{1/2}\, dx^\mu = \int f(y^i)\, (\text{sdet}\, \sigma_{ij})^{1/2}\, dy^i \tag{16.4a}$$

with

$$C_{\alpha\beta} = [\chi_\alpha, \chi_\beta]. \tag{16.4b}$$

In (16.4a), $f(y^i)$ is the function induced on (16.1) by $f(x^\mu)$. The proof of (16.4) is direct and based on the equation

$$dx^\mu = (\text{sdet}\, \sigma_{ij})^{1/2}\, (\text{sdet}\, C_{\alpha\beta})^{-1/2}\, dy^i\, d\chi_\alpha,$$

relating the volume elements in the canonical coordinate system (x^μ) and in the coordinate system (y^i, χ_α) [see (2.22)].

Thus, one has

Path integral

$$= \int [Dx^\mu] \prod_{t,\alpha} \delta(\chi_\alpha) \prod_t (\text{sdet}\, [\chi_\alpha, \chi_\beta])^{1/2} \exp iS[x^\mu(t)]. \tag{16.5}$$

Because of the δ-functions $\delta(\chi_\alpha)$, the ambiguity in $S[x^\mu(t)]$, $S[x^\mu(t)] \rightarrow S[x^\mu(t)] + \int dt\, \mu^\alpha(x^\mu)\, \chi_\alpha(x^\mu)$, is manifestly irrelevant. We assume that a definite choice has been made. By using the Fourier representation of the δ-function, one can transform (16.5) as

$$\text{Path integral} = \int [Dx^\mu]\, [Du] \prod_t (\text{sdet}\, [\chi_\alpha, \chi_\beta])^{1/2}$$

$$\times \exp iS[x^\mu(t), u^\alpha(t)] \tag{16.6a}$$

$$S[x^\mu(t), u^\alpha(t)] = S[x^\mu(t)] - \int u^\alpha \chi_\alpha. \tag{16.6b}$$

The ambiguity in the action amounts then to making the change of integration variables $x^\mu \rightarrow x^\mu$, $u^\alpha \rightarrow u^\alpha - \mu^\alpha(x^\mu)$, whose Jacobian is unity.

16.1.2. Difficulties

The Dirac bracket difficulties analyzed in the operator method are not solved by the path integral. This becomes evident as soon as one tries to define precisely (16.6), which looks deceivingly simple.

For instance, to view (16.6) as the kernel of the evolution operator in the coordinate representation, one first needs to find a complete set of coordinates, *i.e.*, a complete set of commuting variables in the Dirac bracket. The trajectories are then required to take given values of the coordinates at the endpoints. The task of finding a complete set of commuting variables in the Dirac bracket may be untractable.

16.2. REDUCED PHASE SPACE PATH INTEGRAL

16.2.1. Derivation of the Path Integral

We now turn to the problem of the path integral for first-class constraints. For notational simplicity, we assume that there is no second-class constraint. If there were second-class constraints, one would need to include $\prod_t \delta(\chi_\alpha)\,(\text{sdet}\,[\chi_\alpha, \chi_\beta])^{1/2}$ in the formulas below and to replace the Poisson bracket by the Dirac bracket when necessary. The coordinates of the phase space are denoted by z^A and include those extra variables that may be necessary for replacing some of the second-class constraints by equivalent first-class ones (§1.4.3).

Let $z^{*\alpha}(z^A)$ be a complete set of independent observables, *i.e.*, a complete set of gauge-invariant functions,

$$[A, G_a] \approx 0 \Rightarrow A \approx A(z^{*\alpha}). \tag{16.7a}$$

Let $\sigma^{\alpha\beta}$ be the matrix of their Poisson brackets, $\sigma^{\alpha\beta} = [z^{*\alpha}, z^{*\beta}]$. It is an invertible matrix, which depends on $z^{*\alpha}$. The symplectic two-form in the reduced phase space is the inverse $\sigma_{\alpha\beta}$ of $\sigma^{\alpha\beta}$. As analyzed in Chapter 2, the reduced phase space is obtained by factoring out the null surfaces of the 2-form induced on the constraint surface by the phase space symplectic 2-form. The 2-form $\sigma_{\alpha\beta}$ is closed so that one has $\sigma = da$, where a_α is a symplectic potential for $\sigma_{\alpha\beta}$. Because $a_\alpha(z^{*\beta})$ and the symplectic potential $a_A(z^B)$ for the phase space symplectic 2-form σ_{AB} induce the same 2-form on the constraint surface (Appendix 2.A), one has

$$dz^{*\alpha}\, a_\alpha \approx dz^A\, a_A + dM. \tag{16.7b}$$

The reduced phase space quantization method associates operators only to the gauge-invariant functions. It leads accordingly to a path integral that is a sum over trajectories in the reduced phase space,

$$\text{Path integral} = \int [Dz^{*\alpha}] \prod_t (\text{sdet}\,\sigma_{\alpha\beta})^{1/2} \exp iS[z^{*\alpha}(t)]. \tag{16.8a}$$

Here the action $S[z^{*\alpha}(t)]$ is the action induced in the reduced phase space from the original action $S[z^A(t)]$,

$$S[z^{*\alpha}(t)] = \int [a_\alpha(z^*)\,\dot{z}^{*\alpha} - H(z^{*\alpha})]\,dt, \tag{16.8b}$$

where $H(z^{*\alpha})$ is the gauge-invariant Hamiltonian. The action (16.8b) is manifestly gauge invariant in the sense that it involves only $z^{*\alpha}$. Because of (16.7b), it weakly differs from $S[z^A(t)]$ at most by a boundary term that may not be invariant under the transformations generated by G_a (see Exercise 5.16 in this context).

The reduced phase space path integral depends on the reduced phase space variables—and not on the variables of the original phase space. For instance, if the $z^{*\alpha}$ split in conjugate pairs $(q^{*\alpha'}, p^*_{\alpha'})$, the kernel of the evolution operator in the coordinate representation depends on $q^{*\alpha'}$ at the initial and final times.

16.2.2. Faddeev Formula

The path integral (16.8) is gauge invariant by construction and applies provided the reduced phase space exists. It possesses the important feature that it does not require any gauge-fixing condition. In particular, it does not suffer from the Gribov obstruction (if any). However, the path integral (16.8) is not very convenient in that a complete set of gauge-invariant functions is necessary to formulate it, and this may not be possible in practice.

If there exist good canonical gauge conditions, say,

$$\chi_a = 0, \tag{16.9}$$

one can identify the reduced phase space with the slice defined by (16.9) on the constraint surface $G_a = 0$. The system of constraints $\chi_\rho \equiv (G_a, \chi_a)$ is second class. The matrix $[\chi_\rho, \chi_\sigma]$ reads

$$[\chi_\rho, \chi_\sigma] \approx \begin{pmatrix} 0 & [G_a, \chi_b] \\ [\chi_a, G_b] & [\chi_a, \chi_b] \end{pmatrix}, \tag{16.10a}$$

and thus

$$\text{sdet}\,[\chi_\rho, \chi_\sigma] = (\text{sdet}\,[G_a, \chi_b])^2. \tag{16.10b}$$

Furthermore, the Dirac brackets of gauge-invariant functions in the gauge (16.9) coincide with their Poisson brackets (Exercise 1.18). Accordingly, one can rewrite (16.8a) as in (16.5),

$$\text{Path integral} = \int [Dz^A] \prod_{t,a} \delta(\chi_a)\, \delta(G_a) \times \prod_t (\text{sdet}\, [G_a, \chi_b]) \exp iS'[z^A(t)]. \quad (16.11)$$

Here, $S'[z^A(t)]$ contains the appropriate boundary term M appearing in (16.7b), which is necessary to insure that $S'[z^A(t)] = S[z^{*\alpha}(t)]$ on the surface $\chi_a = 0$ and $G_a = 0$.* The boundary conditions on the observables $z^{*\alpha}$ at the endpoints implicitly determine boundary conditions on the z^{*A}'s through $z^{*\alpha} = z^{*\alpha}(z^A)$ and $G_a = 0, \chi_a = 0$.

The formula (16.11) can be further transformed as follows,

$$\text{Path integral} = \int [Dz^A]\, [D\lambda^a]\, [DC^a]\, [D\bar{C}^a] \prod_{t,a} \delta(\chi_a) \times \exp i\Big(S'[z^A(t)] - \int \lambda^a G_a \, dt - \int \bar{C}^b \delta_C \chi_b \, dt\Big), \quad (16.12)$$

where $\bar{C}^a, C^a$ are independent variables of statistics opposite to that of G_a and $\delta_C \chi_b$ stands for the variation of the gauge conditions $\chi_b = 0$ under a gauge transformation of "infinitesimal parameter" C^a. The integration over the multipliers λ^a in (16.12) brings in $\delta(G_a)$, while the integration over the ghosts brings in the superdeterminant of $[G_a, \chi_b]$ (§20.2.4).

16.2.3. Gauge Independence of Path Integral for a Parametrized System Illustrated. Equivalence of the Gauges $t = \tau$ and $t = 0$

We illustrate in this subsection the general analysis of the reduced phase space path integral and its expressions in terms of gauge fixings, the key point being that different choices of gauge conditions are mere rewritings of one and the same reduced phase space amplitude.

We will deal with the parametrized free nonrelativistic particle with constraint

$$\mathcal{H} \equiv p_t + \frac{p^2}{2m} = 0 \quad (16.13a)$$

* If the z^A's are noncanonical in the Poisson bracket, the naive measure $[Dz^A]$ in (16.11) should be replaced by $[Dz^A] \prod_t (\text{sdet}\sigma_{AB})^{1/2}$.

and action

$$S[q, p, t, p_t, N] = \int (p\dot{q} + p_t\dot{t} - N\mathcal{H})\, d\tau. \tag{16.13b}$$

16.2.3a. Reduced Phase Space Transition Amplitude as a Reduced Phase Space Path Integral

For any value of the constant c, a complete set of observables is given by

$$q_c^* = q - \frac{p}{m}(t - c), \qquad p^* = p. \tag{16.14a}$$

These are constants of the motion, which coincide with q and p in the canonical gauge $t = c$. The Heisenberg operators $q_c^*(\tau)$ are independent of τ. For different values of c, they are related by a unitary transformation

$$q_{c_1}^* = \exp\Big[i\,\frac{p^2}{2m}(c_1 - c_2)\Big]\, q_{c_2}^*\, \exp\Big[-i\,\frac{p^2}{2m}(c_1 - c_2)\Big], \tag{16.14b}$$

and the corresponding eigenstates are related by

$$|q_{c_1}^*, \tau_1\rangle = e^{i(p^2/2m)(c_1 - c_2)}\, |q_{c_2}^*, \tau_2\rangle. \tag{16.14c}$$

The mixed transition amplitude

$$(q_{c_2}^* \tau_2 \,|\, q_{c_1}^* \tau_1) = \Big(\frac{m}{2\pi i(c_2 - c_1)}\Big)^{1/2} \exp\Big[-\frac{m(q_{c_2}^* - q_{c_1}^*)^2}{2\,i(c_2 - c_1)}\Big] \tag{16.15a}$$

can be written as a sum over reduced phase paths $q_{c=0}^*(\tau) \equiv q^*(\tau)$, $p_{c=0}^*(\tau) \equiv p^*(\tau)$ of the exponential of the reduced phase space action $S_R[q^*(\tau), p^*(\tau)]$ (see Exercise 16.5). These paths obey the boundary conditions

$$\Big[q^* + \frac{p^*}{m}c_1\Big](\tau_1) = q_{c_1}^*, \tag{16.15b}$$

$$\Big[q^* + \frac{p^*}{m}c_2\Big](\tau_2) = q_{c_2}^*, \tag{16.15c}$$

obtained by using the relation between $q_{c_1}^*$, $q_{c_2}^*$, and q^*, p^*. Since the reduced Hamiltonian $H(q^*, p^*)$ vanishes, the reduced action differs from $\int p^* \dot{q}^*\, d\tau$ by a boundary term adapted to the boundary conditions (16.15b)–(16.15c),

$$S_R[q^*(\tau), p^*(\tau)] = \int p^* \dot{q}^*\, d\tau + \Big[\frac{p^2}{2m}\Big(c_1 + \frac{\tau - \tau_1}{\tau_2 - \tau_1}(c_2 - c_1)\Big)\Big]_{\tau_1}^{\tau_2}. \tag{16.15d}$$

16.2.3b. Canonical Gauge Conditions

The reduced phase space path integral

$$(q^*_{c_2}\tau_2 \,|\, q^*_{c_1}\tau_1) = \int [Dp^*\, Dq^*] \exp iS_R[q^*(\tau), p^*(\tau)] \tag{16.16a}$$

can be written in terms of q, p, t, p_t and gauge conditions. To that end, one observes that the action (16.13b) weakly differs from the reduced action (16.15d) by a surface term,

$$S_R \approx S + \left[\frac{p^2}{2m}\left(c_1 + \frac{\tau - \tau_1}{\tau_2 - \tau_1}(c_2 - c_1) - t\right)\right]_{\tau_1}^{\tau_2}. \tag{16.16b}$$

We shall consider gauge conditions of the form

$$t - f(\tau) = 0 \tag{16.16c}$$

for which the determinant $[t - f(\tau), p_t + H_0]$ is unity. The reduced phase space path integral (16.16a) is, according to our discussion of §16.2.2, equal to

$$(q^*_{c_2}\tau_2 \,|\, q^*_{c_1}\tau_1) = \int [Dp\, Dq\, Dt\, Dp_t] \prod_t \delta\big(t - f(\tau)\big)\, \delta(p_t + H) \exp iS', \tag{16.16d}$$

$$S' = \int \left(p\dot{q} - \frac{p^2}{2m}\frac{df}{d\tau}\right) d\tau + \left[\frac{p^2}{2m}\left(c_1 + \frac{\tau - \tau_1}{\tau_2 - \tau_1}(c_2 - c_1) - f(\tau)\right)\right]_{\tau_1}^{\tau_2}. \tag{16.16e}$$

The paths are subject to the boundary conditions

$$\left(q - \frac{p}{m}(f(\tau) - c_1)\right)(\tau_1) = q^*_{c_1}, \tag{16.16f}$$

$$\left(q - \frac{p}{m}(f(\tau) - c_2)\right)(\tau_2) = q^*_{c_2}. \tag{16.16g}$$

16.2.3c. Gauge $t = 0$

In the gauge $t = 0$, the Hamiltonian vanishes. One finds, nevertheless, from (16.16d)–(16.16g) that the path integral can correctly be described in that gauge as

$$\left(\frac{m}{2\pi i(c_2 - c_1)}\right)^{1/2} \exp - \left[\frac{m(q^*_{c_2} - q^*_{c_1})}{2\,i(c_2 - c_1)}\right]^{1/2} = (q^*_{c_2}\tau_2 \,|\, q^*_{c_1}\tau_1) = \int [Dq\, Dp] \exp\left\{i \int p\,\dot{q}\, d\tau + \left[\frac{p^2}{2m}c\right]_{c_1}^{c_2}\right\} \tag{16.17a}$$

with the boundary conditions

$$q(\tau_1) + \frac{p(\tau_1)}{m}\, c_1 = q^*_{c_1}, \qquad q(\tau_2) + \frac{p(\tau_2)}{m}\, c_2 = q^*_{c_2}. \tag{16.17b}$$

One sees that there is nothing wrong with "the time not flowing."

16.2.3d. Gauge $t \propto \tau$

One may also let the time flow as it is ordinarily done. Although $t = \tau$ is permissible, it is a bit simpler here to adjust the endpoints so that $t(\tau_1) = c_1$ and $t(\tau_2) = c_2$. Thus, we set

$$t = c_1 + \frac{\tau - \tau_1}{\tau_2 - \tau_1}\,(c_2 - c_1). \tag{16.17c}$$

The advantage of this choice is that it makes the surface term in (16.16e) vanish and simplifies the form of the boundary conditions. One gets this time

$$\left(\frac{m}{2\pi i(c_2 - c_1)}\right)^{1/2} \exp - \left[\frac{m(q^*_{c_2} - q^*_{c_1})}{2\,i\,(c_2 - c_1)}\right]^{1/2} = (q^*_{c_2}\tau_2 \,|\, q^*_{c_1}\tau_1)$$
$$= \int [Dq\, Dp]\, \exp i \int \left(p\,\dot{q} - \frac{p^2}{2m}\, \frac{c_2 - c_1}{\tau_2 - \tau_1}\right) d\tau \tag{16.17d}$$

with

$$q^*(\tau_1) = q^*_{c_1}, \qquad q(\tau_2) = q^*_{c_2}. \tag{16.17e}$$

If $c_2 - c_1 \neq 0$, one can further transform the right-hand side of (16.17d) by rewriting the integral as an integral over t. This yields the usual form of the transition amplitude derived in §15.1.1,

$$\int [Dq\, Dp]\, \exp i \int_{t_1}^{t_2} \left(p\frac{dq}{dt} - \frac{p^2}{2m}\right) dt. \tag{16.17f}$$

It is thus quite possible to derive the standard transition amplitudes of the theory by treating the transformations generated by $p_t + H_0 \approx 0$ as ordinary gauge symmetries. This point of view, which treats all the first-class constraints on the same footing, possesses the further advantage that it is not necessary to undig a physical time variable in order to compute physical amplitudes.

16.3. BRST PATH INTEGRAL IN THE FOCK REPRESENTATION

16.3.1. Construction

We now turn to the BRST path integral and first consider the Dirac–Fock quantization in which only half of the constraints are imposed on the physical states. This approach is free from the scalar product difficulties present in the Schrödinger quantization and is particularly useful in field theory, where one associates particles not only to the gauge-invariant degrees of freedom but also to the pure gauge variables and to the ghosts.

The corresponding path integral employs the holomorphic representation in a Hilbert space with indefinite metric. This path integral yields the holomorphic kernel of the evolution operator $\exp[-i(H + [K, \Omega])(t_f - t_i)]$ as a sum over paths in the extended phase space. When written in terms of the q's and the p's, the path integral takes the same form as the path integral in the Schrödinger representation. However, the paths are subject to different boundary conditions (namely, a^* given at t_f and a given at t_i), and, furthermore, the q's and the p's associated with the negative metric bosonic oscillators take imaginary values.

Different choices of the gauge-fixing fermion K define different evolution operators and, hence, different kernels. However, the physical amplitudes are unchanged. This is a consequence of Theorem 14.6 and of formula (14.13). In particular, as the vacuum is BRST invariant, the vacuum expectation value of the evolution operator, obtained by setting $a^* = a^*(t_f) = 0$ and $a = a(t_i) = 0$ in the path integral, does not depend on K.

More generally, the vacuum quantum average $\langle A \rangle_K$ of the BRST-invariant operators $A(z)$ taken at time t does not depend on K,

$$\langle A(t)\rangle_K = \langle A(t)\rangle_{K'}, \quad \text{if} \quad sA \equiv [A, \Omega] = 0. \tag{16.18a}$$

As one also finds

$$\langle\, [B, \Omega](t)\,\rangle_K = 0 \tag{16.18b}$$

for any BRST-exact operator, the vacuum quantum averages (or the quantum averages between any pair of initial and final BRST-invariant states) are defined in cohomology.

16.3.2. Example

The independence of the transition amplitude upon the choice of K can be illustrated in the simple case of a system with two conjugate pairs $(q^1, p_1), (q^2, p_2)$ and constraints $p_1 = 0, p_2 = 0$ quantized in the Fock representation (§13.4.1). This example is pure gauge and has a single physical state, the vacuum. If one takes for K

$$K = \alpha(b^* \bar{c} + \bar{c}^* b), \tag{16.19a}$$

one finds

$$[K, \Omega] = \alpha N, \tag{16.19b}$$

where N is the number operator (14.62). The kernel of the evolution operator is

$$\int [Da\, Da^*\, Db\, Db^*\, Dc\, Dc^*\, D\bar{c}\, D\bar{c}^*] \exp iS_K = \exp\{N \exp -i\alpha(t_f - t_i)\} \tag{16.19c}$$

and possesses an explicit α-dependence. However, the vacuum expectation value of the evolution operator, obtained by setting $a = b = a^* = b^* = c = \bar{c} = c^* = \bar{c}^*$ equal to zero, is equal to one for any choice of α.

16.4. FRADKIN–VILKOVISKY THEOREM—WARD IDENTITIES

16.4.1. Theorem

The proof of independence of the path integral on the choice of gauge-fixing fermion is straightforward when one relies on the operator formalism. It turns out that one can provide an alternative proof based directly on the path integral. This proof, even though formal, is of interest because it also applies to other formulations of the path integral not relying on the Fock representation and for which the operator formalism is more fragile because of scalar product questions.

Consider again the path integral for the evolution operator with Hamiltonian $H + [K, \Omega]$, *i.e.*, the quantum average $\langle 1 \rangle_K$. We assume that this path integral stands for the expectation value between physical states, *i.e.*, that the boundary conditions fulfilled by the trajectories select definite physical states at the endpoints. The boundary conditions are then BRST invariant and the boundary term in the variation of the gauge-fixed action

$$S_K = \int (\dot{z}^\Delta\, a_\Delta(z) - H - [K, \Omega])\, dt, \tag{16.20}$$

evaluated on a trajectory that obeys the boundary conditions, is zero. Here, the $z^\Delta(t)$ stand for all the variables of the extended phase space including the ghosts.

Theorem 16.1. *("Fradkin–Vilkovisky theorem") The quantum average* $\langle 1 \rangle_K$ *does not depend on* K,

$$\langle 1 \rangle_K = \langle 1 \rangle_{K'}. \tag{16.21}$$

Proof. In the path integral

$$\langle 1 \rangle_K = \int [Dz^\Delta] \exp iS_K, \tag{16.22a}$$

let us make the change of integration variables

$$z'^\Delta(t) = z^\Delta(t) + [z^\Delta, \Omega]\,(t)\chi \tag{16.22b}$$

with

$$\chi = -i \int_{t_i}^{t_f} dt\,(K' - K). \tag{16.22c}$$

This change of variables maps the histories over which one integrates in (16.22a) on histories obeying the same boundary conditions. Since Theorem (16.1) is equivalent to $\delta\langle 1 \rangle_K / \delta K = 0$, we can assume $K' - K$ to be infinitesimal.

The infinitesimal parameter χ is time-independent. But because it involves the fields and is written outside the Poisson bracket, (16.22b) is not a canonical transformation in the infinite-dimensional space obtained by taking the product over time of the extended phase space. This means that (16.22b) may change the expression of the integration measure.

This is actually the only thing it changes because the boundary conditions and the action are BRST invariant. [In the proof of the BRST invariance of the action, the fact that the parameter χ involves the histories plays no role. Only $d\chi/dt = 0$ is needed.]

Let us compute the change in the measure $[Dz^\Delta]$, *i.e.*, the superdeterminant of the change of variables (16.22b). One finds

$$\begin{aligned}\frac{\delta z'^\Delta(t)}{\delta z^\Lambda(t')} &= \delta^\Delta_\Lambda\,\delta(t-t') + \frac{\partial[z^\Delta, \Omega]}{\partial z^\Lambda}(t)\,\delta(t-t')\chi \\ &\quad - i\,[z^\Delta, \Omega](t)\,\frac{\partial(K'-K)}{\partial z^\Lambda}(t')\end{aligned}$$

and thus

$$\begin{aligned}\operatorname{sdet}\frac{\delta z'^\Delta(t)}{\delta z^\Lambda(t)} &= 1 + \operatorname{str}\left[\frac{\delta z'^\Delta(t)}{\delta z^\Lambda(t)} - \delta^\Delta_\Lambda\,\delta(t-t')\right] \\ &= i\int_{t_1}^{t_2} [\Omega, K'-K]\,dt\end{aligned}$$

because $\operatorname{str} \partial[z^\Delta, \Omega]/\partial z^\Lambda = 0$ [the BRST transformation is a canonical transformation in the extended phase space (see Exercise 16.6)].

Hence, the transformed measure reads

$$[Dz^{\Delta'}] = [Dz^\Delta] \exp i \int_{t_1}^{t_2} [\Omega, K' - K]\, dt. \tag{16.23}$$

If one substitutes (16.23) back into $\langle 1 \rangle_K$, one gets the desired result

$$\langle 1 \rangle_K = \langle 1 \rangle_{K'},$$

which proves the theorem.

16.4.2. Quantum Averages and BRST Cohomological Classes

The properties (16.18a)–(16.18b) of the quantum averages of BRST-invariant operators can also be given a path integral proof that does not rely directly on the operator formalism.

Theorem 16.2. *The quantum average of any BRST-invariant operator $A(z^\Delta)$ evaluated at time t between physical states does not depend on K,*

$$\langle A(t) \rangle_K = \langle A(t) \rangle_{K'} \quad \textit{if} \quad sA \equiv [A, \Omega] = 0. \tag{16.24}$$

Proof. Follow the same steps as for Theorem 16.1, *i.e.*, perform the same change of integration variables in the path integral and use the further property that A is BRST invariant.

Theorem 16.3. *The quantum average of a BRST-exact operator $B(z^\Delta) = [F, \Omega]$ evaluated at time t between physical states vanishes,*

$$\langle\, [F, \Omega](t) \,\rangle \equiv \langle sF(t) \,\rangle = 0. \tag{16.25}$$

Proof. This theorem is a consequence of the previous one if one rewrites $0 = \langle 1 \rangle_{K'} - \langle 1 \rangle_K$ as the expectation value of the null operator $1 - \exp\left[-i \int dt [K - K', \Omega]\right]$. One can also derive (16.25) more directly, by making the change of integration variables $z^\Delta \to z^\Delta + [z^\Delta, \Omega]\, \varepsilon$ in the path integral $\langle F \rangle_K = \int [Dz^\Delta]\, F \exp iS_K$. The parameter ε is now a constant. Both $[Dz^\Delta]$ and iS_K are invariant, but F need not be; hence, one gets $\langle sF \,\rangle = 0$, as required.

The path integral associates therefore the same, well-defined, quantum average to all the BRST-invariant functions $A(z^\Delta)$ of the extended phase space that belongs to the same BRST cohomological class.

16.4.3. Ward Identities

Although $\langle A \rangle_K$ does not depend on K when A is BRST-invariant, the quantum average of non-BRST-invariant functions depend, in general, on the choice of the gauge-fixing fermion K. In particular, the quantum averages $\langle z^{\Delta_1}(t_1)\, z^{\Delta_2}(t_2) \cdots z^{\Delta_n}(t_n) \rangle_K$ are K-dependent.

The BRST invariance of the action and of the in- and out-states imply, however, interesting relations among the quantum averages of non-BRST-invariant operators. These are known as the Ward identities.

These identities are straightforward generalizations of (16.25) in which the one-time function $F(t)$ is replaced by the multi-time functions $z^{\Delta_1}(t_1)\, z^{\Delta_2}(t_2) \cdots z^{\Delta_n}(t_n)$. They can be derived in the same manner as (16.25) was derived, but rather we shall follow different lines, which are useful in quantum field theory.

To that end, one introduces a classical source j_Δ for each canonical variable. One then considers the "generating functional"

$$Z_K[j] \equiv \langle 1 \rangle_{K,j} = \int [Dz^\Delta] \exp i \Big[S_K + \int dt\, j_\Delta\, z^\Delta \Big]. \tag{16.26}$$

The standard generating functional considered in field theory includes the normalizing factor $(\langle 1 \rangle_{j=0})^{-1}$ so that $Z(j = 0)$ is unity. However, the subsequent identities are insensitive to the normalization, and we have therefore omitted it. The change of integration variables

$$z^\Delta \to z^\Delta + [z^\Delta, \Omega]\, \varepsilon$$

in (16.26) yields the key relation

$$\int j_\Delta(t)\, \langle s z^\Delta(t) \rangle_{K,j}\, dt = 0. \tag{16.27}$$

The Ward identities are obtained by functionally differentiating (16.27) with respect to the sources and setting them equal to zero afterwards. For instance, one differentiation gives

$$\langle s z^\Delta(t) \rangle_{K,j=0} = 0 \tag{16.28a}$$

in agreement with Theorem 16.3. The subsequent Ward identity relates the quantum averages of $\langle z^{\Delta_1}(t_1)\, s z^{\Delta_2}(t_2) \rangle$ and $\langle s z^{\Delta_1}(t_1)\, z^{\Delta_2}(t_2) \rangle$, and reads explicitly

$$\langle z^{\Delta_1}(t_1)\, s z^{\Delta_2}(t_2) \rangle + (-)^{\varepsilon_{\Delta_2}} \langle s z^{\Delta_1}(t_1)\, z^{\Delta_2}(t_2) \rangle = 0, \tag{16.28b}$$

etc.

16.4.4. Zinn–Justin Equation

A useful reformulation of (16.27) is obtained by introducing further independent sources, one for each BRST variation sz^Δ. These sources are denoted by L_Δ and are quite similar to the antifields introduced in Chapters 17 and 18. The new generating functional is

$$Z_K[j,L] \equiv \langle 1\rangle_{K,j,L} = \int [Dz^\Delta] \exp i\Big[S_K + \int dt\,(j_\Delta z^\Delta + L_\Delta\, sz^\Delta\Big]. \tag{16.29}$$

The same change of integration variables $z^\Delta \to z^\Delta + sz^\Delta\,\varepsilon$ yields then

$$\int j_\Delta(t)\,\langle sz^\Delta(t)\rangle_{K,j,L}\,dt = 0 \tag{16.30a}$$

without term proportional to L_Δ because $s^2 = 0$.

The relation (16.30a) is equivalent to

$$\int j_\Delta(t)\,\frac{\delta Z_K[j,L]}{\delta L^*_\Delta(t)}\,dt = 0. \tag{16.30b}$$

If one now defines the effective action $\Gamma_K[z^\Delta_c, L^*_\Delta]$ as the Legendre transformation of $\ell n Z_K$ with respect to the sources j_Δ,

$$\Gamma_K[z^\Delta_c, L^*_\Delta] = \frac{1}{i}\ell n Z_K[j(z_c),L] - \int j_\Delta(t)\,z^\Delta_c(t), \tag{16.31a}$$

where the "classical fields" $z^\Delta_c(t)$ and the sources $j_\Delta(t)$ are related as

$$z^\Delta_c(t) = \frac{1}{i}\,\frac{\delta\,\ell n Z_K}{\delta j^\Delta(t)}, \qquad j^\Delta(t) = -\frac{\delta\,\Gamma_K}{\delta z^\Delta_c(t)}, \tag{16.31b}$$

one can rewrite the Ward identity (16.30a) in terms of the effective action as

$$\int \frac{\delta\,\Gamma_K}{\delta z^\Delta_c(t)}\,\frac{\delta\,\Gamma_K}{\delta L^*_\Delta(t)}\,dt = 0 \tag{16.32a}$$

i.e.,

$$(\Gamma_K, \Gamma_K) = 0, \tag{16.32b}$$

where the antibracket of the classical fields with the BRST sources L^*_Δ is defined in the same way as in Chapter 15,

$$\big(z^\Delta_c(t), L^*_\Gamma(t')\big) = \delta^\Delta_\Gamma\,\delta(t-t'). \tag{16.33}$$

The relation (16.32), known as the Zinn–Justin equation, plays a fundamental role in the study of the renormalization of gauge theories. It was historically through that reformulation of the Ward identities that the antibracket was discovered.

16.5. BRST PATH INTEGRAL IN THE SCHRÖDINGER REPRESENTATION

16.5.1. Projected Kernel of the Evolution Operator

We now turn to the path integral in the Schrödinger (or any equivalent) representation, where, in the absence of ghosts, all the constraints—and not just half of them—are imposed on the physical states.

The projected kernel of the evolution operator in the coordinate representation was defined in §13.3.6 as

$$U_0^P(q',q;t_2-t_1) = \sum_{\alpha,\beta} \psi_\alpha(q')\,(\psi_\alpha \,|\, \exp -iH_0(t_2-t_1) \,|\, \psi_\beta)\,\psi_\beta^*(q),$$

where the sum runs over a complete orthonormal set of solutions of the physical state condition $G_a|\psi\rangle = 0$. In order to be able to impose the ghost number zero condition, it was found necessary to include as a nonminimal sector the Lagrange multipliers λ^a and their momenta b_a, as well as the system of ghosts $(\rho^a, \bar{C}_a)$ associated with the constraint $b_a = 0$ (§14.5.4). This enabled one to construct BRST-closed states at ghost number zero and to rewrite the projected kernel of U_0 as the expectation value between the BRST-closed states $\psi_{q',\eta=0,b=0,\bar{C}=0}$ and $\psi_{q,\eta=0,b=0,\bar{C}=0}$ of the BRST-invariant extension $\exp\left[-i(H+[K,\Omega])(t_2-t_1)\right]$ of U_0 (§14.5.5),

$$U_0^P(q',q;t_2-t_1) = \left(\psi_{q',\eta=0,b=0,\bar{C}=0}, \exp\left[-i(H+[K,\Omega])(t_2-t_1)\right]\psi_{q,\eta=0,b=0,\bar{C}=0}\right). \tag{16.34}$$

The representation (16.34) enables one to rewrite the projected kernel as the BRST path integral

$$U_0^P(q',q;t_2-t_1) = \int [Dq\,Dp\,Db\,D\lambda\,D\eta\,D\mathcal{P}\,D\rho\,D\bar{C}]\,\exp iS_K \tag{16.35a}$$

$$S_K = \int (\dot{q}p + \dot{\eta}\mathcal{P} + \dot{\lambda}b + \dot{\rho}\bar{C} - H - [K,\Omega])\,dt, \tag{16.35b}$$

where the sum extends over all paths obeying

$$q(t_2) = q', \quad q(t_1) = q, \qquad \eta^a(t_2) = \eta^a(t_1) = 0 \tag{16.36a}$$

$$b_a(t_1) = b_a(t_2) = 0, \qquad \bar{C}_a(t_1) = \bar{C}_a(t_2) = 0. \tag{16.36b}$$

The other variables are unrestricted at the endpoints.

Even though the action S_K has no gauge invariance no matter what K is (the equations of motion following from S_K do not contain any arbitrariness), the path integral (16.35) may be ill defined. This is because the states $\psi_{q,\eta=0,b=0,\bar{C}=0}$ are not square-integrable. As we have seen in §14.5.5, a necessary condition for (16.34) to be nonambiguous is that K mixes the minimal and nonminimal sector. This forbids the choice $K = 0$ when the BRST-invariant Hamiltonian H involves only the minimal variables. This difficulty is characteristic of the Schrödinger representation with the boundary conditions (16.36). It does not arise in the Fock representation and, indeed, the choice $K = 0$ in (16.19) ($\alpha = 0$) leads to a path integral without ambiguity.

If one takes in (16.35) the gauge-fixing fermion K of the form

$$K = i\bar{C}_a\chi^a - \mathcal{P}_a\lambda^a, \tag{16.37}$$

one finds that the gauge-fixed action S_K involves the momenta b_a linearly, with coefficient $\dot{\lambda}^a + \chi^a$ (§11.3.2). The integration over b_a yields then $\prod_t \delta(\dot{\lambda}^a + \chi^a)$ and restricts the integration over the Lagrange multiplier to a sum over histories obeying the derivative gauge condition $\dot{\lambda}^a + \chi^a = 0$ This occurs even if the gauge algebra is open and justifies, thus, the use of derivative gauges in that case. For other choices of K, the integral is, however, not restricted to a definite gauge slice but involves, rather, an average over different gauges.

16.5.2. Semiclassical Approximation

The boundary conditions (16.36a)–(16.36b) in the path integral (16.35) are identical to the boundary conditions (11.33) investigated in §11.3.5. Consequently, the path integral (16.35a) is equal, in the classical limit, to

$$A \exp i W(q_f, t_f; q_i, t_i) \tag{16.38}$$

where $W(q_f, t_f; q_i, t_i)$ is the Hamilton principal function (Theorem 11.4) and A is a "slowly varying function."

The fact that the path integral (16.35) obeys the quantum constraint equation becomes, in the classical limit, the statement that W is a solution of the Hamilton–Jacobi equation (5.55a).

16.5.3. Composition Rule

The projected kernel of the evolution operator obeys the composition rule (13.33c)

$$U_0^P(q_3, q_1; t_3 - t_1) = \int U_0^P(q_3, q_2; t_3 - t_2)\, \mu\Big(q_2, \frac{1}{i}\frac{\partial}{\partial q^2}\Big)\, U_0^P(q_2, q_1; t_2 - t_1) \tag{16.39}$$

in which one sums only over physical ("transverse") degrees of freedom (the restriction to the physical degrees of freedom is enforced by μ). On the other hand, the kernel $U_K(q', \eta', b', \bar{C}'; q, \eta, b, \bar{C}; T)$ of the evolution operator $\exp -i(H + [K, \Omega])T$ in the $(q, \eta, b, \bar{C})$-representation obeys the composition rule (see §15.4.4)

$$\begin{aligned} &U_K(q_3, \eta = 0, b = 0, \bar{C} = 0; q_1, \eta = 0, b = 0, \bar{C} = 0, t_3 - t_1) \\ &\quad = \int dq_2\, db_2\, d\eta_2\, d\bar{C}_2\, U_K(q_3, \eta = 0, b = 0, \bar{C} = 0; q_2, \eta_2, b_2, \bar{C}_2; t_3 - t_2) \\ &\qquad \times U_K(q_2, \eta_2, b_2, \bar{C}_2; q_1, \eta = 0, b = 0, \bar{C} = 0; t_2 - t_1), \qquad (16.40) \end{aligned}$$

in which one sums over all intermediate states of the extended Hilbert space of the BRST method. Since $U_K(q', 0, 0, 0; q, 0, 0, 0; T)$ is equal by virtue of (15.89c) to the matrix element (16.34) and, hence, to the projected kernel, one can conclude that the sum over the physical degrees of freedom in (16.39) is equal to the sum (16.40) over all the degrees of freedom, including the pure gauge degrees of freedom and the ghosts. The advantage of (16.40) is that there is no measure μ to insert. The measure is accounted for by the extra variables $b, \eta, \bar{C}$.

16.5.4. Comparison with Reduced Phase Space Path Integral

We have seen in §13.3.6 that the reduced phase space kernel of a gauge-invariant operator and its projected kernel are in different representations. If we assume that the Hilbert space $\mathcal{H}^R$ of the reduced method is isomorphic to the space of solutions of the constraint equations $G_a|\psi\rangle = 0$ (§13.3.1), the reduced phase space kernel is equal to

$$\sum_{\alpha,\beta} \psi_\alpha(q^{*\prime})\, (\psi_\alpha \,|\, \exp -iH_0(t_2 - t_1) \,|\, \psi_\beta)\, \psi^*_\beta(q^*), \qquad (16.41)$$

where $\psi_\alpha(q^*)$ are the wave functions in the Hilbert space $\mathcal{H}^R$ of the reduced method. The projected kernel, on the other hand, is equal to

$$\sum_{\alpha,\beta} \psi_\alpha(q')\, (\psi_\alpha \,|\, \exp -iH_0(t_2 - t_1) \,|\, \psi_\beta)\, \psi^*_\beta(q), \qquad (16.42)$$

where the $\psi_\alpha(q)$ are the wave functions of the physical states in the space of the Dirac method.

In general, $\psi_\alpha(q^*)$ is different from $\psi_\alpha(q)$, so *the reduced phase space path integral and the BRST path integral differ by endpoint wave function factors.* This is not a problem because the kernels (16.41) and (16.42) must be used differently when computing physical amplitudes.

Both (16.41) and (16.42) contain the same physical matrix elements $(\psi_\alpha| \exp -iH_0(t_f - t_i)|\psi_\beta)$, so they are equivalent. The difference between (16.41) and (16.42) is of the same type as—and has no more physical meaning than—the difference between the kernels of the same operator in two different bases. Here, the bases $|q_i^*)$ and $|q_i\rangle$ refer in addition to two different Hilbert spaces.

Although generically different, the expressions (16.41) and (16.42) turn out to be equal in the case of constraints linear and homogeneous in the momenta (internal gauge symmetries). Indeed, the endpoint wave function factors are then equal if one takes the q^*'s to be functions of the q's and if one chooses appropriately the measure in the reduced configuration space.

For instance, for a system with two conjugate pairs $(q^1, p_1), (q^2, p_2)$ and constraint $p_1 = 0$, a complete set of physical states is given by the eigenfunctions of p_1 and q^2 with eigenvalue zero for p_1. Furthermore, one can take $q^* = q^2$. This leads to

$$\langle q^1, q^2 \,|\, p_1 = 0, \tilde{q}^2\rangle = \delta(q^2 - \tilde{q}^2) = (q^2|\tilde{q}^2).$$

A direct path integral proof of the equality of (16.41) and (16.42) for internal gauge symmetries can be given by using the Fradkin–Vilkovisky theorem. It goes as follows. In the BRST path integral (16.35), one can assume that the q's at the endpoints fulfill an arbitrary but fixed canonical gauge condition

$$\chi_a(q) = 0. \tag{16.43}$$

This is because one can move the q's within the same gauge orbit in configuration space without changing $\langle q \,|\, \psi_\beta\rangle$, *i.e.*, without changing the projected kernels of the operators in the q-representation,

$$\begin{aligned} \langle q \,|\, \psi_\beta\rangle &= \langle q|\, e^{i\lambda^a G_a}\, |\psi_\beta\rangle \qquad && (G_a|\psi_\beta\rangle = 0) \\ &= \langle q' \,|\, \psi_\beta\rangle, && q' = \text{gauge transformed of } q. \end{aligned}$$

If one now takes in the BRST path integral the gauge-fixing fermion K to be

$$K = \frac{i}{\varepsilon}\bar{C}_a\chi^a - \mathcal{P}_a\lambda^a, \tag{16.44a}$$

makes the change of integration variables

$$b_a \to \varepsilon b_a, \qquad \bar{C}_a \to \varepsilon \bar{C}_a \tag{16.44b}$$

whose Jacobian is unity, and lets ε go to zero, one precisely gets the reduced phase space path integral (16.12) in the canonical gauge $\chi_a = 0$, with exactly the same action in the exponential. Hence, (16.41) and (16.42) coincide. How to reach canonical gauges in the general case of arbitrary constraints is analyzed in Exercise 16.17.

16.5.5. BRST Path Integral for Generally Covariant Systems—Proper Time Gauge—Causal Propagator

When the Hamiltonian is zero, as normally happens for a generally covariant system, the projected kernel simplifies to

$$U_0^P(q',q) = \sum_\alpha \psi_\alpha(q')\,\psi_\alpha^*(q). \tag{16.45}$$

By folding (16.45) with all possible functions of q, one obtains an (over)-complete set of solutions of the constraint equations in the variable q'. Different choices of the folding function correspond to different boundary conditions for the solution of the constraint equation.

A useful form of the path integral is obtained by using the derivative gauge

$$\dot{\lambda} = 0. \tag{16.46a}$$

This gauge is implemented by taking K to be of the form

$$K = -\mathcal{P}\lambda. \tag{16.46b}$$

(We assume for simplicity that there is only one constraint, namely, the constraint $\mathcal{H} \approx 0$ associated with time reparametrizations. The BRST charge is then $\Omega = \mathcal{H}\eta - ib\rho$, where b is the momentum conjugate to the Lagrange multiplier associated with $\mathcal{H} \approx 0$.) One then finds that the gauge-fixed Hamiltonian is

$$H + [K,\Omega] = \lambda\mathcal{H} - i\bar{\mathcal{P}}\rho \tag{16.46c}$$

and that the path integral becomes

$$\text{Path integral} = (\tau_2 - \tau_1)\int [Dq\,Dp]\prod_t \delta(\dot{\lambda}) \times \exp i\int (p\,\dot{q} - \lambda\mathcal{H})\,dt \tag{16.47a}$$

$$= \int_{-\infty}^{+\infty} dT\,\langle q_2\,|\exp -iT\mathcal{H}\,|\,q_1\rangle. \tag{16.47b}$$

The factor $(\tau_2 - \tau_1)$ comes from the ghost path integral [see (15.94)], while T is defined by

$$T = \lambda(\tau_2 - \tau_1). \tag{16.48a}$$

The variable T is equal to the integral of $\lambda(\tau)$ from τ_1 to τ_2,

$$T = \int_{\tau_1}^{\tau_2} \lambda(\tau)\,d\tau, \tag{16.48b}$$

and is invariant under gauge transformations that vanish at the end-points. It is known as the "proper time" between τ_1 and τ_2, while the gauge $\dot{\lambda} = 0$, in which λ is proportional to T, is known as the proper time gauge. The path integral yields thus a representation of the projected kernel as a sum over all proper times of the matrix elements $\langle q_2 \,|\, \exp -iT\mathcal{H} \,|\, q_1 \rangle$.

In the case of a parametrized system, with $q \equiv (q^i, t)$, the matrix element $(t_2| \exp -iTp_t|t_1\rangle$ yields the δ-function $\delta\big(T-(t_2-t_1)\big)$. The path integral becomes then the standard kernel of the evolution operator

$$(q_2^i \,|\, \exp -iH_0(t_2 - t_1) \,|\, q_1^i)$$

considered in the unparametrized theory. It should be observed that while the path integral (16.47) originally involves paths for which $t(\tau)$ may both increase and decrease as a function of τ, the final amplitude can be written as a sum over paths for which t increases monotonically with the parameter time τ.

It is useful to consider the causal propagator besides the kernel of the evolution operator. This causal propagator $K_F(q_f, t_f; q_i, t_i)$ is equal to zero for $t_f < t_i$ and to $(q_f \,|\, \exp -iH_0(t_f - t_i) \,|\, q_i)$ otherwise,

$$K_F(q_f, t_f; q_i, t_i) = \begin{cases} 0, & t_f < t_i, \quad (16.49a) \\ (q_f \,|\, \exp -iH_0(t_f - t_i) \,|\, q_i), & t_f \geq t_i. \quad (16.49b) \end{cases}$$

The causal propagator is not a solution of the Schrödinger equation because of the jump at $t_f = t_i$. Rather, one has

$$(p_t + H_0)K_F = \delta(t_f - t_i)\, \delta(q_f - q_i). \tag{16.49c}$$

The causality requirement (16.49a) can easily be implemented in the representation (16.47a) by restricting the proper time to positive values only. This has the effect of replacing the δ-function by δ_+ and, hence, $(q_f \,|\, \exp -iH_0(t_f - t_i) \,|\, q_i)$ by (16.49a)–(16.49b). Hence, the causal propagator can also be written as a path integral.

Similar features hold for other generally covariant systems. If one does not restrict the range of the integration variables in the path integral, one gets a solution of the constraint equations. But if one imposes the causality requirement that one should sum over positive proper times only, *i.e.*, over histories going into the future in proper time only, then one gets the analog of the causal propagator (16.49). For instance, in the case of the free relativistic particle, the sum over positive proper times yields the Feynman propagator (Exercise 16.12).

16.5.6. Path Integral in Multiplier Gauges

The BRST formalism enables one to get also a path integral expression for the ordinary (= unprojected) kernel $U_0(q',q)$ of the evolution operator in the Schrödinger representation. This is achieved by taking different boundary conditions for the ghosts, *i.e.*, by folding the BRST-invariant evolution operator between different initial and final states.

Theorem 16.4. *The ordinary kernel $A_0(q',q)$ of the gauge invariant operator $A_0(\hat{q},\hat{p})$ is equal to the matrix element*

$$A_0(q',q) = (\psi_{q',\eta=0,b=0,\bar{C}=0}, A\psi_{q,\mathcal{P}=0,\lambda=0,\rho=0}), \tag{16.50}$$

where A is any BRST-invariant extension of A_0 (with the ordering prescriptions of §14.5.1).

Proof. Take the extension A not to depend on the variables $(\lambda, b, \rho, \bar{C})$ of the nonminimal sector. This is permissible because, as will be seen below, such a choice leads to a well-defined answer. The amplitude (16.50) then factorizes as

$$(\psi_{q',\eta=0}, A\psi_{q,\mathcal{P}=0})\,(\psi_{b=0,\bar{C}=0}, \psi_{\lambda=0,\rho=0}).$$

The second factor is equal to unity (up to a numerical factor), since one has in the b–$\bar{C}$ representation, $\psi_{b=0,\bar{C}=0} = \delta(b)\,\delta(\bar{C})$ and $\psi_{\lambda=0,\rho=0} \sim 1$, so that $(\psi_{b=0,\bar{C}=0}, \psi_{\lambda=0,\rho=0}) \sim \int \delta(b)\,\delta(\bar{C})\,db\,d\bar{C} = 1$. Furthermore, according to the ordering prescriptions of §14.5.1, one finds $A\psi_{q,\mathcal{P}=0} = A_0\psi_{q,\mathcal{P}=0}$. Consequently, one gets

$$\begin{aligned}(\psi_{q',\eta=0}, A\psi_{q,\mathcal{P}=0}) &= \langle q'|A_0|q\rangle \int \delta(\eta)\,d\eta \\ &= \langle q'|A_0|q\rangle.\end{aligned}$$

This proves the theorem.

It follows from Theorem 16.4 that the ordinary kernel of the evolution operator can be written as the path integral

$$U_0(q',q) = \int [Dq\,Dp\,Db\,D\lambda\,D\eta\,D\mathcal{P}\,D\rho\,D\bar{C}]\,\exp iS_K, \tag{16.51a}$$

$$S_K = \int (\dot{q}p + \dot{\eta}\mathcal{P} + \dot{\lambda}b + \dot{\rho}\bar{C} - H - [K,\Omega])\,dt, \tag{16.51b}$$

where the sum extends over all paths obeying

$$q(t_2) = q', \quad q(t_1) = q, \quad \eta^a(t_2) = 0, \quad \mathcal{P}_a(t_1) = 0, \tag{16.52a}$$

$$b_a(t_2) = 0, \quad \lambda^a(t_1) = 0, \quad \bar{C}_a(t_2) = 0, \quad \rho^a(t_1) = 0 \tag{16.52b}$$

(Exercise 16.13). This path integral takes the same form as the path integral (16.35), but the boundary conditions are different.

With the boundary conditions (16.52), it is permissible to take a gauge-fixing fermion that does not mix the minimal and nonminimal sector because the ghost numbers of the initial and final states individually add up to zero in each sector. For instance, one may take $K = 0$. The integral over λ, b, $\bar{C}$, and ρ then factorizes as in the proof of Theorem 16.4, and one gets

$$U_0(q', q) = \int [Dq\, Dp\, D\eta\, D\mathcal{P}] \exp i \int (\dot{q}p + \dot{\eta}\mathcal{P} - H)\, dt, \quad (16.53a)$$

$$H = H_0(q, p) + \text{“}\eta\text{–}\mathcal{P} \text{ terms,”} \quad (16.53b)$$

$$q(t_2) = q', \quad q(t_1) = q, \quad \eta^a(t_2) = 0, \quad \mathcal{P}_a(t_1) = 0. \quad (16.53c)$$

The action in (16.53) can be written as the sum of the extended action S_E in which one has set λ^a equal to zero and of a ghost action S^{gh}. The ghost action itself can be written, up to terms quartic in the ghosts and their momenta, as the integral of the gauge variation of λ^a under a gauge transformation generated by η^a times $\mathcal{P}_a$. Thus, one has

$$S_{K=0} = S_E(\lambda = 0) + S^{\text{gh}}, \quad (16.54a)$$

$$S^{\text{gh}} = \int (\delta_\eta \lambda^a)\Big|_{\lambda=0} \cdot \mathcal{P}_a + \text{quartic ghost terms.} \quad (16.54b)$$

The quartic ghost terms vanish in the case of a closed algebra, and S^{gh} takes then the Faddeev–Popov form in the gauge $\lambda^a = 0$ with $\mathcal{P}_a$ playing the role of $\bar{C}_a$ (§11.3.3). One can accordingly view the path integral (16.53) as the path integral in the multiplier gauge $\lambda^a = 0$. The analysis proceeds in the same way for more general multiplier gauges of the form $\lambda^a = k^a(q, p)$ obtained by taking $K = k^a(q, p)\mathcal{P}_a$ (§11.2.1).

Because the boundary conditions (16.53c) do not select a BRST-closed initial state, the amplitude (16.53a) is not annihilated by the constraints. However, the kernel of U_0 commutes (in the sense of the convolution of the kernels) with the kernel of G_a, since the operator U_0 is gauge invariant. To get an amplitude that is annihilated by the constraints, one needs to use derivative gauges *and the boundary conditions adapted to them.* In that case, the nonminimal sector does not decouple but, rather, brings in the operator $\delta(\hat{G}_a)$ in the amplitude (see proof of Theorem 14.9). One says that in a derivative gauge, the integration over the Lagrange multiplier “averages the ordinary kernel $U_0(q', q)$ over the gauge transformations acting at one endpoint” (Exercise 16.14). This yields the projected kernel.

EXERCISES. CHAPTER SIXTEEN

16.1. Check that the path integral (16.5) is invariant under the allowed redefinitions $\chi_\alpha \to M_\alpha{}^\beta(x)\chi_\beta$ of the second-class constraints, sdet $M_\alpha{}^\beta \neq 0$.

16.2. Verify directly that the path integral (16.11) is invariant under infinitesimal changes of the canonical gauge conditions $\chi_a \to \chi_a + \delta\chi_a$. [Hint: Perform the canonical change of variables $\delta z^A = [z^A, G_a\varepsilon^a]$ (gauge transformation) with ε^a determined by $[\chi_b, G_a]\varepsilon^a \approx \delta\chi_b$.]

16.3. Consider a constrained system (q^i, p_i) with constraints that depend only on q_i, $G_a(q_i) \approx 0$.

(a) Show that there exists a complete set of commuting observables $q^{*\alpha}$ that depend only on q^i and that the change of variables $q^i \to q^{*\alpha}, G_a$ is invertible.

(b) Verify that $\int [p^*_\alpha \dot{q}^{*\alpha} - H(q^*, p^*)]\, dt$ is equal to $\int [p_i\dot{q}^i - H(q,p)]\, dt$ on the constraint surface.

(c) Give a precise time-sliced definition of the path integral (16.11) for this system. Show that the q^i's are restricted by the conditions $G_a(q^i) = 0$ at the endpoints.

16.4. Repeat the analysis of Exercise 16.3 for a first-class constrained system (q^i, p_i) with constraints that are linear in the momenta,

$$G_a(q,p) = \xi^i_a(q)\left(p_i - \frac{\partial V}{\partial q_i}\right).$$

[Show that one can take gauge conditions that depend on q only. Note also that $\int (p_\alpha \dot{q}^{*\alpha} - H)\, dt$ and $\int (p_i \dot{q}^i - H)\, dt$ differ by a nonvanishing boundary term when $V \neq 0$.]

16.5. (a) The path integral $\int [Dq\, Dp] \exp i \int [p\dot{q} - p^2/2m]\, d\tau$ summed over all paths $q(\tau), p(\tau)$ fulfilling $q(\tau_2) = q_2, q(\tau_1) = q_1$ gives the matrix element $\langle q_2 | \exp -i(p^2/2m)(\tau_2 - \tau_1)|q_1\rangle = \langle q_2, \tau_2 | q_1, \tau_1\rangle$, where $|q,\tau\rangle$ is the eigenvector of the Heisenberg operator

$$q(\tau) = [\exp i\, (p^2/2m)\tau] q(0) [\exp -i\, (p^2/2m)\tau]$$

at time τ, $|q,\tau\rangle = [\exp i\,(p^2/2m)\tau]\, |q, 0\rangle$. By making the change of integration variables $q^*(\tau) = q(\tau) - [p(t)/m]\tau$, $p^*(\tau) = p(\tau)$, show that the path integral can be rewritten as

$$\int [Dq^*\, Dp^*] \exp i \left\{ \int p^*\dot{q}^*\, d\tau + \frac{1}{2}\left[\frac{(p^*)^2\tau}{m}\right]_{\tau_1}^{\tau_2} \right\}, \tag{i}$$

where the paths $q^*(\tau), p^*(\tau)$ are subject to the conditions

$$q^*(\tau_1) + \frac{p^*}{m}\tau_1 = q_1, \qquad q^*(\tau_2) + \frac{p^*}{m}\tau_2 = q_2. \tag{ii}$$

Verify explicitly that the value of the action of (i) at the extremum yields correctly the exponential factor $\exp\,(i/2m)[(q_2 - q_1)^2/\tau_2 - \tau_1]$ of the propagation amplitude, even though the Hamiltonian vanishes.

(b) More generally, let $U(c) = \exp iMc$ be a one-parameter family of unitary transformations and let $q(c) = U(c)q(0)U^{-1}(c)$. Set $A \equiv q(c_1), B \equiv q(c_2)$. Write the transformation matrix $\langle b|a\rangle$ connecting the eigenbases of A and B (i) as a path integral with vanishing "Hamiltonian" and non-vanishing boundary term; and (ii) as a path integral with "Hamiltonian" $H = M$ and vanishing boundary term.

16.6. Consider the infinitesimal transformation

$$q'^i = q^i + [q^i, F]\,\varepsilon(q,p),$$
$$p'_i = p_i + [p_i, F]\,\varepsilon(q,p).$$

Show that the Jacobian of the transformation is equal to $1 + [\varepsilon, F]$.

16.7. This exercise shows in yet another way that the path integral (16.35) is annihilated by the constraints. Let $U_K(q', \eta', b', \bar{C}'; q, \eta, b, \bar{C})$ be the kernel of the evolution operator in the $(q, \eta, b, \bar{C})$-representation (see §15.4.4). Show that

$$\int \Omega(q'', \eta'', b'', \bar{C}''; q', \eta', b', \bar{C}')\, U_K(q', \eta', b', \bar{C}'; q, \eta, b, \bar{C})\, dq'\, d\eta'\, db'\, d\bar{C}' =$$

$$\int U_K(q'', \eta'', b'', \bar{C}''; q', \eta', b', \bar{C}')\Omega(q', \eta', b', \bar{C}'; q, \eta, b, \bar{C})\, dq'\, d\eta'\, db'\, d\bar{C}',$$

where $\Omega(q', \eta', b', \bar{C}'; q, \eta, b, \bar{C})$ is the kernel of the BRST operator. Verify that $\Omega(q', \eta', b', \bar{C}'; q, \eta = 0, b = 0, \bar{C} = 0)$ vanishes. Infer from these results that the kernel $U_K(q', \eta', b', \bar{C}'; q, \eta = 0, b = 0, \bar{C} = 0)$ viewed as a wave function of the primed arguments, is annihilated by Ω. Prove then that $U_K(q', \eta' = 0, b' = 0, \bar{C}' = 0; q, \eta = 0, b = 0, \bar{C} = 0)$ is annihilated by the constraints.

16.8. Compute explicitly the path integral (16.35) for a pure gauge system (q, p) with constraint $p - dV/dq = 0$. Show that the semiclassical approximation gives the correct result. Check that the propagator is annihilated by the quantum constraint and, hence, depends on q. Compare with the reduced phase space path integral. (Answer: The projected kernel is equal to $\exp i[V(q_2) - V(q_1)]$; the reduced phase space path integral is equal to one.)

16.9. Let $G_a = 0$ be constraints forming a group $[G_a, G_b] = C_{ab}{}^c G_c$ with $C_{ab}{}^c =$ const. Assume that the group is unimodular, $C_{ab}{}^a = 0$. Show that the projection enforced in the BRST path integral by the integration over the ghosts, the multipliers, and their momenta can be described as an average over the gauge group with the Haar (= right-invariant = left-invariant) measure.

16.10. Consider a bad derivative gauge $\dot{\lambda} + \chi = 0$ of the type discussed in Exercise 3.30. That is, assume that the solution $\lambda(\tau)$ of the equation $\dot{\lambda} = -\chi$ becomes infinite in a finite time in such a way that the proper time $\int_{\tau_1}^{\tau_2} \lambda(\tau)\, d\tau$ does not range over the whole real line as $\lambda(\tau_1)$ goes from $-\infty$ to $+\infty$. Show that if one tries to enforce such a bad gauge in the path integral by means of the gauge-fixing fermion $K = i\bar{C}\chi - \mathcal{P}\lambda$, one gets a path integral (16.35) that is ambiguous. [Hint: The matrix element of the corresponding evolution operator for $\hat{\lambda}$ and its momentum $\hat{b}$ between the eigenstates with zero momentum ($b = 0$) are ambiguous in the absence of boundary conditions at infinity in λ-space. These boundary conditions are necessary to define the quantum system, since wave packets go to infinity in a finite time. Different types of boundary conditions yield different quantum systems. In path integral terms, this means that different manners to define precisely the path integral give different amplitudes.]

16.11. Prove that the projected kernel for a parametrized system in the (t, q)-representation is equal to the kernel of the evolution operator of the unparametrized system. (Take, for instance, for $|\psi_\alpha\rangle$ the solutions of the Schrödinger equation equal to $\delta(q - q_0)$ at time $t = 0$.)

16.12. Show that for a relativistic particle with constraint $p^2 + m^2$ (Exercise 4.1), the path integral (16.47) restricted to positive proper times yields the Feynman propagator.

16.13. Show that the amplitude $\langle q = 0| \exp -iH(q,p)(t_2 - t_1)|p = 0\rangle$ can be written as the sum

$$\int [Dq\, Dp] \exp i \int (\dot{q}p - H)\, dt$$

over paths fulfilling $p(t_1) = 0, q(t_2) = 0$ for either commuting or anticommuting q's and p's.

16.14. Consider the pure gauge system with Hamiltonian $H = p^2/2m$ and constraint $p = 0$. Even though $H \approx 0$, we take a nontrivial form for H in order to get a nontrivial kernel for $\exp -iH(t_2 - t_1)$.

(a) Compute the projected kernel of the evolution operator. [Answer: $U_0^P(q', q) = 1$.]

(b) Compute the ordinary kernel $U_0(q', q; \lambda)$ of

$$\exp\left[-i\left(\frac{p^2}{2m} + \lambda p\right)(t_2 - t_1)\right]$$

where λ is an arbitrary, fixed real member. [Answer:

$$U_0^P(q', q; \lambda) = \left[\frac{m}{2\pi i(t_2 - t_1)}\right]^{1/2} \exp\left[\frac{im(t_2 - t_1)}{2}\left(\frac{q' - q}{t_2 - t_1} - \lambda\right)^2\right].\ ]$$

(c) Verify explicitly that the projected kernel is equal to the average over λ of the ordinary kernel $U_0(q', q; \lambda)$ with measure $(t_2 - t_1)\, d\lambda$,

$$U_0^P(q', q) = \int (t_2 - t_1)\, d\lambda\, U_0(q', q; \lambda).$$

[Note: The amplitude $U_0(q', q; \lambda = 0)$ depends only on $q' - q$ and so is invariant under the same gauge transformation $q' \to q' + \alpha, q \to q + \alpha$ acting at both endpoints. Since $\exp[-i\lambda p(t_2 - t_1)]$ generates a gauge transformation with gauge parameter $\lambda(t_2 - t_1)$, one may view $\langle q'| \exp[-iH(t_2 - t_1)] \exp[-i\lambda p(t_2 - t_1)|q\rangle$ as the amplitude $\langle q'| \exp -iH(t_2 - t_1)|q\rangle$ in which one has gauge-transformed q by the amount $\lambda(t_2 - t_1), q \to q + \lambda(t_2 - t_1)$. The projected kernel is thus obtained by gauge-averaging the ordinary kernel over one endpoint, which yields an amplitude invariant under gauge transformations acting independently at the initial and final times. In the BRST path integral in a derivative gauge, the parameter λ appears as the value of the Lagrange multiplier at one time (the gauge condition $\dot{\lambda} = \cdots$ leaves one constant of integration in λ), while the factor $t_2 - t_1$ in the measure comes from the ghosts; see, for instance, (16.47)–(16.48).]

16.15. Verify explicitly that the reduced phase space path integral is invariant under the redefinitions (1.44) of the first- and second-class constraints.

16.16. Consider the constraint system $(q^1, p_1), (q^2, p_2)$ with Hamiltonian $\frac{1}{2}(p_2)^2$ and constraint $p_1 \approx 0$. Compute the reduced phase space path integral. Verify explicitly that it is equal to the BRST path integral with the representation $p_1 \approx 0$ of the constraint surface, but that it is *not* equal to the BRST path integral if one replaces $p_1 \approx 0$ by $e^{p_2} p_1 \approx 0$. Explain by using the theory of §14.5.2.

16.17. (Path integral representation for U_0^P in a canonical gauge). To apply the trick of §16.5.4 to reach the canonical gauge $\chi_a = 0$, it is necessary that the endpoint data fulfill $\chi_a = 0$. We assume that this can be achieved by means of a gauge transformation. Namely, we assume that $\partial G_a / \partial q^i$ is of maximum rank on $G_a \approx 0$ ("transversality condition") so that one can go from fixed q's to $\chi_a = 0$ by a gauge transformation. For noninternal gauge symmetries, the required gauge transformation induces a surface term in the action and leads to new boundary conditions mixing q^i with p_i. That is, it modifies the class of paths under consideration. Once the boundary conditions obey $\chi_a = 0$, one may take K of the form (16.44a) and take the limit $\varepsilon \to 0$ to get the path integral for the projected kernel U_0^P in the canonical gauge $\chi_a = 0$. Use these observations to derive the path integral (16.16d) for a parametrized particle in the canonical gauge $t = f(\tau)$ from the BRST path integral. [Note that (16.15) coincides with the projected kernel (Exercise 16.11).]

CHAPTER SEVENTEEN

ANTIFIELD FORMALISM: CLASSICAL THEORY

17.1. COVARIANT PHASE SPACE

17.1.1. Path Integral and Spacetime Covariance

The Hamiltonian BRST formalism leads to gauge-fixed actions that can be used for computing quantum-mechanical amplitudes by path integral methods. If the conjugate momenta can be integrated over, one can rewrite the path integral in Lagrangian form. Furthermore, when the original gauge-invariant Lagrangian is Lorentz-invariant, it is in general possible to choose the gauge-fixing fermion K so that the resulting gauge-fixed Lagrangian is also covariant. However, the finding of the appropriate K leading to a manifestly relativistic path integral is not always an easy task within the Hamiltonian formalism. One may thus wonder whether there exists a systematic method for directly writing down the correct covariant gauge-fixed Lagrangians without having to go first through the Hamiltonian formulation.

There exists, indeed, such a method. It is known as the antifield approach to the BRST symmetry. This method is purely Lagrangian and, hence, well adapted to spacetime covariance. It formulates the

BRST symmetry directly in the space of the Lagrangian variables and yields the same gauge-fixed actions as the Hamiltonian formalism. It is, however, less complete in that the only restriction it puts on the measure in the sum-over-paths is that it be gauge invariant. The purpose of this chapter and the next is to expose the Lagrangian antifield method.

17.1.2. Covariant Phase Space in the Absence of Gauge Invariance

The central idea of the antifield formalism is that the usual phase space description of the dynamics is inappropriate to manifest covariance. Rather, one should use the "covariant phase space."

To clearly explain what the covariant phase space is, let us first assume for a moment that there is no gauge invariance. The observables of the system are then usually viewed as the phase space functions $F(q,p)$. As (q,p) at $t=t_0$ completely determines $\big(q(t),p(t)\big)$ through the Hamiltonian equations, one can alternatively view phase space as the space of all solutions of the equations of motion. One can then drop reference to the momenta and consider the solutions $q(t)$ of the equations of motion for q obtained by eliminating p from the Hamiltonian equations. For a covariant theory, it is usually the second-order equations for q that take a manifestly covariant form.

The space of all solutions of the equations of motion is known as the "covariant phase space." In the covariant phase space, one does not take an instantaneous picture of the system at some instant t_0 but, instead, one identifies each phase space point with the entire classical trajectory that it determines.

It should be stressed that the correspondence between q and p at time t_0 and the corresponding solution of the equations of motion depends, in general, on t_0. That is, if one changes t_0, the solution associated with q and p also changes unless the Hamiltonian identically vanishes.

The same ideas apply to field theory, where the observables can be viewed as the functionals $f[\phi^i(x^\mu)]$ of the solutions $\phi^i(x^\mu)$ to the Euler–Lagrange equations of motion $\delta S_0/\delta\phi^i(x)=0$. We shall again use in this chapter and the next the suggestive notations and terminology of finite-dimensional manifold theory and collect the discrete index i and the continuous "index" x^μ into the single index i. The words "functions" and "functionals" will also be used interchangeably and $f[\phi^i(x^\mu)]$ will thus be also denoted by $f(\phi^i)$.

An important feature of the antifield formalism is that its algebraic structure is almost identical to the one of the Hamiltonian BRST formalism provided one looks at the appropriate function spaces. One deals now with the infinite-dimensional functional space I of all possible field histories. A point ϕ^i in I is accordingly an arbitrary entire history that may not solve $\delta S_0/\delta\phi^i$ ($\partial S_0/\partial\phi^i$ in finite-dimensional notations) $= 0$. In I, the equations of motion $\delta S_0/\delta\phi^i = 0$ determine a submanifold Σ, which we call the stationary surface. This submanifold is just the covariant phase space. The space I is not restricted by boundary conditions at the endpoints so that the stationary surface contains all the solutions of the equations of motion. The derivatives $\delta S_0/\delta\phi^i$ are, however, the derivatives of S_0 with fixed endpoints.

The observables are the functions defined on Σ, *i.e.*, the elements of $C^\infty(\Sigma)$ ("smooth" functions on Σ). Any function f on Σ can be extended off Σ to a function $F(\phi^i)$ defined on I, *i.e.*, to an element of $C^\infty(I)$ ("smooth" functions on I). Two different extensions F and F' differ by a function that vanishes on Σ. These functions form an ideal $\mathcal{N}$ as FG vanishes on Σ whenever F or G does. The algebra $C^\infty(\Sigma)$ of the smooth functions on Σ is thus the quotient algebra $C^\infty(I)/\mathcal{N}$ of the smooth functions on I by the functions that vanish on Σ.

17.1.3. Covariant Phase Space in the Presence of Gauge Freedom

If there is gauge freedom, the observables should be, in addition, gauge invariant.

We have seen in Chapter 3 that the gauge transformations

$$\delta_\varepsilon \phi^i = R_\alpha{}^i \varepsilon^\alpha \tag{17.1a}$$

are integrable when the equations of motion hold,

$$R_\alpha{}^j \frac{\delta R_\beta{}^i}{\delta\phi^j} - R_\beta{}^j \frac{\delta R_\alpha{}^i}{\delta\phi^j} = C_{\alpha\beta}{}^\gamma(\phi)\, R_\gamma{}^i + M_{\alpha\beta}{}^{ij}(\phi)\, \frac{\delta S_0}{\delta\phi^j}. \tag{17.1b}$$

Furthermore, they map a solution of the equations of motion on a solution. Accordingly, the gauge transformations generate well-defined orbits on the stationary surface Σ, the dimension of which is equal to the number of independent $R_\alpha{}^i$'s. The gauge-invariant functions are constant along the gauge orbits and, hence, induce definite functions on the quotient space Σ/G of the stationary surface by the gauge orbits.

The covariant phase space in the presence of gauge invariance is Σ/G. This space is isomorphic to the reduced phase space of the Hamiltonian formalism, since the sets of gauge-invariant functions in the Lagrangian and Hamiltonian formalisms are isomorphic (Exercises 3.25 and 17.11).

17.1.4. Lagrangian Homological Perturbation Theory

We thus come to the conclusion that the description of the gauge-invariant functions in terms of the covariant phase space involves the same two steps as in the Hamiltonian formalism: (*i*) first, the restriction to a submanifold Σ; (*ii*) second, the demand of invariance along the gauge orbits. All the ingredients of homological perturbation theory described in Sec. 8.5 are, therefore, met.

What replaces the phase space P of the Hamiltonian formalism is the space I of all histories; what plays the role of the constraint surface Σ is now the stationary surface Σ; in both cases, there are well-defined gauge orbits on Σ. We have thus the following table.

Hamiltonian formalism	Lagrangian formalism
Phase space P	Space I of all histories
Constraint surface Σ ($G_a = 0$)	Stationary surface Σ ($\delta S_0/\delta\phi^i = 0$)
Gauge orbits ($\delta_\varepsilon F = [F, \varepsilon^a G_a]$)	Gauge orbits ($\delta_\varepsilon F = \dfrac{\delta F}{\delta\phi^i} R_\alpha{}^i \varepsilon^\alpha$)

The Hamiltonian part of this table is an instantaneous snapshot of the Lagrangian part: the constraints are what remains on the initial data (q, p) of the equations of motion $\delta S_0/\delta\phi^i = 0$; and the gauge transformations of the Hamiltonian formalism describe the effect on the initial data of the gauge transformations of the Lagrangian formalism (see Chapter 3).

Because the Hamiltonian and Lagrangian formalisms describe the gauge-invariant functions along identical lines, one can also define, in the Lagrangian formalism, a differential s—the (Lagrangian) BRST differential—that yields the observables through its cohomology at ghost number zero,

$$s^2 = 0, \quad s(AB) = A(sB) + (-)^{\varepsilon_B}(sA)B, \tag{17.2a}$$

$$H^0(s) = \{\text{gauge-invariant functions}\}. \tag{17.2b}$$

This differential is obtained by putting together two differentials: first, the (Lagrangian) Koszul–Tate differential δ, which implements the restriction to Σ; second, the (Lagrangian) longitudinal exterior differential d, which picks out the gauge-invariant functions. The BRST differential s is given by

$$s = \delta + d + \text{"more,"} \tag{17.2c}$$

as the main theorem of homological perturbation theory establishes.

There is, however, one important difference between the Hamiltonian and Lagrangian cases: There is no natural Poisson bracket structure in the space I of all histories [even though there is a bracket structure among the on-shell gauge-invariant functions, known as the "Peierls bracket"; see Exercises 17.11, 17.12, and 17.13]. For this reason, the Lagrangian BRST transformation is not a canonical transformation in a standard bracket structure.

It turns out, nevertheless, that there is a natural antibracket structure that can be defined in the Lagrangian BRST complex. This antibracket structure extends to the ghosts the antibracket structure already encountered in the discussion of the path integral for systems without gauge invariance. It plays a crucial role because, as we shall see, the BRST transformation (17.2) is a canonical transformation in the antibracket.

17.1.5. Regularity Conditions

As we already have experience with the ideas of the BRST formalism, we shall develop the antifield analysis without giving all the details. We shall also, for the purpose of notational simplicity, consider explicitly only the case of bosonic reducible gauge theories of the first order. That is, there may be reducibility identities

$$Z_A{}^\alpha R_\alpha{}^i = C_A{}^{ij} \frac{\delta S_0}{\delta \phi^j}, \qquad C_A{}^{ij} = -C_A{}^{ji} \tag{17.3a}$$

on the gauge transformations. The Z's form a complete set,

$$\mu^\alpha R_\alpha{}^i = N^{ij} \frac{\delta S_0}{\delta \phi^j} \Rightarrow \mu^\alpha = A\alpha + M^{\alpha i} \frac{\delta S_0}{\delta \phi^i} \tag{17.3b}$$

and are assumed to be independent ($\rho^A Z_A{}^\alpha \approx 0 \Rightarrow \rho^A \approx 0$ where $\approx$ means now "equal on the stationary surface"). Reducible theories of higher order are treated along identical lines, by adding further ghosts of ghosts and "antifields for antifields." Fermionic variables and fermionic gauge symmetries can also be incorporated by including appropriate sign factors in the intermediate formulas. The final key equation—the master equation (17.27c) below—takes always the same form.

Regularity conditions on the constraint functions played a crucial role in the previous chapters. We shall thus impose similar conditions on the field equations $\delta S_0/\delta\phi^i = 0$. Namely, we shall assume that the derivatives $\delta S_0/\delta\phi^i$ can be locally split on Σ into independent functions y^a and dependent ones z^α, $(\delta S_0/\delta\phi^i) = (y^a, z^\alpha)$ in such a way that (i) the equations $\delta S_0/\delta\phi^i = 0$ are equivalent to $y^a = 0$; *i.e.*, $z^\alpha = 0$ is a consequence of $y^a = 0$; and (ii) the exterior product $\wedge_a dy^a$ does not vanish on Σ. One can then take the y^a's as first coordinates of a new regular coordinate system on I.

When the regularity conditions hold, the following theorem is immediate.

Theorem 17.1. *If the function $F(\phi^i)$ vanishes on Σ, then it can be written as*

$$F(\phi^i) = \lambda^i(\phi)\,\frac{\delta S_0}{\delta\phi^i}$$

with coefficients $\lambda^i(\phi)$ that are regular on Σ.

Proof. The proof formally proceeds as in the finite-dimensional case and will not be repeated here (see proof of Theorem 1.1).

The regularity conditions appear to be fulfilled by all models of direct physical interest. An exception is analyzed in Exercise 17.1.

17.2. KOSZUL–TATE RESOLUTION AND LONGITUDINAL *d*

17.2.1. Koszul–Tate Resolution

The first step in the BRST construction is to implement the restriction from I to Σ. So, one needs to define the Koszul–Tate resolution associated with the stationary surface. This is done by following exactly the same pattern as in the Hamiltonian case (Sec. 10.3).

For each field equation $\delta S_0/\delta\phi^i = 0$, *i.e.*, for each field ϕ^i, one introduces one antifield ϕ^*_i and defines δ on ϕ^i and ϕ^*_i as

$$\delta\phi^i = 0, \tag{17.4a}$$

$$\delta\phi^*_i = -\frac{\delta S_0}{\delta\phi^i}. \tag{17.4b}$$

The minus sign is inserted for later convenience. To preserve the grading properties of δ, one must impose

$$\varepsilon(\phi^*_i) = 1 \tag{17.5a}$$

(as we assume the fields to be bosonic) and

$$\text{antigh}\ \phi_i^* = 1, \qquad \text{antigh}\ \phi^i = 0. \tag{17.5b}$$

With this definition, one finds

$$H_0(\delta) = C^\infty(\Sigma) = \frac{C^\infty(I)}{\mathcal{N}} \tag{17.6}$$

because $(\text{Ker}\ \delta)_0 = C^\infty(I)$ while $(\text{Im}\ \delta)_0 = \mathcal{N}$ (Theorem 17.1).

Since the field equations are not independent, one introduces next further antifields ϕ_α^* at antighost number 2,

$$\text{antigh}\ \phi_\alpha^* = 2, \qquad \varepsilon(\phi_\alpha^*) = 0, \tag{17.7a}$$

and defines

$$\delta\phi_\alpha^* = R_\alpha{}^i\phi_i^*. \tag{17.7b}$$

With this definition, all the Noether identities on the field equations are taken care of. Indeed, the polynomials $R_\alpha{}^i\phi_i^*$, which are closed as a result of the Noether identities

$$R_\alpha{}^i\,\frac{\delta S_0}{\delta\phi^i} = 0 \tag{17.8}$$

and which would not be exact without the introduction of the ϕ_α^*'s, become exact with (17.7b). Hence,

$$H_1(\delta) = 0, \tag{17.9}$$

as in the Hamiltonian case.

Since we have allowed for reducibility identities among the gauge transformations, the construction of δ is not finished yet. One needs antifields ϕ_A^* at antighost number 3, one for each identity (17.3a),

$$\text{antigh}\ \phi_A^* = 3, \qquad \varepsilon(\phi_A^*) = 1. \tag{17.10a}$$

One then defines

$$\delta\phi_A^* = -Z_A{}^\alpha\phi_\alpha^* - \tfrac{1}{2}C_A{}^{ij}\phi_i^*\phi_j^* \tag{17.10b}$$

and finds

$$H_2(\delta) = 0. \tag{17.11}$$

Actually, because the reducibility coefficients $Z_A{}^\alpha$ are independent, one even has

$$H_k(\delta) = 0, \qquad k \neq 0. \tag{17.12}$$

Accordingly, we have proved

Theorem 17.2. *The differential δ defined by (17.4), (17.7), and (17.10) provides a resolution of the algebra of functions on the stationary surface,*

$$H_0(\delta) = C^\infty(\Sigma), \qquad H_k(\delta) = 0, \quad k \neq 0. \tag{17.13}$$

Remarks. (a) the antifields $\phi^*_i, \phi^*_\alpha, \phi^*_A$ play the role of the ghost momenta of the Hamiltonian formalism. However, while the Hamiltonian Koszul–Tate differential associated with the constraint surface is trivial in the absence of gauge freedom (no constraint), there is nevertheless a nontrivial Lagrangian δ in that case because there are nontrivial field equations. This Lagrangian δ is interesting and was already described in §15.4.1 in the context of the Schwinger–Dyson equation.

(b) The requirement that δ be acyclic in degree $k \neq 0$ is sometimes known as the "proper solution" requirement. (See §17.3.2 and §17.4.2 for more on this.)

17.2.2. Any Gauge Transformation that Vanishes On-Shell Is a Trivial Gauge Transformation

Theorem 17.3.

$$\frac{\delta S_0}{\delta \phi^i} \lambda^i = 0 \quad \textit{and} \quad \lambda^i \approx 0 \Rightarrow \lambda^i = \mu^{ij} \frac{\delta S_0}{\delta \phi^j}, \qquad \mu^{ij} = -\mu^{ji}. \tag{17.14}$$

Proof. The theorem, announced in Chapter 3 (Theorem 3.1), is proved exactly as Theorem 10.1, with the mere replacement of G_{a_0} by $\delta S_0 / \delta \phi^i$. We leave the details to the reader. The theorem is connected with the acyclicity of δ for $k \neq 0$ (see Chapter 10).

17.2.3. Longitudinal Exterior Differential *d*

As the orbits generated by the gauge transformations are integrable on the stationary surface Σ, one can define, on Σ, an exterior derivative operator d that acts on longitudinal p-forms along the gauge orbits. This operator only measures how the p-forms change as one moves along the orbits. It contains no information about the transverse directions. The exterior longitudinal d is thus such that $H^0(d) = \{\text{gauge-invariant functions}\}$.

The construction of d proceeds again as in the Hamiltonian case (Sec. 5.3, §9.2.3, and Sec. 10.4), so that we only give the relevant formulas.

In the irreducible case, one can describe d in terms of the gauge transformations by introducing the ghosts C^α,

$$\text{pure gh } C^\alpha = 1, \qquad \varepsilon(C^\alpha) = 1. \tag{17.15a}$$

These can be identified with the one-forms dual to the vector fields defining the infinitesimal gauge transformations. The form degree coincides with the pure ghost number. One has:

$$dF = \frac{\delta F}{\delta \phi^i} R_\alpha{}^i C^\alpha \tag{17.15b}$$

$$dC^\alpha = \tfrac{1}{2} C^\alpha{}_{\beta\gamma} C^\beta C^\alpha . \tag{17.15c}$$

In the reducible case, the gauge transformations are not independent and one considers a "model" D for d by introducing further ghosts of ghosts C^A,

$$\text{pure gh } C^A = 2, \qquad \varepsilon(C^A) = 0, \tag{17.16a}$$

$$DF = \frac{\delta F}{\delta \phi^i} R_\alpha{}^i C^\alpha ,$$

$$DC^\alpha = \tfrac{1}{2} C^\alpha{}_{\beta\gamma} C^\beta C^\alpha + Z_A{}^\alpha C^A + \text{"more,"} \ldots , DC^A = \cdots , \tag{17.16b}$$

One finds

$$H^k(D) = H^k(d). \tag{17.16c}$$

The analysis is identical to the analysis of Sec. 10.4 and need not be repeated here.

17.2.4. δ and Spacetime Locality

We have been a bit cavalier with the functional aspects of the homology of δ and have proceeded as if the space I of all histories were finite-dimensional, so our discussion is rather formal. However, things are not as bad as one may think because spacetime locality comes as a help.

Indeed, the action is a local functional. Therefore, if the equations of motion, the gauge transformations, and the reducibility functions are not only local but also locally complete in the sense of §12.2.2a and §12.2.3, then one can use the tools developed in Chapter 12, but this time in the context of spacetime rather than of space.

In particular, the analog of Theorem 12.5 reads

Theorem 17.4. *(a) Let f be a nonintegrated density with the following properties:*

$$(i)\ \text{antigh} f \geq 1, \quad \text{pure gh } f \geq 1. \tag{17.17a}$$

(ii) f is δ-closed modulo a divergence, i.e.,

$$\delta f = \partial_\mu \, \alpha^\mu . \tag{17.17b}$$

Then f is δ-exact modulo a divergence,

$$f = \delta e + \partial_\mu \, \beta^\mu . \tag{17.17c}$$

Here, e and β^μ are, as are f and α^μ, local functions of the fields ϕ^i, the antifields $\phi^*_i, \phi^*_\alpha, \phi^*_A$, the ghosts C^α, C^A, and their derivatives.

(b) Equivalently, the homology of δ is trivial in the space of local functionals $F = \int f\,dx$ with both positive antighost and pure ghost numbers,

$$\delta F = 0, \qquad \text{pure gh } F \geq 1, \qquad \text{antigh } F \geq 1 \Rightarrow F = \delta E \tag{17.18}$$

(with E also a local functional).

17.3. BRST SYMMETRY—MASTER EQUATION

17.3.1. Antibracket Structure

With the Koszul–Tate differential and the longitudinal exterior derivative at hand, all the building blocks of the BRST symmetry have been constructed. What is required now is to put these ingredients together along the lines of homological perturbation theory,

$$s = \delta + d + \text{“more,”} \qquad s^2 = 0. \tag{17.19}$$

The “main theorem” 8.3 guarantees that this can be done and, furthermore, that the cohomology of s is equal to the cohomology of d,

$$H^g(s) = \begin{cases} 0, & g < 0, \qquad (17.20a) \\ H^g(d), & g \geq 0, \qquad (17.20b) \end{cases}$$

where g in the left-hand side of (17.20) is the (total) ghost number, equal to

$$\text{gh} = \text{pure gh} - \text{antigh}\,. \tag{17.21}$$

In particular, one has the crucial equation

$$H^0(s) = \{\text{gauge-invariant observables}\}. \tag{17.22}$$

In the Hamiltonian formalism, it was found that s was a canonical transformation in an appropriate Poisson bracket structure,

$$sA = [A, \Omega] \qquad \text{(Hamiltonian)}. \tag{17.23a}$$

There is no Poisson bracket structure here but, rather, an antibracket structure, which we denote by $(\ ,\ )$. As we shall show, the BRST transformation of the antifield formalism is a canonical transformation in the antibracket. The equation that replaces (17.23a) reads

$$sA = (A, S) \qquad \text{(antifield formalism)}, \tag{17.23b}$$

where S is the generator of the BRST symmetry. [Although we use the same letter, A in (17.23a) and (17.23b) belongs to different functional spaces.] The notation S is not accidental, since S contains the original classical action S_0 [Eq. (17.28c) below].

What is the antibracket structure of the antifield formalism? The definition of the antibracket suggests itself once it is realized that there is a remarkable symmetry between the fields and the ghosts on the one hand and the antifields on the other hand. This symmetry, in turn, is a consequence of the fact that it is the same functional S_0 that determines both the ghost spectrum (through the gauge symmetries) and the antifield spectrum (through the Noether identities).

Explicitly, one has

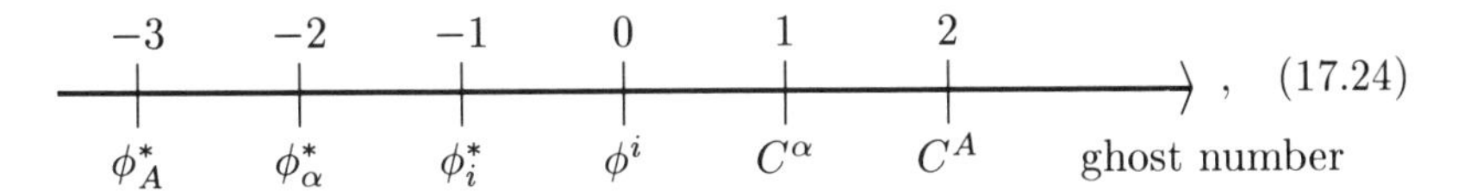

, (17.24)

and so it is natural to declare that the pairs ϕ^i, ϕ^*_i; C^α, ϕ^*_α; and C^A, ϕ^*_A are conjugate,

$$(\phi^i, \phi^*_j) = \delta_j{}^i, \tag{17.25a}$$

$$(C^\alpha, \phi^*_\beta) = \delta_\beta{}^\alpha, \tag{17.25b}$$

$$(C^A, \phi^*_B) = \delta_B{}^A. \tag{17.25c}$$

The difference with the Hamiltonian formalism is that the symmetry in (17.24) is around $(-1/2)$ rather than around 0.

The antibracket (,) is extended to arbitrary functionals A, B of the fields, the ghosts, and the antifields as follows,

$$\begin{aligned}(A, B) = {} & \frac{\delta^R A}{\delta \phi^i}\frac{\delta^L B}{\delta \phi^*_i} - \frac{\delta^R A}{\delta \phi^*_i}\frac{\delta^L B}{\delta \phi^i} \\ & + \frac{\delta^R A}{\delta C^\alpha}\frac{\delta^L B}{\delta \phi^*_\alpha} - \frac{\delta^R A}{\delta \phi^*_\alpha}\frac{\delta^L B}{\delta C^\alpha} \\ & + \frac{\delta^R A}{\delta C^A}\frac{\delta^L B}{\delta \phi^*_A} - \frac{\delta^R A}{\delta \phi^*_A}\frac{\delta^L B}{\delta C^A}.\end{aligned} \tag{17.26}$$

This definition reduces to the definition of §15.5.2 in the absence of ghosts. It is such that the antibracket possesses all the properties (15.101), (15.102), and (15.103) that were derived there. In particular, it is odd, carries ghost number +1, and obeys the Jacobi identity.

17.3.2. Master Equation

Because of the parity and ghost number properties of the antibracket, the generator S of the BRST symmetry in the antibracket,

$$sA = (A, S), \tag{17.27a}$$

should be even and have ghost number zero,

$$\varepsilon(S) = 0, \qquad \mathrm{gh}\,(S) = 0. \tag{17.27b}$$

Furthermore, the nilpotency of s is equivalent to

$$(S, S) = 0, \tag{17.27c}$$

as it follows from the Jacobi identity and the fact that there is no c-number of ghost number one.

The first few terms in S should generate δ and d. This means that in the expansion of S according to antighost number,

$$S = \sum_{n \geq 0} \overset{(n)}{S}, \tag{17.28a}$$

$$\text{antigh}\ \overset{(n)}{S} = n, \tag{17.28b}$$

one should have

$$(i)\ \overset{(0)}{S} = S_0 \tag{17.28c}$$

so that $(\phi_i^*, S) = \delta\phi_i^* +$ "more";

$$(ii)\ \overset{(1)}{S} = \phi_i^*\, R_\alpha{}^i C^\alpha \tag{17.28d}$$

so that $(\phi_\alpha^*, S) = \delta\phi_\alpha^* +$ "more," and $(\phi^i, S) = d\phi^i +$ "more"; and

$$(iii)\ \overset{(2)}{S} = \phi_\alpha^* Z_A{}^\alpha C^A \text{ terms not containing } \phi_\alpha^* C^A, \tag{17.28e}$$

so that the first terms in $\delta\phi_A^*$ and DC^α are appropriately reproduced.

The problem of finding the BRST symmetry as a canonical transformation in the antibracket is thus equivalent to the problem of finding the solution S of (17.27b)–(17.27c) with the boundary conditions (17.28). It is remarkable that the first piece in the BRST generator S is just the original gauge-invariant action.

Equation (17.27c) is named the "master equation" and is the heart of the antifield formalism. The boundary conditions (17.28d) and (17.28e) exclude taking the trivial solution $S = S_0$ of $(S, S) = 0$ in the presence of gauge invariance, *i.e.*, force S to be a "proper solution."

17.3.3. Solution of the Master Equation

The construction of the solution of the master equation $(S, S) = 0$ proceeds perturbatively as the construction of the solution of $[\Omega, \Omega] = 0$. For this reason, we sketch again only the main points. First, one finds that the complete expression for $\overset{(2)}{S}$ in (17.28a), only partially determined by (17.28e), must be

$$\begin{aligned}\overset{(2)}{S} &= \phi^*_\alpha(Z_A{}^\alpha C^A + \tfrac{1}{2}C_{\beta\gamma}{}^\alpha C^\beta C^\gamma) \\ &\quad + \phi^*_i\phi^*_j(-\tfrac{1}{4}M_{\alpha\beta}{}^{ij}C^\alpha C^\beta + \tfrac{1}{2}C_A{}^{ij}C^A)\end{aligned} \tag{17.29}$$

for $(S, S) = 0$ to hold not only at antighost number zero (this is a consequence of the Noether identity) but also at antighost number one.

The next terms in S are then determined recursively by equations that take the form

$$2\,\delta\overset{(n)}{S} + \overset{(n-1)}{D} = 0, \tag{17.30}$$

where $\overset{(n-1)}{D}$ is the component of antighost number $n-1$ of $(\overset{(n-1)}{R}, \overset{(n-1)}{R})$, with

$$\overset{(n-1)}{R} = \sum_{k\leq n-1} \overset{(k)}{S}. \tag{17.31}$$

The local functional $\overset{(n-1)}{D}$ depends on the functionals $\overset{(k)}{S}$ of order $k \leq n-1$. Equations (17.30) have solutions because $\overset{(n-1)}{D}$ is δ-closed as a consequence of the Jacobi identity for the antibracket. These solutions can furthermore be taken to be local functionals, thanks to Theorem 17.4.

We can thus conclude that S exists. Furthermore, it is a local functional provided the expansion (17.28a) stops after a finite number of terms (*i.e.*, $\overset{(n)}{S} = 0$ for $n \geq N$). (The number of derivatives in $\overset{(n)}{S}$ may increase with n. Therefore, if there were an infinite number of $\overset{(n)}{S}$'s, there could be an infinite number of derivatives and S would not be local.)

17.3.4. Canonical Transformation in the Antibracket

The solution of the master equation is not unique, since (17.30) allows for the possibility of adding a δ-exact term to $\overset{(n)}{S}$. Moreover, given the gauge-invariant action $S_0(\phi^i)$, there exist various equivalent descriptions of the gauge symmetries (§3.1.10). These equivalent descriptions define different generating sets and are related as

$$R_\alpha{}^i \to \bar{R}_\alpha{}^i = M_\alpha{}^\beta R_\beta{}^i + T_\alpha{}^{ij} \frac{\delta S_0}{\delta \phi^j}, \qquad T_\alpha{}^{ij} = -T_\alpha{}^{ji}, \tag{17.32a}$$

where $M_\alpha{}^\beta(\phi)$ is invertible. Similarly, one can replace $Z_A{}^\alpha$ by

$$Z_A{}^\alpha \to \bar{Z}_A{}^\alpha = M_A{}^B Z_B{}^\alpha + T_A{}^{\alpha i} \frac{\delta S_0}{\delta \phi^i}. \tag{17.32b}$$

One can even increase the order of reducibility by adding trivial gauge symmetries or reducibility functions. How are the corresponding solutions of the master equation related?

It is easy to see that if one does not increase the reducibility, the ambiguity in S is completely exhausted by the possibility of performing canonical transformations in the antibracket. The verification of this statement proceeds again as in the Hamiltonian formalism (§10.5.5). The enlargement of the generating sets by adding trivial gauge transformations or reducibility equations requires a further concept, that of "nonminimal solution."

17.3.5. Nonminimal Solutions

The requirement that the BRST cohomology at ghost number zero should be given by the gauge-invariant functions does not completely determine the ghost spectrum. For instance, it is always possible to add to a given solution further variables that are cohomologically trivial and, hence, that do not modify $H^k(s)$. This is just as in the Hamiltonian formalism. Solutions of the master equation containing the further variables are known as nonminimal solutions.

Cohomologically trivial variables fulfill

$$s\bar{C} = b, \qquad sb = 0, \qquad \text{gh}\,(\bar{C}) = \text{gh}\,(b) - 1. \tag{17.33}$$

The condition $sF = 0$ eliminates $\bar{C}$, which is not BRST closed. The further passage to the cohomology eliminates b, which is BRST exact. If one requires a canonical action for the BRST symmetry, one must introduce antifields $\bar{C}^*$ and $b*$, respectively conjugate to $\bar{C}$ and b,

$$(\bar{C}, \bar{C}^*) = 1, \qquad (b, b^*) = 1, \tag{17.34a}$$

$$\text{gh}\ \bar{C}^* = -\text{gh}\ C - 1, \qquad \text{gh}\ b^* = -\text{gh}\ b - 1. \tag{17.34b}$$

The term that generates (17.33) through the antibracket reads

$$\bar{C}^* b. \tag{17.35}$$

One then finds $sb^* = \bar{C}^*$, $s\bar{C}^* = 0$, so the pair $b^*, \bar{C}^*$ is also cohomologically trivial.

More complicated nonminimal sectors could be introduced. This will not be done here because their usefulness is not clear. Thus, the solutions of the master equation considered here read

$$\bar{S} = S + \sum \bar{C}^* b, \tag{17.36}$$

where S is the "minimal" solution described above, depending on the minimal set of fields $\phi^i, C^\alpha, C^A, \phi^*_i, \phi^*_\alpha, \phi^*_A$, and where $b, \bar{C}, b^*, \bar{C}^*$ stand for all the trivial variables of the nonminimal sector. The number of required extra pairs depends on the type of gauge-fixing condition that is desired.

The passage from one generating set to another one containing more reducibility becomes now clear. The change of generating sets simply increases the nonminimal sector by cohomologically trivial pairs, since for each new variable that is added, there is another new variable that removes it in cohomology.

17.3.6. Antibracket and BRST Cohomology

Finally, we note that the antibracket structure passes on to the set $H^*(s)$ of BRST cohomological classes, since the BRST transformation is a canonical transformation. However, contrary to what happens with the Poisson bracket, it appears that the induced antibracket in $H^*(s)$ is not of direct physical interest. First of all, the antibracket of two observables is not itself observable, since it possesses ghost number one. Furthermore, in the group case, (A, B) is actually BRST exact, *i.e.*, vanishes in cohomology (Exercise 17.5).

17.4. GAUGE INVARIANCE OF THE SOLUTION OF THE MASTER EQUATION

17.4.1. Abelian Form of S

By a redefinition (17.32a) of the gauge transformations, one can abelianize the gauge transformations. Furthermore, in a region of field space one can redefine the field variables $\phi^i \to \chi^i = \chi^i(\phi^j)$, $\chi^i = (\chi^{\bar{a}}, \chi^{\bar{\alpha}})$ in such a way that (i) the first variables $\chi^{\bar{a}}$ are gauge invariant; and (ii) the gauge transformations are just shifts in the last variables $\chi^{\bar{\alpha}}$. This change of variables is generically nonlocal in spacetime. A proof of these properties is outlined in Exercise 17.6.

The action S_0 depends only on $\chi^{\bar{a}}$, as it is gauge invariant. Together with the boundary conditions, the equations $\delta S_0/\delta\chi^{\bar{a}} = 0$ completely determine $\chi^{\bar{a}}$. The gauge components $\chi^{\bar{\alpha}}$ are, on the other hand, completely arbitrary.

The fields $\chi^{\bar{a}}$ may not be all propagating (the equations $\delta S_0/\delta\chi^{\bar{a}} = 0$ may imply $\chi^{\bar{a}} = 0$ for some $\bar{a}$), so that the number of true degrees of freedom is in general smaller than the number of $\chi^{\bar{a}}$'s (see Exercise 17.7).

A complete set of gauge transformations is given by

$$\delta\chi^{\bar{a}} = 0, \qquad \delta\chi^{\bar{\alpha}} = \varepsilon^{\bar{\alpha}}. \tag{17.37a}$$

The remaining (reducible) gauge transformations can be taken to be

$$\delta\chi^i = R_A{}^i\,\varepsilon^A, \qquad R_A{}^i \equiv 0 \tag{17.37b}$$

$[\alpha = (\bar{\alpha}, A)]$. So, the reducibility equations read

$$Z_B{}^A R_A{}^i = 0, \qquad Z_B{}^A = \delta_B{}^A. \tag{17.37c}$$

The solution of the master equation is given by

$$S = S_0(\chi^{\bar{a}}) + \chi^*_{\bar{\alpha}}\, C^{\bar{\alpha}} + C^*_A C^A, \tag{17.38}$$

where C^*_A are the antifields of ghost number -2 conjugate to the ordinary ghosts associated with the ineffective gauge transformations (17.37b) and not the antifields conjugate to the ghosts of ghosts C^A. The solution S differs from the gauge-invariant action $S_0(\chi^{\bar{a}})$ by manifestly cohomologically trivial pairs that possess exactly the same structure as the nonminimal terms of (17.36).

In practice, the abelian form (17.38) of S is not very useful because it is, in general, impossible to actually find the variables $\chi^{\bar{a}}$ and $\chi^{\bar{\alpha}}$. However, the existence of (17.38) is a useful theoretical tool.

17.4.2. Gauge Transformations of S

We now return to the solution S of the master equation in an arbitrary representation of the gauge symmetry. A remarkable feature of S is that it is invariant under a set of gauge transformations that has an interesting structure.

To display these gauge transformations, we adopt simplified notations. Let ϕ^A $(A = 1, \ldots, N)$ denote collectively the original fields ϕ^i, the ghosts C^α, and all the necessary ghosts of ghosts. These also include the antighosts $\bar{C}$ and the auxiliary fields of the nonminimal sector, if any. We will refer to ϕ^A as the "fields." All the remaining variables, *i.e.*, the antifields $\phi^*_i, \phi^*_\alpha, \ldots, \bar{C}^*, b^*$, etc., will be denoted by ϕ^*_A. From now on, the ϕ^*_A stand thus for all the antifields and not just the antifields of antighost number 3 associated with the reducibility equations. Finally, let us set

$$z^a = (\phi^A; \phi^*_A), \qquad a = 1, \ldots, 2N, \tag{17.39a}$$

and

$$\varepsilon(z^a) = \varepsilon_a. \tag{17.39b}$$

The antibracket can be written as

$$(A, B) = \frac{\delta^R A}{\delta z^a} \zeta^{ab} \frac{\delta^L B}{\delta z^b}, \tag{17.40a}$$

where the (inverse of the) odd symplectic form ζ^{ab} reads

$$\zeta^{ab} = \begin{pmatrix} 0 & \delta_B{}^A \\ -\delta_B{}^A & 0 \end{pmatrix}, \qquad \zeta^{ab} = -\zeta^{ba}. \tag{17.40b}$$

In these notations, the master equation becomes

$$(S, S) = \frac{\delta^R S}{\delta z^a} \zeta^{ab} \frac{\delta^L S}{\delta z^b} = 0, \tag{17.41a}$$

from which one easily derives, upon differentiation with respect to z^c, that

$$\frac{\delta^R S}{\delta z^a} \mathcal{R}_c{}^a = 0. \tag{17.41b}$$

Here, we have set

$$\mathcal{R}_c{}^a = \zeta^{ab} \frac{\delta^L \delta^R S}{\delta z^b \, \delta z^c}. \tag{17.41c}$$

These equations, which express that $s^2 z^c$ is zero, indicate that the functional S is gauge invariant under

$$\delta z^a = \mathcal{R}_c{}^a \varepsilon^c \left(\Leftrightarrow \delta z^a(x) = \int \mathcal{R}_c{}^a(x, y) \, \varepsilon^c(y) \, dy \right), \tag{17.42}$$

where $\varepsilon^c(y)$ are arbitrary spacetime functions.

How many gauge invariances does S possess? Superficially, $2N$, which is the total number of fields and antifields. It actually turns out that the action S has less independent gauge invariances because the matrix $\mathcal{R}_c{}^a$ defining the gauge transformations is nilpotent on-shell,

$$\mathcal{R}_b{}^a \mathcal{R}_c{}^b \approx 0 \qquad \left(i.e., \text{ when } \frac{\delta S}{\delta z^a} = 0 \text{ hold} \right). \tag{17.43}$$

This can be seen by a further differentiation of (17.41b). The gauge transformations (17.42) are thus not independent; there is "on-shell reducibility."

Because $\mathcal{R}^2 \approx 0$, the number of independent gauge transformations in (17.42) is at most equal to N. It is actually precisely equal to N because, as will be seen immediately below, the general solution of $\mathcal{R}v \approx 0$ is given by $v \approx \mathcal{R}t$, so that the nilpotent matrix $\mathcal{R}$ contains only two-dimensional Jordan-blocks $\left(\begin{smallmatrix} 0 & 1 \\ 0 & 0 \end{smallmatrix}\right)$, which have rank one. This makes its total rank equal to $2N/2 = N$. One can further show that (17.42) exhausts all the gauge symmetries of S. So, the solution of the master equation possesses exactly N independent gauge transformations.

The proofs of the statements are most conveniently carried out by making the canonical change of variables $z^a \to \bar{z}^a$ such that $S(\bar{z}^a)$ takes the abelian form of the previous subsection. This is permissible because at an extremum of S the matrix $\mathcal{R}_b{}^a$ transforms as a $\left(\begin{smallmatrix} 1 \\ 1 \end{smallmatrix}\right)$ tensor under canonical transformations, so its rank properties are unchanged in $z^a \to \bar{z}^a$. One easily checks that the gauge transformations of $S(\bar{z}^a)$ are just the arbitrary shifts in the N variables—say, $\bar{z}_2^A$—that do not ocur in $S(\bar{z}^a) \equiv S(\bar{z}_1^A)$, and that these transformations can be written as in (17.42) because the matrix $\delta^2 S/\delta \bar{z}_1^A \, \delta \bar{z}_1^B$ is invertible.

It is remarkable that all the gauge symmetries of the solution of the master equation can be obtained by merely differentiating S. Furthermore, for each field-antifield pair, there is one gauge symmetry. These properties would not hold if S were not a *proper solution* of the master equation, *i.e.*, if it did not obey the boundary conditions (17.28c) and (17.28d) that guarantee that the differential δ in $s = \delta +$ "more" is acyclic in degree $k \neq 0$. For instance, the gauge symmetries of the original gauge-invariant action S_0, which is a nonproper solution of the master equation, cannot be obtained by differentiating S as in (17.41c) (Exercises 17.8 and 17.9).

The fact that S possesses gauge invariances implies that the quantum-mechanical transition amplitude cannot be written as a path integral over both the fields and the antifield histories of $\exp[(i/\hbar)S(\phi,\phi^*)]$,

$$\int [D\phi \, D\phi^*] \exp \frac{i}{\hbar} S(\phi, \phi^*) \qquad \text{(incorrect)} \tag{17.44}$$

because this expression is ill defined. Furthermore, for a system without gauge freedom, the expression (17.44) would not reduce to the path integral of Chapter 15, where one integrates only over the fields. Since there is one gauge invariance for each field-antifield pair, it should be possible to get a well-defined path integral by eliminating the antifields and integrating only over the fields, as in the absence of gauge invariance. How this can be done for an arbitrary gauge system is the subject of the next chapter.

EXERCISES. CHAPTER SEVENTEEN

17.1. Consider the action $S_0 = \int [\frac{1}{2}(\dot{\phi}^2 - \phi'^2) + \lambda(\dot{\phi} - \phi')^2]\, dt\, d\sigma$ for a chiral boson in two dimensions.

(a) Show that the regularity conditions on the equations of motion are not fulfilled. [Hint: $(\dot{\phi} - \phi')^2 = 0$ vanishes quadratically on-shell and cannot be derived from the other equations of motion by local (in spacetime) means.]

(b) Prove that one cannot replace $(\dot{\phi} - \phi')^2$ by $(\dot{\phi} - \phi')$ in the Lagrangian without changing the number of degrees of freedom of the theory.

(c) Show that, by contrast, the Hamiltonian constraints following from S_0 can be chosen so as to obey the regularity conditions.

17.2. An odd bracket structure is said to be real if $(A, B)^* = (B^*, A^*)$ and imaginary if $(A, B)^* = -(B^*, A^*)$. Show that by taking the antifield associated with a real field to be imaginary (and vice versa), the odd bracket structure considered in the text is real.

17.3. Consider a phase space with two bracket structures, one even [,] and one odd (,). Assume that the equations of motion are given by

$$\dot{f} = [f, h] = (f, Q),$$

where h is even and Q is odd. That is, the motion is simultaneously generated by an even generator in the Poisson bracket and an odd generator in the antibracket. Show that the dynamics is invariant under the supersymmetry transformation $\delta_Q f = [f, Q]$ generated by Q in the even bracket. [Hint: Prove that $[h, Q]$ vanishes by using $(Q, Q) = -(Q, Q) = 0$.]

17.4. Construct an explicit example of a supersymmetric dynamics of the type analyzed in the previous exercise. [One simple example is a system with one Bose oscillator (a, a^*) and one Fermi oscillator (A, A^*). The Hamiltonian and supersymmetry charge are, respectively, $h = a^* a + A^* A$ and $Q = a^* A + A^* a$. The nonvanishing brackets are $[a, a^*] = -i = [A, A^*]$, $(a, A^*) = -i = (A, a^*)$.]

17.5. Consider a gauge theory for which the gauge transformations form a group. The gauge orbits are integrable everywhere in I and not just on Σ. Show that the antibracket structure induced in $H^*(s)$ is trivial, *i.e.*, $(A, B) = (K, S)$ if $(A, S) = 0$ and $(B, S) = 0$. (Hint: Show that in each equivalence class of BRST-closed functions, one can find one that does not involve the antifields.)

17.6. (Proof of abelianization of the gauge transformations)

Let $S(q^i)$ be a function of $q^i \in R^n$. Assume that the equations $\partial S/\partial q^i = 0$ are degenerate. Then, $\partial S/\partial q^i = 0$ defines a manifold Σ of dimension m, with $0 < m \leq n$. The gauge transformations are

$$\delta_\varepsilon q^i = R_\alpha{}^i \varepsilon^\alpha, \qquad \delta_\varepsilon S = 0, \qquad \alpha = 1, \ldots, m,$$

where the matrix $R_\alpha{}^i(q)$ is of rank m. Without loss of generality, one can assume that the coordinates $q^i = (q^a, q^\alpha)$ are locally such that $R_\beta{}^\alpha$ is invertible.

(a) Show that $q^\alpha = \mathring{q}^\alpha$ are good gauge conditions and that the equations $\partial S/\partial q^\alpha = 0$ are consequences of the equations $\partial S/\partial q^a = 0$.

(b) Show that the matrix $T_{ab} \equiv \partial/\partial q^a (\partial S/\partial q^b)$ is invertible. [Hint: The $n - m$ functions $\partial S/\partial q^a$ can be used as first coordinates in the vicinity of $\partial S/\partial q^i = 0$, the matrix $T_{ia} \equiv \partial/\partial q^i(\partial S/\partial q^a)$ is thus of rank $n - m$ at the stationary point, and $T_{\alpha a}$ can be expressed in terms of T_{ab} at $\partial S/\partial q^i = 0$ by means of the Noether identity.]

(c) Show that one can make a q^α-dependent, invertible, smooth change of variables

$$q^a \rightarrow x^a = x^a(q^b, q^\alpha)$$

such that $(\partial S/\partial q^a) = 0$ is equivalent to $x^a = 0$ and such that S takes the form

$$S = \eta_{ab}\, x^a\, x^b, \qquad \eta_{ab} = \text{diag}(\pm 1)$$

in the vicinity of the critical point $x^a = 0$. [Hint: Show that for fixed q^α, the stationary problem fixes q^a as a function of q^α. Observe that the stationary point is nondegenerate by (b) and use the fact that a function can always be brought to the quadratic form $\eta_{ab}\, x^a x^b$ in the vicinity of a nondegenerate critical point (Morse lemma).]

(d) What is the form of the gauge transformations in the new coordinate system (x^a, q^α)?

(e) Extend the abelianization theorem to an action $S(q^i, \alpha^A)$ that depends not only on the dynamical variables q^i but also on unvaried extra variables α^A that stand for the boundary conditions.

17.7. (Example of abelianization) Consider the action $S_0 = \frac{1}{2}\int (q - \dot{x})^2\, dt$, invariant under $\delta q = \dot{\varepsilon}$, $\delta x = \varepsilon$.

(a) Show that the change of variables $w = q - \dot{x}$, $x = x$, is permissible and brings the gauge transformations to the canonical form of (17.37a).

(b) Check that S_0 depends only on w and that w is nonpropagating.

(c) Repeat the analysis for $S_0 = \frac{1}{2}\int [(q - \dot{x}^2) + (x - \dot{y})^2]\, dt$.

17.8. Consider the action $S_0[q] = 0$. This is a nonproper solution of the master equation $(S_0, S_0) = 0$.

(a) Show that the gauge transformations of S_0 cannot be obtained by differentiation of S_0.

(b) Write the proper solution of the master equation and check that all its gauge symmetries are given by (17.42).

17.9. Same exercise with the action of Exercise 17.7,

$$S_0[q, x] = \tfrac{1}{2}\int (q - \dot{x})^2\, dt.$$

17.10. Consider the extended Hamiltonian action $S_E(p, q, \lambda) = \int (p_i \dot{q}^i - H_0 - \lambda^a G_a)\, dt$ with first-class constraints $G_a \approx 0$. Show that the passage to the variables χ appearing in the abelianization of the gauge transformations in §17.4.1 amounts to abelianizing the constraints and redefining the Lagrange multipliers as $\lambda^a \rightarrow \lambda^a - \dot{Q}^a$ (in the canonical coordinates where the abelianized constraints are $P_a \approx 0$, *i.e.*, momenta conjugate to $Q^a \approx 0$).

17.11. Let $S[y^i, z^\alpha]$ be an action depending on auxiliary fields z^α. That is, the equations $\delta S/\delta z^\alpha = 0$ can be solved to yield z^α as a function of y^i and its derivatives. Let $\bar{S}[y^i] = S[y^i, z^\alpha(y)]$. Show that the covariant phase space for $S[y^i, z^\alpha]$ is isomorphic to the covariant phase space for $\bar{S}[y^i]$. (Hint: Use the results of Exercise 3.17.) Infer that the covariant phase space of §17.1.3 is isomorphic to the reduced phase space of Chapter 2. [Recall that (*i*) passing to the total Hamiltonian formalism amounts to introducing auxiliary fields, and (*ii*) at a given instant of time, all the first-class constraints—and not just the primary ones—act as gauge generators.]

17.12. (Peierls bracket).
Let $A_0[\phi^i]$ and $B_0[\phi^i]$ be two observables depending on the dynamical variables over a finite time interval. The Peierls bracket $[A_0, B_0]$ is defined as follows. Add to the action an infinitesimal perturbation εA_0; the solutions of the equations of motion then get modified. Let ϕ^i be a given solution of the unperturbed problem and $\phi^i + \varepsilon\delta^-_{A_0}\phi^i$ be a solution of the perturbed problem that reduces to ϕ^i in the remote past (*i.e.*, before the interval). If one replaces ϕ^i by $\phi^i + \varepsilon\delta^-_{A_0}\phi^i$, the on-shell value of B_0 gets modified as $B_0 \to B_0 + \varepsilon\, D_{A_0}B_0$, with

$$D_{A_0}B_0 = \frac{\delta B_0}{\delta\phi^i}\,\delta^-_{A_0}\phi^i.$$

Even though $\delta^-_{A_0}\phi^i$ is defined up to a gauge transformation, $D_{A_0}B_0$ is unambiguous because B_0 is (on-shell) gauge invariant. The Peierls bracket is

$$[A_0, B_0](\phi^i) = D_{A_0}B_0 - D_{B_0}A_0$$

(where ϕ^i is *on-shell*).

(a) Verify explicitly that $[A_0, B_0]$ is on-shell gauge invariant, as well as invariant under the replacements $A_0 \to A_0 + \lambda^i(\delta S_0/\delta\phi^i), B_0 \to B_0 + \mu^i(\delta S_0/\delta\phi^i)$. That is, $[A_0, B_0]$ is defined in the covariant phase space.

(b) Show that the Peierls bracket is invariant under the introduction of auxiliary fields.

(c) Prove that the Peierls bracket coincides with the Hamiltonian Poisson bracket defined in the reduced phase space. [Hint: Use (b) and Exercises 1.4 and 17.11. Compute then the Peierls bracket for the total action.]

17.13. The reduced phase space symplectic structure can be described covariantly as follows. The variation of the action S_0 is given by

$$\delta S_0 = \int \left(\frac{\delta\mathcal{L}_0}{\delta\phi^i}\,\delta\phi^i + \partial_\mu j^\mu\right) d^n x$$

where

$$j^\mu = \frac{\delta\mathcal{L}}{\delta(\partial_\mu\phi^i)}\,\delta\phi^i + \cdots + \frac{\delta\mathcal{L}}{\delta(\partial_{\mu\mu_1\ldots\mu_k}\phi^i)}\,\partial_{\mu_2\ldots\mu_k}\delta\phi^i.$$

The variation δS_0 can be viewed as an exact one-form in the space of all histories.

(a) Using the nilpotency of the exterior derivative δ, $\delta^2 = 0$, show that $\int\partial_\mu(\delta j^\mu)\,d^n x = 0$ on the stationary surface. Hence, the flux of the "symplectic current" $\int \delta j^\mu\,d\varepsilon_\mu$ through a spacelike hypersurface in spacetime does not depend on that hypersurface and defines a closed two-form σ on the stationary surface. This two-form induces a symplectic structure in the quotient space of the stationary surface by the null surfaces of σ.

(b) Show that this symplectic structure is invariant under the introduction of auxiliary fields.

(c) Verify that the quotient space of (a) and its symplectic structure coincides, respectively, with the reduced phase space and its symplectic structure defined in Chapter 2.

CHAPTER EIGHTEEN

ANTIFIELD FORMALISM AND PATH INTEGRAL

The antifield formalism enables one to describe covariantly the observables of a gauge theory in terms of the cohomology of a differential complex that contains spacetime fields, rather than canonical variables at a given instant of time. The description is intrinsic in that no gauge-fixing condition is ever required. It remains to indicate how physical transition amplitudes are computed within the formalism. The purpose of this chapter is to answer that question.

The analysis is organized as follows. First, we write down the central formulas for the path integral in the antifield formalism ("quantum master equation", "quantum BRST symmetry"). This is done somewhat heuristically by making the results plausible. We then justify the formulas by showing that they yield the same answers as the operator formalism.

It should be stated from the outset that the quantum antifield formalism contains infinite expressions that need to be regulated. However, no universal regulation scheme (*i.e.*, a scheme independent of the particular system being dealt with) has been devised at the time of writing. In spite of its mathematical limitations, the formalism is far from being empty, and one feels that those of its central properties already ascertained will survive future developments.

18.1. QUANTUM MASTER EQUATION

18.1.1. Integration of p-Vectors on a Supermanifold

The solution $S(\phi, \phi^*)$ of the master equation and the BRST-invariant functions $A(\phi, \phi^*)$ are polynomials in the antifields. According to the geometric interpretation discussed in §15.5.4, they define multi-vectors on the supermanifold of the fields. The component S independent of ϕ^*_A is a 0-vector (scalar), the component $S^A(\phi)\,\phi^*_A$ is a 1-vector, etc. . . .

To write down the path integral, one needs to know how to integrate expressions involving S or A over the field supermanifold. To that end, it is necessary to generalize the concept of integration from functions to arbitrary p-vectors.

The generalization is guided by an important property of the integral $\int F(\phi)\,[D\phi]$ of a function. If $F(\phi)$ is equal to the divergence ΔV of the vector $V(\phi, \phi^*) = V^A(\phi)\,\phi^*_A$,

$$F = \Delta V, \tag{18.1}$$

$$\Delta = \frac{\delta^R}{\delta\phi^A}\frac{\delta^R}{\delta\phi^*_A}(-)^{\varepsilon_A+1}, \qquad \varepsilon(\Delta) = 1,\ \Delta^2 = 0, \tag{18.2}$$

then

$$\int \Delta V\,[D\phi] = 0 \tag{18.3}$$

(Stokes theorem). As was seen in §15.5.4, this equality follows from the translation invariance of $[D\phi] = \prod_x d\phi$ (nontrivial measure factors, if any, are included in the integrand) and implies the crucial Schwinger–Dyson equation. We shall demand that the searched-for integral of p-vectors preserves the key property (18.3). That is, we shall require that (18.3) holds not just for 1-vectors but also for p-vectors with $p \neq 1$.

One (unsatisfactory) generalization of $\int F(\phi)\,[D\phi]$ that meets this requirement is simply

$$\int \{V_0(\phi) + V_1^A(\phi)\,\phi^*_A + V_2^{AB}(\phi)\,\phi^*_A\,\phi^*_B + \cdots\}\,[D\phi] = \int V_0(\phi)\,[D\phi]. \tag{18.4a}$$

This definition picks the zeroth-order component of V. One can rewrite (18.4a) as

$$\int V\,[D\phi] = \int V(\phi, \phi^* = 0)\,[D\phi], \tag{18.4b}$$

since $V(\phi, \phi^* = 0) = V_0(\phi)$.

The definition (18.4) is quite allright for p-vectors V such that the integral $\int V_0\,[D\phi]$ exists. However, it is not suitable for our purposes because it cannot be used for the solution $S(\phi,\phi^*)$ of the master equation itself. Indeed, the zeroth-order term of $S(\phi,\phi^*)$ in the original gauge-invariant action $S_0(\phi^i)$ and, thus, the integral $\int [D\phi^i\, D(\text{ghosts})] \times \exp{(i/\hbar)}\, S_0(\phi^i)$ is ill defined: One gets $\delta(0)$ for the integration along the gauge directions and 0 for the integration over the ghosts.

One can easily remedy this problem by slightly modifying (18.4b). Rather than setting the antifields equal to zero, one sets them equal to functions $\omega_A(\phi)$ of the fields. These functions should be pure gradients

$$\phi^*_A = \frac{\delta\psi}{\delta\phi^A} \tag{18.5a}$$

for Stokes theorem to hold (Exercise 18.1), so we define the integral of $V = V_0(\phi) + V_1^A(\phi)\phi^*_A + V_2^{AB}(\phi)\,\phi^*_A\,\phi^*_B + \cdots$ as

$$\int V\,[D\phi] = \int V\Big(\phi,\phi^* = \frac{\delta\psi}{\delta\phi}\Big)[D\phi]. \tag{18.5b}$$

The function $\psi(\phi)$ must be of ghost number -1 and Grassmann parity 1 to preserve the grading properties of the fields and antifields,

$$\mathrm{gh}\,\psi = -1, \qquad \varepsilon(\psi) = 1. \tag{18.5c}$$

For the definition (18.5) to be acceptable, the integral should not depend on ψ. This is the case if V is Δ-closed. Indeed, one has

Theorem 18.1. *The integral (18.5b) does not depend on ψ if and only if $\Delta V = 0$.*

Proof. The proof is direct and left to the reader.

It is thus possible to define a satisfactory integral for Δ-closed p-vectors. This integral gives the same value to all Δ-closed p-vectors that differ by a Δ-exact one and so is defined in the space $H^*(\Delta)$ of the Δ-cohomology. The definition of the integral coincides with the original integral $\int F(\phi)\,[D\phi]$ of functions in the case of 0-vectors, since these are automatically Δ-closed and do not depend on ϕ^*_A.

If one sets $\psi = 0$, the integral (18.5) reduces to (18.4). When (18.4) is ill defined but (18.5) well defined for some $\psi \neq 0$, one can view (18.5) as a regularization of (18.4) in which the integrand is modified by a total derivative in field space.

18.1.2. Invariance under Canonical "Phase" Transformations

The integral of a p-vector possesses the property of being invariant under canonical "phase" transformations in the antibracket,

$$\phi^A \to \phi^A, \qquad \phi^*_A \to \phi^*_A + \frac{\delta\chi}{\delta\phi^A}. \tag{18.6}$$

There is a direct consequence of Theorem 18.1, since the transformation (18.6) simply amounts to replacing ψ by $\psi+\chi$. Because of (18.6), one can view the integral (18.5) with a definite choice of ψ, as obtained by (i) making a phase transformation (18.6) with $\chi=\psi$; and (ii) setting $\phi^*_A = 0$ afterwards.

For the purpose of studying the invariance properties of the integral under more general canonical transformation, it is useful to observe that the equations

$$\Omega_A(z^a) \equiv \phi^*_A - \frac{\delta\psi}{\delta\phi^A} = 0, \tag{18.7a}$$

by means of which the antifields are eliminated, are such that

$$(\Omega_A, \Omega_B) = 0. \tag{18.7b}$$

One says that the surface defined by $\Omega_A = 0$ is a "Lagrangian submanifold" (in the antibracket). This definition is invariant under arbitrary canonical transformations, since it is solely formulated in terms of the antibracket. It is easy to verify that any set of independent equations $\Omega_A(z^a) = 0$ that can be solved for the antifields and such that (18.7b) holds is equivalent to $\phi^*_A - \delta\psi/\delta\phi^A = 0$ for some ψ (Exercise 18.2).

18.1.3. Derivation of Quantum Master Equation

We shall postulate that the transition amplitude is given by the integral of $\exp(i/\hbar)W$, where $W(\phi,\phi^*)$ is a function of the fields and of the antifields that differs from $S(\phi,\phi^*)$ by terms of higher order in $\hbar$,

$$W = S + \hbar\, M_1 + \hbar^2 M_2 + \cdots, \qquad \mathrm{gh}\, W = 0, \tag{18.8}$$

$$\langle 1 \rangle = \int [D\phi] \exp\frac{i}{\hbar} W \equiv \int [D\phi] \exp\frac{i}{\hbar} W\Big(\phi, \frac{\delta\psi}{\delta\phi}\Big). \tag{18.9}$$

The leading term in the exponential,

$$S_\psi(\phi) \equiv S\Big(\phi, \frac{\delta\psi}{\delta\phi}\Big), \tag{18.10}$$

is known as the *gauge-fixed action.* The function ψ is known as the (Lagrangian) *gauge-fixing fermion* and must be such that $S_\psi(\phi)$ is nondegenerate, *i.e.*, the equations of motion $\delta S_\psi/\delta\phi^A = 0$ have unique solutions for arbitrary initial conditions. Furthermore, ψ should be taken to be a local functional so that S_ψ itself is a local functional,

$$\psi[\phi^A(x^\mu)] = \int \mu(\phi^A, \partial_\mu\phi^A, \ldots, \partial_{\mu_1}, \ldots, \partial_{\mu_k}\phi^A). \tag{18.11}$$

The higher order terms $M_1, M_2, \ldots$ in (18.8) stand for the integration measure (recall §15.1.5),

$$[D\mu] = [D\phi] \exp i(M_1 + \hbar\, M_2 + \cdots), \tag{18.12a}$$

$$\int [D\phi] \exp \frac{i}{\hbar} W = \int [D\mu] \exp \frac{i}{\hbar} S. \tag{18.12b}$$

The integral extends over histories obeying boundary conditions appropriate to the amplitude at hand.

For (18.9) to be independent of ψ, $\exp(i/\hbar)W$ should be Δ-closed,

$$\Delta \exp \frac{i}{\hbar} W = 0. \tag{18.13}$$

A straightforward calculation shows that this condition is equivalent to

$$i\hbar\, \Delta W - \tfrac{1}{2}(W, W) = 0. \tag{18.14}$$

Equation (18.14) is a cornerstone of the theory and is known as the *quantum master equation.* It guarantees that the integral (18.9) does not depend on the choice of ψ. The antibracket term (W, W) arises in (18.14) because Δ is not a derivation but obeys instead

$$\Delta(\alpha\beta) = \alpha\, \Delta\beta + (-)^{\varepsilon_\beta}(\Delta\alpha)\beta + (-)^{\varepsilon_\beta}(\alpha, \beta). \tag{18.15a}$$

One also has

$$\Delta(\alpha, \beta) = (\alpha, \Delta\beta) - (-)^{\varepsilon_\beta}(\Delta\alpha, \beta). \tag{18.15b}$$

At this point, (18.9) is a postulate. We shall justify this postulate by showing below that the transition amplitude takes, indeed, the form (18.9). To that end, it is necessary to investigate first the general properties of the quantum master equation.

Remarks.

(*i*) If the solution S of the master equation contains nonvanishing higher order terms $\overset{(n)}{S}$, the gauge-fixed action S_ψ will involve the ghosts C^α and the antighosts $\bar{C}_\alpha$ not just quadratically. It will also contain quartic, sixtic, ... ghost interactions.

(*ii*) Exactly as in the Hamiltonian formalism, the gauge-fixed action $S_\psi(\phi)$ depends, in general, on all the original fields ϕ^i. For instance, in electromagnetism, $S_\psi(\phi)$ depends on all the components of the vector potential A_μ. These components are summed over without restriction in the path integral. The terminology "gauge fixed" is used only to convey the fact that $S_\psi(\phi)$ is nondegenerate.

18.1.4. Quantum Averages

Let $A_0(\phi^i)$ be a gauge-invariant function. We shall postulate that the expectation value $\langle A_0 \rangle$ is given by the integral

$$\langle A_0 \rangle = \int \alpha \exp \frac{i}{\hbar} W \, [D\phi] \equiv \int \alpha\Big(\phi, \frac{\delta\psi}{\delta\phi}\Big) \exp \frac{i}{\hbar} W\Big(\phi, \frac{\delta\psi}{\delta\phi}\Big) \, [D\phi] \tag{18.16}$$

where $\alpha(\phi, \phi^*)$ may differ from the BRST-invariant extension $A(\phi, \phi^*)$ solution of $(A, S) = 0$ by correction terms of order at least $\hbar$,

$$\alpha = A + \hbar\,\alpha_1 + \hbar^2\alpha_2 + \cdots, \qquad \mathrm{gh}\,\alpha = 0. \tag{18.17}$$

The function α must be such that the integrand of (18.16) is Δ-closed, so that $\langle A_0 \rangle$ does not depend on ψ,

$$\Delta\Big(\alpha \exp \frac{i}{\hbar} W\Big) = 0. \tag{18.18a}$$

Taking (18.14) into account, one finds that (18.18a) is equivalent to

$$\sigma\alpha \equiv -i\hbar\,\Delta\alpha + (\alpha, W) = 0. \tag{18.18b}$$

Again, (18.16) is at this stage a postulate that will be justified below.

The function $A_\psi(\phi) = A(\phi, \phi^* = \delta\psi/\delta\phi)$ is known as the gauge-fixed form of A. The next terms in (18.17) stand for the "quantum corrections" that may arise from the integration over the momenta if (18.16) is not the phase space path integral (§15.1.5).

18.1.5. Quantum BRST Symmetry—Ward Identity

To zeroth order in $\hbar$, the quantum master equation reduces to the classical master equation

$$\tfrac{1}{2}(W, W) = i\hbar\,\Delta W \overset{\hbar=0}{\longrightarrow} (S, S) = 0. \tag{18.19a}$$

Similarly, Eq. (18.18b) reads

$$\sigma\alpha = 0 \overset{\hbar=0}{\longrightarrow} sA \equiv (A, S) = 0. \tag{18.19b}$$

Equations (18.14) and (18.18b) obtained by demanding that the path integral be independent of ψ become, thus, in the classical limit, the fundamental equations of the classical antifield-BRST formalism that express gauge invariance.

Because σ reduces to s in the limit $\hbar = 0$, it is called the "quantum BRST symmetry." The key feature of σ is its nilpotency,

$$\sigma^2 = 0, \tag{18.20}$$

which follows from (18.2), (18.14), and (18.15b). In the case of a system without gauge invariance, σ reduces to the Schwinger–Dyson operator of §15.5.3. Thus, σ incorporates both the BRST symmetry and the Schwinger–Dyson equation. Although nilpotent, σ fails to be a derivation by terms of order $\hbar$,

$$\sigma(\alpha\beta) = \alpha(\sigma\beta) + (-)^{\varepsilon_\beta}(\sigma\alpha)\beta - i\hbar(-)^{\varepsilon_\beta}(\alpha, \beta). \tag{18.21}$$

The Stokes theorem $\int \Delta V\,[D\phi] = 0$ can be rewritten in terms of σ as

$$\langle\sigma\beta\rangle \equiv \int (\sigma\beta)\,\exp\frac{i}{\hbar}\,W\,[D\phi] = 0. \tag{18.22}$$

The path integral associates, therefore, a well-defined number to a cohomological class of σ, *i.e.,* is defined in the space $H^*(\sigma)$ of the σ-cohomology.

Given a gauge-invariant function on $A_0(\phi^i)$, its BRST extension A is defined up to an s-exact term. That is, if A and A' are two BRST extensions of A_0, then $A' - A = sB$. One can then choose the corresponding solutions of (18.18) so that $\alpha' - \alpha$ is σ-exact, $\alpha' - \alpha = \sigma B$, since this equation reduces to $A' - A = sB$ to first order in $\hbar$ (more on this in §18.2.2). The σ-closed functions α' and α so related yield the same expectation value for A_0.

18.1.6. Zinn–Justin Equation

The Ward identities $\langle\sigma\beta\rangle = 0$ can be conveniently reexpressed in terms of the effective action. The analysis proceeds as in §16.4.4.

The starting point is the generating functional with sources j_A coupled to the fields ϕ^A and further sources L_A introduced through

$$Z_\psi[j, L] = \langle 1\rangle_{\psi,j,L} = \int [D\phi]\,\exp\left[\frac{i}{\hbar}W\left(\phi, L + \frac{\delta\psi}{\delta\phi}\right) + j_A\phi^A\right]. \tag{18.23}$$

The sources j_A occur linearly, but the sources L_A only do so if the solution W of the quantum master equation is linear in the antifields. The change of integration variables $\phi^A \to \phi^A + (\delta^L W/\delta\phi^*)(\phi, L + \delta\psi/\delta\phi)\varepsilon$ yields then the equation

$$j_A \left\langle \frac{\delta^L W}{\delta\phi^*}\left(\phi, L + \frac{\delta\psi}{\delta\phi}\right)\right\rangle = 0, \tag{18.24a}$$

i.e.,

$$j_A \frac{\delta^L Z_\psi}{\delta L_A} = 0. \tag{18.24b}$$

Indeed, by virtue of the quantum master equation, the change in the measure $[D\phi]$, given by $D[\phi + \delta W/\delta\phi^* \varepsilon] = [D\phi](1 + \Delta W \cdot \varepsilon)$ is compensated for by the change in $\exp(iW/\hbar)$, given by $\exp[iW(\phi')/\hbar] = [1 + (i/2\hbar)(W, W)\varepsilon] \exp(iW/\hbar)$, so that the variation of (18.23) reduces to (18.24a).

If one defines the effective action $\Gamma_\psi[\phi_c^A, L_A]$ as the Legendre transformation of $\ell n\, Z_\psi$ with respect to the sources j_A,

$$\phi_c^A = \frac{\hbar}{i}\frac{\delta^L \ell n\, Z_\psi}{\delta j_A}, \qquad j_A = -\frac{\delta^R \Gamma_\psi}{\delta\phi_c^A}, \tag{18.25a}$$

$$\Gamma_\psi[\phi_c^A, L_A] = \frac{\hbar}{i}\,\ell n\, Z_\psi - j_A \phi_c^A, \tag{18.25b}$$

one can rewrite (18.24) as

$$(\Gamma_\psi, \Gamma_\psi) = 0 \tag{18.26a}$$

with

$$(\phi_c^A, L_B) = \delta^A{}_B. \tag{18.26b}$$

The "Zinn–Justin equation" (18.26a) takes exactly the same form as the classical master equation. It plays a fundamental role in the study of the renormalization of gauge theories, where the antibracket structure turns out to be extremely useful.

18.2. SOLUTION OF THE QUANTUM MASTER EQUATION

18.2.1. Ambiguity in W

We shall analyze the quantum master equation order by order in $\hbar$. To zeroth order in $\hbar$, (18.14) becomes the classical master equation $(S, S) = 0$, whose solution, for a given S_0, exists and is unique up to a canonical transformation in the antibracket. Is there an analogous property for W? To analyze this issue, we shall assume that $H^1(s) = H^1(d)$ vanishes, as this greatly simplifies the discussion.

Let us explicitly expand out the quantum master equation order by order in $\hbar$. One finds

$$(M_1, S) = i\,\Delta S \tag{18.27a}$$

and

$$(M_p, S) = N_p \qquad (p \geq 2) \tag{18.27b}$$

with

$$N_p = i\,\Delta M_{p-1} - \tfrac{1}{2} \sum_{n=1}^{p-1} (M_n, M_{p-n}). \tag{18.27c}$$

These equations can be solved recursively if and only if $(\Delta S, S) = 0$ and $(N_p, S) = 0$. These conditions are necessary because the BRST symmetry is nilpotent and sufficient because we assume $H^1(s) = 0$.

If $H^1(s) \neq 0$, the conditions would not be sufficient. One would have to check that ΔS and N_p are in the zeroth cohomological class. If they are not, then there is no solution to (18.27), and the theory is "anomalous." However, checking that ΔS and N_p belong to the zeroth cohomological class is hard to do directly, since ΔS and N_p are actually singular expressions.

It is easy to verify both $(\Delta S, S) = 0$ and $(N_p, S) = 0$ (Exercise 18.6). Accordingly, the solution to the quantum master equation exists. Furthermore, one has the freedom of adding, at each order in $\hbar$, a BRST-closed function. Indeed, given M_n with $n < p$, Eq. (18.27) determines M_p only up to a BRST-invariant function,

$$M_p \to M_p' = M_p + \bar{M}_p, \qquad (\bar{M}_p, S) = 0. \tag{18.28}$$

If $\bar{M}_p$ is exact, $\bar{M}_p = (K_p, S)$, one can choose the next terms $M_{p+1}, M_{p+2}, \ldots$ in W in such a way that the ambiguity $M_p \to M_p + (K_p, S)$ does not modify the path integral. This is obvious if K_p is infinitesimal, since then the change in W can be taken to be such that

$$\exp \frac{i}{\hbar} W' = (1 + \sigma\beta) \exp \frac{i}{\hbar} W. \tag{18.29}$$

The equality $\langle \sigma\beta \rangle = 0$ implies that the path integral is unchanged (Exercise 18.7).

However, $\bar{M}_p$ in (18.28) need not be BRST-exact. In that case, the path integral is modified. Taking into account the isomorphism of $H^0(s)$ with the set of gauge-invariant functions, one sees that *the quantum master equation leaves the freedom of adding a new, independent gauge-invariant function at each order in $\hbar$.* The measure is thus not determined.

That the quantum master equation does not completely determine the measure is not surprising. This equation expresses BRST invariance (*i.e.*,gauge invariance) and nothing more, so it cannot fix the measure in the physical sector. To get the complete expression for the measure, one needs an extra input, such as the operator formalism.

18.2.2. Ambiguity in α

Just as W is not uniquely determined by S and the quantum master equation alone, the quantum corrections to be added to A are not uniquely determined by the equations $\sigma\alpha = 0$. Indeed, the expansion of this equation in powers of $\hbar$ reads

$$(\alpha_p, S) = B_p \qquad (p \geq 1) \tag{18.30a}$$

with

$$B_p = i\,\Delta\alpha_{p-1} - \sum_{k=0}^{p-1} (\alpha_k, M_{p-k}). \tag{18.30b}$$

These equations have a solution if and only if $(B_p, S) = 0$ [assuming again $H^1(s) = 0$]. This condition can easily be checked using the Jacobi identity, the nilpotency of Δ, and the relation (18.15b) (Exercise 18.8). Accordingly, the quantum corrections to A exist and can be found recursively from (18.30), starting from $p = 1$.

However, the solution is not unique. Given A_n with $n < p$, Eq. (18.30a) determines α_p only up to a BRST-invariant function,

$$\alpha_p \to \alpha'_p = \alpha_p + \bar{\alpha}_p, \qquad (\bar{\alpha}_p, S) = 0. \tag{18.30c}$$

If $\bar{\alpha}_p$ is exact, $\bar{\alpha}_p = (B_p, S)$, the resulting ambiguity in α can be taken to be of the form $\alpha \to \alpha + \sigma\beta_p$ and does not change the path integral. But again, $\bar{\alpha}_p$ need not be BRST-exact.

We come thus again to the conclusion that the *quantum BRST invariance equation $\sigma\alpha = 0$ leaves the freedom of adding to α a new, independent gauge-invariant function at each order in $\hbar$*. Again, this is not surprising, since the mere condition of gauge invariance (= BRST invariance) puts no restriction on the gauge-invariant sector. To get the complete expression of α, one needs an extra input, as in the case of the measure. This can again be done through the operator formalism.

18.2.3. Example

One can illustrate the ambiguity in the solution of the master equation in its simplest context, namely, a system without gauge invariance.

In that case, $S_0(\phi^i)$ is a proper solution of the classical master equation, $S = S_0$. Furthermore, since S_0 does not involve the antifields, $\Delta S = \Delta S_0$ identically vanishes. The general solution of the quantum master equation is given by

$$W(\phi^i) = S_0(\phi^i) + \sum_{p\geq 1} \hbar\, M_p(\phi^i), \tag{18.31}$$

where $M_p(\phi^i)$ are arbitrary functions of the fields ϕ^i.

In the case of the phase space path integral with canonical coordinates $\phi^i \equiv (q,p)$, the correct measure is the Liouville measure. It is obtained by setting $M_p = 0$. The master equation alone cannot be used to distinguish this solution from the incorrect measures, which have $M_p \neq 0$ for at least one p. In the case of the configuration space path integral, the correct measure is $\prod_t \sqrt{g}\,[Dq]$ if the Lagrangian is quadratic in the velocities [Eq. (15.30)]. This corresponds to $M_1 = -\int dt\,(i/2)\,\delta(0)\,\ell n\, g$ and $M_p = 0$ for $p \geq 2$. Again, the quantum master equation is not sufficient to pick out the correct solution by itself.

18.2.4. Dimensional Regularization

It must be emphasized that the above manipulations are extremely formal. This is because singular factors of the form $\delta(0), \delta'(0), \ldots$ are generated by Δ when it acts on local functionals, since the second functional derivatives of a local functional evaluated at coincident arguments are singular when they do not vanish. Hence, such singular factors are present in the quantum correction terms.

It is not our purpose here to discuss how these infinities should be regulated. This is an important question that has, unfortunately, not been completely solved at the time of writing. Let us simply indicate that there exist regularization schemes used in field theory (for instance, dimensional regularization) for which the Jacobian of a change of variables local in spacetime is regularized to one. That means that ΔS is set equal to zero, since $(1+\Delta S)$ is the Jacobian of the change of variables defined by the BRST transformation $\phi^A \to \phi^A + (\delta^L S/\delta\phi^*_A)(\phi, \delta\psi/\delta\phi)\,\varepsilon$, which is a local transformation. One finds $D[\phi + \delta S/\delta\phi^*\,\varepsilon] = D[\phi](1 + \Delta S\cdot\varepsilon) = D[\phi]$ $(\Delta S = 0)$. Similarly, $\Delta A = 0$ if A is a local functional.

Setting the singular terms in the quantum master equation equal to zero reduces it to the classical master equation. Similarly, the condition $\sigma\alpha = 0$ becomes $sA = (A,S) = 0$. The path integral is then free from ambiguities, since the only freedom left by the classical formalism is the possibility of adding a BRST-exact term to the integrand. But the expectation value of such a term vanishes.

Accordingly, if it is permissible to drop the singular terms appearing in the quantum master equation, one may take $W = S$ and the measure is irrelevant. This, however, is not always the case (Exercise 18.10).

18.3. INVARIANCE OF THE FORMALISM UNDER CANONICAL TRANSFORMATIONS IN THE ANTIBRACKET

18.3.1. Antifield Formalism Can Only Be Justified up to Quantum Ambiguities in the Measure

We now start with the question of justifying the antifield formalism. The fact that the path integral associates a definite quantum average to the BRST cohomological classes is not sufficient because there is no guarantee so far that this quantum average, even though compatible with gauge invariance, is also compatible with unitarity. For instance, with inappropriate choices of the integration measure, one could find the absurd result $\langle 1 \rangle_\psi = 0$ (see Exercise 18.10).

It is clear that since the corrections $M_1, M_2 \dots$ to W and $\alpha_1, \alpha_2, \dots$ to A are not determined, the best that can be aimed at is to prove that up to these terms, the antifield formalism gives the same formal amplitudes as the operator formalism.

This will be done in two different ways. (i) In the first approach, developed in this section, we shall show that the path integral is invariant under canonical transformations in the antibracket and then show that it reduces, in the abelian representation of the gauge symmetry, to an integral over the gauge-invariant degrees of freedom only. (ii) In the second approach, we shall compare directly the antifield formalism with the Hamiltonian formalism and prove their equivalence.

18.3.2. More on Canonical Transformations

To study the invariance properties of the path integral, we shall need a few results on the superdeterminant of canonical transformations in the antibracket. Let $\phi^A(\phi', \phi'^*), \phi^*_A(\phi', \phi'^*)$ be a canonical transformation with superdeterminant J

$$J = \text{sdet}\frac{\delta(\phi, \phi^*)}{\delta(\phi', \phi'^*)}. \tag{18.32a}$$

Then, one finds that

$$\text{sdet}\frac{\delta\phi}{\delta\phi'} = J^{1/2}, \qquad \text{sdet}\frac{\delta\phi^*}{\delta\phi'^*} = J^{1/2} \tag{18.32b}$$

(Theorem 15.2) and

$$J = 1 + 2\,\Delta F \tag{18.32c}$$

if the transformation is an infinitesimal canonical transformation generated by F (Theorem 15.1).

The operator Δ is not covariant under canonical transformations. Rather, it transforms as

$$\Delta X = \Delta' X + \tfrac{1}{2}(X, \ell n\, J), \tag{18.33}$$

where $X(\phi, \phi^*)$ is an arbitrary function of the fields and the antifields. This property can be checked directly by using the canonical invariance of the antibracket. Or one can first verify (18.33) for infinitesimal transformations and then prove that (18.33) possesses the correct group property, namely, that if (18.33) is true for two transformations, then it is also true for their product. This implies that (18.33) passes on to the finite transformations.

Because $\Delta^2 = 0$, ΔJ vanishes for infinitesimal transformations [see (18.32c)]. This result exponentiates to yield

$$\Delta' \ell n\, J = -\tfrac{1}{4}(\ell n\, J, \ell n\, J) \tag{18.34a}$$

or, equivalently,

$$\Delta' J^{1/2} = 0. \tag{18.34b}$$

Again, this property can be first checked in the case of infinitesimal transformations, where it reduces to $\Delta^2 = 0$. Then, using (18.33), one easily verifies that it obeys the necessary group property to pass on to the finite transformations (Exercise 18.11).

That Δ is not invariant under canonical transformations was not unexpected. Indeed, the concept of divergence of vector fields requires a measure, while Δ is defined without measure. Hence, the identification of Δ with the divergence operator is not covariant. One finds from (18.33) and (18.34b) that the transformation law for Δ can be rewritten in the more natural form

$$\Delta X = J^{-1/2} \Delta' (J^{1/2} X). \tag{18.35}$$

18.3.3. Transformation of W and σ

Theorem 18.2. *Let $W(\phi, \phi^*)$ be a solution of the quantum master equation and let $\phi^A(\phi', \phi'^*), \phi^*_A(\phi', \phi'^*)$ be a canonical transformation in the antibracket with superdeterminant $J = \mathrm{sdet}\, \delta(\phi, \phi^*)/\delta(\phi', \phi')$. Then*

$$W' = W + \frac{\hbar}{i}\, \ell n\, J^{1/2} \tag{18.36a}$$

is a solution of the quantum master equation

$$i\hbar\,\Delta' W' - \tfrac{1}{2}(W', W') = 0 \tag{18.36b}$$

in the primed variables.

Proof. The proof is direct and based on (18.33), (18.34), and the canonical invariance of the antibracket.

Theorem 18.3. *The condition*

$$\sigma\alpha = 0 \tag{18.37}$$

is invariant under canonical transformations in the antibracket.

Proof. It follows from (18.33) and (18.36) that $(\alpha, W) - i\hbar\,\Delta\alpha = (\alpha, W') - i\hbar\,\Delta'\alpha$. Hence, if $\sigma\alpha = 0$, then also $\sigma'\alpha = 0$.

18.3.4. Invariance of the Path Integral

Theorem 18.4. *The path integral (18.16) giving $\langle A_0\rangle$ is invariant under canonical transformations in the antibracket if one transforms W as in (18.36a). [If one replaces W' by another solution of (18.36b), the path integral changes.]*

Proof. The quantum average $\langle A_0\rangle$ can be written as

$$\langle A_0\rangle = \int [D\phi\, D\phi^*]\,\mathrm{sdet}\Big(\frac{\partial\Omega_A}{\partial\phi^*_B}\Big)\,\delta(\Omega_A)\,\alpha\exp\frac{i}{\hbar}W,$$

where the conditions $\Omega_A = 0$ are equivalent to $\phi^*_A - \delta\psi/\delta\phi^A = 0$ [see Eq. (20.31) in Chapter 20]. Because $(\Omega_A, \Omega_B) = 0$, the Ω's can be completed to a canonical coordinate system (μ^A, Ω_A). One can think of $\mathrm{sdet}(\partial\,\Omega_A/\partial\phi^*_B)$ as the square root of the superdeterminant of the canonical transformation connecting μ^A, Ω_A to ϕ^A, ϕ^*_A (Theorem 15.2).

Now make the canonical transformation $\phi, \phi^* \to \phi', \phi^{*\prime}$ in the path integral for $\langle A_0\rangle$. One gets

$$[D\phi\, D\phi^*]\,\mathrm{sdet}\Big(\frac{\partial\,\Omega_A}{\partial\phi^*_B}\Big) = J^{1/2}[D\phi'\, D\phi^{*\prime}]\,\mathrm{sdet}\Big(\frac{\partial\,\Omega_A}{\partial\phi^{*\prime}_B}\Big).$$

Furthermore, because the conditions $(\Omega_A, \Omega_B) = 0$ are canonically invariant, they are equivalent to $\phi'^*_A = \delta\psi'/\delta\phi'_A$ for some $\psi'(\phi')$ that may not be the same function as $\psi(\phi)$. Taking into account the transformation rule (18.36a) of W and the ψ-independence of the path integral, one then finds that the quantum average of BRST-invariant functions is invariant under canonical transformations in the antibracket. This proves the theorem.

It follows from the theorem that different representations of the same gauge symmetry lead to the same path integral, since these different representations amount to making a canonical transformation in field-antifield space. The relevant concept for building the path integral is thus the gauge symmetry itself rather than any particular way to represent it.

Finally, it should be observed that the canonical transformation associated with two locally complete representations of the gauge symmetry is local in spacetime. Within the framework of dimensional regularization, the superdeterminant of its Jacobian matrix can then be taken to be equal to one.

18.3.5. The Path Integral in the Abelian Representation

Because the path integral does not depend on the representation of the gauge symmetry, one can evaluate it in any representation. In particular, one can use the abelian representation (17.37) analyzed in §17.4.1. In that representation, the path integral becomes very simple.

Indeed, it is easy to check that the solution (17.38) of the classical master equation also obeys $\Delta S = 0$. Furthermore, the general solution of $(A, S) = 0$ is given, up to an s-exact term, by an arbitrary function of the gauge-invariant variables $\chi^{\bar{a}}$. Accordingly, W can be assumed to differ from S by correction terms $\hbar M_1, \hbar^2 M_2, \dots$ that are functions of the gauge-invariant variables $\chi^{\bar{a}}$ only. Similarly, α can also be assumed to depend on $\chi^{\bar{a}}$ only and fulfills $(\alpha, W) = 0, \Delta\alpha = 0$.

To evaluate the path integral in the abelian representation (17.37), one takes a gauge-fixing fermion that does not depend on the gauge-invariant variables $\chi^{\bar{a}}$. With that choice, the gauge-fixed action takes the form

$$S_\psi = S_0(\chi^{\bar{a}}) + \bar{S}_\psi, \tag{18.38a}$$

where $\bar{S}_\psi$ involves only the gauge degrees of freedom and the ghosts. There is complete decoupling between the gauge-invariant sector and the gauge/ghost sector. The integration over these latter variables yields a factor independent of $\chi^{\bar{a}}$, so the path integral takes the manifestly gauge-invariant form

$$\langle A_0 \rangle = \int [D\chi^{\bar{a}}]\, \mu(\chi^{\bar{a}})\, \alpha(\chi^{\bar{a}})\, \exp \frac{i}{\hbar}\, S_0(\chi^{\bar{a}}) \tag{18.38b}$$

for some measure $\mu(\chi^{\bar{a}})$. Since (18.38) is the integral over the physical degrees of freedom of the exponential of the gauge-invariant action, it is valid if one chooses correctly the undetermined integration measure $\mu(\chi^{\bar{a}})$. This provides the first justification of the formalism.

The same arguments would go through if we had added cohomologically trivial pairs to the solution (17.38) of the master equation. So, in particular, these do not modify the path integral provided one adjusts again correctly the undetermined quantum corrections.

18.4. EQUIVALENCE OF ANTIFIELD AND HAMILTONIAN FORMALISMS

18.4.1. Gauge-Fixed Form of the BRST Symmetry in the Antifield Formalism

We close this chapter by providing a second proof of the correctness of the antifield formalism. This second proof relies on a comparison with the Hamiltonian formalism. We will only check in the text that the Lagrangian and Hamiltonian gauge-fixed actions are related by the Legendre transformation (lowest order in the stationary phase method). The measure is considered in Exercises 18.17 and 18.18.

First, we determine the form of the BRST symmetry of the gauge-fixed action S_ψ. The "gauge-fixed BRST symmetry" is defined by

$$\bar{s}\phi^A = (s\phi^A)\Big(\phi, \phi^* = \frac{\delta\psi}{\delta\phi}\Big) \equiv \frac{\delta^L S}{\delta\phi^{*A}}\Big(\phi, \phi^* = \frac{\delta\psi}{\delta\phi}\Big). \tag{18.39}$$

When S is linear in the antifields, $\delta^L S/\delta\phi^*_A$ does not depend on the antifields, so $\bar{s}\phi^A$ is independent of the gauge-fixing fermion,

$$S = S_0 + \phi^*_A\, s\phi^A, \tag{18.40a}$$

$$\bar{s}\phi^A = s\phi^A, \tag{18.40b}$$

$$S_\psi = S_0 + s\psi. \tag{18.40c}$$

This occurs, for instance, when the gauge algebra is closed, or when S_0 is the extended Hamiltonian action with the role of the Lagrange multipliers and their antifields exchanged (§18.4.5). In all other cases, however, the gauge-fixed BRST transformation $\bar{s}$ of the antifield formalism *generically depends on the gauge-fixing fermion* ψ. This is in sharp contrast with the Hamiltonian situation, where the BRST transformation takes the same form in the extended phase space irrespectively of the gauge-fixing fermion K.

The following properties are direct (see Exercises 18.13–18.14):

(i) The gauge-fixed action is invariant under the gauge-fixed BRST symmetry,

$$\bar{s}S_\psi = 0. \tag{18.41}$$

(*ii*) The BRST variation $\bar{s}\phi^*_A$ of the antifields viewed as functions of the fields differ from $(s\phi^*_A)(\phi, \phi^* = \delta\psi/\delta\phi)$ by equations of motion terms,

$$\bar{s}\phi^*_A = s\phi^*_A + \frac{\delta^L S_\psi}{\delta\phi^A}. \tag{18.42a}$$

Hence, one has

$$\bar{s}B = sB + \frac{\delta^R B}{\delta\phi^*_A}\frac{\delta^L S_\psi}{\delta\phi^A}. \tag{18.42b}$$

(*iii*) Because of this, the gauge-fixed BRST symmetry of the antifield formalism is, in general, only on-shell nilpotent,

$$\bar{s}^2\phi^A = \text{field equation terms}, \tag{18.43}$$

where the field equations in (18.43) are those of the gauge-fixed action, not those of the original gauge-invariant action. The right-hand side of (18.43) identically vanishes if and only if $\delta^2 S/\delta\phi^*_A\,\delta\phi^*_B$ is zero. For open algebras, $\delta^2 S/\delta\phi^*_A\,\delta\phi^*_B \neq 0$ and $\bar{s}^2$ vanishes only weakly.

18.4.2. Digression. Gauge-Fixed BRST Cohomology

Because the gauge-fixed action S_ψ is BRST invariant, the gauge-fixed stationary surface Σ_ψ defined by

$$\Sigma_\psi : \frac{\delta S_\psi}{\delta\phi^A} = 0 \tag{18.44}$$

is invariant under the BRST symmetry. One can thus define the gauge-fixed BRST cohomology as the space of equivalence classes of weakly BRST-invariant functions $A(\phi)$ modulo weakly BRST-exact ones,

$$\bar{s}A = \lambda^A \frac{\delta S_\psi}{\delta\phi^A}, \tag{18.45a}$$

$$A \sim B \quad \text{iff} \quad A = B + \bar{s}C + \mu^A \frac{\delta S_\psi}{\delta\phi^A}. \tag{18.45b}$$

Here, "weakly" means "on the stationary surface Σ_ψ."

The question, then, is: What is the relationship between the BRST cohomological groups $H^k(s)$ of the BRST symmetry acting in the space of the fields and the antifields on the one hand and the BRST cohomological groups $H^k(\bar{s})$ of the gauge-fixed BRST symmetry on the other hand? The answer is given by the following theorem.

Theorem 18.5.

$$H^k(s) = H^k(\bar{s}). \tag{18.46}$$

In particular, the gauge-fixed BRST cohomology at ghost number zero is isomorphic to the set of gauge-invariant observables.

Proof. To prove the theorem, one defines a Koszul resolution $\bar{\delta}$ for the gauge-fixed stationary surface Σ_ψ. The antifields $\bar{\phi}^*_A \equiv \phi^*_A - \delta\psi/\delta\phi^A$ can be viewed as the generators of this resolution, *i.e.*,

$$\bar{\delta}\bar{\phi}^*_A = -\frac{\delta S_\psi}{\delta\phi^A}, \tag{18.47a}$$

$$\bar{\delta}\phi^A = 0. \tag{18.47b}$$

We denote the resolution degree by r, $r(\bar{\phi}^*_A) = -1$, $r(\phi^A) = 0$. No further generator is needed, as the equations $\delta S_\psi/\delta\phi^A = 0$ are independent (the gauge is fixed). With the definition (18.47), one easily finds [use Eq. (18.42a)],

$$s\bar{\phi}^*_A = \bar{\delta}\bar{\phi}^*_A + \text{"higher order,"} \tag{18.48a}$$

$$s\phi^A = \bar{\delta}\phi^A + \bar{s}\phi^A + \text{"higher order,"} \tag{18.48b}$$

where the expansion is made according to the resolution degree r. Because $\bar{s}^2 \approx 0$, one can apply the techniques of homological perturbation theory (Sec. 8.4), which guarantee that (18.46) indeed holds. This proves the theorem.

It also follows from the main theorem of homological perturbation theory that one can go from $H^k(s)$ to $H^k(\bar{s})$ simply by setting $\bar{\phi}^*_A$ equal to zero, *i.e.*, ϕ^*_A equal to $\delta\psi/\delta\phi^A$,

$$A_\psi(\phi) = A\Big(\phi, \phi^* = \frac{\delta\psi}{\delta\phi}\Big), \tag{18.49a}$$

and, indeed, one finds from $(A, S) \equiv sA = 0$ that

$$\bar{s}A_\psi = \lambda^A \frac{\delta^L S_\psi}{\delta\phi^A}, \qquad \lambda^A = \frac{\delta^R A}{\delta\phi^*_A}. \tag{18.49b}$$

Conversely, given an element of $H^k(\bar{s})$, one finds the corresponding elements of $H^k(s)$ by a recursive procedure based on the expansion in powers of $\bar{\phi}^*_A$. This procedure works because of the acyclicity of $\bar{\delta}$ and the Jacobi identity for the antibracket.

18.4.3. Equivalence of Antifield BRST Symmetry and Hamiltonian BRST Symmetry

One can now provide a general argument showing the equivalence of the antifield and Hamiltonian formalisms. This argument is based on the uniqueness of the BRST symmetry and goes as follows. The antifield gauge-fixed action S_ψ is a local functional and is not degenerate. Hence, it can be rewritten in Hamiltonian form without difficulty—if it is not already. There can be, at most, second-class constraints, and these can be eliminated by means of the Dirac bracket method. When this is done, the Hamiltonian formalism is constraint-free. (If the gauge-fixed action S_ψ contains higher order time derivatives, one can still develop a Hamiltonian formalism. This formalism simply requires more conjugate pairs.)

To any local-in-time functional A of the fields and their time derivatives, one can associate, by using the equations of motion of S_ψ if necessary, a well-defined phase space function. In particular, if one expresses the BRST Noether charge Ω_ψ following from the BRST invariance of S_ψ, in terms of the canonical variables, one gets a phase space function with the following features: (a) Ω_ψ is off-shell nilpotent because the (Dirac) bracket $[\Omega_\psi, \Omega_\psi]$, which must be zero on-shell, does not contain the time derivatives, *i.e.*, cannot involve the equations of motion. Thus, it must identically vanish, $[\Omega_\psi, \Omega_\psi] = 0$. (b) The canonical transformation generated by Ω_ψ starts like a gauge transformation because $\bar{s}\phi^i = R_\alpha{}^i C^\alpha +$ "more". (The Lagrangian and Hamiltonian gauge transformations are equivalent; see Chapter 3.)

The properties (a) and (b) are just the defining properties of the Hamiltonian BRST charge. From the general theorems on the existence and uniqueness of the BRST charge in the Hamiltonian formalism, one can thus infer that Ω_ψ differs from the gauge-independent BRST charge Ω constructed along Hamiltonian lines at most by a canonical change of variables and the possible addition of cohomologically trivial pairs. The transformation relating Ω_ψ to Ω may have a complicated, ψ-dependent structure, but it is canonical.

Similarly, the Hamiltonian H_ψ resulting from S_ψ is BRST invariant, $[H_\psi, \Omega_\psi] = 0$, and of the form $H_\psi = H_0 +$ "more", since $S = S_0 +$ "more". We can thus conclude from this general argument that the Lagrangian and Hamiltonian BRST formalisms are equivalent (more on this in Exercise 18.15).

We shall now present a different argument that makes the equivalence more concrete. The analysis follows the detailed procedure by which one goes from the action S_L in Lagrangian form to the extended Hamiltonian action S_E and consists of establishing the following three assertions: (*i*) the antifield path integrals based on the original action S_L and the total action S_T are equal; (*ii*) the antifield path integrals based on S_T and S_E are also equal; and (*iii*) the path integrals obtained from S_E by either the antifield method or the Hamiltonian BRST method are identical.

18.4.4. The Antifield Path Integral Based on S_L, S_T, and S_E Are the Same

Because the conjugate momenta and the Lagrange multipliers associated with the primary constraints are auxiliary fields (Exercise 1.4), the equality of the path integrals based on S_T and S_L is a direct consequence of the following:

Theorem 18.6. *Let $S_0[y,z]$ be an action depending on auxiliary fields z; that is, the equations $\delta S_0/\delta z = 0$ can be used to express the z's as functions of the y's and their derivatives. Let $\bar{S}_0[y]$ be the action obtained from $S_0[y,z]$ by elimination of the auxiliary fields, $\bar{S}_0[y] = S_0[y, z(y)]$. Then, the gauge-fixed actions and gauge-fixed BRST symmetries obtained by the antifield method and respectively based on $S_0[y,z]$ and $\bar{S}_0[y]$ coincide after the z's are eliminated from $S_\psi[y,z]$ by means of their own equations of motion. (The z's remain auxiliary for choices of ψ that do not involve them.)*

Proof. The proof is direct and outlined in Exercise 18.16.

The next step is to compare the path integrals based on S_T and S_E. To that end, one assumes that the consistency algorithm takes the form of §3.3.2; this guarantees that the Dirac conjecture holds, so that S_L and S_T manifestly exhibit all the gauge invariances of the theory. The following theorem then holds.

Theorem 18.7. *The gauge-fixed actions and gauge-fixed BRST symmetries obtained by the antifield method and respectively based on S_T and S_E coincide (after the extra variables appearing in S_E^ψ are eliminated by means of their own equations of motion).*

Proof. As second-class constraints and their multipliers can be thought of as auxiliary fields (Exercise 3.16), one can assume that the second-class constraints have been eliminated. The proof is then direct and outlined in Exercise 18.18. Note that the nonminimal sectors for S_T in Theorems 18.6 and 18.7 may be different, so that the gauge-fixed actions S_T^ψ in Theorems 18.6 and 18.7 may involve different sets of variables.

This does not matter because the path integral does not depend on the choice of nonminimal sector.

18.4.5. Antifield Formalism for the Extended Hamiltonian Action

We now complete the argument proving the equivalence of the antifield method applied to the original Lagrangian action S_L with the Hamiltonian BRST method. This is done by establishing that the antifield method for S_E and the Hamiltonian BRST method lead to identical path integrals.

For notational simplicity, we assume the constraints to be irreducible. The analysis of the general case proceeds along the same lines (Exercise 18.19).

Theorem 18.8. *Let Ω be the BRST charge for the first-class constraint system $G_a(z^A) = 0$ and H be a BRST-invariant extension of the Hamiltonian H_0.*

(a) The solution of the master equation for the extended Hamiltonian action

$$S_E = \int (\dot{z}^A a_A(z) - H_0 - \lambda^a G_a)\, dt \tag{18.50a}$$

is given by

$$\begin{aligned} S[z^A, z^*_A, \lambda^a, \lambda^*_a, \eta^a, \eta^*_a] \\ = \int \big(\dot{z}^A a_A + \dot{\eta}^a \mathcal{P}_a - H + z^*_A[z^A, \Omega] + \lambda^a[\mathcal{P}_a, \Omega] \\ + \eta^*_a[\eta^a, \Omega]\big)\, dt \end{aligned} \tag{18.50b}$$

where the identification

$$\lambda^*_a = -\mathcal{P}_a \tag{18.50c}$$

has been made.

(b) The functional W,

$$W = S + \frac{\hbar}{2i}\delta(0) \int \ell n\, (\operatorname{sdet} \sigma_{AB})\, dt, \tag{18.50d}$$

solves the quantum master equation. Here, σ_{AB} is the symplectic 2-form associated with $a_A(z)$, $\sigma = da$.

Proof. (a) First, it is easy to check that $(S, S) = 0$. This is a consequence of $[\Omega, \Omega] = 0$ and $[H, \Omega] = 0$. Furthermore, (18.50b) reduces to (18.50a) when the antifields are set equal to zero. As S contains the correct spectrum of fields and antifields (the gauge transformations are irreducible) and obeys the "boundary conditions" (17.28d), it is a proper solution of the master equation.

(b) The equation $\frac{1}{2}(W, W) = i\hbar \, \Delta W$ follows from a direct calculation that uses $d\sigma = 0$.

Note that the z's can include the Lagrange multipliers associated with the original constraints and their momenta b. The G's contain then $b = 0$.

With Theorem 18.8 at hand, the path integral equivalence of the antifield method applied to (18.50a) with the Hamiltonian BRST method is immediate. The Hamiltonian gauge-fixing fermion $K = K(z, \eta, \mathcal{P})$ leads to the gauge-fixed action

$$S_K = \int dt \big(\dot{z}^A a_A + \dot{\eta}^a \mathcal{P}_a - H - [K, \Omega]\big). \tag{18.51}$$

The integration measure is $[Dz][D\eta][D\mathcal{P}] \prod_t (\mathrm{sdet}\, \sigma_{AB})^{1/2}$ (product over time of Liouville measure). The elimination of the antifields z^*_A and η^*_a and of the fields λ^a by means of the formula

$$z^*_A(t) = \frac{\delta \psi}{\delta z^A(t)}, \qquad \eta^*_a = \frac{\delta \psi}{\delta \eta^a(t)}, \qquad \lambda^a(t) = -\frac{\delta \psi}{\delta \lambda^*_a(t)}, \tag{18.52a}$$

$$\psi = -\int K \, dt, \tag{18.52b}$$

leads to the same gauge-fixed action and integration measure. [The singular term in W yields the local measure $\prod_t (\mathrm{sdet}\, \sigma_{AB})^{1/2}$; this term is absent in canonical coordinates.]

Observe further that S is linear in the variables that are eliminated. Therefore, the gauge-fixed BRST symmetry does not actually depend on ψ and coincides with the Hamiltonian BRST symmetry. There is complete agreement between the antifield and Hamiltonian BRST approaches.

Remarks.

(*i*) In (18.52a), one views λ^a as an "antifield" and $-\lambda^*_a$ as a "field." This is permissible because $-\lambda^*_a$ and λ^a are conjugate,

$$(-\lambda^*_a, \lambda^b) = \delta_a{}^b.$$

More generally, one can replace some fields by the corresponding antifields if there exist good gauge-fixing fermions in term of the new "fields."

(*ii*) The antifield method allows for gauge-fixing fermions that depend on the time derivatives of the canonical variables and, hence, that are more general than (18.52b). The resulting gauge-fixed actions lead to a bigger phase space, since the equations of motion may no longer be first order. This amounts to increasing the Hamiltonian nonminimal sector.

(*iii*) When working with observables that depend on the canonical variables only and not on their time derivatives, the freedom of adding to the observables terms that vanish with the equations of motion is frozen. This is why the Schwinger–Dyson equation was not needed to prove that $\langle A(q,p,\mathcal{P},\eta)\rangle_K$ does not depend on K in §16.4.2.

EXERCISES. CHAPTER EIGHTEEN

18.1. Let $S = S^{AB}\,\phi^*_A\,\phi^*_B$ be an arbitrary 2-vector. Show that $\int \Delta S(\phi,\phi^* = \omega)\,[D\phi]$ vanishes for all S's if and only if $d\omega = 0$, *i.e.*, ω is a gradient ($\omega_A = \delta\psi/\delta\phi^A$ for some ψ. (Hint: Make an integration by parts and use Stokes theorem.) Prove then that $\int \Delta V(\phi,\phi^* = \delta\psi/\delta\phi)[D\phi] = 0$ for an arbitrary p-vector, *i.e.*, a divergence remains a divergence after the replacement of ϕ^* by $\delta\psi/\delta\phi$.

18.2. Show that if the equations $\Omega_A(z^a) = 0$ can be solved for ϕ^*_A and obey $(\Omega_A,\Omega_B)=0$, then they are equivalent to $\phi^*_A = \delta\psi/\delta\phi^A$ for some ψ. Prove, furthermore, that there is a canonical transformation (in the antibracket) $(\phi,\phi^*) \to (\mu,\Omega)$ in which the Ω's are the new antifields.

18.3. Derive (18.14) from (18.13).

18.4. Derive (18.18b) from (18.18a).

18.5. Prove the ψ-independence of the expectation value (18.16) by making the change of integration variables

$$\phi^A \to \phi^A - \frac{i}{\hbar}\left(\frac{\delta^L W}{\delta\phi^*_A}\left(\phi,\frac{\delta\psi}{\delta\phi}\right)\right)(\psi'-\psi).$$

[Hints: (*i*) By using the quantum master equation, show that the variation in the measure and in W combine to shift ψ to ψ' in $W(\phi,\phi^* = \delta\psi/\delta\phi)$. (*ii*) By using $\sigma\alpha = 0$ and the Schwinger–Dyson equation, show that the change in α amounts to replacing ψ by ψ' in $\alpha(\phi,\phi^* = \delta\psi/\delta\phi)$.]

18.6. Prove that $(\Delta S, S) = 0$ and $(N_p, S) = 0$ in (18.27). [Hint: Use the Jacobi identity, $\Delta^2 = 0$ and (18.15b).]

18.7. (a) Show that if one modifies M_p $(p \geq 1)$ as $M_p \to M_p + (K_p, S)$ (with M_k, $k < p$ unchanged), one can choose the subsequent terms in the measure so that the solution of the master equation is modified as

$$W \to W' = W + \frac{\hbar}{i}\ell n(1 + i\hbar^{p-1}\sigma\bar{K}_p)$$

with an appropriate $\bar{K}_p$ (that can be taken to be equal to K_p if $p \geq 2$. (Hint: Check that this equation yields a solution of the quantum master equation that coincides with W up to the required order $\hbar^p$.)

(b) Verify that the path integral $\langle 1\rangle_W$ is equal to the path integral $\langle 1\rangle_{W'}$.

18.8. Prove $(B_p, S) = 0$ where B_p is given by (18.30b). (Hint: Proceed as in Exercise 18.6.)

18.9. Assume that S is linear in the antifields, with $\Delta S = 0$. Take $W = S$. Show that a complete set of BRST observables depending only on the fields exists. [Hint: The gauge algebra closes off-shell, so one can construct strongly gauge-invariant functions $A_0(\phi^i)$. For these, one has $A = A_0$.]

18.10. This exercise shows that the measure may be quite important even if $\Delta S = 0$. Consider the Lagrangian $L_0 = \frac{1}{2}\dot{q}^2$. The minimal solution of the master equation is $S[q(t), q^*(t)] = \frac{1}{2}\int \dot{q}^2\, dt$. Add the nonminimal sector $\bar{C}, b$ to get the full solution $S[q(t), q^*(t), \bar{C}(t), \bar{C}^*(t), b(t), b^*(t)] = \int [\frac{1}{2}\dot{q}^2 + \bar{C}^* b]\, dt$.

(a) Show that $\Delta S = 0$. So, S is also a solution of the quantum master equation.

(b) Prove, however, that taking $W = S$ leads to nonsensical results: The path integral $\langle 1 \rangle_\psi = \int [Dq\, D\bar{C}\, Db] \exp(i/\hbar) S_\psi$ vanishes no matter what ψ is.

(c) Show that the source of the difficulty lies in the measure: $[D\bar{C}\, Db]$ should be replaced by $\prod_t \delta(b)\, \delta(\bar{C})\, d\bar{C}\, db$, which is also invariant under the BRST transformation $b \to b$, $\bar{C} \to \bar{C} + b\varepsilon$.

(d) Because the coefficient of $d\bar{C}\, db$ in the correct measure contains δ-functions, it cannot be exponentiated in a meaningful way [no meaning to $\ell n\, \delta(b)$ or $\ell n\, \delta(\bar{C})$]. One can, however, develop a smoother formalism as follows. The "gauge-fixed action S_ψ" of (b) cannot properly be gauge fixed because it possesses the fermionic gauge invariance $\bar{C} \to \bar{C} + \varepsilon(t)$ no matter what ψ is. This is the explanation for the vanishing result found in (b) (so S_ψ really "screams"). To fully gauge-fix the system (when q, b and $\bar{C}$ are taken as the "fields"), one needs to enlarge the nonminimal sector, with one new variable of ghost number $+1$ balancing the ghost number -1 of $\bar{C}$. (*i*) Add to S the term $\int x^* C\, dt$. (*ii*) Show that $\psi = \int dt\, \bar{C}[q + \lambda x]$ is acceptable. (*iii*) Study the λ-(in)dependence of the path integral. (*iv*) Show that one gets the measure of (c) by integrating over x and C.

18.11. (a) Verify that (18.33) is invariant under the simultaneous replacements of Δ by Δ' and J by J^{-1}.

(b) Prove the equivalence of (18.34a) with (18.34b).

(c) Check directly that (18.34b) obeys the group property by using (18.15a) and (18.33).

18.12. Prove (18.33) and (18.34) directly by using the fact that any canonical transformation in the antibracket possesses a fermionic generating function G, which can be taken to be functions of ϕ and $\phi^{*\prime}$ if $\phi \leftrightarrow \phi'$ is invertible and which is such that $\phi'^A = \partial G/\partial \phi'^*_A$, $\phi^*_A = \partial G/\partial \phi^A$ (see Exercise 15.30).

18.13. Check (18.41). [Hint: Start from $(S, S) = 0$ and use the fact that ϕ^*_A is equal to a gradient.]

18.14. Verify (18.42) and compute explicitly the coefficient of the variational derivative of S_ψ in $s^2\phi^A$ [Eq. (18.43)].

18.15. Let L be a Lagrangian leading to second-class constraints only (no gauge invariance), and let $\delta_\varepsilon q^i$ be a rigid transformation leaving L invariant up to a total time derivative. Let $A(q, \dot{q}, \ddot{q}, \ldots)$ be a function that is on-shell invariant under $\delta_\varepsilon q^i$.

(a) Show that $\delta_\varepsilon q^i$ can be extended to the momenta and the multipliers so as to be a symmetry transformation of the total action (see Exercise 3.17). Prove that the corresponding conserved Noether charge Q can be redefined by on-shell vanishing terms so as to depend on q and p only and to obey $[Q, H_0] \approx 0, [Q, \chi_\alpha] \approx 0$ (for all the second-class constraints and not just the primary ones). Conclude that Q defines a symmetry transformation of the action in which all the second-class constraints have been eliminated.

(b) Show that the phase space function $A(q,p)$ induced by $A(q, \dot{q}, \ddot{q}, \ldots)$ is such that $[A, Q]^* = 0$.

(c) Let $C(q, \dot{q}, \ddot{q}, \ldots) = \delta_\varepsilon D(q, \dot{q}, \ddot{q}, \ldots)$ (on-shell). Show that $C(q,p) = [D(q,p), Q]^*$. This gives another proof that the Hamiltonian and antifield BRST cohomology are isomorphic if one takes for L the gauge-fixed Lagrangian and for $\delta_\varepsilon q$ the gauge-fixed BRST symmetry.

18.16. Prove Theorem 18.6. [Hints: (*i*) Reformulate the gauge transformations $\delta_\varepsilon y, \delta_\varepsilon z$ of $S_0[y, z]$ so that $\delta_\varepsilon y = \bar{\delta}_\varepsilon y$, where $\bar{\delta}_\varepsilon y$ are the gauge transformations of $\bar{S}_0[y] \equiv S_0(y, z(y))$ (see Exercise 3.17). (*ii*) Show that the solution S of the master equation (for S_0) is related to the solution $\bar{S}$ of the master equation for $\bar{S}_0$ as follows,

$$S(y, y^*, z, z^*) = \bar{S}(y, y^*) + S_0(y, z) - \bar{S}_0(y) + \text{“higher orders,''}$$

where the higher order terms vanish when the antifields z^* are set equal to zero. Conversely, $\bar{S}$ is obtained from S by setting z^* equal to zero and replacing z by the solution of $\delta S / \delta z|_{z^*=0} = \delta S_0 / \delta z = 0$. (*iii*) Show that gauge-fixing fermions that involve only y, $\psi = \psi(y)$, enable one to transform the path integral with the auxiliary fields included into the path integral without the auxiliary fields. (*iv*) Observe that the same result holds if one adds the same nonminimal sectors to both S and $\bar{S}$.]

18.17. (a) Consider gauge transformations forming a group. Show from the master equation that the measure terms M_p in W can be assumed to depend only on the fields and that one can take $M_p = 0$ for $p \geq 2$.

(b) Let $S_0[y, z]$ be an action with quadratic auxiliary fields, $S_0[y, z] = \bar{S}_0[y] + \int f_{\alpha\beta}(y)\,(z^\alpha - Z^\alpha(y))\,(z^B - Z^b(y))\,dx$. Assume that the gauge transformations form a group and that $f_{\alpha\beta}$ is gauge invariant. By a redefinition of the z's, one can assume $Z^\alpha = 0$. Verify explicitly that one can choose the respective solutions $\bar{W}$ and W of the quantum master equation for $\bar{S}_0$ and S_0 in such a way that the path integrals $\int [Dz\, Dy] \exp(i/\hbar)\, W$ and $\int [Dy] \exp(i/\hbar)\, \bar{W}$ are equal (and not just the gauge-fixed actions).

(c) Consider now a general action $S_0[y, z]$ with auxiliary fields z^α. Show that $\bar{W}[y, y^*]$ defined by $\exp(i/\hbar)\, \bar{W} = \int [Dz] \exp(i/\hbar)\, W[y, y^*, z, z^* = 0]$ is a solution of the quantum master equation for $\bar{S}_0$. [Hints: The integral $\int Dy \exp(i/\hbar)\, \bar{W}[y, y^* = \delta\psi/\delta y]$ does not depend on ψ.]

18.18. Prove Theorem 18.7. [Hints: Assume first that the constraints are irreducible and that the number of primary, secondary, ... constraints at each generation are equal, *i.e.*, $m_1 = m_2 = \cdots = m_L$. Under these conditions: (*i*) Prove that the solution of the master equation for S_T is obtained from the solution S of the master equation for S_E by (a) setting the Lagrange multipliers $u^{m_2}, \ldots, u^{m_L}$ and the antifields $\eta^*_{m_1}, \eta^*_{m_L}, \ldots, \eta^*_{m_{L-1}}$ associated with the ghosts $\eta^{m_1}, \ldots, \eta^{m_{L-1}}$ equal to zero; (b) solving $\delta S/\delta u^*_{m_2} = 0, \ldots, \delta S/\delta u^*_{m_L} = 0, \delta S/\delta \eta^{m_1} = 0, \ldots, \delta S/\delta \eta^{m_{L-1}} = 0$ for $u^*_{m_2}, u^*_{m_3}, \ldots, u^*_{m_L}, \eta^{m_1}, \ldots, \eta^{m_{L-1}}$ (the analysis of the solution proceeds as in §3.3.4); and (c) identifying η^{m_L} with the ghosts of the total formalism. (*ii*) Take gauge-fixing fermions that depend only on q, p, $u^*_{m_1}$, and η^{m_L}, and eliminate q^*, p^*, η^*_a and $\lambda^a \equiv (u^{m_1}, u^{m_2}, \ldots)$ in favor of q, p, η^a, $\lambda^*_a \equiv (u^*_{m_1}, u^*_{m_2}, \ldots)$. Once the theorem is established for irreducible constraints such that $m_1 = \cdots = m_L$, the general case is easily covered.] By using the argument of Exercise 18.17(c), show then that the measures in the path integrals for S_E and S_T can also be chosen to agree.

18.19. Extend the analysis of §18.4.5 to the reducible case. [Hint: Take into acount the additional gauge symmetry given by (3.37a)].

18.20. Consider the nonminimal sector $\int dt(\alpha^* \beta + \pi_\alpha \pi^{*\beta})$ added to the solution (18.50b) of the master equation for the extended acton S_E. Show that by a canonical transformation in the antibracket, this nonminimal sector can be rewritten as $\int dt(-\dot{\pi}_\beta + \alpha^*)\beta + \pi_\alpha(\dot{\alpha} + \pi^*_\beta)$. Prove that the extension of the antifield nonminimal sector is equivalent to adding the nonminimal term $\pi_\alpha \beta$ to the BRST charge Ω.

18.21. Write down explicitly the gauge transformations (17.42) of S in the abelian representation (17.38) of the gauge symmetry. Show that one can devise good gauge conditions $\phi^*_A = \delta\psi/\delta\phi^A$ for (17.42) only if the gauge and ghost modes come in quartets of the form $q^* C + \bar{C}^* b$. If this does not happen in the minimal sector, one needs to add nonminimal variables to form the quartets.

18.22. Consider the action $S_0 = \int (p\dot{q} - H - u^\alpha \chi_\alpha)\, dt$ with only second-class constraints $\chi_\alpha \approx 0, \det[\chi_\alpha, \chi_\beta] \neq 0$. Show that $S = S_0$ is a proper solution of the master equation. Verify that $\Delta S = 0$ but that the correct solution W of the quantum master equation to be used in the path integral differs from S by quantum corrections if $\det[\chi_\alpha, \chi_\beta]$ depends on the dynamical variables.

18.23. Consider the Lagrangian $\bar{L}_0 = L_0(q^i, \dot{q}^i) - \frac{1}{2}\Sigma(Q^a)^2$ where $L_0(q^i, \dot{q}^i)$ is regular. The system has primary constraints ($P_a \approx 0$), as well as secondary constraints ($Q^a \approx 0$), which are second class. The Hamiltonian path integral is

$$\int [Dq^i\, Dp_i\, DQ^a\, DP_a] \prod_t \delta(Q^a)\, \delta(P_a) \exp i\, S_H.$$

By making the integration over the momenta, using the Fourier representation of $\delta(Q^a) = 1/\sqrt{2\pi} \int d\lambda_a \exp i\lambda_a Q^a$, making the change of integration variables $Q_a \to Q_a + \lambda_a$ and integrating over λ^a, show that the path integral takes the Lagrangian form

$$\int [Dq^i\, DQ^a]\, \mu[q^i] \exp i \int \bar{L}_0\, dt$$

with some local measure $\mu[q^i]$ no longer containing $\delta(Q^a)$, in agreement with the antifield formalism.

18.24. (Nonminimal sector without gauge invariance)
The introduction of a nonminimal sector for a system without gauge invariance enables one to make formal "changes of variables" $q^i = f^i(Q^j, \dot{Q}^j, \ddot{Q}^j, \ldots, \overset{(k)}{Q}{}^j)$ involving time derivatives of the Q's—and, hence, such that the Q's obey higher order equations of motion. This exercise indicates how.

(a) Add the nonminimal sector $\int dt(\bar{C}^{*i}b_i + Q_i^* C^i)$ to the action $S_0 = \int L_0(q, \dot{q})\, dt$, with $\text{gh}\, Q^i = \text{gh}\, b_i = 0$. Show that the gauge-fixing fermion $\psi = \int dt\, \bar{C}_i(q^i - f^i)$ yields the path integral

$$\int [DQ\, DC\, D\bar{C}] \times \exp i \int \{L_0(f(Q, \dot{Q}, \ddot{Q}, \ldots), \dot{f}(Q, \dot{Q}, \ddot{Q}, \ldots)) + \text{ghost terms}\}\, dt$$

in which the substitution $q^i = f^i(Q, \dot{Q}, \ddot{Q}, \ldots)$ has been made.

(b) Perform the Hamiltonian analysis and show that the physical subspace coincides with the one defined by $L(q, \dot{q})$. The extra degrees of freedom contained in the Q's are thus compensated for by the ghosts.

CHAPTER NINETEEN

FREE MAXWELL THEORY. ABELIAN TWO-FORM GAUGE FIELD

In this chapter we illustrate the general principles on two simple examples. The first is Maxwell theory without sources, which is an irreducible gauge theory. The second is the theory of a noninteracting two-form gauge field, for which the gauge transformations are reducible. The great technical simplicity of these models enables one to exhibit explicitly the cohomological mechanism by which the ghost and gauge modes decouple, both at the classical and quantum levels. Even though the two models of this chapter are free, they are relevant to the description of more realistic interacting theories, for which they provide a description of some of the asymptotic fields.

19.1. FREE MAXWELL FIELD

19.1.1. Hamiltonian Analysis

The action for the free electromagnetic field reads

$$S_0^L[A_\mu] = \int d^4x(-\tfrac{1}{4} F^{\mu\nu}F_{\mu\nu}), \tag{19.1}$$

where the field strength $F_{\mu\nu}$ is given by

$$F_{\mu\nu} = \partial_\mu A_\nu - \partial_\nu A_\mu. \tag{19.2}$$

The action is invariant under the gauge transformation

$$\delta A_\mu = \partial_\mu \varepsilon. \tag{19.3}$$

The transformation (19.3) involves the gauge parameter ε together with its first time derivatives. Hence, the "gauge strikes twice" and there are 4 (components of vector potential) minus 2 (independent number of gauge parameters at a given time) = 2 physical degrees of freedom per space point (Sect. 3.3). These correspond to the two transverse helicity states of the photon.

The canonical momenta are

$$\pi^0 = 0, \qquad \pi^i = F^{i0}, \tag{19.4}$$

$$[A_\mu(\vec{x}\,), \pi^\nu(\vec{x}\,')] = \delta_\mu{}^\nu\, \delta(\vec{x} - \vec{x}\,'), \tag{19.5}$$

and there is thus one primary constraint,

$$\phi_1 \equiv \pi^0 = 0. \tag{19.6}$$

The canonical Hamiltonian H_C is found to be

$$H_C = H_0 + \int d^3x\, A_0(-\partial_i\, \pi^i) \tag{19.7a}$$

with

$$H_0 = \int d^3x(\tfrac{1}{2}\,\pi^i\,\pi_i + \tfrac{1}{4}\,F^{ij}\,F_{ij}). \tag{19.7b}$$

The consistency condition $\dot{\pi}^0 \approx 0$ leads to the secondary constraint

$$\phi_2 \equiv -\pi^i{}_{,i} = 0, \tag{19.8}$$

which is recognized as Gauss law ($-\pi^i{}_{,i} \equiv \vec{\nabla}\cdot\vec{E}$, $E^i \equiv F^{0i}$). There is no further constraint because $\pi^i{}_{,i}$ commutes with the Hamiltonian. The algebra of the constraints and of the Hamiltonian is given by

$$[\phi_1(\vec{x}\,), \phi_1(\vec{x}\,')] = 0, \quad [\phi_1(\vec{x}\,), \phi_2(\vec{x}\,')] = 0,$$
$$[\phi_2(\vec{x}\,), \phi_2(\vec{x}\,')] = 0, \tag{19.9a}$$

$$[H_C, \phi_1(\vec{x}\,)] = \phi_2(\vec{x}\,), \qquad [H_C, \phi_2(\vec{x}\,)] = 0, \tag{19.9b}$$

since

$$[H_0, \phi_1(\vec{x}\,)] = [H_0, \phi_2(\vec{x}\,)] = 0. \tag{19.9c}$$

The extended action reads

$$S_0^E[\pi^\mu, A_\mu, \lambda^1, \lambda^2] = \int d^4x(\pi^i \dot{A}_i + \pi^0 \dot{A}_0 - H_C - \lambda^1\phi_1 - \lambda^2\phi_2) \tag{19.10}$$

and is invariant under the gauge transformations generated by $\int(\varepsilon^1\phi_1 + \varepsilon^2\phi_2)\,dx$ with independent ε^1 and ε^2 (§3.2.2)

$$\delta A_0 = \varepsilon^1, \qquad \delta\pi^0 = 0, \tag{19.11a}$$
$$\delta A_i = \partial_i\,\varepsilon^2, \qquad \delta\pi^i = 0, \tag{19.11b}$$
$$\delta\lambda^1 = \dot{\varepsilon}^1, \qquad \delta\lambda^2 = \dot{\varepsilon}^2 - \varepsilon^1, \tag{19.11c}$$

The total action is obtained by imposing the gauge condition $\lambda^2 = 0$. It is

$$S_0^T[\pi^\mu, A_\mu, \lambda^1] = \int d^4x(\pi^i\dot{A}_i + \pi^0\dot{A}_0 - H_C - \lambda^1\phi_1). \tag{19.12}$$

The residual gauge invariances of (19.12) are such that $\delta\lambda^2 = 0$, *i.e.*, from (19.11c),

$$\delta A_0 = \dot{\varepsilon}, \qquad \delta A_i = \partial_i\,\varepsilon, \tag{19.13a}$$
$$\delta\lambda^1 = \ddot{\varepsilon} \tag{19.13b}$$

($\delta\pi^0 = \delta\pi^i = 0$) with $\varepsilon \equiv \varepsilon^2$. If one eliminates π^0, π^i, and λ^1 from (19.12) by means of their own equations of motion, one gets the original action (19.1). The gauge transformations (19.13a) coincide with (19.3).

It is customary to eliminate from (19.12) only the auxiliary fields λ^1 and π_0 by means of their equations of motion, which leads to

$$S_0^H[A_0, A_i, \pi^i] = \int d^4x(\pi^i\dot{A}_i - H_0 - A_0\phi_2). \tag{19.14}$$

This action is invariant under the gauge transformation (19.13a) of S_0^T and could have been obtained directly from (19.1) by making the Legendre transformation only on $\dot{A}_i$ and π^i.

If one considers A_0 as a Lagrange multiplier, the action (19.14) has the form of an extended action $\int(p\,\dot{q} - H_0 - \lambda G)\,dt$ but in the smaller phase space of the A_i and the π^i. For this reason, one says that in electromagnetism, the dynamical variables are the spatial components of the vector potential and their conjugate momenta and that the temporal component A_0 is the Lagrange multiplier for Gauss' law.

19.1.2. Classical BRST Cohomology

The distinction between (A_i, π^i) and (A_0, π^0) is conventional and depends on where one draws the line between the minimal and nonminimal sectors. If one starts with the constraint $\phi_2(x) \approx 0$ in the phase space (A_i, π^i) and adds a ghost pair $\eta(\vec{x}\,), \mathcal{P}(\vec{x}\,)$ with

$$[\mathcal{P}(\vec{x}\,), \eta(\vec{x}\,')] = -\delta(\vec{x} - \vec{x}\,'),$$

the BRST charge is

$$\Omega^{\text{Min}} = \int \eta\, \phi_2\, d^3x, \tag{19.15}$$

since $[\phi_2(x), \phi_2(x')] = 0$. Adding the multiplier $A_0(x)$ and its momentum $\pi^0(x) \equiv b$, with antighost $\bar{C}(x)$ and momentum $\rho(x)$,

$$[\rho(\vec{x}\,), \bar{C}(\vec{x}\,')] = -\delta(\vec{x} - \vec{x}\,'),$$

one gets the nonminimal solution [see §11.3.2 and Eq. (11.22)]

$$\Omega = \int [-i\rho b + \eta\phi_2]\, d^3x. \tag{19.16}$$

But (19.16) is just the minimal solution for the description based on the extended action (19.10). One could further increase the nonminimal sector by adding λ^1, λ^2, their momenta, and the corresponding system of ghosts. This will not be necessary for dealing with the standard Lorentz gauge condition $\partial^\mu A_\mu = 0$ but is useful when considering gauge conditions involving time derivatives of A_0 of order higher than the first (see §19.1.7 below). We shall thus, in a first step, base the discussion on the BRST charge (19.16), which is sufficient for dealing with the Lorentz gauge, and refer to A_0 as a multiplier.

The classical cohomology of the BRST differential is easily worked out if one introduces the transverse and longitudinal components of $A_i(x)$ and $\pi^i(x)$ according to

$$A^{Ti}(x) = A^i(x) - \partial^i \Delta^{-1}(\vec{\nabla} \cdot \vec{A}\,), \quad A^{L'}(x) = \Delta^{-1}(\vec{\nabla} \cdot \vec{A}\,), \tag{19.17a}$$

$$\pi^{Ti}(x) = \pi^i(x) - \partial^i \Delta^{-1}(\vec{\nabla} \cdot \vec{\pi}), \qquad \pi^{L'}(x) = \vec{\nabla} \cdot \vec{\pi}, \tag{19.17b}$$

where Δ^{-1} is the inverse Laplace operator. The BRST symmetry takes then the simple form

$$sA^{L'} = \eta, \qquad s\eta = 0, \qquad s\mathcal{P} = \pi^{L'}, \qquad s\pi^{L'} = 0, \tag{19.18a}$$

$$sA_0 = -i\rho, \qquad s\rho = 0, \qquad s\bar{C} = ib, \qquad sb = 0, \tag{19.18b}$$

$$sA^{Ti} = 0, \qquad s\pi^{Ti} = 0. \tag{19.18c}$$

The variables η, $\pi^{L'}$, ρ, and b are the BRST variations of $A^{L'}$, $\mathcal{P}$, A_0, and $\bar{C}$, respectively. Only the transverse fields are BRST closed without being BRST exact. One can thus apply the method of Theorem 8.2 (see also §9.A.2) to show that the classical BRST cohomology is exhausted by the functionals of $A^{Ti}(\vec{x}\,)$ and $\pi^{Ti}(\vec{x}\,)$. The contracting homotopy σ defined by

$$\sigma\eta = A^{L'}, \qquad \sigma A^{L'} = 0, \qquad \sigma\pi^{L'} = \mathcal{P}, \qquad \sigma\mathcal{P} = 0, \tag{19.19a}$$

$$\sigma(-i\rho) = A_0, \qquad \sigma A_0 = 0, \qquad \sigma(ib) = \bar{C}, \qquad \sigma\bar{C} = 0, \tag{19.19b}$$

$$\sigma A^{Ti} = 0, \qquad \sigma\pi^{Ti} = 0, \tag{19.19c}$$

and satisfies

$$\sigma s + s\sigma = \tilde{N}, \tag{19.19d}$$

where $\tilde{N}$ counts the degree in the unphysical variables $A_{L'}, \pi_{L'}, A_0, b, \eta, \mathcal{P}, \rho, \bar{C}$, *i.e.*, $\tilde{N} = A_{L'}(\partial/\partial A_{L'}) + \pi_{L'}(\partial/\partial \pi_{L'}) + A_0(\partial/\partial A_0) + b(\partial/\partial b) + \eta(\partial/\partial \eta) + \mathcal{P}(\partial/\partial \mathcal{P}) + \rho(\partial/\partial \rho) + \bar{C}(\partial/\partial \bar{C})$. It follows that if a BRST-closed functional F is of degree $n \neq 0$ in the unphysical variables,

$$sF = 0, \qquad \tilde{N}F = nF,$$

then it is BRST exact, $F = s[(1/n)\sigma F]$. The only nonexact BRST-closed functionals are those of degree zero in the unphysical variables, *i.e.*, the functionals $F[A^{Ti}(\vec{x}\,), \pi^{Ti}(\vec{x}\,)]$ of the transverse fields.

The BRST cohomology is thus isomorphic to the space of functionals on the reduced phase space, since the two conjugate pairs $\big(A^{Ti}(\vec{x}\,), \pi^{Ti}(\vec{x}\,)\big), (\partial_i A^{Ti} \equiv 0, \partial_i \pi^{Ti} \equiv 0)$ provide a complete set of observables. These variables may be thought of as being the gauge-invariant extensions of A^i and π^i from the transverse gauge $\vec{\nabla} \cdot \vec{A} = 0$. When one introduces the further nonminimal variables λ^1, λ^2, their momenta, and the corresponding set of ghosts, the isomorphism of the BRST cohomology with the space of observables is explicitly verified in the same fashion.

19.1.3. Antifield Formalism

The derivation of the minimal solution of the master equation is straightforward because the gauge transformations (19.3) are irreducible and abelian. Hence, one needs only one ghost field C and the antifields A^*_μ, C^*. The minimal solution reads explicitly

$$S^{\text{Min}} = -\tfrac{1}{4}\int F_{\mu\nu}F^{\mu\nu}\, d^4x + \int A^{*\mu}\partial_\mu C\, d^4x. \tag{19.20}$$

It will be necessary to add the nonminimal term $\int d^4x\, \bar{C}^* b$ to S in order to implement covariant gauge conditions. The complete (nonminimal) solution of the master equation is given by

$$S = -\tfrac{1}{4}\int F_{\mu\nu}F^{\mu\nu}\, d^4x + \int A^{*\mu}\partial_\mu C\, d^4x + i\int \bar{C}^* b\, d^4x. \tag{19.21}$$

(The factor i is included in front of the nonminimal term because we take the antifields to be imaginary. This makes S real.)

19.1.4. Path Integral—Gauge-fixed Action

19.1.4a. Hamiltonian Treatment

To implement the Lorentz gauge $\partial_\mu A^\mu = 0$, it is necessary to take a gauge-fixing fermion K of the form

$$K = \int (i\bar{C}\chi - \mathcal{P}A_0)\, d^3x \tag{19.22a}$$

where χ is such that $\dot{A}_0 + \chi = 0$ is equivalent to $\partial^\mu A_\mu = 0$. This is accomplished by choosing

$$\chi = -\partial^k A_k \tag{19.22b}$$

(see §11.3.2 and §11.3.3). Since H_0 is BRST invariant, the gauge-fixed Hamiltonian reads

$$\begin{aligned} H_K &= H_0 + [K, \Omega] \\ &= H_0 + \int d^3x\, (-i\bar{C}\,\Delta\eta - \chi b - A_0\, \vec{\nabla}\cdot\vec{\pi} + i\mathcal{P}\rho). \end{aligned} \tag{19.22c}$$

Integration over the momenta $\mathcal{P}$, ρ, and π^i in the path integral

$$\int [DA_\mu\, D\pi^\mu\, D\eta\, D\bar{C}\, D\mathcal{P}\, D\rho]\, \exp i \int d^4x\, (\dot{A}_\mu \pi^\mu + \dot{\eta}\mathcal{P} + \dot{\bar{C}}\rho - H_K) \tag{19.23}$$

gives the Lagrangian path integral

$$\int DA_\mu\, Db\, D\eta\, D\bar{C}\, \exp i\, S_K \tag{19.24a}$$

with

$$S_K = \int (-\tfrac{1}{4}\, F_{\mu\nu}F^{\mu\nu} - b(\partial^\mu A_\mu) + i\,\bar{C}\,\Box\eta)\, d^4x \tag{19.24b}$$

($\Box\eta = -\ddot{\eta} + \Delta\eta$). The integration over the b-field yields then a δ-function of the gauge condition in the integrand, *i.e.*, restricts the integration over the potential A_μ to those obeying $\partial_\mu A^\mu = 0$.

One can alternatively consider gauge-fixing fermions of the form (11.24a)

$$K_\alpha = i\,\bar{C}\chi - \mathcal{P}\lambda + \frac{i}{2}\,\alpha\bar{C}b \tag{19.25}$$

with $\chi = -\partial^k A_k$ and where α is an arbitrary real number. After integration over the momenta π^i, ρ, and $\mathcal{P}$, these gauge-fixing fermions lead to the path integral

$$\int [DA_\mu\, Db\, D\bar{C}\, D\eta]\, \exp i\, S'_{K_\alpha}, \tag{19.26a}$$

$$S'_{K_\alpha} = \int \Big[-\frac{1}{4} F_{\mu\nu}F^{\mu\nu} - \Big(\partial^\mu A_\mu - \frac{\alpha}{2}\, b\Big) b + i\bar{C}\Box\eta\Big)\Big] d^4x. \qquad (19.26b)$$

This time, the integration over b is Gaussian so that the integral over A_μ is not restricted to a definite gauge. One says that one has a Gaussian average over the gauge conditions $\partial^\mu A_\mu - \alpha b = 0$ (see Exercise 19.3). The equality of the path integrals (19.26) and (19.24) follows from the independence of the transition amplitude on the choice of gauge-fixing fermions (Theorems 14.5 or 16.1).

When dealing with the family of gauge-fixing fermions (19.25), one often refers to different choices of the parameter α as being different "choices of gauges." The choice $\alpha = 1$ is known as the Feynman gauge and is particularly simple because the equations of motion for A_μ are then simply

$$\Box A_\mu = 0 \qquad (\alpha = 1). \qquad (19.27)$$

The choice $\alpha = 0$ is known as the "Landau gauge" and yields the gauge-fixed action (19.24).

In terms of the variables A_μ, b, $\bar{C}$, and η, the BRST transformation reads

$$sA_\mu = \partial_\mu \eta, \qquad s\eta = 0, \qquad s\bar{C} = ib, \qquad sb = 0. \qquad (19.28)$$

If one replaces b by its on-shell value, $b = (1/\alpha)\partial_\mu A^\mu$, one finds that $s\bar{C} = ib$ is replaced by $s\bar{C} = (i/\alpha)\partial_\mu A^\mu$. The BRST transformation is then no longer nilpotent on $\bar{C}$.

19.1.4b. Antifield Treatment

One can reach the gauge-fixed actions (19.24) and (19.26) more directly within the antifield formalism. Elimination of the antifields by means of the gauge-fixing fermion

$$\psi = \int d^4x\, i\,\bar{C}\Big(\partial^\mu A_\mu - \frac{\alpha}{2}\, b\Big) \qquad (19.29a)$$

yields

$$S_\psi = \int d^4x \Big[-\frac{1}{4} F_{\mu\nu}F^{\mu\nu} - \Big(\partial^\mu A_\mu - \frac{\alpha}{2}\, b\Big) b + i\,\bar{C}\Box C\Big], \qquad (19.29b)$$

i.e., exactly (19.26) with the identification of C with η (besides the identifications $b_{\text{Ham.}} = b_{\text{Lag.}}$, $\bar{C}_{\text{Ham.}} = \bar{C}_{\text{Lag.}}$ anticipated by the notations). Hence, the gauge-fixed action of the antifield formalism agrees with the gauge-fixed action of the Hamiltonian formalism (as otherwise known on general grounds).

Furthermore, ΔS vanishes so that S solves the quantum master equation. One can thus take $S = W$, which amounts to adopting the naive integration measure $[DA_\mu\, DC\, D\bar{C}]$ in the path integral. Again, this agrees with the Hamiltonian path integral, since the integration over the momenta yields just a numerical factor that can be absorbed in the measure.

19.1.5. Faddeev–Popov Determinant

The ghost term in (19.24) and (19.26) is quadratic in the ghosts and of the Faddeev–Popov form $\bar{C}\,\delta_\eta$(gauge condition) (see §11.3.3). It can be rewritten as the determinant of the operator $\Box$ (in the functional space defined by the boundary conditions on $\bar{C}$ and η),

$$\int D\bar{C}\, D\eta \, \exp \int i\,\bar{C}\Box\eta\, dx = \det \Box. \tag{19.30}$$

In the present case, this determinant does not depend on the gauge field A_μ and may be absorbed in the normalization. In the case of Yang–Mills theory, however, the ghosts are coupled to the gauge fields (Exercise 19.4). The determinant defined by the ghost action—and known as the Faddeev–Popov determinant—cannot be ignored.

The traditional route leading to the ghost action is actually opposite to the method followed in this book. Namely, one first shows that the integration over the gauge field in Yangs–Mills theory must be performed with a nontrivial determinant to secure gauge independence of the physical amplitudes. The ghosts are then regarded as a trick used to replace the determinant by an integration over new fields with a local action. This alternative route is sketched in Exercise 19.5. It cannot be applied, however, to gauge transformations not forming a closed algebra.

Note that the alternate method gives no clue as to what reality properties should be assigned to the ghosts. The choice adopted throughout this book (here, $\eta^* = \eta$, $\bar{C}^* = \bar{C}$) leads to a hermitian BRST charge, making possible the consistent decoupling of the null states in the operator formalism.

19.1.6. Operator Quantization

To analyze the quantum BRST cohomology, it is useful to Fourier-analyze the fields and to pass to oscillator variables. We shall make the transformation in such a way that the oscillators do oscillate $[a(t) \sim e^{-i\omega t}a(t_0)]$ in the Feynman gauge ($\alpha = 1$).

We define an invertible transformation of the (A_μ, π^μ) to the (a_0, a_m, a_0^*, a_m^*) as follows:

$$A_0(\vec{x}\,) = \frac{1}{(2\pi)^{3/2}} \int \frac{d^3k}{\sqrt{2\omega(\vec{k}\,)}} \times [a_0(\vec{k}\,)\, e^{i\vec{k}\vec{x}} + a_0^*(\vec{k}\,)\, e^{-i\vec{k}\vec{x}}], \tag{19.31a}$$

$$A_i(\vec{x}\,) = \frac{1}{(2\pi)^{3/2}} \int \frac{d^3k}{\sqrt{2\omega(\vec{k}\,)}} \sum_{m=1,2,3} \times [a_m(\vec{k}\,)\, e_i^m(\vec{k}\,)\, e^{i\vec{k}\vec{x}} + a_m^*(\vec{k}\,)\, e_i^m(\vec{k}\,)\, e^{-i\vec{k}\vec{x}}], \tag{19.31b}$$

$$b(\vec{x}\,) \equiv \pi^0(\vec{x}\,) = \frac{i}{(2\pi)^{3/2}} \int d^3k \sqrt{\frac{\omega(\vec{k}\,)}{2}} \big[\big(a_0(\vec{k}\,) + a_L(\vec{k}\,)\big)\, e^{i\vec{k}\vec{x}} - \big(a_0^*(\vec{k}\,) + a_L^*(\vec{k}\,)\big)\, e^{-i\vec{k}\vec{x}}\big], \tag{19.31c}$$

$$\pi_i(\vec{x}\,) = -\frac{i}{(2\pi)^{3/2}} \int d^3k \sqrt{\frac{\omega(\vec{k}\,)}{2}} \times \big[\big(a_m(\vec{k}\,)\, e_i^m(\vec{k}\,) + a_0(\vec{k}\,)\, e_i^3(\vec{k}\,)\big)\, e^{i\vec{k}\vec{x}} - \big(a_m^*(\vec{k}\,)\, e_i^m(\vec{k}\,) + a_0^*(\vec{k}\,)\, e_i^3(\vec{k}\,)\big)\, e^{-i\vec{k}\vec{x}}\big], \tag{19.31d}$$

where $\omega(\vec{k}\,) = |\vec{k}|$. Here, $e_i^m \vec{k}p$ are three vectors chosen as follows: $e_i^3(\vec{k}\,)$ is a unit vector along $\vec{k}$, $e_i^3(\vec{k}\,) = k_i/\omega = -e_i^3(-\vec{k})$, which define the longitudinal direction ($a_L = a_3$), whereas $e_i^a(\vec{k}\,)$, $a = 1, 2$, are two orthonormal (spatial) vectors orthogonal to $\vec{k}$. The oscillators $a_a^*(\vec{k}\,)$ are thus the transverse oscillators, creating transverse photons, and the oscillators $a_L(\vec{k}\,)$ create longitudinal photons, while the oscillators $a_0(\vec{k}\,)$ create temporal (or "scalar") photons. The nonzero Poisson brackets are:

$$[a_0(\vec{k}\,), a_0^*(\vec{k}\,)] = i\,\delta(\vec{k} - \vec{k}'\,) \qquad \text{(P.B.)}, \tag{19.32a}$$

$$[a_L(\vec{k}\,), a_L^*(\vec{k}\,)] = -i\,\delta(\vec{k} - \vec{k}'\,) \qquad \text{(P.B.)}, \tag{19.32b}$$

$$[a_a(\vec{k}\,), a_b^*(\vec{k}\,)] = -i\,\delta(\vec{k} - \vec{k}'\,)\,\delta_{ab} \qquad \text{(P.B.)}. \tag{19.32c}$$

We extend the transformation (19.31) to the ghosts and their momenta as follows:

$$\eta(\vec{x}\,) = -\frac{1}{(2\pi)^{3/2}} \int \frac{d^3k}{2\omega^{3/2}} [c(\vec{k}\,)\, e^{i\vec{k}\vec{x}} + c^*(\vec{k}\,)\, e^{-i\vec{k}\vec{x}}], \tag{19.33a}$$

$$\bar{C}(\vec{x}\,) = -\frac{i}{(2\pi)^{3/2}} \int d^3k\, \omega^{1/2} [\bar{c}(\vec{k}\,)\, e^{i\vec{k}\vec{x}} - \bar{c}^*(\vec{k}\,)\, e^{-i\vec{k}\vec{x}}], \tag{19.33b}$$

$$\mathcal{P}(\vec{x}\,) = \frac{i}{(2\pi)^{3/2}} \int d^3k\, \omega^{3/2}\, [\bar{c}(\vec{k}\,)\, e^{i\vec{k}\vec{x}} + \bar{c}^*(\vec{k}\,)\, e^{-i\vec{k}\vec{x}}\,], \tag{19.33c}$$

$$\rho(\vec{x}\,) = -\frac{1}{(2\pi)^{3/2}} \int \frac{d^3k}{2\omega^{1/2}}\, [c(\vec{k}\,)\, e^{i\vec{k}\vec{x}} - c^*(\vec{k}\,)\, e^{-i\vec{k}\vec{x}}\,]. \tag{19.33d}$$

The nonvanishing brackets are

$$[c(\vec{k}\,), \bar{c}^*(\vec{k'}\,)] = -i\,\delta(\vec{k} - \vec{k'}\,),\ [\bar{c}(\vec{k}\,), c^*(\vec{k'}\,)] = -i\,\delta(\vec{k} - \vec{k'}\,) \quad \text{(P.B.)}. \tag{19.33e}$$

The BRST charge expressed in terms of oscillators is

$$\Omega = \int d^3k\, \left[c^*(\vec{k}\,)\big(a_0(\vec{k}\,) + a_L(\vec{k}\,)\big) + \big(a_0^*(\vec{k}\,) + a_L^*(\vec{k}\,)\big)\, c(\vec{k}\,)\right] \tag{19.34}$$

and involves only the null combination $a_0(\vec{k}\,) + a_L(\vec{k}\,)$ and its complex conjugate. We thus redefine $a_0, a_L \to a, b$ as follows:

$$a(\vec{k}\,) = a_L(\vec{k}\,) + a_0(\vec{k}\,), \tag{19.35a}$$

$$b(\vec{k}\,) = \tfrac{1}{2}\big(a_L(\vec{k}\,) - a_0(\vec{k}\,)\big). \tag{19.35b}$$

These null oscillators are such that

$$[a(\vec{k}\,), b^*(\vec{k}\,)] = -i\,\delta(\vec{k} - \vec{k'}\,),\ [b(\vec{k}\,), a^*(\vec{k'}\,)] = -i\,\delta(\vec{k} - \vec{k'}\,) \quad \text{(P.B.)} \tag{19.35c}$$

(the unwritten brackets vanish).

Note that the Fourier component of $b(x)$ is $a(\vec{k}\,)$, not $b(\vec{k}\,)$! This unfortunate notation is due to our wish to comply with the traditional use of $b(x)$ for π^0 on the one hand and to have notations that agree with those of Chapter 14 on the other hand.

In terms of these variables, the BRST charge and gauge-fixed Hamiltonian become

$$\Omega = \int d^3k\, [c^*(\vec{k}\,)\, a(\vec{k}\,) + a^*(\vec{k}\,)\, c(\vec{k}\,)], \tag{19.36}$$

$$\begin{aligned} H_\alpha = &\int d^3k\, \omega(\vec{k}\,) \sum_{a=1,2} a_a^*(\vec{k}\,)\, a_a(\vec{k}\,) \\ &+ \int d^3k\, \omega(\vec{k}\,)\, \big(\bar{c}^*(\vec{k}\,)\, c(\vec{k}\,) + c^*(\vec{k}\,)\, \bar{c}(\vec{k}\,) + a^*(\vec{k}\,)\, b(\vec{k}\,) + b^*(\vec{k}\,)\, a(\vec{k}\,)\big) \\ &+ \tfrac{1}{4}(\alpha - 1) \int d^3k\, \omega(k) \big[a^*(\vec{k}\,)\, a^*(-\vec{k}) + a(\vec{k}\,)\, a(-\vec{k}) - 2\, a^*(\vec{k}\,)\, a(\vec{k}\,)\big]. \end{aligned} \tag{19.37}$$

The first term in the Hamiltonian is the physical transverse energy. The second and third terms are individually BRST exact and so do not contribute to physical expectation values. The second term has the structure of the occupation number operator for the unphysical modes

$$N = \int d^3k \, [\bar{c}^*(\vec{k}\,) \, c(\vec{k}\,) + c^*(\vec{k}\,) \, \bar{c}(\vec{k}\,) + a^*(\vec{k}\,) \, b(\vec{k}\,) + b^*(\vec{k}\,) \, a(\vec{k}\,)], \quad (19.38)$$

weighted with the frequency $\omega(\vec{k}\,)$. The last term does not have a nice interpretation but vanishes in the Feynman gauge. The constraints and the BRST charge for free electromagnetism have the form already investigated in §14.4.2. The cohomology is thus established by observing that—as one should have expected—the counting operator (19.38) for the unphysical modes is BRST exact,

$$N = \tilde{K}\Omega + \Omega\tilde{K}, \qquad \tilde{K} = \int d^3k \, \big(\bar{c}^*(\vec{k}\,) \, b(\vec{k}\,) + b^*(\vec{k}\,) \, \bar{c}(\vec{k}\,)\big).$$

Accordingly, a physical state with a nonvanishing number of unphysical particles (= longitudinal and temporal photons, and antighost and ghost particles) is necessarily BRST exact,

$$N|\psi\rangle = n|\psi\rangle, \quad n \neq 0, \qquad \Omega|\psi\rangle = 0 \Rightarrow |\psi\rangle = \Omega\Big(\frac{1}{n}\tilde{K}|\psi\rangle\Big). \quad (19.39)$$

Therefore, the only nontrivial physical states are those containing solely transverse photons. The modes $a(\vec{k}\,), b(\vec{k}\,), c(\vec{k}\,), \bar{c}(\vec{k}\,)$ form "quartets" and disappear in cohomology. The BRST-state cohomology is isomorphic with the Fock space of the reduced method.

The Hamiltonian (19.37) does not annihilate the vacuum, except when $\alpha = 1$ (Feynman gauge). For other values of α, the vacuum develops with time into the vacuum plus a nonzero null state. If one makes the Bogoliubov transformation,

$$b'^*(\vec{k}\,) = b^*(\vec{k}\,) + \tfrac{1}{4}(\alpha - 1) \, a(-\vec{k}), \quad (19.40a)$$

$$b'(\vec{k}\,) = b(\vec{k}\,) + \tfrac{1}{4}(\alpha - 1) \, a^*(-\vec{k}), \quad (19.40b)$$

the new vacuum annihilated by the $b'(\vec{k}\,)$ and $a(\vec{k}\,)$ is also annihilated by the Hamiltonian and is therefore time-independent.

19.1.7. Gauge $\Box\partial_\mu A^\mu = 0$

We shall end our analysis of the free Maxwell theory by considering gauge conditions that involve time derivatives of A_0 of order higher than the first. Such gauge conditions can be imposed in the Hamiltonian formalism if one increases the nonminimal sector. This illustrates the flexibility of the framework.

19.1.7a. Antifield Treatment

A covariant gauge condition involving higher derivatives is $\Box\partial_\mu A^\mu = 0$. It can be implemented in the antifield formalism by taking as gauge-fixing fermion

$$\psi' = \int d^4x\, i\,\bar{C}\Box\partial_\mu A^\mu. \tag{19.41a}$$

The gauge-fixed action reads

$$S_{\psi'} = \int d^4x\,[-\tfrac{1}{4}F_{\mu\nu}F^{\mu\nu} - (\Box\partial_\mu A^\mu)b + i\,\bar{C}\Box\Box C]. \tag{19.41b}$$

The equations of motion for the bosonic fields are $\Box A^\mu - \partial^\mu(\partial^\rho A_\rho + \Box b) = 0$, $\Box\partial_\mu A^\mu = 0$. They can be solved to express $\dddot{b}$, $\dddot{A}^0$, and $\ddot{A}^k$ as functions of the derivatives of b, A^0, and A^k of lower order with respect to time. Hence, the number of required bosonic initial data is $12\big(A_\mu, \dot{A}_\mu, (\partial_\mu A^\mu)^{\cdot}, b, \dot{b}, \ddot{b}\big)$. Similarly, since the equations for the ghosts are $\Box\Box C = 0$ and $\Box\Box\bar{C} = 0$, the number of required fermionic data is $8(C, \dot{C}, \ddot{C}, \dddot{C}, \bar{C}, \dot{\bar{C}}, \ddot{\bar{C}}, \dddot{\bar{C}})$. The difference between the numbers of bosonic and fermionic initial data $(12 - 8 = 4)$ is equal to the number of independent initial data for the transverse degrees of freedom.

19.1.7b. Hamiltonian Treatment

To implement the gauge $\Box\partial^\mu A_\mu = 0$ in the canonical formalism, the extended phase space should have the same number of dimensions as the number of initial data, which is equal to 20, while keeping the number of physical degrees of freedom unchanged. This implies that one should enlarge the nonminimal sector with two bosonic conjugate pairs and two fermionic conjugate pairs. This is precisely the extra variables needed to write down the BRST charge associated with the extended formulation (19.10). Denoting by B_1 and B_2 the variables, respectively, conjugate to λ^1 and λ^2, and by $(\eta^1, \mathcal{P}_1), (\eta^2, \mathcal{P}_2)$ the corresponding ghost pairs, one finds that the full BRST charge with this new nonminimal sector included reads

$$\Omega = \int [\eta^1 B_1 + \eta^2 B_2 - i\,\rho\,\pi^0 + \eta\,\phi_2]\,d^3x. \tag{19.42a}$$

Once the appropriate nonminimal sector has been identified, the derivation of the gauge-fixed action within the canonical formalism is straightforward. We leave it to the reader to check that, after Gaussian integration over $\lambda^1, B_2, \pi^0, \lambda^2, \rho, \mathcal{P}, \mathcal{P}_2, \eta^1, \eta^2, \bar{C}$, and π^k, the gauge-fixing fermion

$$K = \int d^3x\,[i\bar{C}(-\partial^k A_k + \lambda^2) - \mathcal{P}A_0 - \lambda^2\,\Delta\mathcal{P}_1 - \lambda^1\mathcal{P}_2] \tag{19.42b}$$

and gauge-fixed Hamiltonian

$$H = H_0 + [K, \Omega] \tag{19.42c}$$

lead to the gauge-fixed action (19.41b). To see this, one must change the identifications of the Lagrangian and Hamiltonian variables and set $b = B_1$ and $\bar{C} = i\mathcal{P}_1$ (Exercise 19.7).

We also leave it to the reader to verify that the new variables $\lambda^1, \lambda^2, B^1, B^2, \eta^1, \eta^2, \mathcal{P}_1$, and $\mathcal{P}_2$ form oscillator quartets that disappear in cohomology (§14.4.2 and Exercise 19.8).

One could go up and consider higher order derivative gauges, such as $\Box\Box\partial^\mu A_\mu = 0$. This would require a further enlargement of the nonminimal sector with a coupling through the gauge-fixing term $[K, \Omega]$ between the new variables and the previous ones. The BRST charge in Hamiltonian form is always the same up to nonminimal terms forming null quartets. How many of these null quartets are relevant to the dynamics depends on the form of the gauge conditions, which may couple more and more of these null quartets to the original variables through the (BRST-exact) gauge-fixed term $[K, \Omega]$.

19.1.8. Temporal Gauge

One could also go down and consider multiplier gauges involving no derivative of A_0 at all (§3.4.2)

$$A_0 - \chi(A_i) = 0, \tag{19.43a}$$

e.g., $A_0 = 0$ ("temporal gauge"). To enforce multiplier gauges, one takes Hamiltonian gauge-fixing fermions of the form

$$K = -\chi(A_i)\mathcal{P} \tag{19.43b}$$

that do not mix the minimal sector $(A_k, \pi^k, \eta, \mathcal{P})$ with the nonminimal sector $(A_0, \pi^0, \bar{C}, \rho)$ (§11.2.1 and §16.5.6). The gauge-fixed action reads then,

$$S_K = S_{\mathrm{I}} + S_{\mathrm{II}}, \tag{19.43c}$$

$$S_{\mathrm{I}} = \int [\dot{A}_k \pi^k - H_0 - \chi(A_i)\phi_2 - \mathcal{P}(\dot{\eta} - \delta_\eta \chi)]\, d^4x, \tag{19.43d}$$

$$S_{\mathrm{II}} = \int (\dot{A}_0 \pi^0 + \dot{\bar{C}}\rho)\, d^4x. \tag{19.43e}$$

The first term, S_{I}, depends only on the variables $(A_k, \pi^k, \mathcal{P}, \eta)$ of the minimal sector. The second term, S_{II}, involves only the variables $(A_0, \pi^0, \bar{C}, \rho)$ of the nonminimal sector. The ghost term in S_{I} can be rewritten in the Faddeev–Popov form "antighost times variation of the gauge condition with the gauge parameter replaced by the ghost," provided one regards now $\mathcal{P}$ as being the Lagrangian antighost. This shows again that the identification of the Lagrangian and Hamiltonian variables is not universal (*i.e.*, independent of gauge fixing).

After integration over the momenta π^k, the path integral can be rewritten as

$$\int [DA_k \, D\mathcal{P} \, D\eta] \exp i \int [-\tfrac{1}{4} F_{\mu\nu} F^{\mu\nu}|_{A^0=\chi} - \mathcal{P}\, \delta_\eta (A_0 - \chi)] \, d^4x$$

$$\times \int [D\lambda \, D\pi^0 \, D\bar{C} \, D\rho] \exp i \int (\dot{\lambda}\pi^0 + \dot{\bar{C}}\rho) \, d^4x, \qquad (19.43\text{f})$$

where we have replaced the integration variable A_0 of the nonminimal sector by λ in the second line of (19.43f). Although the minimal and nonminimal sectors are now completely uncoupled in the dynamical equations of motion, they are still coupled through the boundary conditions in the path integral. Indeed, the oscillator variables of the Fock representation mix the longitudinal and temporal components, as well as ρ with η and $\mathcal{P}$ with $\bar{C}$. This feature is characteristic of the Fock representation defined by (19.31) and (19.33). One could have defined the oscillator variables without mixing the minimal and nonminimal sectors by combining together the gauge and ghost fields at $\vec{k}$ and $-\vec{k}$ (Exercise 19.10). In that case, the path integral over λ, π^0, $\bar{C}$, and ρ entirely decouples and is equal to unity. The path integral (19.43f) reduces then to

$$\int [DA_k \, D\mathcal{P} \, D\eta] \exp i \int [-\tfrac{1}{4} F_{\mu\nu} F^{\mu\nu}|_{A^0=\chi} - \mathcal{P}\, \delta_\eta (A_0 - \chi)] \, d^4\alpha. \qquad (19.43\text{g})$$

A similar feature holds in the Schrödinger representation with the boundary conditions of §16.5.6. The antifield treatment of the gauge $A_0 = 0$ is outlined in Exercise 19.14.

19.2. ABELIAN 2-FORM GAUGE FIELDS

The key difference between the theories of a 1-form gauge field A_μ and a 2-form $A_{\mu\nu}$ is that in the latter case the gauge transformations are reducible.

The action is

$$S_0^L[A_{\mu\nu}] = -\tfrac{1}{12} \int d^4x \, F_{\mu\nu\rho} \, F^{\mu\nu\rho}, \qquad (19.44)$$

where the field strength $F_{\mu\nu\rho}$ is given by

$$F_{\mu\nu\rho} = \partial_\mu A_{\nu\rho} + \partial_\nu A_{\rho\mu} + \partial_\rho A_{\mu\nu}. \qquad (19.45)$$

The action is invariant under the gauge transformations

$$\delta A_{\mu\nu} = \partial_\mu \varepsilon_\nu - \partial_\nu \varepsilon_\mu. \tag{19.46}$$

These gauge transformations are reducible, since the choice of gauge parameters,

$$\varepsilon_\mu = \partial_\mu \Lambda \tag{19.47}$$

leads to $\delta A_{\mu\nu} = 0$. There is no further reducibility identity.

The counting of physical degrees of freedom proceeds as indicated in Sec. 3.3. The gauge transformations at a given time—say, $t = 0$—read

$$\delta A_{0k} = \partial_0 \bar{\varepsilon}_k, \qquad \delta A_{mn} = \partial_m \bar{\varepsilon}_n - \partial_n \bar{\varepsilon}_m,$$

where we have redefined $\bar{\varepsilon}_k = \varepsilon_k - \partial_k \int_0 dt\, \varepsilon_0$ $\big(\bar{\varepsilon}_k(t=0) = \varepsilon_k(t=0)$; $\dot{\bar{\varepsilon}}_k(t=0) = (\dot{\varepsilon}_k - \partial_k \varepsilon_0)(t=0)\big)$. They involve effectively the three independent components $\partial_0 \bar{\varepsilon}_k$, but only two of the $\bar{\varepsilon}_n$'s, since $\delta A_{mn} = 0$ if $\bar{\varepsilon}_n = \partial_n \Lambda$. Hence, there are five independent gauge parameters in the gauge transformations at a given time. Since there is no second-class constraint (see below), the number of degrees of freedom is 6 (number of field components $A_{\mu\nu}$) $-\,5$ (number of independent gauge parameters at a given time) $= 1$.

19.2.1. Hamiltonian Analysis

The canonical analysis is direct (Exercise 19.13): There are three primary constraints,

$$\pi^{0j} \approx 0, \tag{19.48}$$

and three secondary constraints,

$$G^j = -2\,\partial_i\,\pi^{ij} \approx 0, \tag{19.49}$$

which are not independent but obey instead the reducibility identity,

$$\partial_j G^j = 0. \tag{19.50}$$

The gauge-invariant Hamiltonian H_0 is given by

$$H_0 = \int d^3x\,(\pi_{ij}\,\pi^{ij} + \tfrac{1}{12} F_{ijk}\,F^{ijk}). \tag{19.51}$$

The action in which the temporal components A_{0i} are treated as Lagrange multipliers [as in (19.14)] reads

$$S_0^H[A_{ij}, \pi^{ij}, A_{0i}] = \int d^3x\,(\pi^{ij}\partial_0 A_{ij} - H_0 - A_{0i}G^i). \tag{19.52}$$

19.2.2. Classical BRST Cohomology

The constraints (19.49) are abelian. The minimal solution of the BRST equation $[\Omega, \Omega] = 0$ is given by

$$\Omega^{\text{Min}} = \int (\eta_i G^i + \eta \mathcal{P}^i_{,i})\, d^3x, \tag{19.53a}$$

where $(\eta^i, \mathcal{P}_i)$ are the ghost pair for $G^i \approx 0$ and where $(\eta, \mathcal{P})$ is the ghost-of-ghost pair associated with the reducibility equation (19.50) (see Chapter 10),

$$[\mathcal{P}^i(\vec{x}\,), \eta_j(\vec{x}\,')] = -\delta^i{}_j\, \delta(\vec{x} - \vec{x}\,'), \tag{19.54a}$$

$$[\mathcal{P}(\vec{x}\,), \eta(\vec{x}\,')] = -\delta(\vec{x} - \vec{x}\,'), \tag{19.54b}$$

$$\text{gh}\, \eta_i = 1 = -\text{gh}\, \mathcal{P}^i, \qquad \varepsilon(\eta_i) = \varepsilon(\mathcal{P}_i) = 1, \tag{19.54c}$$

$$\text{gh}\, \eta = 2 = -\text{gh}\, \mathcal{P}, \qquad \varepsilon(\eta) = \varepsilon(\mathcal{P}) = 0. \tag{19.54d}$$

The spatial 2-form A_{ij} can be decomposed into longitudinal and transverse components,

$$A_{ij} = \partial_i A_j{}^T - \partial_j A_i{}^T + \varepsilon_{ijk}\, \partial_k A^L \tag{19.55}$$

$(\partial^i A_i{}^T = 0)$. The two transverse components are pure gauge, while A^L is gauge invariant. The BRST transformation pairs the two transverse components of A_i with those of the ghosts η_i,

$$sA_i{}^T = \eta_i{}^T, \qquad s\eta_i{}^T = 0. \tag{19.56a}$$

For the momenta, one has

$$s\mathcal{P}_i{}^T = \pi_i{}^T, \qquad s\pi_i{}^T = 0, \tag{19.56b}$$

while the longitudinal modes A^L and π^L are BRST invariant,

$$sA^L = 0, \qquad s\pi^L = 0. \tag{19.56c}$$

If one had not included the ghost-of-ghost term $\int \eta \mathcal{P}_{,i}{}^i\, dx$ in (19.53a), one would have found $s\eta^L = 0$, $s\mathcal{P}^L = 0$ and the BRST cohomology would have been wrongly given by the functionals $F[A^L(\vec{x}\,), \pi^L(\vec{x}\,), \eta^L(\vec{x}\,), \mathcal{P}^L(\vec{x}\,)]$. The role of the ghosts of ghosts is to remove $\eta^L(\vec{x}\,)$ and $\mathcal{P}^L(\vec{x}\,)$ in cohomology by pairing η^L with η and $\mathcal{P}^L$ with $\mathcal{P}$,

$$s\eta^L = \eta, \qquad s\eta = 0, \tag{19.56d}$$

$$s\mathcal{P} = \mathcal{P}_L, \qquad s\mathcal{P}_L = 0. \tag{19.56e}$$

With the ghosts of ghosts in, the BRST cohomology is correctly given by the functionals $F[A^L(\vec{x}\,), \pi^L(\vec{x}\,)]$ of the longitudinal modes, *i.e.*, by the functionals on the reduced phase space. The verification of these assertions is most easily carried out by means of a homotopy as in §19.1.2.

19.2.3. Nonminimal Sector—Operator Formalism

To discuss the nonminimal sector, it is convenient to introduce some terminology. The nonminimal variables consist of groups of two cohomologically trivial conjugate pairs ("blocks") of the form $(x, p), (y, \pi)$ with $sy = p, sp = 0, sx = \pi, s\pi = 0, \Omega \sim p\pi$. One has

$$\operatorname{gh} x = -\operatorname{gh} p,\ \varepsilon(x) = \varepsilon(p); \quad \operatorname{gh} \pi = -\operatorname{gh} y = -\operatorname{gh} p + 1; \\ \varepsilon(y) = \varepsilon(\pi) = \varepsilon(p) + 1. \tag{19.57}$$

Accordingly, either $\operatorname{gh} p \leq 0$ and $\operatorname{gh} \pi > 0$; or $\operatorname{gh} \pi \leq 0$ and $\operatorname{gh} p > 0$. By changing names, if necessary, we can assume $\operatorname{gh} p \leq 0$. We shall denote a block by $[p]$ and assign to it the ghost number and Grassmann parity of p. Knowing the ghost number and parity of a block, one can reconstruct the ghost number and parity of each of its elements by means of (19.57).

The BRST charge (19.53a) has the form $\Omega \sim {\pi_T}^i {\eta_T}^i + \eta \mathcal{P}^L$ and contains thus three blocks: the two even blocks $[{\pi_T}^i]$ of ghost number zero and the odd block $[\mathcal{P}^L]$ of ghost number minus one. To define oscillator variables and to apply the quartet mechanism, one must combine two blocks with the same ghost number in pairs (§14.4.1). Although permissible from the point of view of ghost numbers, it is not suitable to pair the two transverse components with each other, since this would spoil manifest rotation invariance. Hence, one must introduce at least three new blocks. However, as will be seen below, the need to preserve also locality will bring in even other additional blocks.

If the constraints $G^i \approx 0$ were irreducible, one would simply introduce three blocks $[b_i]$ containing the Lagrange multipliers λ^i and combine them with the blocks defined by the constraints, as in electromagnetism. This does not work here because one can only combine the two transverse ${b_i}^T$'s with the two independent constraints. One of the b_i's, namely, b^L, is left over. Furthermore, one needs to take care of the block $[\mathcal{P}^L]$ containing the ghost of ghost. Thus, it is necessary to introduce, besides the standard three blocks $[b_i]$ with ghost number 0 and parity 0, one further block $[b_0]$ with the same ghost number and parity and one block $[b]$ with ghost number -1 and parity 1. The extra block $[b_0]$ may be thought of as necessary to preserve locality, since this requirement forces one to describe the transverse fields ${b_i}^T$ by means of the three independent components b_i.

Once these variables are introduced, one can combine the blocks in the BRST charge in pairs and express the variables of the combined blocks into null oscillators $[(p_1, p_2) \to (a, a^*), (\pi_1, \pi_2) \to (c, c^*)]$. The BRST charge takes then the form of §14.4.1, $\Omega \sim \sum_\alpha (a^*_\alpha c_\alpha + c^*_\alpha a_\alpha)$.

The oscillators a^*_α, c^*_α, together with their non-BRST-invariant commutator conjugates, form quartets that do not contribute to the cohomology (§14.4.2 and §19.1.6). Therefore, the nontrivial BRST-closed states contain only the longitudinal particles of the 2-form gauge field. The BRST cohomology is isomorphic to the Fock space of the reduced method.

19.2.4. Generalization: Hamiltonian Nonminimal Sector for Arbitrary Reducible Theories

One can extend the analysis to more general reducible constraints $G_{a_0} \approx 0$ with reducibility functions $Z_{a_1}{}^{a_0}, Z_{a_2}{}^{a_1}, \dots$ (Chapter 10). Since our purpose is only to count the number of nonminimal blocks required in this case, without explicitly defining the null oscillators, one may assume that the constraints are pure momenta and that the Z's take the canonical form of §10.2.5 ($Z_{A_0}{}^{a_{k-1}} = (0, \delta_{A_k}{}^{a_{k-1}}), Z_{\alpha_k}{}^{a_{k-1}} = 0$).

The minimal BRST charge reads

$$\Omega = \eta^{a_0} G_{a_0} + \eta^{a_1} Z_{a_1}{}^{a_0} \mathcal{P}_{a_0} + \cdots + \eta^{a_L} Z_{a_L}{}^{a_{L-1}} \mathcal{P}_{L-1} \tag{19.58}$$

and contains $m_0 - m_1 + m_2 \cdots + (-)^L m_L$ independent blocks $[G_{a_0}]$ of ghost number zero, $m_1 - m_2 + \cdots + (-)^{L-1} m_L$ independent blocks $[Z_{a_1}{}^{a_0} \mathcal{P}_{a_0}]$ of ghost number minus one, ..., and m_L blocks of ghost number minus L.

Let us first take care of the blocks $[G_{a_0}]$. We introduce m_0 blocks $[b_{a_0}]$ with $\varepsilon(b_{a_0}) = \varepsilon_{a_0}$ and $\text{gh}\, b_{a_0} = 0$. If $m_1 = 0$, we combine all the $[b_{a_0}]$'s with the $[G_{a_0}]$'s and this is finished. If $m_1 \neq 0$, we cannot combine all the $[b_{a_0}]$ s with the $[G_{a_0}]$'s, since $Z_{a_1}{}^{a_0} G_{a_0}$ vanishes. The $Z_{a_1}{}^{a_0} b_{a_0}$ are left over. We thus introduce m_1 blocks $[b_{a_1}]$ with $\varepsilon(b_{a_1}) = \varepsilon(Z_{a_1}{}^{a_0} b_{a_0}) = \varepsilon_{a_1}$, $\text{gh}\, b_{a_1} = 0$. If $m_2 = 0$, we combine all the $[b_{a_1}]$'s with the $[Z_{a_1}{}^{a_0} b_{a_0}]$'s and the remaining $m_0 - m_1$ $[b_{a_0}]$'s with the constraints, and the process ends. If $m_2 \neq 0$, this cannot be done because $Z_{a_2}{}^{a_1}(Z_{a_1}{}^{a_0} b_{a_0})$ vanishes, thus the $Z_{a_2}{}^{a_1} b_{a_1}$ are left over. So, one introduces m_2 blocks $[b_{a_2}]$ with $\varepsilon(b_{a_2}) = \varepsilon(Z_{a_2}{}^{a_1} b_{a_1}) = \varepsilon_{a_2}$, $\text{gh}\, b_{a_2} = 0$. If $m_3 = 0$, one combines $[b_{a_2}]$ with $[Z_{a_2}{}^{a_1} b_{a_1}]$, the remaining $m_1 - m_2$ $[b_{a_1}]$ with the independent $[Z_{a_1}{}^{a_0} b_{a_0}]$ and the remaining $m_0 - m_1 + m_2$ $[b_{a_0}]$ with the independent $[G_{a_0}]$. If $m_3 \neq 0$, the process goes on at the higher levels in the same fashion.

One must then take care of the blocks $[Z_{a_1}{}^{a_0} \mathcal{P}_{a_0}]$. This leads along the same lines to a tower of blocks of ghost number minus one and of parity $\varepsilon_{a_i} + 1$. We thus get the following nonminimal blocks:

$$[b_{s a_s}] \qquad s = 0, \dots, L;\ a_s = 1, \dots, m_s, \tag{19.59a}$$

$$[b_s{}^{s'}{}_{a_s}] \qquad s' = 1, \dots, L;\ s = s', s'+1, \dots, L;\ a_s = 1, \dots, m_s, \tag{19.59b}$$

$$\varepsilon(b_{sa_s}) = \varepsilon_{a_s} + s \text{ (mod 2)}; \ \text{gh}\, b_{sa_s} = -s, \tag{19.59c}$$

$$\varepsilon(b_s{}^{s'}{}_{a_s}) = \varepsilon_{a_s} + s - s' \text{ (mod 2)}; \ \text{gh}\,(b_s{}^{s'}{}_{a_s}) = s' - s, \tag{19.59d}$$

$$\Omega = \Omega^{\text{Min}} + \sum_{s=0}^{L} b_{sa_s} \rho_s{}^{a_s} + \sum_{s'=1}^{L} \sum_{s=s'}^{L} b_s{}^{s'}{}_{a_s}\, \rho_s{}^{s' a_s}, \tag{19.59e}$$

where the ρ's are the other BRST-invariant variables in the blocks $[b]$. The blocks associated with G_{a_0} are $[b_{0a_0}]$ and $[b_s{}^s{}_{a_s}]$, $s = 1, \dots, L$; those associated with $Z_{a_i}{}^{a_{i-1}} \mathcal{P}_{a_{i-1}}$ are $[b_{ia_i}]$ and $[b_s{}^{s-i}{}_{a_s}]$ with $s \geq i+1$. The above nonminimal sector is the standard nonminimal sector introduced in the Hamiltonian formalism. It is, however, not the only possible one, since by combining the constraints among each other, fewer nonminimal variables are needed (see Exercise 19.10).

19.2.5. Path Integral

Since the theory is free, the integration over the momenta in the Hamiltonian path integral can only yield a numerical factor that may be absorbed in the normalization. Thus, the measure can be taken to be one. For that reason, we shall skip the somewhat cumbersome Hamiltonian derivation of the path integral in covariant gauges and proceed instead with the more direct antifield method.

Because the gauge transformations are reducible, the minimal set of fields and antifields is given by

$$\begin{array}{cccccc} -3 & -2 & -1 & 0 & 1 & 2 \\ \hline C^* & C^{*\mu} & A^{*\mu} & A_\mu & C_\mu & C \end{array} \longrightarrow . \tag{19.60}$$

The minimal solution of the master equation reads

$$S = \int [-\tfrac{1}{12} F_{\mu\nu\rho} F^{\mu\nu\rho} + A^{*\mu\nu}(\partial_\mu C_\nu - \partial_\nu C_\mu) + C^{*\mu} \partial_\mu C]\, d^4x \tag{19.61a}$$

and is such that

$$\Delta S = 0. \tag{19.61b}$$

A covariant gauge condition that generalizes the Lorentz gauge is

$$\partial^\mu A_{\mu\nu} = 0. \tag{19.62}$$

The conditions (19.62) are redundant ($\partial^\nu \partial^\mu A_{\mu\nu}$ identically vanishes no matter what $A_{\mu\nu}$ is) and are thus three conditions for the three independent transformations contained in (19.46).

To mimic the electromagnetic case, one first tentatively introduces antighosts $\bar{C}^\mu$ for the gauge fixing of $A_{\mu\nu}$ and considers the nonminimal solution

$$S^{\text{Nonmin}} = S + \int \bar{C}^*_\mu b^\mu \, d^4x \qquad \text{(incorrect)}. \tag{19.63a}$$

To impose $\partial^\mu A_{\mu\nu} = 0$, one may attempt a gauge-fixing fermion of the form

$$\psi = \int \bar{C}^\mu \partial^\nu A_{\nu\mu} \, d^4x \qquad \text{(incorrect)} \tag{19.63b}$$

to find

$$S_\psi = \int [-\tfrac{1}{12} F_{\mu\nu\rho} F^{\mu\nu\rho} - \tfrac{1}{2}(\partial^\mu \bar{C}^\nu - \partial^\nu \bar{C}^\mu)(\partial_\mu C_\nu - \partial_\nu C_\mu) + \partial^\nu A_{\nu\mu} \cdot b^\mu] \qquad \text{(incorrect)}. \tag{19.63c}$$

This cannot be the correct answer because (i) the integration over b_μ yields $\prod_{x,\nu} \delta(\partial^\mu A_{\nu\mu})$ in the path integral. This product of delta-functions contains $\delta(0)$ because the arguments $\partial^\nu A_{\nu\mu}$ are not independent; (ii) the action (19.63c) is gauge invariant under gauge transformations affecting the ghost and antighost fields, $C_\mu \to C_\mu + \partial_\mu \varepsilon$, $\bar{C}_\mu \to \bar{C}_\mu + \partial_\mu \bar{\varepsilon}$. This formally yields a "compensating zero" in the path integral.

To remedy these problems, one extends the nonminimal sector by adding terms that enable one to gauge fix the ghost gauge freedom. So, one adds $\int \bar{C}^* b \, d^4x$ for the gauge freedom of the ghosts C_μ and $\int \eta^* \pi \, d^4x$ for the gauge freedom of the antighosts $\bar{C}^\mu$,

$$S^{\text{Nonmin}} = S + \int (\bar{C}^*_\mu b^\mu + \bar{C}^* b + \eta^* \pi) \, d^4x \tag{19.64}$$

with

$$\text{gh}\, b^\mu = 0 = \text{gh}\, \bar{C}^*_\mu, \qquad \text{gh}\, b^*_\mu = -1 = \text{gh}\, \bar{C}^\mu, \tag{19.65a}$$

$$\text{gh}\, b = -1 = -\text{gh}\, \bar{C}^*, \qquad \text{gh}\, b^* = 0, \qquad \text{gh}\, \bar{C} = -2, \tag{19.65b}$$

$$\text{gh}\, \pi = 1 = -\text{gh}\, \eta^*, \qquad \text{gh}\, \pi^* = -2, \qquad \text{gh}\, \eta = 0. \tag{19.65c}$$

An appropriate gauge-fixing fermion is then

$$\psi = \int [\bar{C}^\mu (\partial^\nu A_{\nu\mu}) + \bar{C} \partial^\nu C_\nu + \bar{C}^\nu \partial_\nu \eta], \tag{19.66a}$$

which yields the gauge-fixed action

$$S_\psi = \int [-\tfrac{1}{12} F_{\mu\nu\rho} F^{\mu\nu\rho} + \tfrac{1}{2}(\partial^\mu \bar{C}^\nu - \partial^\nu \bar{C}^\mu)(\partial_\mu C_\nu - \partial_\nu C_\mu) - \partial^\nu \bar{C} \partial_\nu C + (\partial^\nu A_{\nu\mu} + \partial_\mu \eta) b^\mu + (\partial^\nu C_\nu) b - (\partial_\nu \bar{C}^\nu) \pi] \, d^4x. \tag{19.66b}$$

The gauge freedom of the ghosts and the antighosts is now fixed, since the integration over b enforces $\partial^\nu C_\nu = 0$, while the integration over π enforces $\partial_\nu \bar{C}^\nu = 0$. Furthermore, the integration over b^μ yields $\delta(\partial^\nu A_{\nu\mu} + \partial_\mu \eta)$ instead of $\delta(\partial^\mu A_{\nu\mu})$, and the arguments of the δ-functions are independent. The removal of the gauge invariances of the ghost and antighost fields automatically solves the problem of the redundancy of the gauge conditions. The gauge-fixed action (19.66b) is thus acceptable. Note that $\partial^\nu A_{\nu\mu} + \partial_\mu \eta = 0$ implies $\Box \eta = 0$, *i.e.*, $\eta = 0$ in virtue of the boundary conditions on the unphysical variable η (vacuum to vacuum). Hence, the gauge conditions $\partial^\nu A_{\nu\mu} = 0$ are indeed enforced by (19.66b).

To reach a Gaussian average representation, one must add to (19.66a) the term

$$\int (\alpha \, \bar{C}^\mu \, b_\mu + \beta \, \bar{C} b + \gamma \, \eta \, \pi) \, d^4x \tag{19.67}$$

linear in the auxiliary fields b_μ, b, and π.

19.2.6. Generalization: Antifield Nonminimal Sector for Arbitrary Reducible Theories

The gauge-fixed action in Hamiltonian form has no gauge invariance no matter how K is taken, since the equations of motion are $\dot{q} = \partial H_K / \partial p$, $\dot{p} = -\partial H_K / \partial q$ with a definite H_K involving no arbitrary function. This is not true for the action $S_\psi(\phi) = S(\phi, \phi^* = \delta\psi/\delta\phi)$ obtained within the antifield formalism when ψ is incorrectly chosen. As explained in Chapter 18, the gauge-fixing fermion ψ must be such that the equations of motion following from S_ψ are nondegenerate, *i.e.*, possess a unique solution for arbitrary physical initial conditions. This may not be possible if the nonminimal sector is inappropriately chosen. The guidelines for constructing the nonminimal sectors in the Hamiltonian and antifield cases are thus different.

The fields appearing in the path integral for the abelian 2-form gauge field can be put in two pyramids as follows:

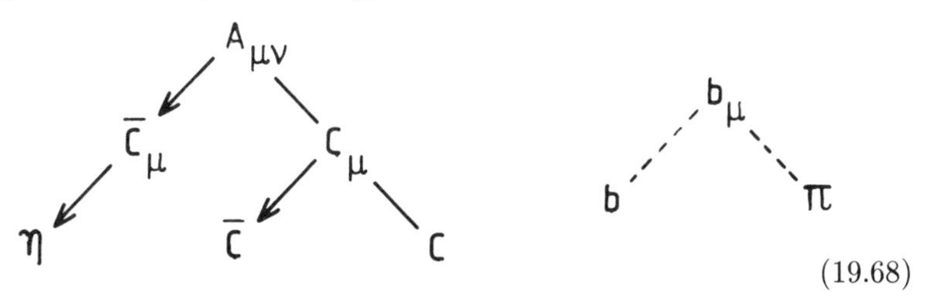

(19.68)

There are corresponding pyramids for the antifields. The fields $A_{\mu\nu}$, C_μ, and C connected by solid lines (right branch of the bigger pyramid) are the minimal fields. All the other fields belong to the nonminimal sector. The ghost number and parity of b_μ, b, and π appearing in the small pyramid can be recovered from those of $\bar{C}^\mu$, η, and $\bar{C}$, which appear in the nonminimal piece of the large pyramid. They must make the nonminimal term $\bar{C}^{*\mu}\, b_\mu + \eta^*\pi + \bar{C}^* b$ in S even and of ghost number zero. Therefore, it is sufficient to analyze the bigger pyramid.

The nonminimal part of that pyramid is formed as follows. The antighost $\bar{C}^\mu$ is introduced for freezing the gauge freedom of $A_{\mu\nu}$. It has ghost number -1 and parity $+1$. Because of the reducibility of the gauge transformations, the equations of motion for $\bar{C}^\mu$ and C_μ are degenerate. To gauge-fix this degeneracy, one introduces one even antighost $\bar{C}$ of ghost number -2 for the ghost C_μ and one even ghost η of ghost number 0 for the antighost $\bar{C}^\mu$. So, the variables $\bar{C}_\mu$, η, and $\bar{C}$ appear in order to remove a gauge degeneracy in the equations of motion for the fields $A_{\mu\nu}$, $\bar{C}_\mu$ and C_μ. This is the meaning of the arrow in the diagram (19.68). The sum of the ghost numbers of two variables connected by an arrow is -1 so that the terms $\bar{C}^\mu(sA_{\mu\nu})$, $\bar{C}(sC_\mu)$ and $\eta(s\bar{C}^\mu)$ have ghost number zero.

If there were higher order reducibility identities, the variables η, $\bar{C}$, and C would obey degenerate field equations. To freeze that degeneracy would require further nonminimal variables. The bigger pyramid of (19.68) is thus generalized to an arbitrary irreducible gauge theory as

$$
\begin{array}{ccccccccc}
 & & & & \phi^i & & & & \\
 & & & \swarrow & & \searrow & & & \\
 & & \bar{C}_{0\alpha_0} & & & & C_0^{\alpha_0} & & \\
 & \swarrow & & & & \swarrow & & \searrow & \\
 & C_1'^{\alpha_1} & & & \bar{C}_{1\alpha_1} & & & & C_1^{\alpha_1} \\
 \swarrow & & & \swarrow & & & & \swarrow & & \searrow \\
\bar{C}'_{2\alpha_2} & & C_2'^{\alpha_2} & & & \bar{C}_{2\alpha_2} & & & C_2^{\alpha_2} \\
\swarrow & & \swarrow & & & \swarrow & & \swarrow & & \searrow
\end{array}
\tag{19.69}
$$

The right branch without an arrow contains the minimal fields. The other variables are introduced to remove degeneracies in the equations of motion $\delta S_\psi = 0$. In $A \to B$, B is there to freeze the ambiguity in the field equations for A. The sum of the ghost numbers of two fields connected by an arrow is equal to -1. This fixes the ghost number of each field in the pyramid, since the ghost numbers of the minimal fields are known. The fields on a horizontal line have the parity of the minimal ghost of ghost ... of ghost $C_s^{\alpha_s}$ belonging to the same line.

The complete nonminimal solution reads

$$S = S^{\text{Min}} + \bar{C}_0^{*\alpha_0}\, b_{0\alpha_0} + \bar{C}_1^{*\alpha_1}\, b_{1\alpha_1} + C'^{*}_{1\alpha_1} b'^{\alpha_1}_1 + \cdots \tag{19.70}$$

where the b's, which stand for b^μ, b, and π of (19.68), are the elements of the smaller pyramid associated with (19.69) according to the rule that each nonminimal variable in (11.69) has a b such that the right-hand side of (19.70) is even and of ghost number zero. The gauge-fixing fermion ψ must be such that $\delta S_\psi = 0$ is a nondegenerate variational principle. The simplest possibility is to include in ψ, for each arrow $A \to B$, a term of the form $Bf(A)$ such that $f(A) = 0$ are (redundant) gauge conditions for A. Also, one must impose $\text{gh}\,\big(f(A)\big) = \text{gh}\,(A)$ so that $\text{gh}\,\psi = -1$.

This completely determines the nonminimal spectrum that permits the elimination of the antifields in the path integral, as appropriate for "relativistic gauges" such as $\partial^\mu A_{\mu\nu} = 0$. It should be mentioned that if one were to exchange the role of some of the fields and antifields, as appropriate to gauges such as $A_0 = 0$, then the required nonminimal sector would, in general, be different. (See in this context §18.4.5 on the antifield treatment of the extended action S_E as well as Exercise 19.14.)

EXERCISES. CHAPTER NINETEEN

19.1. Consider the action $\bar{S}[A_\mu, \phi] = \int [-\frac{1}{4} F_{\mu\nu}F^{\mu\nu} + \frac{1}{2}(\phi - \partial_\mu A^\mu)^2]\, d^4x$. Show that upon elimination of ϕ by means of its equations of motion, one gets back the standard Maxwell action. Work out the gauge transformations of $\bar{S}$. Prove without going to the Hamiltonian that the Hamiltonian formalism for $\bar{S}$ contains not only primary and secondary constraints but also a tertiary constraint. Then verify this assertion by deriving explicitly the Hamiltonian.

19.2. Derive the equations of motion following from the gauge-fixed action (19.26b) and check that all the fields obey the wave equation in the Feynman gauge ($\alpha = 1$).

19.3. The path integral (19.26a) can be obtained from the path integral (19.24a) by averaging over different gauges as follows. Integrating (19.24a) over the b-field, one gets the path integral

$$\int [DA_\mu\, D\eta\, D\bar{C}] \prod \delta(\partial^\mu A_\mu) \exp i \left[-\tfrac{1}{4} \int F_{\mu\nu}F^{\mu\nu}\, d^4x + \int i\bar{C}\Box\eta\, d^4x \right].$$

This integral does not depend on the choice of gauge conditions, so one can substitute for $\delta(\partial^\mu A_\mu)$ the quantity $\delta(\partial^\mu A_\mu - \mu)$. One can then average over μ with a Gaussian weight (the integral does not depend on μ), *i.e.*, replace $\delta(\partial^\mu A_\mu - \mu)$ by $\int [D\mu] \exp i\lambda \int \mu^2\, d^4x \prod \delta(\partial^\mu A_\mu - \mu)$. Evaluate the integral and show that the resulting integrand in the path integral is equal to what becomes the integrand of (19.26a) after the b-field has been integrated over.

19.4. Consider the Yang–Mills action $S_0 = -\frac{1}{4}\int F^a_{\mu\nu}F_a^{\mu\nu}\, d^4x$, with $F^a_{\mu\nu} = \partial_\mu A^a_\nu - \partial_\nu A^a_\mu - C_{bc}{}^a A^b_\mu A^c_\nu$. Here, $C_{ab}{}^c$ are the structure constants of a compact Lie group.

(a) Derive the Hamiltonian and constraints. [Answer: The primary constraints are $\pi^0_a \approx 0$. The secondary constraints are $-\partial_i \pi_a{}^i + C^b{}_{ac} A_i{}^c \pi_b{}^i \approx 0$. The Hamiltonian is $H_0 = \int d^3x \frac{1}{2}(\vec{\pi}^2 + \vec{\mathcal{B}}^2)$, $\mathcal{B}^{ai} = \frac{1}{2}\varepsilon^{ijk}\partial_j A_k{}^a$.]

(b) Derive the path integral in the Lorentz gauge $\partial^\mu A^a_\mu = 0$ and show that the ghosts η^a and the antighosts $\bar{C}_a$ are now coupled to the gauge fields. [Answer: The gauge-fixed action is $\int[-\frac{1}{4}F^a_{\mu\nu}F_a^{\mu\nu} - b_a(\partial^\mu A^a_\mu - (\alpha/2)b^a) - i\,\partial_\mu \bar{C}_a\, D^\mu \eta^a$, with $D_\mu \eta^a = \partial_\mu \eta^a + C^a{}_{bc}\,\eta^b A^c_\mu$.]

(c) Derive the same path integral through the antifield method. [Hint: The minimal solution of the antifield formalism is $S^{\text{Min}} = \int d^4x(-\frac{1}{4}F^a_{\mu\nu}F_a^{\mu\nu} + A_a^{*\mu}\, D_\mu C^a + \frac{1}{2}C^*_a C^a{}_{bc} C^b C^c)$.]

(d) Show that $\Delta S^{\text{Min}} = 0$. How crucial is the identity $C^a{}_{ac} = 0$ (valid for a compact group)?

19.5. (Faddeev–Popov determinant for Yang–Mills field A^a_μ)

(a) Let $\mathcal{F}^a(A)(x) = 0$ be a local gauge condition. Define $\Delta^{-1}_{\mathcal{F}}(A) \equiv \int[Dg \prod_x \delta[\mathcal{F}({}^gA)]$ where $[Dg]$ is the product over x of the invariant measure for the (compact) Yang–Mills group and ${}^gA$ is the gauge transformed of A, ${}^gA = gAg^{-1} - g\,dg^{-1}$. Show that $\Delta_{\mathcal{F}}({}^gA) = \Delta_{\mathcal{F}}(A)$.

(b) Let $I(A)$ be a gauge-invariant functional. Show that $\int[DA]\, I(A)\, \Delta_{\mathcal{F}}(A)\, \delta[\mathcal{F}(A)]$ does not depend on $\mathcal{F}$. (Hint: Insert $1 = \Delta_{\mathcal{F}'}(A)\int[Dg]\, \delta[\mathcal{F}'({}^gA)]$ in the integrand, interchange the order of integrations and use the invariance of $[Dg]$ and of $[DA]$.)

(c) Show that if A fulfills the gauge condition $\mathcal{F}(A) = 0$, then $\Delta_{\mathcal{F}}(A)$ is equal to the determinant of the operator $(\delta/\delta\alpha^a(x)\ \mathcal{F}^b(A))|_{\alpha=0}$ defined by the ghost action $\bar{C}_a \delta_\eta \mathcal{F}^b$ [see Exercise 19.4 and §11.3.3 ("Faddeev–Popov determinant")]. Here, $\alpha^b(x)$ is an infinitesimal gauge transformation. [Hint: If A fulfills $\mathcal{F}(A) = 0$, only the g's close to the identity contribute to the integral. For those g's, $[Dg] = \prod_x d\alpha$.]

19.6. Verify the Poisson brackets (19.32). [Hint: Use $e^3_i(\vec{k}) = -e^3_i(-\vec{k})$.]

19.7. Verify that the gauge-fixed Hamiltonian (19.42b) is the Hamiltonian for the gauge-fixed action (19.41b). (Hint: Start from the gauge-fixed Hamiltonian action. Perform the Gaussian integrals over $\lambda^1, B_2, \pi^0, \lambda^2, \rho, \mathcal{P}, \mathcal{P}_2, \eta^1, \eta^2, \bar{C}$, and π^k. Identify $b_{\text{Lag.}}$ with B_1 and $\bar{C}_{\text{Lag.}}$ with $i\,\mathcal{P}_1$.)

19.8. Verify that the nonminimal variables $\lambda^1, \lambda^2, B^1, B^2, \eta^1, \eta^2, \mathcal{P}_1$, and $\mathcal{P}_2$ in (19.42) define oscillator quartets that do not contribute to the BRST cohomology. (Hints: Expand B^1 and B^2 in terms of null oscillators α

$$B_1(\vec{x}) \sim \int d^3k\,[\alpha(\vec{k})\, e^{-i\vec{k}\vec{x}} + \alpha^*(\vec{k})\, e^{-i\vec{k}\vec{x}}],$$

$$B_2(\vec{x}) \sim i\int d^3k\,[\alpha(\vec{k})\, e^{i\vec{k}\vec{x}} - \alpha^*(\vec{k})\, e^{i\vec{k}\vec{x}}],$$

etc., and similarly for the ghosts η^1 and η^2. Apply then the homotopy argument of §19.1.6 and §14.4.2.) Show that the Hamiltonian in the gauge $\Box \partial^\mu A_\mu = 0$ leaves the vacuum invariant up to a null state.

19.9. Verify formulas (19.43c)–(19.43f).

19.10. The BRST charge (19.15) is equal to $\Omega^{\text{Min}} = \int d^3x\, \eta(\vec{x})\, E^{L'}(\vec{x})$. Show that one can write it in the standard null oscillator form (14.60a) without introducing a nonminimal sector by defining

$$E^{L'}(\vec{x}) = \frac{i}{(2\pi)^{3/2}} \int d^3k \sqrt{2\omega}\, \theta(\vec{k}\cdot\vec{n})\, [e^{i\vec{k}\cdot\vec{x}}\, A(\vec{k}) - e^{-i\vec{k}\cdot\vec{x}}\, A^*(\vec{k})]$$

$$A^{L'}(\vec{x}) = \frac{1}{(2\pi)^{3/2}} \int \frac{d^3k}{\sqrt{2\omega}}\, \theta(\vec{k}\cdot\vec{n})\, [e^{i\vec{k}\cdot\vec{x}}\, B(\vec{k}) + e^{i\vec{k}\cdot\vec{x}}\, B^*(\vec{k})],$$

and similarly for the ghosts. Here, $\vec{n}$ is an arbitrary constant vector, θ is the Heaviside step function, and the variables $A(\vec{k})$, $A^*(\vec{k})$, $B(\vec{k})$, and $B^*(\vec{k})$ are defined only in the half-plane $\vec{k}\cdot\vec{n} > 0$ where they satisfy the null oscillator commutation relations. The BRST charge reads explicitly

$$\Omega^{\text{Min}} = \int d^3k\, \theta(\vec{k}\cdot\vec{n})\, (\eta^*(\vec{k})\, A(\vec{k}) + A^*(\vec{k})\, \eta(\vec{k})).$$

Note that the vector $\vec{n}$ breaks manifest rotational invariance.

19.11. Perform the canonical analysis of massive electrodynamics,

$$\mathcal{L} = -\tfrac{1}{4} F_{\mu\nu} F^{\mu\nu} - \tfrac{1}{2} m^2\, A_\mu\, A^\mu.$$

(a) Show that there are second-class constraints.
(b) Show that the process of replacing the system by an equivalent first-class one leads to the Lagrangian density $\mathcal{L} = -\frac{1}{4} F_{\mu\nu}\, F^{\mu\nu} - \frac{1}{2}(m\, A_\mu - \partial_\mu \phi)^2$ with an extra scalar field ϕ ("Stuckelberg field").

19.12. Show that in four spacetime dimensions, the abelian 2-form gauge theory is equivalent to the theory of a massless scalar field. (Hint: Rewrite the equations of motion in terms of the dual ${}^*F^\mu = \frac{1}{3!}\varepsilon^{\mu\nu\rho\sigma} F_{\nu\rho\sigma}$ of the field strength.)

19.13. Derive explicitly the total action and the extended action for the 2-form gauge field. Show how one can recover the gauge transformations (19.46) of the original action from those of the extended formalism. [Hint: ε_0 appears as the gauge parameter of the extra gauge symmetry (3.37a) associated with the reducibility of the constraints.] Verify that elimination from the total action of π^{0i} and of the multiplier associated with the primary constraint $\pi^{0i} \approx 0$ by means of their equations of motion yield (19.52), just as in electromagnetism.

19.14. Show that the temporal gauge $A_0 = 0$ of electromagnetism can be obtained by eliminating A^{*k} and A_0 from the minimal solution of the master equation according to $A^{*k} = \delta\psi/\delta A_k$, $A_0 = -\delta\psi/\delta A^{*0}$ with $\psi = 0$. Show that A^{*0} becomes identified with the antighost, and observe that no nonminimal sector is required.

19.15. Show that the "axial gauge" $A_0 = A_z = 0$ can be imposed in the Maxwell theory. Compute the Dirac brackets of $A_\mu(\vec{x})$ and $\pi^\mu(\vec{x})$ in the axial gauge.

19.16. Find the extended Hamiltonian for the Maxwell field interacting in a gauge-invariant manner with a charged scalar field. The Lagrangian density is

$$\mathcal{L} = -\tfrac{1}{4}F_{\mu\nu}F^{\mu\nu} - \tfrac{1}{2}(D_\mu\phi^*)(D^\mu\phi) - V(\phi,\phi^*)$$

where

$$D_\mu\phi = (\partial_\mu - ieA_\mu)\phi.$$

19.17. Same question for the Maxwell–Dirac system with Lagrangian density

$$\mathcal{L} = -\tfrac{1}{4}F_{\mu\nu}F^{\mu\nu} + \bar{\psi}\gamma^\mu D_\mu\psi$$

where

$$D_\mu\psi = (\partial_\mu - ieA_\mu)\psi.$$

19.18. Perform the Hamiltonian–BRST analysis for a 3-form abelian gauge field with Lagrangian density

$$\mathcal{L} = -\frac{1}{2\cdot 4!}F_{\mu\nu\rho\sigma}F^{\mu\nu\rho\sigma}$$

where

$$F_{\mu\nu\rho\sigma} = \tfrac{1}{3}\partial_{[\mu}A_{\nu\rho\sigma]}.$$

(Hints: There are now "ghosts of ghosts of ghosts.")

19.19. Consider a system with Hamiltonian H and BRST charge Ω. Assume that the asymptotic dynamics is determined by the free Hamiltonian H_0 and free BRST charge Ω_0. The S-matrix is defined by

$$S = \lim_{t''\to\infty,\, t'\to-\infty} e^{iH_0t''}e^{-iH(t''-t')}e^{-iH_0t'},$$

and one has

$$\Omega_0 = \lim_{|t|\to\infty} e^{iH_0t}\Omega e^{-iH_0t}.$$

Verify that the S-matrix commutes with the free BRST charge Ω_0, $[S,\Omega_0]=0$, and so maps the physical subspace determined by Ω_0 on itself.

CHAPTER TWENTY

COMPLEMENTARY MATERIAL

We collect in this chapter general information not directly connected with the theory of gauge systems but necessary for following some of the developments made in this book. The chapter is organized as follows. First, we briefly review some aspects of the exterior calculus on a supermanifold (Sec. 20.1). Next, we discuss the Berezin integral (Sec. 20.2). Finally, we turn to the representation theory of the anticommutation relations for fermionic operators (Sec. 20.3).

20.1. EXTERIOR CALCULUS ON A SUPERMANIFOLD: CONVENTIONS

Our conventions concerning the exterior calculus possess the following key features.

(i) The commutation properties are governed by a single Z_2-grading called the Grassmann parity. That is, each object (superfunctions, vectors, differential forms) is assigned a definite Grassmann parity equal to 0 or 1, and whenever an object of Grassmann parity ε_1 is passed through an object of Grassmann parity ε_2, one picks up the sign factor $(-)^{\varepsilon_1 \varepsilon_2}$. This rule is the most convenient one because of its economy.

(*ii*) Derivations act from the right. This is to conform with the conventional Hamiltonian conventions, where the generator of a transformation is put to the right in the Poisson bracket.

(*iii*) The exterior derivative d has Grassmann parity $+1$,

$$\varepsilon(d) = 1, \tag{20.1}$$

and by duality, $\varepsilon(\vec{\partial}) = 1$.

In particular, if X is a vector field of Grassmann parity ε_X and f, g are superfunctions of respective Grassmann parity ε_f and ε_g, the action of X on f, denoted by $f\overleftarrow{\partial}_X$ [or just $X(f)$ when no confusion can arise], is a superfunction with the properties

$$\varepsilon(f\overleftarrow{\partial}_X) = \varepsilon_f + \varepsilon_X + \varepsilon(\overleftarrow{\partial}) = \varepsilon_f + \varepsilon_X + 1, \tag{20.2a}$$

$$(fg)\overleftarrow{\partial}_X = f(g\overleftarrow{\partial}_X) + (f\overleftarrow{\partial}_X)g(-)^{\varepsilon_g(\varepsilon_X+1)}. \tag{20.2b}$$

The vector fields e_A "tangent to the coordinate lines z^A" have Grassmann parity $\varepsilon_A + 1$ and are defined by

$$f\overleftarrow{\partial}_{e_A} \equiv \frac{\partial^r f}{\partial z^A}. \tag{20.3}$$

Any vector field can be expanded in terms of the e_A's as

$$X = e_A X^A, \tag{20.4}$$

where the X^A are superfunctions of Grassmann parity $\varepsilon_X + \varepsilon_A + 1$.

Because we assign Grassmann parity $+1$ to $\overleftarrow{\partial}$, the vector fields turn out to be all odd in the pure bosonic case. This choice agrees with the commutativity properties of p-vectors on an ordinary manifold. With the convention, the contraction of a 1-form by a 1-vector, which yields a function, carries no Grassmann parity, $\varepsilon[\alpha(X)] = \varepsilon_\alpha + \varepsilon_X = 0$ (bosonic case). Note that the antisymmetry of an ordinary 2-form β

$$\beta(X,Y) = -\beta(Y,X) \qquad (\text{"ordinary" case})$$

is compatible with this rule, since

$$(-)^{\varepsilon_X \varepsilon_Y} = -1 \qquad (\text{"ordinary" case})$$

for ordinary vector fields.

More generally, p-forms on a supermanifold are such that

$$\omega(X_1, \ldots, X_i, X_{i+1}, \ldots, X_p) = (-)^{\varepsilon_{X_i}\varepsilon_{X_{i+1}}}\, \omega(X_1, \ldots, X_{i+1}, X_i, \ldots, X_p) \tag{20.5a}$$

and

$$\varepsilon[\omega(X_1, \ldots, X_p)] = \varepsilon_\omega + \sum_{i=1}^{p} \varepsilon_{X_i}. \tag{20.5b}$$

Because of (20.5a), the components $\omega_{A_1\dots A_p}$ of a p-form are completely antisymmetric in the following sense

$$\omega_{A_1\dots A_i A_{i+1}\dots A_p} = (-)^{(\varepsilon_{A_i}+1)(\varepsilon_{A_{i+1}})}\omega_{A_1\dots A_{i+1}A_i\dots A_p}, \tag{20.5c}$$

since the tangent vectors e_A have parity $\varepsilon_A + 1$. The exterior product compatible with convention (20.5a) is

$$(\omega_1\omega_2)(X_1,\dots,X_{p+q}) = \frac{1}{p!}\frac{1}{q!}\sum_P \omega_1(X_{P(1)},\dots,X_{P(p)}) \\ \omega_2(X_{P(p+1)},\dots,X_{P(p+q)})\,(-)^{\phi(P)}. \tag{20.6}$$

The sum in (20.6) extends over all $(p+q)!$ permutations of $1,2,\dots p+q$. The sign factor $(-)^{\phi(P)}$ associated with the permutation P in (20.6) is the sign that one obtains by putting $\omega_1\omega_2X_1\dots X_{p+q}$ in the order $\omega_1 X_{P(1)}\dots X_{P(p)}\,\omega_2 X_{P(p+1)}\dots X_{P(p+q)}$ according to the rules of (i).

The Lie bracket $[X_1, X_2]$ of two vector fields X_1 and X_2 is defined by

$$f\overleftarrow{\partial}_{[X_1,X_2]} = f(\overleftarrow{\partial}_{X_1}\overleftarrow{\partial}_{X_2} - (-)^{(\varepsilon_{X_1}+1)(\varepsilon_{X_2}+1)}\overleftarrow{\partial}_{X_2}\overleftarrow{\partial}_{X_1}). \tag{20.7}$$

One has

$$\varepsilon([X_1,X_2]) = \varepsilon(X_1) + \varepsilon(X_2) + 1, \tag{20.8a}$$

$$[X_1,X_2] = -(-)^{(\varepsilon_{X_1}+1)(\varepsilon_{X_2}+1)}[X_2,X_1]. \tag{20.8b}$$

The Jacobi identity reads

$$\big[[X_1,X_2],X_3\big] + (-)^{(\varepsilon_{X_1}+1)(\varepsilon_{X_2}+\varepsilon_{X_3})}\big[[X_2,X_3],X_1\big] \\ + (-)^{(\varepsilon_{X_3}+1)(\varepsilon_{X_1}+\varepsilon_{X_2})}\big[[X_3,X_1],X_2\big] = 0. \tag{20.9}$$

The exterior derivative is defined by

$$(d\alpha)(X_0,\dots,X_p) = (-)^{\sum_{i=0}^{p}\varepsilon_{X_i}} \\ \times\Big[\sum_{j=0}^{p}(-)^{\rho_j+\varepsilon_{X_j}}\,\alpha(X_0,\dots,\hat{X}_j,\dots,X_p)\,\overleftarrow{\partial}_{X_j} + \\ \sum_{0\le i<j<p}(-)^{\rho_{ij}+\varepsilon_{X_j}+1}\,\alpha(X_0,\dots,\hat{X}_i,\dots\hat{X}_j,\dots,X_p,[X_i,X_j])\Big], \tag{20.10}$$

where the caret ˆ denotes omission. The sign factors $(-)^{\rho_j}$ and $(-)^{\rho_{ij}}$ are those appearing in

$$X_jX_{j+1}\dots X_p = (-)^{\rho_j}X_{j+1}\dots X_pX_j,$$
$$X_i\dots X_j\dots X_p = X_{i+1}\dots X_{j-1}X_{j+1}\dots X_pX_iX_j(-)^{\rho_{ij}}.$$

The exterior derivative d is nilpotent and acts as a right derivation,

$$d^2 = 0, \tag{20.11a}$$

$$d(\alpha\beta) = \alpha\, d\beta + (-)^{\varepsilon_\beta}\,(d\alpha)\beta. \tag{20.11b}$$

Any closed form α ($d\alpha = 0$) is locally exact ($\alpha = d\beta$) (Poincaré lemma).

Finally, we turn to the important concept of submanifolds. A submanifold in the supermanifold of the z's is a set of points defined by equations

$$f_m(z) = 0 \tag{20.12}$$

that have the following property: One can cover (20.12) by superdomains, on each of which the superfunctions f_m can be split into independent f_u's ($u = 1, \ldots, U$) and dependent f_γ's, $f_m = (f_u, f_\gamma)$, in such a way that (i) $f_\gamma = 0$ is a consequence of $f_u = 0$; and (ii) the matrix

$$\frac{\partial f_u}{\partial z^A} \tag{20.13}$$

is of maximal rank U. By "maximal rank U," we mean that there is a square $U \times U$ submatrix that is invertible. In that case, there exists locally a new coordinate system in which the f_u are coordinate functions. If the functions f_u do not fulfill these regularity conditions—and cannot be redefined so as to fulfill them, (*e.g.*, $f_u \to \sqrt{f_u}$)—then Eqs. (20.12) do not define a submanifold.

An example of a submanifold is $z^1 = 0$ where z^1 is a fermionic (or bosonic) coordinate. A subset that is *not* a submanifold is $z^1 z^2 = 0$ where z^1 and z^2 are both fermionic. In that case, indeed, the matrix $(\partial f_u / \partial z^A)$ reads $(0, 0, \ldots, 0, z^2, -z^1, \ldots)$ and there is no 1-by-1 submatrix that can be inverted, since neither z^1 nor z^2 possess an inverse. The regularity condition is thus not fulfilled and cannot be fulfilled by redefinitions of the f's.

A vector field such that

$$f_m \overleftarrow{\partial}_X = 0 \tag{20.14}$$

for all m is said to be tangent to the submanifold, just as in the pure bosonic case.

The definition of the longitudinal exterior derivative associated with the foliation of a manifold by "gauge orbits" proceeds exactly as in the bosonic case (Sec. 5.3).

If X_i is a basis of vector fields tangent to the orbits they close in the Lie bracket,

$$[X_i, X_j] = X_k C_{ij}{}^k \tag{20.15}$$

for some structure functions $C_{ij}{}^k$. Conversely, the condition (20.15) is sufficient for the vectors X_i to be locally surface-forming if, in addition, the vectors X_i are such that one can locally split them as $(X_i) = (X_a, X_\alpha)$ where (i) the matrix ${X_a}^A$ of the components of the first set is of maximum rank; (ii) the vectors of the second set are combinations of the vectors of the first set. The dimension of the surfaces to which the X_i are tangent is equal to the number of the independent X_a's.

20.2. INTEGRATION ON A SUPERMANIFOLD

20.2.1. Definition

The Berezin integral of a function $f(\theta)$ of Fermi variables is defined by the formula

$$\int f(\theta)\, d\theta_1\, d\theta_2 \ldots d\theta_n \equiv \int f(\theta)\, d\theta = f_{n\ldots 1} \tag{20.16}$$

where $f_{n\ldots 1}$ is the coefficient of $\theta_n \theta_{n-1} \ldots \theta_1$ in the expansion of $f(\theta)$ ("top component"). The definition (20.16) can be generalized to functions $f(x,\theta)$ defined on a supermanifold with both Bose and Fermi variables. The integral of f over the superdomain $\mathcal{V}$ with coordinates x, θ is given by

$$\int_{\mathcal{V}} f(x,\theta)\, dx\, d\theta = \int_U f_{n\ldots 1}(x)\, dx, \tag{20.17}$$

where U is the real domain underlying $\mathcal{V}$ and where $\int_U$ is the ordinary Riemann integral. The integral over the supermanifold itself is obtained by patching together the superdomains by means of a partition of unity , as in the bosonic case [for the details, see De Witt (1984) or Berezin (1987)].

Because the top component of $\partial f/\partial\theta^k$ vanishes, one gets

$$\int \frac{\partial f}{\partial \theta^k}\, dx\, d\theta = 0. \tag{20.18a}$$

This shows (i) that the integral (20.17) is invariant under translations of θ^k, $\theta^k \to \theta^k + \alpha^k$; and (ii) that an integration-by-parts formula holds without surface term. By contrast, there is a surface term for the bosonic variables

$$\int_{\mathcal{V}} \frac{\partial f}{\partial x^i}\, dx\, d\theta = \int_{\partial U} f_{n\ldots 1}(x)\, \delta_i{}^j\, d\Sigma_j. \tag{20.18b}$$

The θ-integral is analogous to a bosonic integral always extended from $-\infty$ to $+\infty$ for functions that vanish fast enough at infinity. There is no concept of boundary of the θ-integration range.

The purpose of this section is to investigate how (20.17) transforms under changes of variables and to show that there is a natural measure in any superphase space, the "Liouville measure."

20.2.2. Supertrace–Superdeterminant

A matrix

$$M = \begin{pmatrix} A & B \\ C & D \end{pmatrix}, \tag{20.19}$$

is called a supermatrix if its Bose–Bose part A and Fermi–Fermi part D have even elements and its Bose–Fermi part B and Fermi–Bose part C have odd elements. The Jacobian matrix $\partial(x,\theta)/\partial(x',\theta')$ of the change of variables $x = x(x',\theta')$, $\theta = \theta(x',\theta')$ is a supermatrix (see §6.3.2), and this is the main reason that we are interested here in supermatrices.

The ordinary matrix multiplication of two supermatrices is again a supermatrix. Hence, supermatrices form a subalgebra of the algebra of all matrices.

The supertrace $\operatorname{str} M$ of the supermatrix M is defined by

$$\operatorname{str} M = \operatorname{tr} A - \operatorname{tr} D \tag{20.20}$$

with a minus sign for the fermionic part D. One easily checks

$$\operatorname{str}(M_1 + M_2) = \operatorname{str} M_1 + \operatorname{str} M_2, \tag{20.21a}$$

$$\operatorname{str}(M_1 M_2) = \operatorname{str}(M_2 M_1), \tag{20.21b}$$

and hence the supertrace of a commutator is zero,

$$\operatorname{str}[M_1, M_2] = 0. \tag{20.21c}$$

It should be stressed that (20.21b) holds only because a minus sign has been included in front of D in (20.20). This minus sign compensates the minus sign arising from the commutation of the elements of B with those of C.

The superdeterminant of an invertible supermatrix is defined by

$$\operatorname{sdet} M = \exp \operatorname{str} \ell n\, M \tag{20.22a}$$

with

$$\ell n\, M = \ell n(I + M - I) = \sum_{k \geq 1} \frac{(M - I)^k}{k} (-)^{k-1}. \tag{20.22b}$$

The definition (20.22a) is such that for matrices differing infinitesimally from the identity

$$M = I + \varepsilon, \tag{20.23a}$$

one has

$$\text{sdet}\, M = 1 + \text{str}\, \varepsilon. \tag{20.23b}$$

Since $\ell n(M_1 M_2)$ differs from $\ell n M_1 + \ell n M_2$ by a commutator, one gets from (20.21c),

$$\text{sdet}\,(M_1 M_2) = (\text{sdet}\, M_1)(\text{sdet}\, M_2). \tag{20.23c}$$

Furthermore, the decompositions of M into triangular matrices, given by

$$M = \begin{pmatrix} A & 0 \\ C & I \end{pmatrix} \begin{pmatrix} I & A^{-1}B \\ 0 & D - CA^{-1}B \end{pmatrix}$$

$$= \begin{pmatrix} I & B \\ 0 & D \end{pmatrix} \begin{pmatrix} A - BD^{-1}C & 0 \\ D^{-1}C & I \end{pmatrix},$$

yields the following expressions for sdet M,

$$\text{sdet}\, M = \det\, A\,[\det\,(D - CA^{-1}B)]^{-1} \tag{20.24a}$$

$$= (\det\, D)^{-1} \det\,(A - BD^{-1}C). \tag{20.24b}$$

If M is noninvertible (*i.e.*, if one of the matrices A or D is noninvertible), then the expression (20.24) may be ill defined (*e.g.*, of the form 0/0) and sdet$M$ then does not exist.

The supertranspose of M is defined by

$$M^T = \begin{pmatrix} A^T & C^T \\ -B^T & D^T \end{pmatrix} \tag{20.25a}$$

and is such that*

$$(M_1 M_2)^T = {M_2}^T {M_1}^T, \tag{20.25b}$$

$$\text{str}\, M^T = \text{str}\, M, \tag{20.25c}$$

$$\text{sdet}\, M^T = \text{sdet}\, M, \tag{20.25d}$$

$$(M^{-1})^T = (M^T)^{-1}. \tag{20.25e}$$

The same properties would have held if we had defined the supertranspose by inserting the minus sign in front of C^T rather than in front of B^T in (20.25a).

It follows from (20.21b) and (20.23c) that the supertrace and the superdeterminant are invariant under

$$M \to M' = NMN^{-1}, \tag{20.26}$$

where N is an invertible supermatrix.

---

* Note that for matrices α, β with odd elements, the usual transposition is such that $(\alpha\beta)^T = -\beta^T \alpha^T$.

20.2.3. Change of Variables. Superdensities

The importance of the superdeterminant lies in the following theorem.

Theorem 20.1. *Let $(x,\theta) \to (x',\theta')$ be an invertible change of variables and $f(x,\theta)$ be a function of compact support contained in U. Then*

$$\int_{\mathcal{V}} f(x,\theta)\, dx\, d\theta = \int_{\mathcal{V}} f\big(x(x',\theta'),\theta(x',\theta')\big)\, \mathrm{sdet}\Big(\frac{\partial^R(x,\theta)}{\partial(x',\theta')}\Big)\, dx'\, d\theta'. \tag{20.27}$$

[We assume that the coordinate transformation $x \to x'$ induced in U preserves the orientation. The function $f(x,\theta) = \sum f_{i_1\dots i_k}\theta^{i_1}\dots\theta^{i_k}$ is of compact support if each of its components $f_{i_1\dots i_k}(x)$ is of compact support.]

Proof. We shall only prove here the theorem for changes of variables close to the identity,

$$x^i = x'^i + \xi^i(x',\theta'), \tag{20.28a}$$
$$\theta^k = \theta'^k + \eta^k(x',\theta'). \tag{20.28b}$$

We refer to Berezin 1987 for the general case.

One has

$$f\big(x(x',\theta'),\theta(x',\theta')\big)\, \mathrm{sdet}\Big(\frac{\partial^R(x,\theta)}{\partial(x',\theta')}\Big) = f(x',\theta') + \frac{\partial(f\xi^i)}{\partial x'^i} - \frac{\partial(f\eta^k)}{\partial\theta'^k},$$

where we have used

$$\mathrm{sdet}\Big(\frac{\partial^R(x,\theta)}{\partial(x',\theta')}\Big) = 1 + \frac{\partial\xi^i}{\partial x'^i} - \frac{\partial\eta^k}{\partial\theta'^k}$$

and

$$-\frac{\partial(f\eta^k)}{\partial\theta'^k} = \frac{\partial^R f}{\partial\theta'^k}\eta^k - f\frac{\partial\eta^k}{\partial\theta'^k}.$$

So in order to prove (20.27), one must show that

$$\int_{\mathcal{V}} \Big[\frac{\partial(f\xi^i)}{\partial x^i} - \frac{\partial(f\eta^k)}{\partial\theta^k}\Big]\, dx\, d\theta = 0.$$

But this is an immediate consequence of (20.18a) and (20.18b) and of the fact that f vanishes on the boundary ∂U.

Remarks. (a) It is the right derivative $\partial^R(x,\theta)/\partial(x',\theta')$ that appears in (20.27) because we adopt the convention that the index A in $\partial z^A/\partial z'^B$ is a row index, while the index B is a column index (see discussion in §6.3.2). Had we adopted the opposite convention, the left derivative would have appeared. This actually only matters for $\partial x^i/\partial\theta'^k$. The chain rule implies that the super-Jacobian matrix obeys $\partial^R(x,\theta)/\partial(x'',\theta'') = [\partial^R(x,\theta)/\partial(x',\theta')]\,[\partial^R(x',\theta')/\partial(x'',\theta'')]$ (as a matrix product), so the superdeterminant $\operatorname{sdet}\partial^R(x,\theta)/\partial(x',\theta')$ obeys the group property necessary for exponentiating the infinitesimal transformation (20.28).

(b) If f is not of compact support contained in U, (20.27) is, in general, not true. There is an extra boundary term. In the pure bosonic case, this boundary term just corresponds to the change of the integration range of the bosonic variables. This is no longer true when fermions are present, as the following example due to Leites shows:

$$\int x\,dx\,d\theta_1\,d\theta_2$$

with $U = [0,1]$. One makes the change of variables $x = x' + \theta_1'\theta_2'$, $\theta_1' = \theta_1$, $\theta_2 = \theta_2'$. The components along unity of x and x' both belong to $[0,1]$, and there is no change of integration domain. Yet,

$$\int x\,dx\,d\theta_1\,d\theta_2 = \int_0^1 0\,dx = 0,$$

but

$$\int (x' + \theta_1'\theta_2')\,\frac{\partial(x,\theta)}{\partial(x',\theta')}\,dx'\,d\theta' = \int \theta_1'\theta_2'\,dx'\,d\theta'$$
$$= -\int_0^1 dx' = -1.$$

(c) The theorem remains true when $x^i \in R^n$, provided f vanishes fast enough at infinity (if f is oscillatory, one must insert appropriate convergence factors).

(d) It follows from (20.27) that $d\theta^1\,d\theta^2 = d\theta^2\,d\theta^1 \operatorname{sdet}\partial(\theta^1,\theta^2)/\partial(\theta^2,\theta^1) = -d\theta^2\,d\theta^1$, *i.e.*, the $d\theta$'s should be regarded as anticommuting in the Berezin integral. This is to be contrasted with their behavior in the exterior calculus, where they were found to be commuting in order to achieve $d^2 = d\theta^i\,d\theta^j\,\partial_j\,\partial_i = 0$. It is sometimes convenient to pass the $d\theta$'s through other variables within a Berezin integral. This is done by taking the $d\theta$'s to anticommute also with the θ's and with any other odd variable (or constant) that may appear in the Berezin integral.

Because $dx\, d\theta = \text{sdet}\,[\partial^R(x,\theta)/\partial(x',\theta')]\, dx'\, d\theta'$, one defines a superdensity $\rho(x,\theta)$ (of weight one) to be an object that transforms as

$$\rho(x,\theta)\text{sdet}\Big(\frac{\partial^R(x,\theta)}{\partial(x',\theta')}\Big) = \rho'(x',\theta'), \tag{20.29}$$

so that $\rho\, dx\, d\theta = \rho'\, dx'\, d\theta'$. In order to define an invariant integration theory on a supermanifold, one needs a superdensity ("supermeasure").

As has already been put in practice, we will often drop the prefix "super" where there is no possible confusion.

In the bosonic case, a density defines by duality a form of maximum rank. This property no longer holds in the presence of fermions. This is not surprising, since there is then no form of maximum rank. For example, all the powers $(d\theta)^k$ differ from zero. Because there is no sensible concept of duality, one cannot reformulate the integration theory in terms of the exterior forms defined in Sec. 20.1, and one cannot view the "volume element" $d\theta^1\, d\theta^2 \ldots d\theta^n$ as the exterior product of the one-forms $d\theta^i$.

We finish this subsection with the following result, which is needed in the analysis of the antibracket formalism.

Theorem 20.2. *Let $z = (y, w)$ be coordinates of a supermanifold where the y's and the w's can be either even or odd. If the change of coordinates $z = z(z')$ is such that $y = y(y', w')$ can be inverted for y', then*

$$\text{sdet}\,\frac{\partial z}{\partial z'} = \text{sdet}\,\frac{\partial y}{\partial y'}\left(\text{sdet}\,\frac{\partial w'}{\partial w}\right)^{-1}. \tag{20.30}$$

[The superdeterminants are computed by reordering, if necessary, the rows and their corresponding columns simultaneously in such a way that the bosonic indices come first.]

Proof. One has

$$\frac{\partial z}{\partial z'} = \begin{pmatrix} \alpha & \beta \\ \gamma & \delta \end{pmatrix}$$

with

$$\alpha = \frac{\partial y}{\partial y'}, \qquad \beta = \frac{\partial y}{\partial w'}, \qquad \gamma = \frac{\partial w}{\partial y'}, \qquad \delta = \frac{\partial w}{\partial w'}.$$

Therefore,

$$\frac{\partial z}{\partial z'} = \begin{pmatrix} \alpha & 0 \\ \gamma & I \end{pmatrix} \begin{pmatrix} I & \alpha^{-1}\beta \\ 0 & \delta - \gamma\alpha^{-1}\beta \end{pmatrix}$$

and

$$\frac{\partial z'}{\partial z} = \left(\frac{\partial z}{\partial z'}\right)^{-1} = \begin{pmatrix} I & -\alpha^{-1}\beta(\delta - \gamma\alpha^{-1}\beta)^{-1} \\ 0 & (\delta - \gamma\alpha^{-1}\beta)^{-1} \end{pmatrix} \begin{pmatrix} \alpha^{-1} & 0 \\ -\gamma\alpha^{-1} & I \end{pmatrix},$$

so that $\partial w'/\partial w$ is invertible and given by

$$\frac{\partial w'}{\partial w} = (\delta - \gamma\alpha^{-1}\beta)^{-1}.$$

The theorem follows immediately by using (20.23c).

20.2.4. Delta Function—Gaussian Integrals

The fermionic delta function is defined by

$$\int f(\theta)\,\delta(\theta - \theta_0)\,d\theta = f(\theta_0). \tag{20.31a}$$

It is an ordinary function explicitly given by

$$\delta(\theta - \theta_0) = \theta - \theta_0 = \int e^{i(\theta-\theta_0)\pi}\,\frac{d\pi}{i}. \tag{20.31b}$$

The definition (20.31a) can be extended directly to an arbitrary number of bosonic and fermionic variables,

$$\int f(x,\theta) \prod_i \delta(x^i - x_0^i) \prod_k \delta(\theta^k - \theta_0^k)\,dx\,d\theta = f(x_0^i, \theta_0^k). \tag{20.31c}$$

It follows from Theorem 20.1 that

$$\prod_i \delta\big(x^i(x',\theta') - x_0^i\big) \prod_k \delta\big(\theta^k(x',\theta') - \theta_0^k\big) =$$

$$\left(\mathrm{sdet}\,\frac{\partial(x,\theta)}{\partial(x',\theta')}\bigg|_{x'_0,\theta'_0}\right)^{-1} \prod_i \delta(x'^i - x'^i_0) \prod_k \delta(\theta'^k - \theta'^k_0), \tag{20.31d}$$

where

$$x_0^i = x^i(x'^i_0, \theta'^k_0), \qquad \theta_0^k = \theta^k(x'^i_0, \theta'^k_0). \tag{20.31e}$$

Another useful integral is the Gaussian integral. Let M be a symmetric supermatrix,

$$M^T = M. \tag{20.32a}$$

Then one easily gets (see Exercise 20.4)

$$\int \exp\left[-\frac{1}{2} z^A M_{AB}\, z^B\right] dz = (2\pi)^{n/2}\,(\mathrm{sdet}\,M)^{-1/2} \tag{20.32b}$$

with

$$z^A = (x^i, \theta^k), \quad dz = dx\, d\theta,$$
$$n = \text{number of bosonic degrees of freedom.} \qquad (20.32c)$$

If one adds $N_A\, z^A + C$ to the power of the exponential in (20.32b), with $\varepsilon(N_A) = \varepsilon_A$, $\varepsilon(C) = 0$, the right-hand side of (20.32b) is multiplied by the value of the exponential at the extremum.

A straightforward application of (20.32) is

$$\int \exp\left[-\theta_1^i\, D_{ij}\, \theta_2^j\right] d\theta_1\, d\theta_2 = \det D, \qquad (20.33)$$

where D is an arbitrary even matrix. The rule for the Gaussian integrals over fermionic variables is thus that the determinant come upstairs rather than downstairs as is the case for bosonic variables.

20.2.5. Liouville Measure

As in the pure bosonic case, the even symplectic structure associated with the Poisson bracket defines a measure on super-phase space, the "Liouville measure." The construction of this measure starts with the observation that the Poisson bracket matrix

$$\sigma^{AB} = [z^A, z^B] \qquad (20.34a)$$

is a supermatrix that transforms as

$$\sigma'^{AB} \equiv [z'^A, z'^B] = \frac{\partial^R z'^A}{\partial z^C}\, [z^C, z^D]\, \frac{\partial^L z'^B}{\partial z^D}. \qquad (20.34b)$$

In matrix notations, this equation reads

$$\sigma' = J \sigma J^T, \qquad J = \frac{\partial^R z'}{\partial z}. \qquad (20.34c)$$

By taking the superdeterminant of (20.34c), one gets

$$(\text{sdet}\, \sigma'^{AB})^{1/2} = \text{sdet} \frac{\partial^R (z')}{\partial z}\, (\text{sdet}\, \sigma^{AB})^{1/2}, \qquad (20.35a)$$

and hence, for the inverse matrix defining the symplectic 2-form,

$$(\text{sdet}\, \sigma_{AB})^{1/2}\, \text{sdet} \frac{\partial^R z}{\partial z'} = (\text{sdet}\, \sigma'_{AB})^{1/2}. \qquad (20.35b)$$

This shows that $(\text{sdet}\, \sigma_{AB})^{1/2}$ defines a measure [cf. (20.29)]. This measure is constant in canonical coordinates.

20.3. QUANTIZATION OF FERMI DEGREES OF FREEDOM: CLIFFORD ALGEBRAS

20.3.1. Introduction

For anticommuting degrees of freedom, the graded commutator is equal to the anticommutator. In the case of c-number (Dirac) brackets,

$$[\theta_A, \theta_B]^* = -i\eta_{AB}, \qquad A = 1, \ldots, n, \tag{20.36a}$$

the anticommutation relations thus read

$$\hat{\theta}_A \hat{\theta}_B + \hat{\theta}_B \hat{\theta}_A = i[\theta_A, \theta_B]^* = \eta_{AB}. \tag{20.36b}$$

We have set $\hbar = 1$. Without loss of generality, one can assume that η_{AB} is diagonal, with diagonal elements equal to ± 1. The variables θ_A are classically real and, hence, should be represented by operators that are hermitian, $\theta^*_A = \theta_A$.

The relation (20.36b) defines a Clifford algebra with n generators. The irreducible representation of (20.36b) is unique and $2^{n/2}$-dimensional when n is even. Furthermore, on the space of the irreducible representation of (20.36b), one can define two inequivalent scalar products, making the θ_A's hermitian. These scalar products simply differ by an overall sign. There are two irreducible representations of (20.36b), both of dimension $2^{(n-1)/2}$, when n is odd. We will not give a proof of these statements here. Rather, we shall give an explicit construction of the irreducible representations of (20.36b).

As the ghost phase space is even dimensional (the ghosts come with their momenta), the case of interest in the BRST quantization of gauge systems is when n is even.

20.3.2. Clifford Algebras with an Even Number of Generators

With even n, the matrix η_{AB} can be decomposed in two-dimensional blocks of one of the following types:

$$\begin{pmatrix} 1 & 0 \\ 0 & 1 \end{pmatrix}, \tag{20.37}$$

$$\begin{pmatrix} 1 & 0 \\ 0 & -1 \end{pmatrix}, \tag{20.38}$$

or

$$\begin{pmatrix} -1 & 0 \\ 0 & -1 \end{pmatrix}. \tag{20.39}$$

We first construct the Clifford algebras associated with the elementary blocks (20.37)–(20.39).

20.3.2a. Clifford Algebra Associated with (20.37)

The change of variables

$$a = \frac{1}{\sqrt{2}}(\theta_1 + i\theta_2), \qquad a^* = \frac{1}{\sqrt{2}}(\theta_1 - i\theta_2) \tag{20.40}$$

brings the Clifford algebra to the familiar Fermi oscillator form,

$$a^2 = 0, \qquad (a^*)^2 = 0, \qquad aa^* + a^*a = 1. \tag{20.41}$$

The representation space is two-dimensional and contains the "vacuum state" anihilated by a,

$$a|0\rangle = 0, \tag{20.42a}$$

as well as the "filled state" $a^*|0\rangle$,

$$|0\rangle, \quad a^*|0\rangle. \tag{20.42b}$$

If one assumes that the vacuum state has positive norm normalized to plus one, then the scalar products are diagonal in the basis (20.42b), and their diagonal elements are equal to plus one. The norms are thus positive. (An inequivalent scalar product is obtained by taking $\langle 0 \,|\, 0\rangle = -1$. The norms are then all negative.)

In the holomorphic representation, the states are represented as functions of a^*. In particular, $|0\rangle$ is given by the function 1 and $a^*|0\rangle$ by the function a^*. The operators a and a^* are then

$$a = \frac{\partial}{\partial a^*}, \qquad a^* = \text{multiplication by } a^*. \tag{20.43}$$

The scalar product of two states $f(a^*)$ and $g(a^*)$ is given by

$$\langle f \,|\, g\rangle = \int (\exp -a^*a)\, \big(f(a^*)\big)^*\, g(a^*)\, da^*\, da \tag{20.44}$$

where

$$\big(f(a^*)\big)^* = f_0^* + f_1^*\, a \tag{20.45a}$$

if f is given by

$$f(a^*) = f_0 + f_1\, a. \tag{20.45b}$$

The integral in (20.44) stands for the Berezin integral (20.16)

$$\int a^* \, da^* = 1, \quad \int a \, da = 1, \quad \int da^* = 0, \quad \int da = 0. \tag{20.46}$$

20.3.2b. Clifford Algebra Associated with (20.38)

The change of variables

$$\eta = \frac{1}{\sqrt{2}} (\theta_1 + \theta_2), \qquad \pi = \frac{1}{\sqrt{2}} (\theta_1 - \theta_2), \tag{20.47a}$$

yields

$$\eta^2 = 0, \qquad \pi^2 = 0, \qquad \eta \, \pi + \pi \, \eta = 1, \tag{20.47b}$$

$$\eta^* = \eta, \qquad \pi^* = \pi. \tag{20.47c}$$

The irreducible representation space is again two-dimensional and can be taken to be the space of functions of η ("Schrödinger representation"). One has

$$\eta = \text{multiplication by } \eta, \qquad \pi = \frac{\partial}{\partial \eta}, \tag{20.48a}$$

$$(f, g) = \int f^*(\eta) \, g(\eta) \, d\eta, \tag{20.48b}$$

with

$$f^*(\eta) = f_0^* + f_1^* \eta \tag{20.48c}$$

if f is given by

$$f(\eta) = f_0 + f_1 \eta. \tag{20.48d}$$

The scalar products are explicitly

$$(1,1) = 0, \qquad (1, \eta) = (\eta, 1) = 1, \qquad (\eta, \eta) = 0. \tag{20.49}$$

In the basis $e_1 = (1/\sqrt{2})\,(1+\eta)$, $e_1 = (1/\sqrt{2})\,(1-\eta)$, the scalar products become

$$(e_i, e_j) = \begin{pmatrix} 1 & 0 \\ 0 & -1 \end{pmatrix}, \tag{20.50}$$

which explicitly shows the presence of negative norm states.

This case corresponds to the ghost representation space, with $\pi = -i\mathcal{P}$. One may define an inequivalent scalar product simply by changing the signs of all the norms (20.49) (see Exercise 20.14).

20.3.2c. Clifford Algebra Associated with (20.39)

The change of variables

$$a = \frac{1}{\sqrt{2}}(\theta_1 + i\theta_2), \qquad a^* = \frac{1}{\sqrt{2}}(\theta_1 - i\theta_2) \tag{20.51a}$$

leads to a Fermi oscillator algebra with the "wrong sign,"

$$a^2 = 0, \qquad a^{*2} = 0, \qquad a\,a^* + a^*a = -1. \tag{20.51b}$$

If the vacuum $|0\rangle$ annihilated by a is taken to have positive norm $+1$, then the filled state $a^*|0\rangle$ has negative norm -1. In the holomorphic representation where the states are functions of a^*, one finds

$$a = -\frac{\partial}{\partial a^*}, \qquad a^* = \text{multiplication by } a^*, \tag{20.52a}$$

while the scalar product is given by

$$(f, g) = -\int \exp a^* a \left(f(a^*)\right)^* g(a^*)\, da^*\, da. \tag{20.52b}$$

One can again change the sign of the scalar product without violating the hermiticity properties of a and a^*.

Note that the representation space for (20.51) is isomorphic to the representation space for (20.47). The change of variables relating a and a^* to η and π is

$$a = \tfrac{1}{2}(1 - \eta + \pi - 2\,\pi\,\eta), \tag{20.53a}$$
$$a^* = \tfrac{1}{2}(1 - \eta + \pi - 2\,\eta\,\pi). \tag{20.53b}$$

This change of variables is nonlinear and does not preserve the Grassmann parity.

20.3.2d. Combining the Representations of (20.37)–(20.39)

The irreducible representation of the Clifford algebra with arbitrary even n is easily constructed by combining the previous building blocks. If there are k_1 oscillators a_i with positive norms, k_2 oscillators a_α with negative norms, and k_3 conjugate pairs (η_M, π_M), the states can be described as functions of the anticommuting variables a_i^*, a_α^*, η_M,

$$f = f(a_i^*, a_\alpha^*, \eta_M) \tag{20.54a}$$

with

$$a_i = \frac{\partial^L}{\partial a_i^*}, \qquad a_\alpha = -\frac{\partial^L}{\partial a_\alpha^*}, \qquad \pi_M = \frac{\partial^L}{\partial \eta_M} \tag{20.54b}$$

and

$$a_i^* = \text{multiplication by } a_i^*, \tag{20.54c}$$

$$a_\alpha^* = \text{multiplication by } a_\alpha^*, \tag{20.54d}$$

$$\eta_M = \text{multiplication by } \eta_M. \tag{20.54e}$$

In (20.54b), one takes left derivatives, as is conventional in quantum mechanics where the operators act from the left on the states.

The formula for the scalar product is

$$(f, g) = (i)^{\phi(k_3)} \int f^* g \, e^{-N} \prod da^* \, da \, d\eta, \tag{20.55}$$

where N is the occupation number for the oscillators

$$N = \sum_i a_i^* \, a_i - \sum_\alpha a_\alpha^* \, a_\alpha. \tag{20.56}$$

The phase $(i)^{\phi(k_3)}$ is equal to i for $k_3 = 2, 3 \bmod 4$ and to 1 otherwise. It is included so that the "measure" $(i)^{\phi(k_3)} \prod d\eta$ is real.

A particularly interesting case arises when one has two blocks (20.38), *i.e.*, two sets of real pairs (η_1, π_1) and (η_2, π_2) obeying (20.47). In that instance, it is convenient to go to a new basis, the "Fock basis." This is done as follows.

First, one defines

$$C = \frac{1}{\sqrt{2}} (\eta_1 + i\eta_2), \qquad C^* = \frac{1}{\sqrt{2}} (\eta_1 - i\eta_2), \tag{20.57a}$$

$$\bar{C} = \frac{1}{\sqrt{2}} (\pi_1 + i\pi_2), \qquad \bar{C}^* = \frac{1}{\sqrt{2}} (\pi_1 - i\pi_2). \tag{20.57b}$$

One finds

$$C^2 = 0, \qquad (C^*)^2 = 0, \qquad CC^* + C^*C = 0, \tag{20.57c}$$

$$\bar{C}^2 = 0, \qquad (\bar{C}^*)^2 = 0, \qquad \bar{C}\bar{C}^* + \bar{C}^*\bar{C} = 0, \tag{20.57d}$$

$$C\bar{C}^* + \bar{C}^*C = 1, \qquad C^*\bar{C} + \bar{C}C^* = 1, \tag{20.57e}$$

$$C\bar{C} + \bar{C}\,C = 0, \qquad C^*\bar{C}^* + \bar{C}^*C^* = 0. \tag{20.57f}$$

These relations define two conjugate null Fermi oscillators. A representation of the relations (20.57c)–(20.57f) is obtained by acting on the vacuum $|0\rangle$ annihilated by C and $\bar{C}$,

$$C\,|0\rangle = 0, \qquad \bar{C}\,|0\rangle = 0, \tag{20.57g}$$

with the creation operators C^* and $\bar{C}^*$. This yields four independent states,

$$|0\rangle, \quad C^* |0\rangle, \quad \bar{C}^* |0\rangle, \quad C^* \bar{C}^* |0\rangle. \tag{20.57h}$$

Because C^* ($\bar{C}^*$) anticommutes with C ($\bar{C}$), $C^* |0\rangle$ ($\bar{C}^* |0\rangle$) is a null state. If one represents the states as functions of C^* and $\bar{C}^*$, one gets

$$C^* = \text{multiplication by } C^*, \tag{20.57i}$$

$$\bar{C}^* = \text{multiplication by } \bar{C}^*, \tag{20.57j}$$

$$C = \frac{\partial^L}{\partial \bar{C}^*}, \qquad \bar{C} = \frac{\partial^L}{\partial C^*}, \tag{20.57k}$$

and

$$(f, g) = \int e^{-N} f^* g \, dC \, d\bar{C} \, dC^* \, d\bar{C}^* \tag{20.57l}$$

with

$$N = \bar{C}^* C + C^* \bar{C}. \tag{20.57m}$$

The Fock representation (20.57) of the Clifford algebra with $\eta_{AB} = \text{diag}\,(1, -1, 1, -1)$ is completely equivalent to the product of two representations (20.48), or of one representation (20.43)–(20.45) with one representation (20.52) (Exercise 20.10).

20.3.2e. Grassmann Parity

A Grassmann parity can be consistently assigned to the states in such a way that even operators do not change the parity of the states, while odd operators do. One possibility is to assign even parity to the vacuum. The parity of a state (of definite parity) is then equal to the parity of the function representing it in the holomorphic representation. This is unambiguous because that function is unique. Alternatively, one can decree the vacuum to be of odd parity, in which case the parities of the states are all reversed.

In the first instance, the parity of the states is equal to $(-)^N$, where N is the occupation number. In the second instance, it is equal to $(-)^{N+1}$. A similar construction can be worked out in the Schrödinger representation.

The Grassmann parity can be related to the eigenvalues of $\gamma_{n+1} \equiv (\sqrt{2})^n \theta_1 \theta_2 \ldots \theta_n$ as follows. Because $\gamma_{n+1}^2 = 1$ (if γ_{n+1}^2 is found to be equal to -1, one should include an i in its definition), its eigenvalues are ± 1. One has

$$\gamma_{n+1} = \prod_{\text{oscillators}} (2N_i - 1) \tag{20.58}$$

where N_i is the occupation number of the i-th oscillator. As N_i equals 0 or 1, $2N_i - 1$ is equal to ± 1. States for which an even number of oscillators are unoccupied, are of the same parity, and are also of "positive chirality," *i.e.*, eigenstates of γ_{n+1} with eigenvalue $+1$. Similarly, the states of "negative chirality" have the opposite Grassmann parity.

20.3.3. Clifford Algebra with an Odd Number of Generators

20.3.3a. Irreducible Representations of the Clifford Algebra

When the number n of generators is even, there are two inequivalent representations of the Clifford algebra (20.36). These irreducible representations are characterized as follows. Let Γ be the product of the θ's, normalized so that $\Gamma^2 = \pm 1$,

$$\Gamma = (\sqrt{2})^n \, \theta_1 \ldots \theta_n \qquad (n \text{ odd}). \tag{20.59a}$$

[The sign of Γ^2 is fixed for a definite η_{AB} in (20.36).] The operator Γ commutes with each generator θ_A. Hence, in any irreducible representation of the Clifford algebra, it must be a multiple of the identity. Thus, one has:

$$\Gamma = +1 \qquad \text{or} \qquad \Gamma = -1 \qquad \text{if} \qquad \Gamma^2 = +1, \tag{20.59b}$$

and

$$\Gamma = i \qquad \text{or} \qquad \Gamma = -i \qquad \text{if} \qquad \Gamma^2 = -1. \tag{20.59c}$$

For a given sign of Γ^2, there are thus two possibilities, which define different representations.

By using (20.59), one can express the last generator θ_n in terms of the $(n-1)$ first ones. One finds that θ_n is proportional to what we called γ_{n+1} above. The irreducible representations of the algebra are then obtained by first constructing the irreducible representation of the Clifford algebra defined by $\theta_1, \theta_2, \ldots$ up to θ_{n-1}; and then using (20.59) to get θ_n.

20.3.3b. Reality Conditions

The construction that we have just outlined yields a representation of the algebraic relations (20.36). However, it may fail to give the correct reality properties to θ_n.

The difficulty appears most transparently in the case of an algebra with one generator θ_1 such that ${\theta_1}^2 = -1$. In that case, the representation space is one-dimensional and $\theta_1 = \pm i$. But then θ_1 is necessarly imaginary no matter how the scalar product is defined in the representation space. This is in conflict with $\theta_1^* = \theta_1$. Hence, the irreducible representations of the algebra (20.36) may not yield representations of the reality conditions.

To get the correct realization of the reality conditions $\theta_A^* = \theta_A$, one needs to consider reducible representations of the algebra. The simplest possibility is to add just one generator θ_{n+1}. The total algebra has then an even number of generators and can be treated along the lines of §20.3.2.

Note that if one does not add an extra generator, there is no well-defined Grassmann parity in the irreducible representation space of a Clifford algebra with an odd number of generators. This is because Γ is even by (20.59b)–(20.59c) (it is a multiple of the identity), but it is also odd as the product of an odd number of θ's. Similarly, θ_n is at the same time odd and even (as the product $\theta_1\theta_2\ldots\theta_{n-1}$). The absence of a well-defined Grassmann parity is not a problem in itself.

Some remarks on the reducible representations of the algebra ${\theta_1}^2 = -1$ are gathered in Exercise 20.11.

EXERCISES. CHAPTER TWENTY

20.1. Sometimes "antisymmetry" is defined as

$$T^{a_1\ldots a_i a_{i+1}\ldots a_k} = (-)^{\varepsilon_{a_i}\varepsilon_{a_{i+1}}} T^{a_1\ldots a_{i+1}a_i\ldots a_k}.$$

Show that one can redefine $T^{a_1\ldots a_k} \longrightarrow \tilde{T}^{a_1\ldots a_k}$ by an appropriate sign factor

$$\tilde{T}^{a_1\ldots a_k} = (-)^{f(\varepsilon_{a_1},\ldots,\varepsilon_{a_k})} T^{a_1\ldots a_k}$$

so that

$$\tilde{T}^{a_1\ldots a_i a_{i+1}\ldots a_k} = (-)^{(\varepsilon_{a_i}+1)(\varepsilon_{a_{i+1}}+1)} T^{a_1\ldots a_{i+1}a_i\ldots a_k},$$

as in (20.5c).

20.2. A matrix with even and odd elements is said to be odd if the parities of its elements are given by $\varepsilon(M_{AB}) = \varepsilon_A + \varepsilon_B + 1$. [It is even if $\varepsilon(M_{AB}) = \varepsilon_A + \varepsilon_B$. A supermatrix is thus an even matrix.] Let M and N be two odd matrices. Show that $\mathrm{str}\,(MN + NM) = 0$.

20.3. Show directly from definition (20.22) that $\mathrm{sdet}\begin{pmatrix} A & 0 \\ C & I \end{pmatrix} = \det A$ and $\mathrm{sdet}\begin{pmatrix} I & B \\ 0 & D \end{pmatrix} = (\det D)^{-1}$. [Hint: Use $\begin{pmatrix} A & 0 \\ C & I \end{pmatrix} = \begin{pmatrix} A & 0 \\ 0 & I \end{pmatrix}[I + N]$ with $N = \begin{pmatrix} 0 & 0 \\ C & 0 \end{pmatrix}$. Check that $\log(I + N) = N$ and is thus traceless.]

20.4. Let $D_{ij} = -D_{ji}$, $i, j = 1, 2, \ldots, N = 2N'$, be an antisymmetric, even matrix. The "Pfaffian" is defined by

$$Pf(D) = \frac{1}{2^{N'}}\frac{1}{N'!}\varepsilon^{i_1 i_2\ldots i_{N-1}i_N} D_{i_1 i_2} D_{i_3 i_4}\ldots D_{i_{N-1}i_N}.$$

(a) Prove that $[Pf(D)]^2 = \det D$.

(b) Show by direct expansion of the exponential that

$$\int \exp\left[-\frac{1}{2}\theta^i D_{ij}\theta_j\right] d\theta_1 d\theta_2 \ldots d\theta_N = Pf(D).$$

(c) Prove (20.32b) by integrating first over the fermionic variables and using the results of (b), as well as the known Gaussian formula for bosonic variables.

(d) Supply a different proof of (20.32b) based on the invariance of the superdeterminant under the transformation (20.26). (Hint: Diagonalize M.)

20.5. Let $X(t), M(t)$ be supermatrices such that

$$\frac{dX}{dt} = MX, \qquad X(0) = I.$$

Show that $\operatorname{sdet} X(t) = \exp\left[\int_0^t \operatorname{str} M(s)\, ds\right]$.

20.6. (a) Check that as a supermatrix the Poisson bracket matrix σ^{AB} defined by (20.34a) is neither symmetric nor antisymmetric, $(\sigma^{AB})^T \neq \pm\sigma^{AB}$. Rather, $A = -A^T$, $B = -C^T$, $D = D^T$.

(b) Verify explicitly that the inverse matrix σ_{AB} fulfills $\sigma_{AB} = (-)^{(\varepsilon_A+1)(\varepsilon_B+1)}\sigma_{BA}$, *i.e.*, has blocks $\alpha = -\alpha^T$, $\beta = \gamma^T$, $\delta = \delta^T$. (Compare with Exercise 20.1.)

20.7. Verify explicitly formula (20.27) in the case of the change of variables $\theta'_1 = \theta_2$, $\theta'_2 = \theta_1$, whose superdeterminant is equal to minus one.

20.8. Let f^1 be a function of $\theta_1, \ldots, \theta_k$ and f^2 be a function of $\theta_{k+1}, \ldots, \theta_n$. Show that $\int f^2 f^1\, d\theta_1 \ldots d\theta_k\, d\theta_{k+1} \ldots d\theta_n = (-)^\varepsilon \int f^2\, d\theta_{k+1} \ldots d\theta_n \int f^1\, d\theta_1 \ldots d\theta_k$, where $(-)^\varepsilon$ is the sign factor obtained by passing $d\theta_{k+1} \ldots d\theta_n$ through $f^1_{k\ldots 1}\theta_k \ldots \theta_1\, d\theta_1 \ldots d\theta_k$ according to the rules of §20.2.3, remark (d). (This sign factor is equal to unity if $f^1_{k\ldots 1}$ is even.)

20.9. Check explicitly that the scalar product (20.55) is such that $a_i^* = (a_i)^*$, $a_\alpha^* = (a_\alpha)^*, \eta_L^* = \eta_L, \pi_L^* = \pi_L$.

20.10. Show that the fermionic Fock representation (20.57) of the Clifford algebra with $\eta_{AB} = \operatorname{diag}(1, -1, 1, -1)$ is completely equivalent to the Schrödinger representation in which the states are functions of η^1 and η^2. Write explicitly the wave function of the Fock vacuum in the Schrödinger representation.

20.11. Consider the Clifford algebra ${\theta_1}^2 = -1$.

(a) Add one generator $\theta_2, {\theta_2}^2 = 1$ and consider the Schrödinger representation (20.48), $\psi = \psi(\eta), \eta = (1/\sqrt{2})(\theta_1 + \theta_2)$. Verify that θ_1 is hermitian.

(b) Show that the eigenvalues of θ_1 are $\pm i$.

(c) Check that the eigenstates of θ_1 (*i*) have zero norm and (*ii*) have no definite Grassmann parity.

20.12. Consider the Clifford algebra ${\theta_1}^2 = 1$. Add one generator θ_2 such that ${\theta_2}^2 = 1$ and discuss the realization of θ_1 in the holomorphic representation (20.43).

20.13. Show that there is no representation of the commutation relation $aa^* + a^*a = -1$ in a positive definite Hilbert space. (Hint: Let ψ be a nonzero state with $\langle\psi \mid \psi\rangle > 0$. Show that either $\langle a\psi \mid a\psi\rangle$ or $\langle a^*\psi \mid a^*\psi\rangle$ is strictly negative.)

20.14. Show that the scalar products in the representation space of (20.36b) with n even are completely determined by (*i*) the algebra (20.36b); (*ii*) the condition $\theta_A^* = \theta_A$; and (*iii*) the norm $\langle 0 \,|\, 0 \rangle$ of the vacuum (oscillator case) or the scalar product $(1, \eta^1 \ldots \eta^m)$ (Schrödinger case). Prove that in the case of (20.38), the transformation $\eta \to \eta' = -\eta, \pi \to \pi' = -\pi$ is not unitarily implementable (one goes from one choice of sign for the scalar product to the other).

BIBLIOGRAPHY

The following is a list of some of the works that we have found useful when writing this book. They range from general developments to analysis of particular issues. The list of references has, of course, no pretense whatsoever of completeness.

Chapters 1 and 2

Arnold, V. 1978. *Mathematical Methods of Classical Mechanics, Graduate Texts in Math.*, n°60. Berlin: Springer.

Batalin, I. A., and Fradkin, E. S. 1987. "Operational quantization of dynamical systems subject to second class constraints," *Nucl. Phys.* **B279**:514.

Bergmann, P. G., and Goldberg, I. 1955. "Dirac bracket transformation in phase space," *Phys. Rev.* **98**:531.

Dirac, P.A.M. 1950. "Generalized Hamiltonian Dynamics," *Can. J. Math.* **2**:129.

——— . 1951. "The Hamiltonian form of field dynamics," *Can. J. Math.* **3**:1.

——— . 1958, "Generalized Hamiltonian Dynamics," *Proc. Roy. Soc. (London)* **A246**:326.

——— . 1967, *Lectures on Quantum Mechanics*. Yeshiva University, New York: Academic Press.

Faddeev, L. D., and Jackiw, R. 1988. "Hamiltonian reduction of unconstrained and constrained systems," *Phys. Rev. Lett.* **60**:1692.

Flato, M., Lichnerowicz, A., and Sternheimer, D. 1976. "Deformation of Poisson brackets, Dirac brackets and applications," *J. Math. Phys.* **17**:1754.

Gitman, D. M., and Tyutin, I. V. 1990. *Quantization of fields with constraints*. Berlin: Springer.

Gribov, V. N. 1978. "Quantization of non-Abelian gauge theories," *Nucl. Phys.* **B139**:1.

Hanson, A., Regge, T., and Teitelboim, C. 1976. *Constrained Hamiltonian Systems*. Rome: Accad. Naz. dei Lincei.

Lichnerowicz, A. 1975. "Variété symplectique et dynamique associée à une sous-variété," *C. R. Acad. Sc. Paris* **280**:523.

Marsden, J. E., and Weinstein, A. 1974. "Reduction of symplectic manifolds with symmetry," *Rep. Math. Phys.* **5**:121.

Sniatycki, J. 1974. "Dirac brackets in geometric dynamics," *Ann. Inst. Henri Poincaré* **A20**:365.

Sudarshan, E.D.G., and Mukunda, N. 1974. *Classical Dynamics: a modern perspective*. New York: Wiley.

Chapter 3

Anderson, J. L., and Bergmann, P. G. 1951. "Constraints in covariant field theories," *Phys. Rev.* **83**:1018.

Castellani, L. 1982. "Symmetries in constrained Hamiltonian systems," *Ann. Phys. (N.Y.)* **143**:357.

DeWitt, B. S. 1964. "Dynamical theory of groups and fields." In *Relativity, groups and topology*, ed. C. g and B. S. DeWitt. New York: Gordon and Breach.

Fradkin, E. S., and Vilkovisky, G. A. 1977. "Quantization of relativistic systems with constraints: equivalence of canonical and covariant formalisms in quantum theory of gravitational field," CERN report TH-2332.

Géhéniau, J. 1959. "Remarques sur les identités relatives à un Lagrangien," *Bull. Cl. Sc. Acad. Roy. Belg.* **45**:447.

Henneaux, M. 1990. "Elimination of the auxiliary fields in the antifield formalism," *Phys. Lett.* **238B**:299.

Henneaux, M., Teitelboim, C., and Zanelli, J. 1990. "Gauge invariance and degree of freedom count," *Nucl. Phys.* **B332**:169.

Mukunda, N. 1980. "Generators of symmetry transformations for constrained Hamiltonian systems," *Physica Scripta* **21**, 783.

Teitelboim, C. 1977. "Supergravity and the square root of constraints," *Phys. Rev. Lett.* **38**:1106.

Chapter 4

Carlip, S. 1990. "Observables, Gauge Invariance and Time in 2 + 1 Dimensional Quantum Gravity," *Phys. Rev.* **D42**:2647.

Kuchař, K. 1973. "Canonical quantization of gravity." In *Relativity, Astrophysics and Cosmology*, ed. W. Israel, p. 238. Dordrecht: Reidel.

Rovelli, C. 1991. "Time in quantum gravity: an hypothesis," *Phys. Rev.* **D43**:442.

Teitelboim, C. 1982. "Quantum mechanics of the gravitational field," *Phys. Rev.* **D25**:3159.

Chapter 5

Bergmann, P. G. 1966. "Hamilton–Jacobi and Schrödinger theory in theories with first-class Hamiltonian constraints," *Phys. Rev.* **144**:1078.

Eisenhart, L. P. 1933. *Continuous groups of transformations*. Princeton: Princeton University Press.

Gomis, J., Henneaux, M., and Pons, J. M. 1990. "Existence theorem for gauge symmetries in Hamiltonian constrained systems," *Class. Quant. Grav.* **7**:1089.

Lanczos, C. 1970. *The variational principles of mechanics*, Fourth edition. Toronto: University of Toronto Press.

Schouten, J. A., and van der Kulk, W. 1949. *Pfaff's problem and its generalizations*. Oxford: Clarendon Press.

Shanmugadhasan, S. 1973. "Canonical formalism for degenerate Lagrangians," *J. Math. Phys.* **14**:677.

Sudarshan, E.D.G., and Mukunda, N. 1974. *Classical Dynamics: a modern perspective*. New York: Wiley.

Chapters 6 and 7

Berezin, F. A. 1966. *The method of second quantization*. New York: Academic Press.

——— . 1987. *Introduction to superanalysis, Mathematical Physics and Applied Mathematics*, n°9. Dordrecht: Reidel.

Berezin, F. A., and Leites, D. A. 1975. "Supermanifolds," *Sov. Math. Dokl.* **16**:1218.

Berezin, F. A., and Marinov, M. S. 1977. "Particle spin dynamics as the Grassmann variant of classical mechanics," *Ann. Phys. (N.Y.)* **104**:336.

Casalbuoni, R. 1976a. "On the quantization of systems with anticommuting variables," *Nuovo Cimento* **33A**:115.

——— . 1976b. "The classical mechanics for Bose–Fermi systems," *Nuovo Cimento* **33A**:389.

DeWitt, B. S. 1984. *Supermanifolds*. Cambridge: Cambridge University Press.

Galvão, C.A.P., and Teitelboim, C. 1980. "Classical supersymmetric particles," *J. Math. Phys.* **21**:1863.

Leites, D. A. 1980. "Introduction to the theory of supermanifolds," *Russ. Math. Surv.* **35**:1.

Martin, J. L. 1959. "Generalized classical dynamics and the 'classical analogue' of a Fermi oscillator," *Proc. Roy. Soc.* **A251**:536.

Chapters 8, 9, 10, 11, and 12

Batalin, I. A., and Vilkovisky, G. A. 1977. "Relativistic S-Matrix of dynamical systems with boson and fermion constraints," *Phys. Lett.* **69B**:309.

Baulieu, L. 1985. "Perturbative gauge theories," *Phys. Rep.* **129**:3.

Baulieu, L., and Thierry–Mieg, J. 1982. "The principle of BRS symmetry: an alternative approach to Yang–Mills theories," *Nucl. Phys.* **B197**:477.

Becchi, C., Rouet, A., and Stora, R. 1975. "Renormalization of the abelian Higgs–Kibble model," *Commun. Math. Phys.* **42**:127.

———. 1976. "Renormalization of gauge theories," *Ann. Phys.* **98**:287.

Bonora, L., and Cotta–Ramusino, P. 1983. "Some remarks on BRS transformation, anomalies and the cohomology of the Lie algebra of the group of gauge transformations," *Commun. Math. Phys.* **87**:589.

Bott, R., and Tu, L. W. 1982. *Differential forms in algebraic topology.* New York: Springer.

Brandt, F., Dragon, N., and Kreuzer, M. 1990. "Completeness and nontriviality of the solutions of the consistency conditions," *Nucl. Phys.* **B332**:224.

Browning, A. D., and McMullan, D. 1987. "The Batalin, Fradkin, Vilkovisky formalism for higher order theories," *J. Math. Phys.* **28**:438.

Cartan, H., and Eilenberg, S. 1956. *Homological algebra.* Princeton: Princeton University Press.

DeWitt, B. S. 1967. "Quantum Theory of Gravity II. The manifestly covariant theory," *Phys. Rev.* **162**:1195.

Dixon, J. A. 1975. "Field redefinitions and renormalization of gauge theories," *Nucl. Phys.* **B99**:420.

———. 1989. "Calculating BRST cohomology using spectral sequences." In *Proc. XVII International Conference on Differential Geometric Methods in Theoretical Physics*, ed. Solomon et al. Singapore: World Scientific.

Doubrovine, B., Novikov, S., and Fomenko, A. 1982. *Géométrie contemporaine*, vol 2, Editions MiR (Moscow).

Dubois–Violette, M. 1987. "Systèmes dynamiques constraints: l'approche homologique," *Ann. Inst. Fourier* **37**:45.

Dubois–Violette, M., Henneaux, M., Talon, M., and Viallet, C. M. 1991. "Some results on local cohomologies in field theory," *Phys. Lett.* **267B**:81.

Dubois–Violette, M., Talon, M., and Viallet, C. M. 1985. "BRS Algebras. Analysis of the Consistency Equations of Gauge Theory," *Commun. Math. Phys.* **102**:105.

Faddeev, L. D., and Popov, V. N. 1967. "Feynman diagrams for the Yang–Mills field," *Phys. Lett.* **25B**:30.

Feynman, R. P. 1963. "Quantum theory of gravitation," *Acta Phys. Polon.* **XXIV**:697.

Figueroa–O'Farrill, J. M. 1990. "A topological characterization of classical BRST cohomology," *Commun. Math. Phys.* **127**:181.

Fisch, J., Henneaux, J., Stasheff, J., and Teitelboim, C. 1989. "Existence, uniqueness and cohomology of the classical BRST charge with ghosts of ghosts," *Commun. Math. Phys.* **120**:379.

Fradkin, E. S., and Fradkina, T. E. 1977. "Quantization of relativistic systems with boson and fermion first- and second-class constraints," *Phys. Lett.* **72B**:343.

Fradkin, E. S., and Vilkovisky, G. A. 1977. "Quantization of relativistic systems with constraints: equivalence of canonical and covariant formalisms in quantum theory of gravitational field," CERN report TH-2332.

Frölicher, A., and Nijenhuis, A. 1956. "Theory of vector-valued differential forms I," *Proc. Kon. Ned. Ak. Wet. Amsterdam* **A59**:338.

Gugenheim, V.K.A.M., and Stasheff, J. D. 1986. "On perturbations and A_∞-structures," *Bull. Soc. Math. Belgique* **38**:237.

Henneaux, M. 1985. "Hamiltonian form of the path integral for theories with a gauge freedom," *Phys. Rep.* **126**:1.

Henneaux, M., and Teitelboim, C. 1988. "BRST cohomology in classical mechanics," *Commun. Math. Phys.* **115**:213.

Huebschmann, J. 1989. "Constraints, symmetries, homological perturbations and the BRST-transformation," Heidelberg preprint.

Kostant, B., and Sternberg, S. 1987. "Symplectic reduction, BRS cohomology and infinite-dimensional Clifford algebras," *Ann. Phys. (N.Y.)* **176**:49.

Koszul, J.–L. 1950. "Sur un type d'algèbres différentielles en rapport avec la transgression," Colloque de Topologie, Bruxelles, CBRM, Liège.

Stasheff, J. D. 1988. "Constrained Poisson algebras and strong homotopy representations," *Bull. Amer. Math. Soc.* **19**:287.

———. 1989. "Homological reduction of constrained algebras," preprint.

Stora, R. 1977. "Continuum Gauge Theories." In *New Developments in Quantum Field Theory and Statistical Mechanics*, ed. M. Levy and P. Mitter. New York: Plenum.

Tate, J. 1957. "Homology of Noetherian rings and local rings," *Illinois J. Math.* **1**:14.

Thierry–Mieg, J. 1990. "BRS structure of the antisymmetric tensor gauge theories," *Nucl. Phys.* **B335**:334 (original version: preprint 1980).

Tyutin, I. V. 1975. "Gauge invariance in field theory and statistical mechanics,"Lebedev preprint FIAN, n°39.

van Holten, J. W. 1989. "Aspects of BRST quantization." In *Functional integration, geometry and strings; Proc. XXVth Karpacz Winter School in Theor. Physics*, ed. Z. Haba and J. Sobczyk. Basel: Birkhäuser.

Wess, J., and Zumino, B. 1971. "Consequences of anomalous Ward identities," *Phys. Lett.* **37B**:95.

Chapter 13

Christ, N., and Lee, T. D. 1980. "Operator ordering and Feynman rules in gauge theories," *Phys. Rev.* **D22**:939.

Dirac, P.A.M. 1930. *The Principles of Quantum Mechanics*. Oxford: Clarendon Press.

——. 1967, *Lectures on Quantum Mechanics*. Yeshiva University, New York: Academic Press.

Faddeev, L. D. 1984. "Operator anomaly for the Gauss law," *Phys. Lett.* **145B**:81.

Gotay, M. J., and Sniatycki, J. 1981. "On the quantization of presymplectic dynamical systems via coisotropic imbeddings," *Commun. Math. Phys.* **82**:377.

Gribov, V. N. 1978. "Quantization of non-Abelian gauge theories," *Nucl. Phys.* **B139**:1.

Jackiw, R. 1984. "Topological Investigations of Quantized Gauge Theories." In *Relativity, Groups and Topology II*, ed. B. S. DeWitt and R. Stora. Amsterdam: Elsevier.

Nagy, K. L. 1966. *State vector spaces with indefinite metric in quantum field theory*. Groningen: P. Noordhoff.

Nakanishi, N. 1972. "Indefinite-Metric Quantum Field Theory," *Prog. Theor. Phys. (Suppl.)* **51**:1.

Tuynman, G. M. 1990. "Reduction, quantization and non-unimodular groups," *J. Math. Phys.* **31**:83.

Van Hove, L. 1951. "Sur le problème des relations entre les transformations unitaires de la mécanique quantique et les transformations canoniques de la mécanique classique," *Bull. Cl. Sc. Acad. Roy. Belg.* **37**:610.

Chapter 14

Bes, D. R., and Kurchan, J. 1990. *The Treatment of Collective Coordinates in Many Body Systems. An Application of the BRST Invariance, World Scientific Lecture Notes in Physics*, n°34. Singapore: World Scientific.

Curci, G., and Ferrari, R. 1976. "An alternative Approach to the Proof of Unitarity for Gauge Theories," *Nuovo Cim.* **A35**:273.

Figueroa–O'Farrill, J. M., and Kimura, T. 1988. "The cohomology of BRST complexes," Stony Brook preprint ITP-SB-88-34 (revised version).

Freeman, M. D., and Olive, D. I. 1986. "BRS cohomology in string theory and the no-ghost theorem," *Phys. Lett.* **175B**:151.

Frenkel, I. B., Garland, H., and Zuckerman, G. J. 1986. "Semi-infinite cohomology and string theory," *Proc. Natl. Acad. Sci.* **83**:8442.

Frolov, S. A., and Slavnov, A. A. 1989a. "Physical Subspace Norm in Hamiltonian BRST-quantization," *Phys. Lett.* **218B**:461.

Fujikawa, K. 1980. "Non-Trivial Realization of the BRS Supersymmetry," *Prog. Theor. Phys.* **63**:1364.

Hwang, S. 1983. "Covariant quantization of the string in dimensions $D \leq 26$ using a Becchi–Rouet–Stora formulation," *Phys. Rev.* **D28**: 2614

Kato, M., and Ogawa, O. 1983. "Covariant quantization of strings based on BRS invariance," *Nucl. Phys.* **B212**:443.

Kostant, B., and Sternberg, S. 1987. "Symplectic reduction, BRS cohomology and infinite-dimensional Clifford algebras," *Ann. Phys. (N.Y.)* **176**:49.

Kugo, T., and Ojima, I. 1979. "Local covariant operator formalism of non-Abelian gauge theories and quark confinement problem," *Prog. Theor. Phys. (Suppl.)* **66**:1.

Marnelius, R. 1981. "A simple relation between covariant and noncovariant quantization of gauge theories," *Phys. Lett.* **99B**:467.

Nagy, K. L. 1966. *State vector spaces with indefinite metric in quantum field theory.* Groningen: P. Noordhoff.

Nakanishi, N. 1972. "Indefinite-Metric Quantum Field Theory," *Prog. Theor. Phys. (Suppl.)* **51**:1.

Nishijima, K. 1984."Representations of BRS algebra,"*Nucl. Phys.* **B238**: 601.

Rybkin, G. N. 1991. "State space in BRST-quantization and Kugo–Ojima quartets," *J. Mod. Phys. A* **6**:1675.

Schwarz, A. S. 1989. "Lefschetz trace formula and BRST," *Mod. Phys. Lett.* **A4**:1891.

Slavnov, A. A. 1989. "Unitarity condition in covariant quantum field theory with indefinite metric," *Theor. Math. Phys.* **79**:579.

Spiegelglas, M. 1987. "Q_{BRST} Cohomology: a mechanism for getting rid of negative norm states, with an application to the bosonic string," *Nucl. Phys.* **B283**:205.

Teitelboim, C. 1982. "Quantum mechanics of the gravitational field," *Phys. Rev.* **D25**:3159.

Tuynman, G. M. 1990. "Reduction, quantization and non-unimodular groups," *J. Math. Phys.* **31**:83.

Chapter 15

Berezin, F. A., and Marinov, M. S. 1977. "Particle spin dynamics as the Grassmann variant of classical mechanics," *Ann. Phys. (N.Y.)* **104**:336.

Boulware, D., and Gross, D. J. 1984. "Lee–Wick Indefinite Metric Quantization: a Functional Integral Approach," *Nucl. Phys.* **B233**:1.

Carreau, M., Farhi, E., and Gutman, S. 1990. "The functional integral for a free particle in a box," *Phys. Rev.* **D42**:1194.

Cohen, L. 1976. "Correspondence rules and path integrals," *J. Math. Phys.* **17**:597.

Dirac, P.A.M. 1933. "The Lagrangian in quantum mechanics," *Phys. Z. Sowjetunion* **Band 3**:Heft. 1

Faddeev, L. D., and Slavnov, A. A. 1980. *Gauge fields: Introduction to quantum theory*. Reading: Benjamin/Cummings.

Feynman, R. P. 1948. "Space-time approach to non-relativistic quantum mechanics," *Rev. Mod. Phys.* **20**:267.

——. 1951. "An Operator Calculus Having Applications in Quantum Electrodynamics," *Phys. Rev.* **84**:108.

Feynman, R. P., and Hibbs, A. R. 1965. *Quantum Mechanics and Path Integrals*. New York: McGraw–Hill.

Fradkin, E. S. 1973. "Hamilton Formalism in Covariant Gauge and the Measure in Quantum Gravity." In *Proc. Xth Winter School of Theoretical Physics in Karpacz (Poland) Acta Universitatis Wratislaviensis*, n°207, p. 93.

Galvão, C.A.P., and Teitelboim, C. 1980. "Classical supersymmetric particles," *J. Math. Phys.* **21**:1863.

Marinov, M. S. 1980. "Path integrals in quantum theory: an outlook of basic concepts," *Phys. Rep.* **60**:1.

Nijenhuis, A. 1955. "Jacobi-type identities for bilinear differential concomitants of certain tensor fields," *Proc. Kon. Ned. Akad. Wet. Amsterdam* **A58**:390.

Pauli, W. 1943. "On Dirac's New Method of Field Quantization," *Rev. Mod. Phys.* **15**:175.

Schouten, J. A. 1954. "On the differential operators of first order in tensor calculus," *Convegno di Geom. Diff.* **1**:7.

Schwinger, J. 1951. "The Theory of Quantized Fields. I.," *Phys. Rev.* **82**:914.

van Holten, J. W. 1987. "Quantum noncompact σ-models," *J. Math. Phys.* **28**:1420.

Witten, E. 1990. "A note on the antibracket formalism," *Mod. Phys. Lett.* **A5**:487.

Chapter 16

Batalin, I. A., and Vilkovisky, G. A. 1977. "Relativistic S-Matrix of dynamical systems with boson and fermion constraints," *Phys. Lett.* **69B**:309.

Christ, N., and Lee, T. D. 1980. "Operator ordering and Feynman rules in gauge theories," *Phys. Rev.* **D22**:939.

Faddeev, L. D. 1969. "The Feynman integral for singular Lagrangians," *Theor. Math. Phys.* **1**:1.

Fradkin, E. S. 1973. "Hamilton Formalism in Covariant Gauge and the Measure in Quantum Gravity." In *Proc. Xth Winter School of Theoretical Physics in Karpacz (Poland) Acta Universitatis Wratislaviensis*, n°207, p. 93.

Fradkin, E. S., and Fradkina, T. E. 1977. "Quantization of relativistic systems with boson and fermion first- and second-class constraints," *Phys. Lett.* **72B**:343.

Fradkin, E. S., and Vilkovisky, G. A. 1977. "Quantization of relativistic systems with constraints: equivalence of canonical and covariant formalisms in quantum theory of gravitational field," CERN report TH-2332.

Halliwell, J. J., and Hartle, J. B. 1991. "Wave functions constructed from an invariant sum-over-histories satisfy constraints," *Phys. Rev.* **D4**:1170.

Henneaux, M. 1985. "Hamiltonian form of the path integral for theories with a gauge freedom," *Phys. Rep.* **126**:1.

Hirschfeld, P. 1979. "Strong evidence that Gribov copying does not affect the gauge theory functional integral," *Nucl. Phys.* **B157**:37.

Teitelboim, C. 1982. "Quantum mechanics of the gravitational field," *Phys. Rev.* **D25**:3159.

Zinn–Justin, J. 1975. "Renormalization of gauge theories." In *Trends in Elementary Particle Theory, Lecture Notes in Physics*, n°37, ed. H. Rollnik and K. Dietz. Berlin: Springer.

Chapters 17 and 18

Ashtekar, A., Bombelli, L., and Koul, R. K. 1987. "Phase space formulation of general relativity without a $3+1$ splitting." In *The Physics of Phase Space*, ed. Y. S. Kim and W. W. Zachari. Berlin: Springer–Verlag.

Barnich, G., Henneaux, M., and Schomblond, C. 1991. "Covariant description of the canonical formalism," *Phys. Rev.* **D44**:939.

Batalin, I. A., and Vilkovisky, G. A. 1983. "Quantization of gauge theories with linearly dependent generators," *Phys. Rev.* **D28**:2567 [erratum: 1984. *Phys. Rev.* **D30**:508].

———. 1984. "Closure of the gauge algebra, generalized Lie equations and Feynman rules," *Nucl. Phys.* **B234**:106.

———. 1985. "Existence theorem for gauge algebra," *J. Math. Phys.* **26**:172.

Batlle, C., Gomis, J., París, J., and Roca, J. 1990. "Field-antifield formalism and Hamiltonian BRST approach," *Nucl. Phys.* **B329**:139.

Crnković, Č., and Witten, E. 1987. "Covariant description of canonical formalism in geometrical theories." In *Three Hundred Years of Gravitation*, ed. S. W. Hawking and W. Israel, p. 676. Cambridge: Cambridge University Press.

de Wit, B., and van Holten, J. W. 1978. "Covariant quantization of gauge theories with open gauge algebra," *Phys. Lett.* **79B**:389.

Dresse, A., Fisch, J.M.L., Grégoire, P., and Henneaux, M. 1991. "Equivalence of the Hamiltonian and Lagrangian path integrals for gauge theories," *Nucl. Phys.* **B354**:191.

Fisch, J.M.L., and Henneaux, M. 1990. "Homological perturbation theory and the algebraic structure of the antifield-antibracket formalism for gauge theories," *Commun. Math. Phys.* **128**:627.

Grigoryan, G. V., Grigoryan, R. P., and Tyutin, I. V. 1991. "Equivalence of Lagrangian and Hamiltonian BRST quantizations: systems with first-class constraints." In *Sov. Nucl. Phys.*, **53**, n°6, p. 1058.

Henneaux, M. 1990. "Elimination of the auxiliary fields in the antifield formalism," *Phys. Lett.* **238B**:299.

Kallosh, R. E. 1978. "Modified Feynman rules in supergravity," *Nucl. Phys.* **B141**:141.

Kijowski, J., and Szczyrba, W. 1976. "A canonical structure for classical field theories," *Commun. Math. Phys.* **46**:183.

Peierls, R. E. 1952. "The commutation laws of relativistic field theory," *Proc. Roy. Soc. (London)* **A214**:143.

Siegel, W. 1989. "Batalin–Vilkovisky from Hamiltonian BRST," *Int. J. Mod. Phys.* **A4**:3951.

Slavnov, A.A. 1991. "Equivalence Theorem for Spectrum Changing Transformations," *Phys. Lett.* **258B**:391.

Troost, W., van Nieuwenhuizen, P., and Van Proeyen, A. 1990. "Anomalies and the Batalin–Vilkovisky Lagrangian formalism," *Nucl. Phys.* **B333**:727.

Volkov, D. V., Pashnev, A. I., Soroka, V. A., and Tkach, V. I. 1989. "Hamiltonian dynamical systems with even and odd Poisson brackets," *Theor. Math. Phys.* **79**:424.

Voronov, B. L., and Tyutin, I. V. 1982a. "Formulation of gauge theories of general form I," *Theor. Math. Phys.* **50**:218.

Witten, E. 1990. "A note on the antibracket formalism," *Mod. Phys. Lett.* **A5**:487.

Zinn–Justin, J. 1975. "Renormalization of gauge theories." In *Trends in Elementary Particle Theory, Lecture Notes in Physics*, n°37, ed. H. Rollnik and K. Dietz. Berlin: Springer.

Chapter 19

Baulieu, L. 1985. "Perturbative gauge theories," *Phys. Rep.* **129**:3.

Becchi, C., Rouet, A., and Stora, R. 1975. "Renormalization of the abelian Higgs–Kibble model," *Commun. Math. Phys.* **42**:127.

——— . 1976. "Renormalization of gauge theories," *Ann. Phys.* **98**:287.

DeWitt, B. S. 1967. "Quantum Theory of Gravity II. The manifestly covariant theory," *Phys. Rev.* **162**:1195.

Dirac, P.A.M. 1967, *Lectures on Quantum Mechanics.* Yeshiva University, New York: Academic Press.

Faddeev, L. D., and Popov, V. N. 1967. "Feynman diagrams for the Yang–Mills field," *Phys. Lett.* **25B**:30.

Faddeev, L. D., and Slavnov, A. A. 1980. *Gauge fields: Introduction to quantum theory.* Reading: Benjamin/Cummings.

Feynman, R. P. 1963. "Quantum theory of gravitation," *Acta Phys. Polon.* **XXIV**:697.

Frolov, S. A. 1989. "BRST-quantization of gauge theories in temporal like gauges," Steklov Mathematical Institute preprint.

Gitman, D. M., and Tyutin, I. V. 1990. *Quantization of fields with constraints.* Berlin: Springer.

Hanson, A., Regge, T., and Teitelboim, C. 1976. *Constrained Hamiltonian Systems.* Rome: Accad. Naz. dei Lincei.

Henneaux, M. 1985. "Hamiltonian form of the path integral for theories with a gauge freedom," *Phys. Rep.* **126**:1.

Itzykson, C., and Zuber, J.–B. 1980. *Quantum Field Theory.* New York: McGraw–Hill.

Jackiw, R. 1984. "Topological Investigations of Quantized Gauge Theories." In *Relativity, Groups and Topology II*, ed. B. S. DeWitt and R. Stora. Amsterdam: Elsevier.

Kugo, T., and Ojima, I. 1979. "Local covariant operator formalism of non-Abelian gauge theories and quark confinement problem," *Prog. Theor. Phys. (Suppl.)* **66**:1.

Siegel, W. 1980. "Hidden ghosts," *Phys. Lett.* **93B**:170.

Stora, R. 1977. "Continuum Gauge Theories." In *New Developments in Quantum Field Theory and Statistical Mechanics*, ed. M. Levy and P. Mitter. New York: Plenum.

Thierry–Mieg, J. 1990. "BRS structure of the antisymmetric tensor gauge theories," *Nucl. Phys.* **B335**:334 (original version:. preprint 1980)

Wess, J., and Zumino, B. 1971. "Consequences of anomalous Ward identities," *Phys. Lett.* **37B**:95.

Chapter 20

Berezin, F. A. 1966. *The method of second quantization.* New York: Academic Press.

——— . 1987. *Introduction to superanalysis, Mathematical Physics and Applied Mathematics*, n°9. Dordrecht: Reidel.

Berezin, F. A., and Leites, D. A. 1975. "Supermanifolds," *Sov. Math. Dokl.* **16**:1218.

DeWitt, B. S. 1984. *Supermanifolds.* Cambridge: Cambridge University Press.

Leites, D. A. 1980. "Introduction to the theory of supermanifolds," *Russ. Math. Surv.* **35**:1.

INDEX